FUNDAMENTALS OF
DOMINATION IN GRAPHS

PURE AND APPLIED MATHEMATICS

A Program of Monographs, Textbooks, and Lecture Notes

MONOGRAPHS AND TEXTBOOKS IN
PURE AND APPLIED MATHEMATICS

1. *K. Yano*, Integral Formulas in Riemannian Geometry (1970)
2. *S. Kobayashi*, Hyperbolic Manifolds and Holomorphic Mappings (1970)
3. *V. S. Vladimirov*, Equations of Mathematical Physics (A. Jeffrey, ed.; A. Littlewood, trans.) (1970)
4. *B. N. Pshenichnyi*, Necessary Conditions for an Extremum (L. Neustadt, translation ed.; K. Makowski, trans.) (1971)
5. *L. Narici et al.*, Functional Analysis and Valuation Theory (1971)
6. *S. S. Passman*, Infinite Group Rings (1971)
7. *L. Dornhoff*, Group Representation Theory. Part A: Ordinary Representation Theory. Part B: Modular Representation Theory (1971, 1972)
8. *W. Boothby and G. L. Weiss, eds.*, Symmetric Spaces (1972)
9. *Y. Matsushima*, Differentiable Manifolds (E. T. Kobayashi, trans.) (1972)
10. *L. E. Ward, Jr.*, Topology (1972)
11. *A. Babakhanian*, Cohomological Methods in Group Theory (1972)
12. *R. Gilmer*, Multiplicative Ideal Theory (1972)
13. *J. Yeh*, Stochastic Processes and the Wiener Integral (1973)
14. *J. Barros-Neto*, Introduction to the Theory of Distributions (1973)
15. *R. Larsen*, Functional Analysis (1973)
16. *K. Yano and S. Ishihara*, Tangent and Cotangent Bundles (1973)
17. *C. Procesi*, Rings with Polynomial Identities (1973)
18. *R. Hermann*, Geometry, Physics, and Systems (1973)
19. *N. R. Wallach*, Harmonic Analysis on Homogeneous Spaces (1973)
20. *J. Dieudonné*, Introduction to the Theory of Formal Groups (1973)
21. *I. Vaisman*, Cohomology and Differential Forms (1973)
22. *B.-Y. Chen*, Geometry of Submanifolds (1973)
23. *M. Marcus*, Finite Dimensional Multilinear Algebra (in two parts) (1973, 1975)
24. *R. Larsen*, Banach Algebras (1973)
25. *R. O. Kujala and A. L. Vitter, eds.*, Value Distribution Theory: Part A; Part B: Deficit and Bezout Estimates by Wilhelm Stoll (1973)
26. *K. B. Stolarsky*, Algebraic Numbers and Diophantine Approximation (1974)
27. *A. R. Magid*, The Separable Galois Theory of Commutative Rings (1974)
28. *B. R. McDonald*, Finite Rings with Identity (1974)
29. *J. Satake*, Linear Algebra (S. Koh et al., trans.) (1975)
30. *J. S. Golan*, Localization of Noncommutative Rings (1975)
31. *G. Klambauer*, Mathematical Analysis (1975)
32. *M. K. Agoston*, Algebraic Topology (1976)
33. *K. R. Goodearl*, Ring Theory (1976)
34. *L. E. Mansfield*, Linear Algebra with Geometric Applications (1976)
35. *N. J. Pullman*, Matrix Theory and Its Applications (1976)
36. *B. R. McDonald*, Geometric Algebra Over Local Rings (1976)
37. *C. W. Groetsch*, Generalized Inverses of Linear Operators (1977)
38. *J. E. Kuczkowski and J. L. Gersting*, Abstract Algebra (1977)
39. *C. O. Christenson and W. L. Voxman*, Aspects of Topology (1977)
40. *M. Nagata*, Field Theory (1977)
41. *R. L. Long*, Algebraic Number Theory (1977)
42. *W. F. Pfeffer*, Integrals and Measures (1977)
43. *R. L. Wheeden and A. Zygmund*, Measure and Integral (1977)
44. *J. H. Curtiss*, Introduction to Functions of a Complex Variable (1978)
45. *K. Hrbacek and T. Jech*, Introduction to Set Theory (1978)
46. *W. S. Massey*, Homology and Cohomology Theory (1978)
47. *M. Marcus*, Introduction to Modern Algebra (1978)
48. *E. C. Young*, Vector and Tensor Analysis (1978)
49. *S. B. Nadler, Jr.*, Hyperspaces of Sets (1978)
50. *S. K. Segal*, Topics in Group Kings (1978)
51. *A. C. M. van Rooij*, Non-Archimedean Functional Analysis (1978)
52. *L. Corwin and R. Szczarba*, Calculus in Vector Spaces (1979)
53. *C. Sadosky*, Interpolation of Operators and Singular Integrals (1979)

54. *J. Cronin*, Differential Equations (1980)
55. *C. W. Groetsch*, Elements of Applicable Functional Analysis (1980)
56. *I. Vaisman*, Foundations of Three-Dimensional Euclidean Geometry (1980)
57. *H. I. Freedan*, Deterministic Mathematical Models in Population Ecology (1980)
58. *S. B. Chae*, Lebesgue Integration (1980)
59. *C. S. Rees et al.*, Theory and Applications of Fourier Analysis (1981)
60. *L. Nachbin*, Introduction to Functional Analysis (R. M. Aron, trans.) (1981)
61. *G. Orzech and M. Orzech*, Plane Algebraic Curves (1981)
62. *R. Johnsonbaugh and W. E. Pfaffenberger*, Foundations of Mathematical Analysis (1981)
63. *W. L. Voxman and R. H. Goetschel*, Advanced Calculus (1981)
64. *L. J. Corwin and R. H. Szczarba*, Multivariable Calculus (1982)
65. *V. I. Istrățescu*, Introduction to Linear Operator Theory (1981)
66. *R. D. Järvinen*, Finite and Infinite Dimensional Linear Spaces (1981)
67. *J. K. Beem and P. E. Ehrlich*, Global Lorentzian Geometry (1981)
68. *D. L. Armacost*, The Structure of Locally Compact Abelian Groups (1981)
69. *J. W. Brewer and M. K. Smith, eds.*, Emmy Noether: A Tribute (1981)
70. *K. H. Kim*, Boolean Matrix Theory and Applications (1982)
71. *T. W. Wieting*, The Mathematical Theory of Chromatic Plane Ornaments (1982)
72. *D. B. Gauld*, Differential Topology (1982)
73. *R. L. Faber*, Foundations of Euclidean and Non-Euclidean Geometry (1983)
74. *M. Carmeli*, Statistical Theory and Random Matrices (1983)
75. *J. H. Carruth et al.*, The Theory of Topological Semigroups (1983)
76. *R. L. Faber*, Differential Geometry and Relativity Theory (1983)
77. *S. Barnett*, Polynomials and Linear Control Systems (1983)
78. *G. Karpilovsky*, Commutative Group Algebras (1983)
79. *F. Van Oystaeyen and A. Verschoren*, Relative Invariants of Rings (1983)
80. *I. Vaisman*, A First Course in Differential Geometry (1984)
81. *G. W. Swan*, Applications of Optimal Control Theory in Biomedicine (1984)
82. *T. Petrie and J. D. Randall*, Transformation Groups on Manifolds (1984)
83. *K. Goebel and S. Reich*, Uniform Convexity, Hyperbolic Geometry, and Nonexpansive Mappings (1984)
84. *T. Albu and C. Năstăsescu*, Relative Finiteness in Module Theory (1984)
85. *K. Hrbacek and T. Jech*, Introduction to Set Theory: Second Edition (1984)
86. *F. Van Oystaeyen and A. Verschoren*, Relative Invariants of Rings (1984)
87. *B. R. McDonald*, Linear Algebra Over Commutative Rings (1984)
88. *M. Namba*, Geometry of Projective Algebraic Curves (1984)
89. *G. F. Webb*, Theory of Nonlinear Age-Dependent Population Dynamics (1985)
90. *M. R. Bremner et al.*, Tables of Dominant Weight Multiplicities for Representations of Simple Lie Algebras (1985)
91. *A. E. Fekete*, Real Linear Algebra (1985)
92. *S. B. Chae*, Holomorphy and Calculus in Normed Spaces (1985)
93. *A. J. Jerri*, Introduction to Integral Equations with Applications (1985)
94. *G. Karpilovsky*, Projective Representations of Finite Groups (1985)
95. *L. Narici and E. Beckenstein*, Topological Vector Spaces (1985)
96. *J. Weeks*, The Shape of Space (1985)
97. *P. R. Gribik and K. O. Kortanek*, Extremal Methods of Operations Research (1985)
98. *J.-A. Chao and W. A. Woyczynski, eds.*, Probability Theory and Harmonic Analysis (1986)
99. *G. D. Crown et al.*, Abstract Algebra (1986)
100. *J. H. Carruth et al.*, The Theory of Topological Semigroups, Volume 2 (1986)
101. *R. S. Doran and V. A. Belfi*, Characterizations of C*-Algebras (1986)
102. *M. W. Jeter*, Mathematical Programming (1986)
103. *M. Altman*, A Unified Theory of Nonlinear Operator and Evolution Equations with Applications (1986)
104. *A. Verschoren*, Relative Invariants of Sheaves (1987)
105. *R. A. Usmani*, Applied Linear Algebra (1987)
106. *P. Blass and J. Lang*, Zariski Surfaces and Differential Equations in Characteristic $p > 0$ (1987)
107. *J. A. Reneke et al.*, Structured Hereditary Systems (1987)
108. *H. Busemann and B. B. Phadke*, Spaces with Distinguished Geodesics (1987)
109. *R. Harte*, Invertibility and Singularity for Bounded Linear Operators (1988)

Additional Volumes in Preparation

FUNDAMENTALS OF DOMINATION IN GRAPHS

Teresa W. Haynes
East Tennessee State University
Johnson City, Tennessee

Stephen T. Hedetniemi
Clemson University
Clemson, South Carolina

Peter J. Slater
University of Alabama in Huntsville
Huntsville, Alabama

MARCEL DEKKER, INC.   NEW YORK · BASEL · HONG KONG

ISBN 0-8247-0033-3

The publisher offers discounts on this book when ordered in bulk quantities. For more information, write to Special Sales/Professional Marketing at the address below.

This book is printed on acid-free paper.

MARCEL DEKKER, INC.
270 Madison Avenue, New York, New York 10016
http://www.dekker.com

Current printing (last digit):
10 9 8 7 6 5 4 3 2 1

PRINTED IN THE UNITED STATES OF AMERICA

Dedicated to our friend and colleague,

Ernie Cockayne
University of Victoria

a pioneer and prolific researcher in domination

$$ir(G) \leq \gamma(G) \leq i(G) \leq \beta_0(G) \leq \Gamma(G) \leq IR(G)$$

To Elizabeth (Nell) Hamilton Gilliam,
my grandmother, mother, and friend
who has always been there for me.

Teresa Haynes

To Sandee, my wife, my colleague,
and my best friend.

Steve Hedotniemi

To Mary Elizabeth Fay Slater
who left a trail of kindness.

For Paul and Meghan
that their mistakes
be ones of commission,
not of omission.

Pete Slater

Preface

Within the last twenty-five years, concurrent with the growth of such areas as computer science, electrical and computer engineering, and operations research, graph theory has seen explosive growth. Perhaps the fastest-growing area within graph theory is the study of domination and related subset problems, such as independence, covering, and matching.

This book provides the first comprehensive treatment of domination in graphs. Because the vast majority of results in this important area of research have been very recently obtained, the comprehensive treatment that we provide is unavailable elsewhere. Part of the reason for the increased interest in these subset problems is their many and varied applications in such fields as linear algebra and optimization, design and analysis of communication networks, social sciences, computational complexity, and algorithm design. Consequently, this book will be of interest to mathematicians, computer scientists, operations researchers, economists, social scientists, electrical and computer engineers, chemists, systems engineers, and many others.

Our goal is to provide a general treatment of the theoretical, algorithmic, and application aspects of domination in graphs. Our objectives are:

- to present results in a manner that allows the reader to quickly learn the mathematical tools and methods used to conduct research in this field

- to consolidate and organize much of the material in the more than 1200 papers already published on domination in graphs

- to explain the most fundamental results on domination in graphs in an easy-to-understand style

- to provide a comprehensive bibliography on domination

- to provide a unified terminology and notation that is consistent with the majority of papers in the field

- to identify many unsolved problems in the area

- to present many of the applications of domination and related subset problems

- to provide a representative sample of domination algorithms and NP-completeness results

- to illustrate many different mathematical perspectives from which one can study and generalize the concept of domination in graphs.

v

The result of these objectives is a book that is appropriate for use at several different levels. Containing the only existing comprehensive bibliography on domination, it is a valuable reference book. It is suitable as a textbook for graduate level courses in graph theory and could also be used by advanced undergraduates. As a textbook, it is complete with exercises ranging in level from introductory and easy to advanced and difficult. The book is self-contained, with the necessary basic definitions and preliminary graph theoretic results presented in the prolegomenon. A second course in graph theory can treat the prolegomenon as a reference and begin with Chapter 1. A majority of the readers of this text will have had some exposure —if not a full course— to graph theory but it could well be used for a first course in graph theory. Instructors who want to use it in a first course can expand beyond subset problems by developing the topics in the prolegomenon. It can also be used as a supplemental text for courses in graph theory, combinatorics, and related areas. As a definitive reference guide on domination in graphs, it will be a valuable tool to researchers and advanced graduate students. They can use the book to familiarize themselves with the subject, the research techniques, and major research accomplishments in the field. Advanced graduate students will find many topics that can be developed into masters theses and Ph.D. dissertations.

We note that this area is rapidly being developed and the bibliography constantly needs updating. We plan to maintain the bibliography in the future and would appreciate updates from you. We have tried to eliminate mistakes from this text, but surely some remain. We would appreciate being informed of any that you may find and/or your suggestions and comments. Our e-mail addresses are Teresa W. Haynes at haynes@etsu.etsu-tn.edu, Stephen T. Hedetniemi at hedet@cs.clemson.edu, and Peter J. Slater at slater@math.uah.edu.

Teresa W. Haynes
Stephen T. Hedetniemi
Peter J. Slater

Acknowledgments

Many people have contributed to the development of this field. We acknowledge the fundamental efforts of Berge and Ore, who mathematically formulated the concept of domination in graphs. The bibliography lists more than 300 authors, many of whom have made significant contributions. Among the most prolific are Ernie Cockayne, Michael Henning, Renu Laskar, Christine Mynhardt, and Bohdan Zelinka.

We acknowledge and sincerely thank the following people who graciously read individual chapters and offered many helpful suggestions: Ashok Amin, Tony Barkauskas, Julie Carrington, Gerard Chang, Gary Chartrand, Jean Dunbar, Odile Favaron, Gerd Fricke, Dana Grinstead-Tanner, Frank Harary, Bert Hartnell, Johan Hattingh, Sandee Hedetniemi, Mike Henning, David Jacobs, Mike Jacobson, Debra Knisley, Dieter Kratsch, Renu Laskar, Linda Lawson, Lisa Markus, Debbie Meiers, Christine Mynhardt, Doug Rall, Brooks Reid, Lorna Stewart, Jan Arne Telle, Lucas Vandermerwe, and Lutz Volkmann.

We appreciate the support and help from the Mathematics Department of East Tennessee State University, the Computer Science Department of Clemson University, and the Mathematical Sciences Department of the University of Alabama at Huntsville. Specifically, we thank George Poole for providing a congenial working environment and Sherri Renfro for all her help.

We give special thanks to Ben Phillips for his invaluable assistance in the preparation of this book. His tireless efforts are greatly appreciated.

Many people, each in his/her own way, have influenced the writing of this book. Although I don't attempt to list them all, I sincerely appreciate everyone. In particular, I thank the following, my friends and colleagues, who have given me support and encouragement: Jean Dunbar for the fun times, constant encouragement, and friendship; Gayla Domke, Debra Knisley, and Linda Lawson for their support and caring; Mike Henning for his friendship and prayers; Frank Harary for the encouragement and all he has taught me; Lisa Markus for the kind words and hours in the gym; Renu Laskar for opening her home to me; Christine Mynhardt for encouraging me to vasbyt, thanks Kieka; Ernie Cockayne for reminding me to relax and assuring me that I didn't have to write this book; and Gary Chartrand for his friendship and the algorithm for finding all errors in the book ("find all errors on page x"). Of course, I must mention the joy of working with my two coauthors, Steve and Pete, our friendship endured the domination. Again I thank Ben Phillips for the tremendous amount of work he did in preparation of this book and for always being available when I needed his help. Also, I thank Mark Buckner who spent much of one summer helping compile the bibliography.

Surely, my part in writing this book could not have been completed without the support of my family. I thank all of them. Foremost, I thank my husband,

Ben Buckner, for his love, encouragement, and understanding about all the times we had to be apart physically while I was off "playing with the boys," for his ability and willingness to understand when I was so often preoccupied with domination, and generally for believing in me. Finally, I want to thank God and acknowledge that He is the source of all wisdom and of my strength and joy.

Teresa W. Haynes

As the bibliography at the end of this book attests, hundreds of people have made contributions to the field of domination in graphs. It is because of these researchers that this book is written, and we acknowledge their many excellent contributions. In particular, I acknowledge the many and creative contributions of Ernie Cockayne. If ever a field of study had a leading researcher, it is Ernie, my frequent coauthor and very good friend. This one's for you, Ern!

For the past 15 years at Clemson University, we have held a research seminar, whose members have made many contributions to the study of domination in graphs. I am honored to acknowledge each of them: Sandee Hedetniemi, Renu Laskar, Jean Dunbar, Doug Rall, Lisa Markus, David Jacobs, Eleanor Hare, Alice McRae, James Knisely, Charles Wallis, Fred Harris, Gerd Fricke our good friend and colleague at Wright State University, Aniket Majumdar, Srini Madella, Dan Pillone, Ken Peters, John Pfaff, Gayla Domke, Johan Hattingh and Mike Henning, both of whom visited us from South Africa, Bert Hartnell, who visited us several times from Nova Scotia, and Tom Wimer, who started this seminar many years ago. Thanks to all.

Finally, permit me to add a personal comment. In 1962 Oystein Ore first defined the domination number of a graph. When Ernie Cockayne and I first started to study dominating sets in graphs in 1975, we accepted Ore's terminology. I have often wondered, however, if we should have changed this term. It has been said "fear is to domination, as love is to dominion." Thus, I can't help thinking that we would have sent a more positive message to researchers in this field had we changed Ore's terminology to the dominion number of a graph.

Steve Hedetniemi

I must first thank Sandy Reeder, who endured many moments of my being distracted by ideas for this book, and many other things.

Those of us who have wisely chosen to do our research in graph theory have been fortunate. There is a worldwide network of wonderful people in this research area. My personal good fortune includes having worked with scores of them, especially Teresa and Steve. Several years of collaborative efforts on these books leave me with no doubt that, in her and his own ways, they are even weirder than I am.

It is probably from our students that we learn the most. So thanks to DLG, for the joy and the agony; to CBS, for the interest and industry; to DEL, for marching to the beat of his choice; to RG, for being the paradigm of a scholar; and to WJS, for sharing some beer, for appropriately turning pale and saying "But that means ...," and thereby saving what little is left of my sanity.

PJS

Contents

Prolegomenon

This book concerns dominating sets and other related subsets in graphs. Our goal is to present a self-contained, comprehensive treatment of this subject - an area of graph theory that is rich in history, applications, interesting questions and results, and unsolved research questions.

This introduction contains most of the necessary definitions and some preliminary graph theoretic results. (However, the formal definition of "domination" is not presented until Chapter 1.) Experienced graph theorists can skip to the section titled Theme and Plot.

1 Graph Theory: Terminology and Concepts

As we shall see, there are many useful applications of graph theory. However, similar to the development of probability theory, where much of the original work was motivated by efforts to understand games of chance, portions of graph theory have been motivated by the study of games and recreational mathematics. Nevertheless, even in these instances, this had led to important areas of graph theory. It is within this game theoretic context that we present many of the definitions and basic concepts of graph theory in this section, with some mention of applications.

We begin by considering a four-by-four chessboard with 16 squares available for the placement of chess pieces, each of the 16 squares being represented by a point labelled with one of the letters in $\{a, b, c, ..., n, o, p\}$. To indicate that a chess piece can move from one square to another in a single move, we draw a line segment joining points corresponding to these letters. In Figures 1(a) and 1(b) we have assumed the piece is a rook or a knight, respectively.

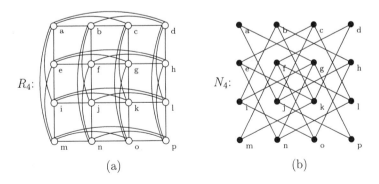

Figure 1: The rooks graph R_4 and knights graphs N_4.

Some of the questions one asks are : The sequence $(a, b, c, d, h, g, f, e, i, j, k, l,$ $p, o, n, m, a)$ of moves takes a rook from point a to each other point and back to a with every point entered exactly once. Is there such a sequence for knight moves in N_4? What is the minimum number of points for R_4 or N_4 on which we can place rooks or knights, respectively, so that every (other) point is within one move of one of these points? What is the maximum number of points for R_4 or N_4 so that rooks or knights, respectively, placed on these points have the property that no two can capture each other? In order to formulate problems of these types in more general settings, some definitions are useful.

A *graph* $G = (V, E)$ consists of a (finite) set denoted by V, or by $V(G)$ if one needs to make clear which graph is under consideration, and a collection E, or $E(G)$, of unordered pairs $\{u, v\}$ of distinct elements from V. Each element of V is called a *vertex* (or a point, or a node), and each element of E is called an *edge* (or a line, or a link). The number of vertices, the cardinality of V, is called the *order* of G and is denoted by $|V|$, and $|E|$ is called the *size* of G. We usually use n to denote the order and m the size and typically have $V(G) = \{v_1, v_2, ..., v_n\}$. We write $v_i v_j \in E(G)$ to mean $\{v_i, v_j\} \in E(G)$, and if $e = v_i v_j \in E(G)$, we say that v_i and v_j are *adjacent* and that e and v_i are *incident*. For example, $V(R_4) = V(N_4) = \{a, b, , c, ..., p\}$, $fg \in E(R_4)$, $fg \notin E(N_4)$, and so f and g are adjacent in R_4 but not in N_4.

The *open neighborhood* $N(v)$ of the vertex v consists of the set of vertices adjacent to v, that is, $N(v) = \{w \in V : vw \in E\}$, and the *closed neighborhood* of v is $N[v] = N(v) \cup \{v\}$. For example, in the graph N_4, we have $N(h) = \{b, j, o\}$. For a set $S \subseteq V$, the *open neighborhood* $N(S)$ is defined to be $\cup_{v \in S} N(v)$, and the *closed neighborhood* of S is $N[S] = N(S) \cup S$. The *degree* $deg(v)$ of v is the number of edges incident with v or, equivalently, $deg(v) = |N(v)|$. We have $deg_{N_4}(h) = 3$ and $deg_{R_4}(h) = 6$. The *degree sequence* of G is $(deg(v_1), deg(v_2), \cdots, deg(v_n))$, typically written in nondecreasing or nonincreasing order. The *minimum* and *maximum degrees* of vertices in $V(G)$ are denoted by $\delta(G)$ and $\Delta(G)$, respectively. If $\delta(G) = \Delta(G) = r$, then G is said to be *regular* of degree r, or simply r-regular. A 3-regular graph is also called a *cubic graph*. The graph N_4 has $\delta(N_4) = 2$ and $\Delta(N_4) = 4$, and R_4 is regular of degree six. Almost everyone's first theorem in graph theory is the following.

Theorem 1 *For a graph G of size $|E| = m$,*

$$\sum_{v \in V} deg(v) = 2m.$$

Proof. One can simply count the number of incidences in two ways. First, each vertex v is in $deg\ v$ incidences. Alternatively, each of the m edges has two incidences. $\square$

A *walk* of length k is an alternating sequence $W = u_0, e_1, u_1, e_2, u_2, e_3, ..., u_{k-1},$ e_k, u_k of vertices and edges with $e_i = u_{i-1} u_i$. Because every two distinct vertices

	a	b	c	d	e	f	g	h	i	j	k	l	m	n	o	p	ecc	dist
a	0	3	2	5	3	4	1	2	2	1	4	3	5	2	3	2	5	42
b	3	0	3	2	2	3	2	1	1	2	1	4	2	3	2	3	4	34
c	2	3	0	3	1	2	3	2	4	1	2	1	3	2	3	2	4	34
d	5	2	3	0	2	1	4	3	3	4	1	2	2	3	2	5	5	42
e	3	2	1	2	0	3	2	3	3	2	1	2	2	1	4	3	4	34
f	4	3	2	1	3	0	3	2	2	3	2	1	1	2	1	4	4	34
g	1	2	3	4	2	3	0	3	1	2	3	2	4	1	2	1	4	34
h	2	1	2	3	3	2	3	0	2	1	2	3	3	4	1	2	4	34
i	2	1	4	3	3	2	1	2	0	3	2	3	3	2	1	2	4	34
j	1	2	1	4	2	3	2	1	3	0	3	2	4	3	2	1	4	34
k	4	1	2	1	1	2	3	2	2	3	0	3	1	2	3	4	4	34
l	3	4	1	2	2	1	2	3	3	2	3	0	2	1	2	3	4	34
m	5	2	3	2	2	1	4	3	3	4	1	2	0	3	2	5	5	42
n	2	3	2	3	1	2	1	4	2	3	2	1	3	0	3	2	4	34
o	3	2	3	2	4	1	2	1	1	2	3	2	2	3	0	3	4	34
p	2	3	2	5	3	4	1	2	2	1	4	3	5	2	3	0	5	42

Figure 2: The distance matrix for the graph N_4.

are either nonadjacent or are incident with exactly one common edge, W can be denoted more simply as $W = u_0, u_1, u_2, ..., u_k$ (then, implicitly, $u_i u_{i+1} \in E$ for $0 \leq i \leq k - 1$). If all k edges are distinct, then W is called a *trail*. If $u_0 = u_k$, then W is said to be *closed*, and a closed trail is a *circuit*. A walk with $k+1$ distinct vertices $u_0, u_1, ..., u_k$ is a *path*, and if $u_0 = u_k$ but $u_1, u_2, ..., u_k$ are distinct, then the trail is a *cycle* (more specifically a k-*cycle*). In N_4, the walk $W = g, a, j, p, g, i, b, h, o, f, d, k, m, f, l, n, e, c$ is a trail but not a path. It is *spanning* because it contains every vertex at least once.

If $u_0 = v$ and $u_k = x$, then W is said to be a $v - x$ *walk* of length k. A graph G is *connected* if for every pair v, x of vertices there exists a $v - x$ path; otherwise, G is *disconnected*. If there exists at least one $v - x$ walk, then the *distance* $d(v, x)$ between v and x is the minimum length of a $v - x$ walk; if no $v - x$ walk exists, then we write $d(v, x) = \infty$. The *ith open neighborhood* of v for $i \geq 0$ consists of those vertices at distance i from v, and is denoted by $N^i(v) = \{w \in V : d(v, w) = i\}$. The *ith closed neighborhood* is $N^i[v] = \{w \in V : d(v, w) \leq i\}$.

In particular, $N^0(v) = N^0[v] = \{v\}$, $N^1(v) = N(v)$, and $N^1[v] = N[v]$. For the graph N_4, $N^0(a) = \{a\}$, $N^1(a) = \{j, g\}$, $N^2(a) = \{i, n, p, c, h\}$, $N^3(a) = \{b, o, e, l\}$, $N^4(a) = \{k, f\}$, and $N^5(a) = \{d, m\}$. The *eccentricity* of vertex v is $ecc(v) = \max\{d(v, w) : w \in V\}$. The *radius* of G is $rad(G) = \min\{ecc(v) : v \in V\}$ and the *diameter* of G is $diam(G) = \max\{ecc(v) : v \in V\}$. Clearly every $v \in V(R_4)$ has $ecc(v) = 2$, so $rad(R_4) = diam(R_4) = 2$. Note that $rad(N_4) = ecc(b) = ecc(f) = 4$, and $diam(N_4) = ecc(a) = 5$. The matrix in

Figure 2 lists the distances for all pairs of vertices in N_4. In addition to the eccentricity of a vertex, another measure of the centrality of a vertex v is its *distance*, $dist(v) = \sum_{w \in V(G)} d(v, w)$.

There are, in fact, several matrices commonly used to represent a graph. For a graph G with $V = \{v_1, v_2, \cdots, v_n\}$ and $E = \{e_1, e_2, \cdots, e_m\}$, the *adjacency matrix* $A = [a_{i,j}]$ is the n-by-n binary matrix with $a_{i,j} = 1$ if and only if $v_i v_j \in E$. Note that the *ith* row and the *ith* column each contain $deg(v_i)$ ones and $n - deg(v_i)$ zeros, and A is symmetric. We can equivalently define A by $a_{i,j} = 1$ if and only if $v_i \in N(v_j)$. Let I_n be the n-by-n identity matrix. The n-by-n binary *closed neighborhood matrix* $N = [n_{i,j}] = A + I_n$ has $n_{i,j} = 1$ if and only if $v_i \in N[v_j]$. The *ith* row and column of N have $1 + deg(v_i)$ ones, and N is also symmetric. The n-by-m *incidence matrix* $H = [h_{i,j}]$ is the binary matrix with $h_{i,j} = 1$ if and only if v_i is incident with edge e_j. The n-by-n *distance matrix* $D = [d_{i,j}]$ has $d_{i,j} = d(v_i, v_j)$. See Figure 3.

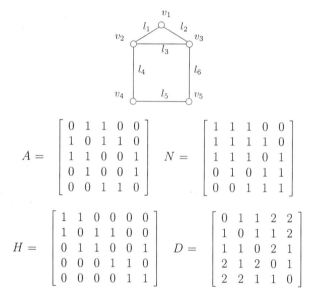

Figure 3: Adjacency, closed neighborhood, incidence, and distance matrices.

As one might expect, there are many classes of graphs in which we are particularly interested. For example, we have the class of cycles. The *cycle C_n* of order $n \geq 3$ has size $m = n$, is connected and is 2-regular. A tree T is a connected graph with no cycles. The trees of order five and six are presented in Figure 4. The path P_n is the tree of order n with diameter $n - 1$, and the *star* $K_{1,n-1}$ has one vertex v of degree $n - 1$ and $n - 1$ vertices of degree one. In any graph, a vertex of degree one is called an *endvertex*, and an edge incident with an endvertex is called a *pendant edge*.

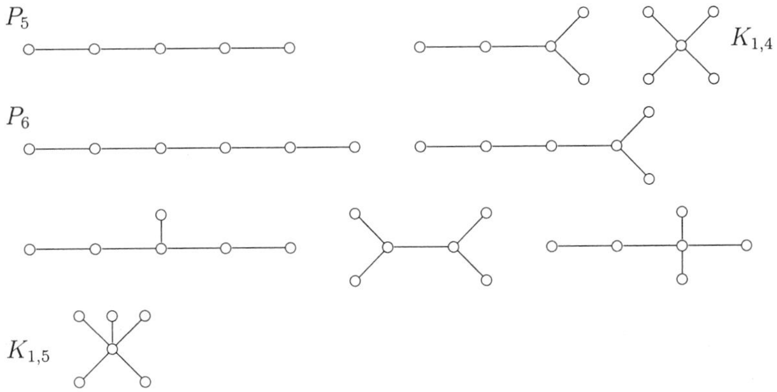

Figure 4: The trees of order five and six.

Theorem 2 *For any graph T, the following are equivalent.*

(a) *T is a tree (that is, T is connected and acyclic).*

(b) *T is connected and $m = n - 1$.*

(c) *T is acyclic and $m = n - 1$.*

(d) *For any two vertices v and w in $V(T)$, there exists a unique $v - w$ path.*

Theorem 3 *Every tree of order $n \geq 2$ has at least two endvertices.*

For the tree T in Figure 5, the vertex v is most central with respect to the eccentricity measure, namely, $rad(T) = ecc(v) = 4 \leq ecc(u)$ for every $u \in V(T)$. Note that $diam(T) = ecc(x) = 8$. The most central vertex relative to the distance measure, however, is w with $dist(w) = 35 \leq dist(u)$ for every $u \in V(T)$.

Theorem 4 *For any connected graph G, $rad(G) \leq diam(G) \leq 2\,rad(G)$.*

The *center* $C(G)$ of G is the set of vertices of minimum eccentricity, namely, $C(G) = \{v \in V(G) : ecc(v) \leq ecc(u) \text{ for all } u \in V(G)\}$. The *median* $M(G)$ of G is the set of vertices of minimum distance, that is, $M(G) = \{v \in V(G) : dist(v) \leq dist(u) \text{ for all } u \in V(G)\}$. In 1869 Jordan [751] proved a theorem equivalent to the following result.

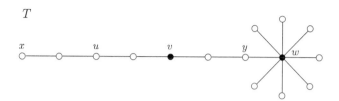

Figure 5: A tree T with $rad(T) = ecc(v) = 4$ and $dist(w) = 35$.

Theorem 5 *For any tree T, both the center $C(T)$ and the median $M(T)$ consist of one vertex or two adjacent vertices.*

Note that if v is an endvertex of a tree T, then the graph $T - v$, obtained from T by deleting v from $V(T)$ and the edge incident with v from $E(T)$, remains connected and is acyclic, so $T - v$ is also a tree. (This simple observation forms the basis for many inductive proofs about trees.) More generally, we are frequently interested in the substructure of a given graph G. We say that H is a *subgraph* of G, denoted $H < G$, if $V(H) \subseteq V(G)$ and $uv \in E(H)$ implies $uv \in E(G)$. If a subgraph H satisfies the added property that for every pair u, v of vertices, $uv \in E(H)$ if and only if $uv \in E(G)$, then H is called an *induced subgraph* of G. The induced subgraph H with $S = V(H)$ is called the *subgraph induced by S* and is denoted by $\langle S \rangle$. A *component* of a graph G is a connected subgraph not properly contained in any other connected subgraph. Clearly, G has exactly one component if and only if G is connected. In the graph R_4, the induced subgraph $\langle \{a, c, k, i, b, d, l, j\} \rangle$ consists of two components, each of which is a 4-cycle C_4. For $S \subseteq V$, we write $G - S$ to denote the induced subgraph $\langle V - S \rangle$.

Among all connected graphs of order n, trees have the minimum number of edges and the *complete graph*, denoted K_n, has the maximum possible $n(n-1)/2$ edges. The *complement* $\overline{G}$ of graph G has $V(\overline{G}) = V(G)$ and $uv \in E(\overline{G})$ if and only if $uv \notin E(G)$. In particular, $\overline{K}_n$ has n vertices and no edges. In N_4 each of $\langle \{a, c, f, h, i, k, n, p\} \rangle$ and $\langle \{b, d, e, g, j, l, m, o\} \rangle$ is a $\overline{K}_8$. The graph N_4 is therefore an example of a *bipartite graph*, that is, a graph G for which V can be partitioned as $V = V_1 \cup V_2$ with no two adjacent vertices in the same V_i. For example, trees are bipartite graphs. The *complete multipartite graph* $K_{n_1, n_2, \ldots, n_t}$ has $n = n_1 + n_2 + \ldots + n_t$ vertices and $V(K_{n_1, n_2, \ldots n_t}) = S_1 \cup S_2 \cup \ldots \cup S_t$, where $|S_i| = n_i$ for $1 \le i \le t$, $\{u, v\} \subseteq S_i$ implies u and v are not adjacent, and $u \in S_i$ and $v \in S_j$ with $i < j$ implies u and v are adjacent. Of particular interest are the complete bipartite graphs $K_{r,s}$, and, specifically, $K_{1, n-1}$ is called a *star*.

Note that $K_{2,2}$ and C_4 can therefore be considered to be the same graph. Technically, graphs G and H are *identical* if $V(G) = V(H)$ and $E(G) = E(H)$, or equivalently $G < H$ and $H < G$. We say that G and H are *isomorphic*, denoted

$G = H$, if there exists a bijection $\psi : V(G) \to V(H)$ such that $uv \in E(G)$ if and only if $\psi(u)\psi(v) \in E(H)$. The graphs P and R in Figure 6 are isomorphic. For example, we can let $\psi(1) = a$, $\psi(2) = i$, $\psi(3) = d$, $\psi(4) = e$, $\psi(5) = f$, $\psi(6) = g$, $\psi(7) = j$, $\psi(8) = h$, $\psi(9) = b$, and $\psi(10) = c$.

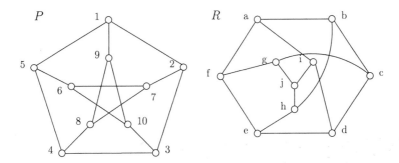

Figure 6: Isomorphic copies of the Petersen graph.

We next describe methods of constructing a third graph from two given graphs G and H. The *union* $K = G \cup H$ has $V(K) = V(G) \cup V(H)$ and $E(K) = E(G) \cup E(H)$. For disjoint graphs G and H, the *join* $K = G + H$ has $V(K) = V(G) \cup V(H)$ and $E(K) = E(G) \cup E(H) \cup \{uv : u \in V(G)$ and $v \in V(H)\}$. The *(cartesian) product* $K = G \times H$ has $V(K) = V(G) \times V(H)$ and vertices (u_1, v_1) and (u_2, v_2) in $V(K)$ are adjacent if and only if either $u_1 = u_2$ and $v_1 v_2 \in E(H)$ or $v_1 = v_2$ and $u_1 u_2 \in E(G)$. The *grid* $G_{j,k}$ is $P_j \times P_k$; the *cylinder* $C_{j,k}$ for $j \geq 3$ is $C_j \times P_j$; and the *torus* $T_{j,k}$ for $j \geq 3$ and $k \geq 3$ is $C_j \times C_k$. See Figure 7.

For a graph G with edges, the *line graph* $L(G)$ is the graph whose vertices correspond to the edges (lines) of G, and two vertices in $L(G)$ are adjacent if and only if the corresponding edges in G are adjacent (that is, are incident with a common vertex). See Figure 8 for the line graph of the Petersen graph P from Figure 6.

2   Two Applications

2.1   Centrality and domination for facilities location

Suppose that each vertex in a graph represents a site where customers are located, and we can choose one or more sites at which to locate facilities to serve these customers optimally. Measures of optimality typically involve centrality measures such as choosing centers, medians, or centroids. For example, suppose we have a fixed number p of facilities to locate. If we want to minimize the

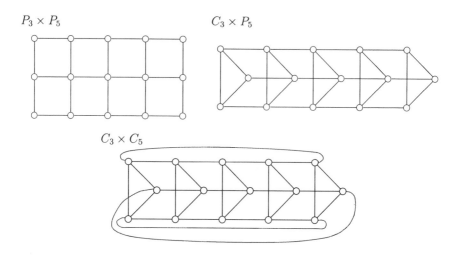

Figure 7: A grid, a cylinder, and a torus.

maximum distance a customer has to travel in order to get to the facility vertex closest to him/her, then we have the well-studied *p-center problem*. If, on the other hand, we wish to minimize the average distance of a customer to his/her nearest facility vertex, then we have the also well-studied *p-median problem*. There is also a *centroidal* measure of centrality that is used in competitive facility location problems. (See P. J. Slater, Maximin facility location, *J. Res. Nat. Bur. Standards 79B* (1975) 107–115.) For these three problems one does as well as possible, given the constraint that only p facility vertices can be chosen. For the tree T in Figure 5, a 2-center solution is $\{u, y\}$, and $\{u, w\}$ is a 2-median solution.

Suppose we have a p-center type of problem in that we are interested in minimizing the maximum distance from a vertex to a set S of p facility vertices. For $v \in V$ and $S \subseteq V$, the *distance from v to S* is defined as $d(v, S) = \min\{d(v, s) : s \in S\}$, and the *eccentricity* of S is $ecc(S) = \max\{d(v, S) : v \in V\}$. The p-center problem is to find a set S of cardinality p in V so that the $ecc(S)$ is as small as possible. On the other hand, the value of p might not be given, but rather a positive integer k is given, and we require that $ecc(S) \leq k$. The problem now might be to minimize the cardinality of S, that is, to find $\min\{|S| : ecc(S) \leq k\}$. The *kth power graph* G^k of a graph G has the same vertex set as G with two vertices adjacent in G^k if and only if they are at a distance at most k in G. That is, $V(G^k) = V(G)$ and $E(G^k) = \{uv : d_G(u, v) \leq k\}$. Note that a vertex set S has eccentricity at most k in G if and only if it has eccentricity one in G^k, that is, $ecc_G(S) \leq k$ if and only if $ecc_{G^k}(S) \leq 1$. This provides one motivation for the basic definition given in Chapter 1 of a dominating set.

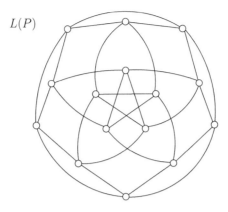

$L(P)$

Figure 8: The line graph $L(P)$ of the Petersen graph P.

2.2 Independence and chromatic number for scheduling

A college is currently offering classes $C_1, C_2, ..., C_n$, and needs to schedule final examinations so that whenever there is a student taking classes C_i and C_j, the examinations in these two classes must be given at different time periods. What is the least number of time periods required? We can construct a graph $G = (V, E)$ with $V = \{v_1, v_2, ..., v_n\}$, and let $v_i v_j \in E$ if and only if there is at least one student enrolled in both classes C_i and C_j. Note that the set of classes assigned to any one time period has the property that no two of its corresponding vertices are adjacent. The minimum time scheduling problem involves partitioning V into sets of this type.

For a graph $G = (V, E)$, a set $S \subset V$ is *independent* if no two vertices in S are adjacent. The *independence number* $\beta_0(G)$ is the maximum cardinality of an independent set in G. A maximum independent set is called a β_0-set. The minimum k such that we can partition $V = S_1 \cup S_2 \cup ... \cup S_k$, where each S_i is independent, is the *chromatic number* $\chi(G)$.

For the graph G in Figure 9, the set $\{1, 2, 5, 6, 8, 9\}$ is independent and $\beta_0(G) = 6$. Note that the only way to extend this to a partition of V into independent sets is $\{\{1, 2, 5, 6, 8, 9\}, \{3\}, \{4\}, \{7\}\}$. However, $\chi(G) = 3$ and a suitable partitioning of V is $\{\{1, 2, 7\}, \{3, 8, 9\}, \{4, 5, 6\}\}$.

An important concept in the study of graph subset problems is the distinction between "minimal" and "minimum" sets and between "maximal" and "maximum" sets with a given property. Consider the problem of actually computing the (maximum) independence number $\beta_0(G)$ and finding a β_0-set. Note that every subset of an independent set is independent. To attempt to find a β_0-set, one might start with $S = \emptyset$ and iterate the following procedure. Choose any

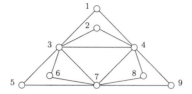

Figure 9: $\beta_0(G) = 6$ and $\chi(G) = 3$.

vertex $v \in V(G) - S$ that is not adjacent to any vertex in S, and replace S by
$S \cup \{v\}$. When no more vertices can be added to S, we have a maximal independent set. For example, consider the cycle C_8 in Figure 10, and assume we
examine the vertices in the order $(5,8,1,6,4,3,7,2)$. Vertices 5 and 8 are added to
S, and the next possible vertex one can add to S is the vertex 3. Then $\{5,8,3\}$ is
a *maximal independent set*, namely, an independent set S with the property that
any vertex set properly containing S is not independent. However, $\beta_0(C_8) = 4$
with maximum independent sets $\{1,3,5,7\}$ and $\{2,4,6,8\}$. The *lower independence number* $i(G)$ is the minimum cardinality of a maximal independent set of
G. Clearly, $i(C_8) = 3$. Also, for the star, $\beta_0(K_{1,n-1}) = n - 1$ and $i(K_{i,n-1}) = 1$.
For paths, $\beta_0(P_n) = \lceil n/2 \rceil$ and $i(P_n) = \lceil n/3 \rceil$. For grids, $\beta_0(P_j \times P_k) = \lceil jk/2 \rceil$
and, for j and k large, $i(P_j \times P_k)$ is approximately $jk/5$.

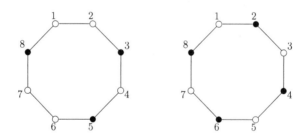

Figure 10: $\beta_0(C_8) = 4$ and $i(C_8) = 3$.

3   Theme and Plot

In this section we illustrate the type of problems which will be the primary
focus of this book: local subset problems. In the main, the phrase "(closed)
neighborhood subset problems" is accurate. A problem of this type typically
involves finding as a solution a subset S of V or E. Further, verification that S is

actually a solution can be accomplished, for example, by sequentially examining each $N[v]$ independent of the other closed neighborhoods, for $1 \leq i \leq n$. For example, if the problem is to decide whether $\beta_0(G) \geq k$, then we can verify that a given set S of cardinality $|S| \geq k$ is a solution by seeing if $|N(v_i) \cap S| = 0$ for each $v_i \in S$. To further illustrate what we mean, consider the following two problems.

Problem 1. The *girth* $g(G)$ of a graph G is the length of a shortest cycle in G, while its *circumference* $c(G)$ is the length of a longest cycle. Given a graph G and a value k, we might want to know if $c(G) \geq k$. In particular, if each vertex represents a city and each edge represents a direct link between two cities it joins, we might be interested in knowing whether there is a route for a sales representative that starts at a given city, visits each city exactly once, and returns to the starting city. That is, we want to know if $c(G) = n$ (or equivalently, if there exists a spanning cycle). If this is possible, that is, if $c(G) = n$, then G is said to be *hamiltonian*, and a spanning cycle is called a *hamiltonian cycle*. Starting with ordered lists $(v_1, v_2, ..., v_n)$ of V and $(e_1, e_2, ..., e_m)$ of E, a proposed solution can be defined by deciding, for $1 \leq i \leq m$, whether or not e_i is an edge to be used in the spanning cycle. How do we then verify that we have a valid solution? Certainly it is easy enough to do, but can we do it by a simple scan over one or both of the ordered lists? By examining each $N[v_i]$ one can easily verify if each v_i is incident with exactly two of the proposed edges, but a verification that we have one cycle rather than two or more requires a "global" tracing of the edges.

Problem 2. Again assume that each vertex represents a city and now an edge joining cities u and v indicates that a transmission from a tower at one of these cities will reach the other. Our goal might be to establish a smallest set of transmission towers so that every city either has a tower or is adjacent to a city that has one. For a given value k, we might want to know if there is such a set of k or fewer towers. For the ordered list $(v_1, v_2, ..., v_n)$, we can decide, for each i, $1 \leq i \leq n$, whether city v_i gets one of the k available towers. Having decided on a set S of cities to receive towers, with $|S| \leq k$, how do we verify that we have a valid solution? In this case, there is a "local" procedure. Namely, for $1 \leq i \leq n$, we can check to see if $N[v_i] \cap S \neq \emptyset$.

It is the kind of problem illustrated by Problem 2 with which we will be primarily concerned. Clearly, we are not suggesting that every graph theory problem can be categorized as either "global" or "local". For example, one might seek a circular bus route (a cycle in G) such that every vertex is adjacent to at least one vertex of the cycle. However, we think that the idea of a "local subset problem" is a unifying concept.

Any good novel ought to have a worthwhile theme worked out in a carefully constructed plot. We choose to state our theme as follows. In terms of the amount and significance of the results in the area, and in the interest and applications of the results, the work on graph theoretic subset problems rivals the work on structural problems.

We focus on subset problems in this book and have attempted to present an underlying framework for such problems. We certainly hope the readers experience as much joy in investigating these problems as we have.

EXERCISES

1 (a) For each of the following, find, if possible, a graph with the given degree sequence:

  (i) $(1, 2, 3, 4, 5, 6, 7, 8)$.

  (ii) $(1, 2, 3, 4, 4, 5, 6, 7)$.

  (iii) $(1, 2, 3, 3, 4, 5, 5, 6, 7)$.

  (b) Which sequences $(d_1, d_2, ..., d_n)$ can be realized as the degree sequence of at least one graph? of exactly one graph?

2 Prove: If G contains two distinct $u - v$ paths, then G contains a cycle.

3 Show that among any six people there are three mutual acquaintances or three pairwise strangers. Equivalently, if each edge of a K_6 is colored red or blue, then there is either a red triangle or a blue triangle.

4 Find, if possible, hamiltonian cycles in the graphs N_4 and N_8.

5 (a) Prove Theorem 4: $rad(G) \leq diam(G) \leq 2\, rad(G)$.

  (b) For each positive integer k, find graphs G_k and H_k such that $rad(G_k) = diam(G_k) = k$ and $rad(H_k) = k$ and $diam(H_k) = 2k$.

6 The tree T in Figure 5 has $n = 15$ and the distance from the center to the median is three. Construct a tree on $4k + 3$ vertices whose distance from its center to its median is k.

7 Jog from the center to the median of your town/campus.

8 Prove Theorem 2 which presents three characterizations of trees.

9 (a) Find an induced cycle C_6 in the rook graph R_4.

  (b) What is the largest length of an induced cycle in R_4?

  (c) What is the largest order of an induced subtree in R_4?

10 (a) Show that there are 11 nonisomorphic trees of order seven and 23 of order eight.

 (b) How many nonisomorphic graphs of order five are there?

 (c) Draw all cubic graphs of order at most eight.

11 For the tree T in Figure 5,

 (a) find three disjoint closed neighborhoods (is four possible?), and

 (b) find five disjoint open neighborhoods (is six possible?).

12 (a) What is the minimum number of open (respectively, closed) neighborhoods whose union is $V(N_4)$?

 (b) What is the minimum number of open (respectively, closed) neighborhoods in N_4 such that every edge in $E(N_4)$ is in the subgraph induced by at least one of them?

13 (a) Does the Petersen graph P have a spanning path? a spanning cycle?

 (b) For which k does P have a cycle of length k?

14 Find k-centers and k-medians for the graphs in Figure 7 and for the knight's graph N_8 for $k = 1, 2, 3$.

15 (a) Determine the independence number β_0 for $P_j \times P_k$, $P_j \times C_k$, and $C_j \times C_k$. Determine the chromatic number χ for these graphs.

 (b) Determine $\chi(L(P))$ for the graph in Figure 8.

16 Prove Brooks' Theorem [171],

 (a) $\chi(G) \le \Delta(G) + 1$, and

 (b) $\chi(G) = \Delta(G) + 1$ if and only if G is K_n or C_{2k+1}.

17 Show that $\lceil n/\beta_0(G) \rceil \le \chi(G) \le n - \beta_0(G) + 1$.

18 Prove: A graph G is bipartite if and only if it contains no odd cycles.

19 (a) Prove Dirac's Theorem: $\delta(G) \ge n/2$, $n \ge 3$, implies G is hamiltonian.

 (b) Prove Ore's Theorem: If $deg(u) + deg(v) \ge n \ge 3$ for every pair u, v of nonadjacent vertices, then G is hamiltonian.

 (c) Which complete multipartite graphs $K_{n_1, n_2, \dots, n_t}$ are hamiltonian?

20 Show that if $1 \le j \le k$, then there exists a graph $G_{j,k}$ such that $i(G_{j,k}) = j$ and $\beta_0(G_{j,k}) = k$.

21 Prove Turan's Theorem: The maximum number of edges in a triangle-free graph of order n is $\lfloor n^2/4 \rfloor$.

22 Prove: If G is disconnected or $diam(G) \geq 3$, then $diam(\overline{G}) \leq 2$.

23 Find a graph G of order eight such that $G = \overline{G}$.

24 (a) Which trees T of order six have the minimum/maximum values of $i(T)$ and $\beta_0(T)$?

   (b) Which of these trees T have $i(T) = \beta_0(T)$?

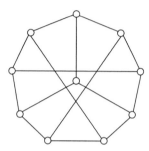

25 Is the graph shown above isomorphic to the Petersen graph?

26 Prove that any two longest paths in a connected graph must have a vertex in common.

Chapter 1

Introduction

As we said in the Prolegomenon, this is a book about dominating sets in graphs. Having defined graphs and many fundamental terms in graph theory, we are now in a position to define and discuss the concept of a dominating set in a graph. Through a series of applications, we will also present several different types of dominating sets. We will consider the difficulty of designing algorithms for finding dominating sets of minimum cardinality in a graph, which leads to a discussion of computational complexity and NP-completeness. We conclude this chapter with a brief review of the historical development of the study of domination in graphs.

1.1   Dominating Queens

The following problem can be said to be the origin of the study of dominating sets in graphs. Figure 1.1 illustrates a standard 8×8 chessboard on which is placed a queen. According to the rules of chess a queen can, in one move, advance any number of squares horizontally, vertically, or diagonally (assuming that no other chess piece lies in its way). Thus, the queen in Figure 1.1 can move to (or attack, or dominate) all of the squares marked with an 'x'. In the 1850s, chess enthusiasts in Europe considered the problem of determining the minimum number of queens that can be placed on a chessboard so that all squares are either attacked by a queen or are occupied by a queen. Figure 1.1 illustrates a set of six queens which together attack, or dominate, every square on the board. It was correctly thought in the 1850s, that five is the minimum number of queens that can dominate all of the squares of an 8×8 chessboard (see Exercise 1.1). The Five Queens Problem is to find a dominating set of five queens.

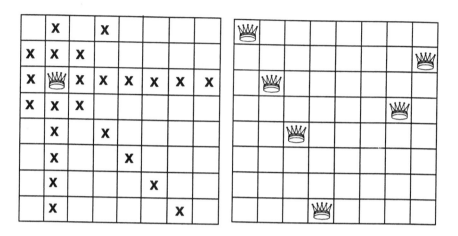

Figure 1.1: Queens.

1.2   Dominating Sets in Graphs

The problem of dominating the squares of a chessboard can be stated more generally as a problem of dominating the vertices of a graph.

Definition. A set $S \subseteq V$ of vertices in a graph $G = (V, E)$ is called a *dominating set* if every vertex $v \in V$ is either an element of S or is adjacent to an element of S.

Definitions. For $S \subseteq V$, a vertex $v \in S$ is called an *enclave* of S if $N[v] \subseteq S$, and $v \in S$ is an *isolate* of S if $N(v) \subseteq V - S$. A set is said to be *enclaveless* if it does not contain any enclaves.

There are several different ways to define a dominating set in a graph, each of which illustrates a different aspect of the concept of domination. Consider the following equivalent definitions. A set $S \subseteq V$ of vertices in a graph $G = (V, E)$ is a *dominating set* if and only if:

(i) for every vertex $v \in V - S$, there exists a vertex $u \in S$ such that v is adjacent to u;

(ii) for every vertex $v \in V - S$, $d(v, S) \le 1$;

(iii) $N[S] = V$;

(iv) for every vertex $v \in V - S$, $|N(v) \cap S| \ge 1$, that is, every vertex $v \in V - S$ is adjacent to at least one vertex in S;

(v) for every vertex $v \in V$, $|N[v] \cap S| \geq 1$;

(vi) $V - S$ is enclaveless.

Notice that if S is a dominating set of a graph G, then every superset $S' \supseteq S$ is also a dominating set. On the other hand, not every subset $S'' \subseteq S$ is necessarily a dominating set. We will be interested in studying minimal dominating sets in graphs, where a dominating set S is a *minimal dominating set* if no proper subset $S'' \subset S$ is a dominating set.

The set of all minimal dominating sets of a graph G is denoted by $MDS(G)$. Figure 1.2 illustrates a graph having minimal dominating sets of cardinality three (the set $\{1,3,5\}$), four (the set $\{3,6,7,8\}$), and five (the set $\{2,4,6,7,8\}$).

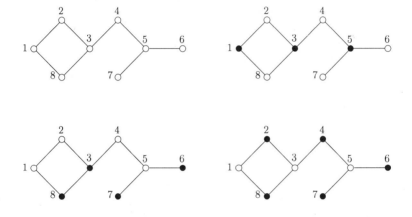

Figure 1.2: Minimal dominating sets.

The first three theorems about dominating sets in graphs were given by Ore in his 1962 book, *Theory of Graphs*, as follows:

Theorem 1.1 [924] *A dominating set S is a minimal dominating set if and only if for each vertex $u \in S$, one of the following two conditions holds:*

(a) *u is an isolate of S,*

(b) *there exists a vertex $v \in V - S$ for which $N(v) \cap S = \{u\}$.*

Proof. Assume that S is a minimal dominating set of G. Then for every vertex $u \in S$, $S - \{u\}$ is not a dominating set. This means that some vertex v in $V - S \cup \{u\}$ is not dominated by any vertex in $S - \{u\}$. Now either $v = u$, in which case u is an isolate of S, or $v \in V - S$. If v is not dominated by $S - \{u\}$, but is dominated by S, then vertex v is adjacent only to vertex u in S, that is, $N(v) \cap S = \{u\}$.

Conversely, suppose that S is a dominating set and for each vertex $u \in S$, one of the two stated conditions holds. We show that S is a minimal dominating set. Suppose that S is not a minimal dominating set, that is, there exists a vertex $u \in S$ such that $S - \{u\}$ is a dominating set. Hence, u is adjacent to at least one vertex in $S - \{u\}$, that is, condition (a) does not hold. Also, if $S - \{u\}$ is a dominating set, then every vertex in $V - S$ is adjacent to at least one vertex in $S - \{u\}$, that is, condition (b) does not hold for u. Thus neither condition (a) nor (b) holds, which contradicts our assumption that at least one of these conditions holds. $\square$

Theorem 1.1 suggests the following definition. Let S be a set of vertices, and let $u \in S$. We say that a vertex v is a *private neighbor* of u (with respect to S) if $N[v] \cap S = \{u\}$. Furthermore, we define the *private neighbor set of* u, with respect to S, to be $pn[u, S] = \{v : N[v] \cap S = \{u\}\}$. Notice that $u \in pn[u, S]$ if u is an isolate in $\langle S \rangle$, in which case we say that u is its own private neighbor. Given this terminology, we can say that a dominating set S is a minimal dominating set if and only if every vertex in S has at least one private neighbor, that is, for every $u \in S$, $pn[u, S] \neq \emptyset$. For example, consider the minimal dominating set $\{3,6,7,8\}$ in the graph in Figure 1.2. Vertex 3 has vertices 2 and 4 as private neighbors, vertex 8 has vertex 1 as a private neighbor, while vertices 6 and 7 are their own private neighbors.

Theorem 1.2 [924] *Every connected graph G of order $n \geq 2$ has a dominating set S whose complement $V - S$ is also a dominating set.*

Proof. Let T be any spanning tree of G, and let u be any vertex in V. Then the vertices in T fall into two disjoint sets S and S' consisting, respectively, of the vertices with an even and odd distance from u in T. Clearly, both S and $S' = V - S$ are dominating sets for G. $\square$

Theorem 1.3 [924] *If G is a graph with no isolated vertices, then the complement $V - S$ of every minimal dominating set S is a dominating set.*

Proof. Let S be any minimal dominating set of G. Assume vertex $u \in S$ is not dominated by any vertex in $V - S$. Since G has no isolated vertices, u must be dominated by at least one vertex in $S - \{u\}$, that is, $S - \{u\}$ is a dominating set, contradicting the minimality of S. Thus every vertex in S is dominated by at least one vertex in $V - S$, and $V - S$ is a dominating set. $\square$

Definitions. The *domination number* $\gamma(G)$ of a graph G equals the minimum cardinality of a set in $MDS(G)$, or equivalently, the minimum cardinality of a dominating set in G. The *upper domination number* $\Gamma(G)$ equals the maximum cardinality of a set in $MDS(G)$, or equivalently, the maximum cardinality of a minimal dominating set of G. It is easy to see that for the graph G in Figure 1.2, $\gamma(G) = 3$, while $\Gamma(G) = 5$. Notice that the set $S = \{1, 3, 5\}$ is a dominating set of minimum cardinality; this is called a *γ-set* of G. Notice further that S is an independent set. This is also called an *independent dominating set* of G. The minimum cardinality of an independent dominating set of G is the *independent domination number $i(G)$*.

One of the earliest and most basic theorems about the domination number of a graph is the following, due to Nieminen. Let $\varepsilon_F(G)$ denote the maximum number of pendant edges in a spanning forest of G (a *forest* is an acyclic graph). We leave the proof as an exercise.

Theorem 1.4 [920] *For any graph G, $\gamma(G) + \varepsilon_F(G) = n$.*

Now that we have defined and illustrated dominating sets, minimum dominating sets, and the domination number of a graph, we describe, in Sections 1.3 through 1.10, a variety of situations in which dominating sets naturally occur.

1.3 Sets of Representatives

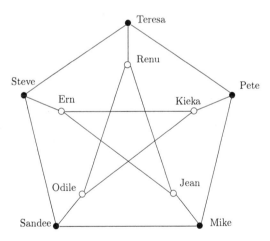

Figure 1.3: A total dominating set.

Let the vertices of a graph (Figure 1.3) represent a group of people. An edge between two people means that they have something in common. We wish to form a committee (a subset of the vertices), with as few members as possible, such that everyone not on the committee has something in common with at least one person on the committee. Thus, we seek a minimum cardinality

dominating set. Notice that the set $S = \{$Teresa, Pete, Mike, Sandee, Steve$\}$ is
a minimal dominating set, while the set $D = \{$Steve, Renu, Pete$\}$ is a minimum
dominating set (or a γ-set) for this graph. The set S has another property of
interest. Every person is adjacent to a member of S, including the members of
S themselves, that is, no member of S is an isolate in $\langle S \rangle$. Since each committee
member might feel more comfortable knowing at least one other member of the
committee, a set with this property could enhance committee performance. A
set S is a *total dominating set* if $N(S) = V$, or equivalently, if for every vertex
$v \in V$, there exists a vertex $u \in S$, $u \neq v$, such that u is adjacent to v. The *total
domination number* $\gamma_t(G)$ equals the minimum cardinality of a total dominating
set of G. The total domination number of the graph G in Figure 1.3 is four,
that is, $\gamma_t(G) = 4$; a γ_t-set is $\{$Steve, Teresa, Renu, Pete$\}$.

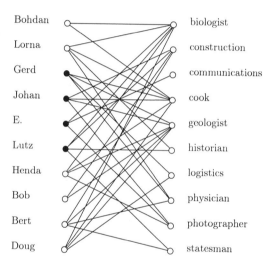

Figure 1.4: Bipartite domination.

Next, let the vertices of a graph be of two kinds: (i) those representing the
members of an exploration society, and (ii) those representing types of skills
(see Figure 1.4). In this graph, an edge between a person and a skill means that
person has expertise in that skill area. We wish to send an expedition to set
up a camp and explore some remote part of the world. In order to save costs,
we want to send as few people as possible, but we must be sure that every skill
is possessed by at least one member of the expedition. Thus, in this bipartite
graph, we seek a minimum subset of the vertices in the left column (people)
that dominates all of the vertices in the right column (skills) (for example, the
subset $\{$Gerd, Johan, E., Lutz$\}$ in Figure 1.4).

One real world application of this type of domination (*bipartite domination*) occurs in making personnel assignments. For example, at one time the U. S. Navy had approximately 500,000 personnel, many of whom periodically qualified for reassignments. In any given month, from 30,000 to 50,000 personnel would be reassigned to fill a similar number of available jobs requiring differing types of expertise. Needless to say, this monthly assignment problem is quite complex.

1.4 School Bus Routing

Most school districts in the country provide school buses for transporting children to and from school. Most also operate under certain rules, one of which usually states that no child shall have to walk farther than, say, one quarter mile to a bus pickup point. Thus, the school district must construct a route for each bus that gets within one quarter mile of every child in its assigned area. Another rule might stipulate that no bus ride can take more than some specified number of minutes, and another rule puts limits on the number of children that a bus can carry at any one time.

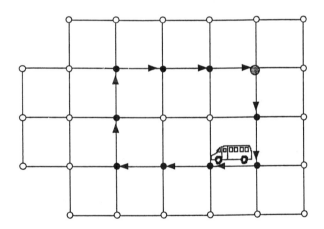

Figure 1.5: Distance-2 cycle domination.

Consider the graph in Figure 1.5. Let us say that this represents a street map of part of a city, where each edge represents one city block. The school is located at the large vertex. Let us assume that the school district has decided that no child shall have to walk more than two blocks in order to be picked up by a school bus. Therefore, we must construct a route for a school bus that leaves the school, gets within two blocks of every child and returns to the school. One such simple route is indicated by the directed edges in Figure 1.5. Notice that some of the children live close enough to walk to school.

A second possible route is indicated in Figure 1.6. With this route the school bus can turn around and drive back down a street. Both routes define what are called *distance-2 dominating sets* in the sense that every vertex not on the route (not in the set) is within distance two (two edges) of at least one point on the route. These routes also define what are called *connected dominating sets* in the sense that the set of shaded vertices on the route forms a connected subgraph of the entire graph. The *connected domination number* $\gamma_c(G)$ equals the minimum cardinality of a dominating set S such that $\langle S \rangle$ is connected. Finally, the route in Figure 1.5 defines a *cycle dominating set* in the sense that the vertices on this route form a cycle.

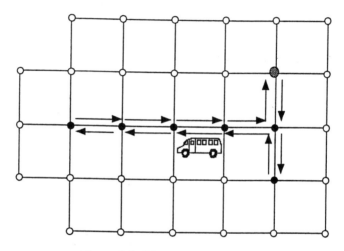

Figure 1.6: Distance-2 domination.

1.5 Computer Communication Networks

Consider a computer network modeled by a graph $G = (V, E)$, for which vertices represent computers and edges represent direct communication links between pairs of computers. Let the vertices in Figure 1.7 represent an array, or network, of 16 computers, or processors. Each processor can pass information to the processors to which it is directly connected. Assume that from time to time we need to collect information from all processors. We do this by having each processor route its information to one of a small set of collecting processors (a dominating set). Since this must be done relatively often and relatively fast, we cannot route this information over too long a path. Thus we need to identify a small set of processors which are close to all other processors. Let us say that we will tolerate at most a two-unit delay between the time a processor sends its information and the time it arrives at a nearby collector. In this case we seek a *distance-2 dominating set* among the set of all processors. The two shaded vertices form a distance-2 dominating set in the hypercube network in Figure 1.7.

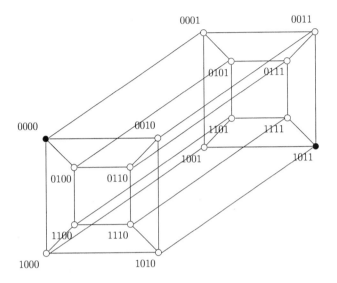

Figure 1.7: Distance-2 domination in the hypercube Q_4.

1.6 (r, d)-Configurations

Consider again a computer network modeled by a graph $G = (V, E)$. In such a distributed computing environment, we are often faced with the problem of allocating copies of a scarce or valuable hardware or software resource, such as a high-speed printer, a high-performance graphics workstation, or a very large database or fileserver, to certain computers in such a way that every computer not having such a resource has efficient access to a copy of this resource.

We say that a computer at vertex u can *effectively access* a resource at vertex v if $d(u, v) \leq d$ for some suitably small distance d. If $d = 1$, then the minimum number of copies of a given resource, which can be allocated to computers in a network so that all computers either have the resource or have efficient access to a copy of the resource, is simply the domination number of the network.

Suppose, furthermore, that we need to allocate many different resources, but all computers have a fixed capacity r, which prevents us from assigning more that r resources to any computer. We wish to determine the maximum number

$R_{(r,d)}(G)$ of different resources we can allocate to the computers of a network G such that (i) no more than r different resources are allocated to any one computer and (ii) every computer has efficient access to every resource. An allocation that achieves this maximum is called an (r, d)-*configuration*. In Figure 1.8 we show a (2,1)-configuration of the 5-cycle, where $R_{(2,1)}(C_5) = 5$. Here we allocate five different resources, numbered 1, 2, 3, 4, and 5. Notice that for each value of i, $1 \leq i \leq 5$, the pair of vertices labelled i is a dominating set. When $r = d = 1$, $R_{(1,1)}(G)$ is the *domatic number* of G (defined and discussed in Chapter 8).

For more information on (r, d)-configurations, the interested reader is referred to the work of Fujita, Yamashita, and Kameda [503, 504].

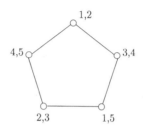

Figure 1.8: An $R_{(2,1)}$-configuration.

1.7 Radio Stations

Suppose that we have a collection of small villages in a remote part of the world, for example, Alaska, the Northwest Territories, Siberia, the Outback, the Andes, the Himalayas, or the Serengetti. We would like to locate radio stations in some of these villages so that messages can be broadcast to all of the villages in the region. Since each radio station has a limited broadcasting range, we must use several stations to reach all villages. But since radio stations are costly, we want to locate as few as possible which can reach all other villages.

Let each village be represented by a vertex. An edge between two villages is labelled with the distance, say in kilometers, between the two villages (see Figure 1.9(a)). Let us assume that a radio station has a broadcast range of fifty kilometers. What is the least number of stations in a set which dominates (within distance 50) all other vertices in this graph? A set $\{B, F, H, J\}$ of cardinality four is indicated in Figure 1.9(b). Notice in this case that since we have assumed that a radio station has a broadcast range of only fifty kilometers, we can essentially remove all edges in the graph in Figure 1.9(a) which represent a distance of more than fifty kilometers. This gives us the graph in Figure 1.9(b). We need only to find a dominating set in this graph. Notice that if we could

afford radio stations which have a broadcast range of seventy kilometers, three
radio stations would suffice.

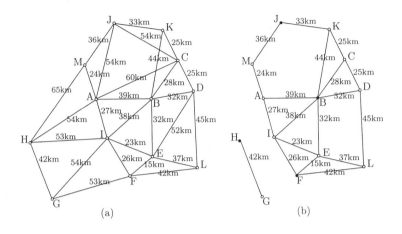

Figure 1.9: Radio stations.

1.8 Social Network Theory

In her Ph.D. dissertation, Kelleher presented research on dominating sets in
social network graphs [771]. The ideas in this section are based on this work;
the interested reader is referred also to Kelleher and Cozzens [772].

In social network theory one studies relationships that exist among members
of a group. The people in such a group are called *actors*. These relationships
are typically defined in terms of one or more dichotomous properties, that is, a
property that for any two actors unambiguously either holds or does not hold.
Given such a property, one can construct a *social network graph*, in which the
vertices represent the actors, and an edge between two vertices indicates that the
property in question holds between the corresponding actors. A *social network
clique* is a set of actors, between any two of whom a given property either always
holds or never holds (that is, the set of vertices in a social network clique either
induces a complete graph or an independent set).

A *status* is a set S of vertices in a social network graph which has the property
that for any two vertices $u, v \in S$, $N(u) \cap V - S = N(v) \cap V - S$, that is, the
set of vertices in $V - S$ dominated by u equals the set of vertices in $V - S$
dominated by v. Thus, all of the vertices in a status dominate the same set of
vertices outside of the status. Note that it is assumed that every status must
have at least two vertices.

We say that two vertices u and v in a social network graph are *structurally equivalent* if either $N(u) = N(v)$ or $N[u] = N[v]$. A set S is called a *structurally equivalent set* if every two vertices in S are structurally equivalent.

Social network theorists are interested in finding every maximal structurally equivalent set and every status in a social network graph, and they have developed algorithms to do so. Kelleher and Cozzen's work shows that these sets can be found using the properties of dominating sets in graphs. In particular, they proved the following.

Theorem 1.5 [772] *If a γ-set S of a connected graph G of order $n \geq 2$ is a status of G, then S is an independent dominating set of cardinality two.*

Proof. Let S be a γ-set of G which is a status. Since G is connected and has no isolated vertices, there must be at least one vertex $v \in V - S$. Since S is a γ-set, v must be adjacent to at least one vertex in S. But since S is a status, every vertex of S must be adjacent to v. Furthermore, every vertex in S must be adjacent to every vertex in $V - S$. Now since S is a status, $|S| \geq 2$.

Assume that $|S| \geq 3$, and let $u \in S$ and $v \in V - S$. Since S is a status, u is adjacent to every vertex in $V - S$, and v is adjacent to every vertex in S. Thus, $\{u, v\}$ is a dominating set, contradicting the minimality of S. Therefore, $|S| = 2$. If u is adjacent to v, then clearly $\{u\}$ is a dominating set of G, again contradicting the minimality of S. Therefore, $|S| = 2$ and S is an independent set. $\square$

Corollary 1.6 [772] *If a γ-set S of a connected graph G is also a structurally equivalent set, then S consists of two independent vertices each of which has degree $n - 1$.*

1.9 Land Surveying

In the field of surveying, a typical task is to compile a topographic map of a tract of land by determining the positions and elevations of a carefully selected set of control points. A grid method can be used in areas where the topography is fairly regular. The area is divided uniformly into squares or rectangles, by two sets of lines running in perpendicular directions and spaced uniformly apart, for example, 100 feet (as in Figure 1.10).

One way the surveyor identifies the grid points to be mapped is to set stakes at the intersections of these lines. It is then necessary to determine the elevations of all grid points. This is done with the use of a total station instrument which is an electronic theodolite (or transit) containing an integral EDM (electronic distance measuring instrument) and angle measuring capability. An EDM sends out light from one point on a line to another point. At the other point, a retroprism (a reflector or a transmitter-receiver), reflects the light back to the

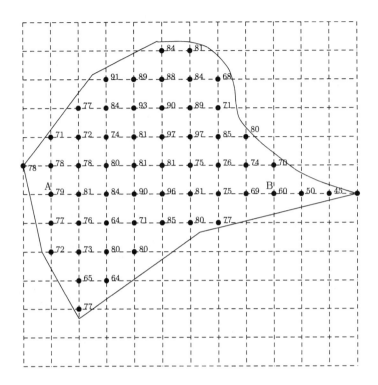

Figure 1.10: Elevations for a topographic map.

EDM, where it is analyzed electronically to give the horizontal and vertical distances between the two points and the horizontal and vertical angles to the survey point. These measurements are stored in electronic field books, and later downloaded into automatic drafting instruments which generate the contour lines and other planimetric features.

The total station instrument is controlled by a surveying party chief, who selects the control points on the tract from which measurements are to be taken. A rodman then holds a prism pole, or reflector rod, in a vertical position at each of the grid points. Since these are line-of-sight measurements, any obstruction, such as a building structure, a tree, a steep ravine, or gully, can prevent a measurement of a grid point from being taken. For example, Figure 1.10 indicates that a hill obstructs the line-of-sight measurement from point A to point B. In this case, the total station instrument must be moved to another control point from which no line-of-sight obstruction exists to the given grid point.

Unfortunately, it is time-consuming to move and change the initial reading in the total station instrument. For instance, one must carry the total station instrument to a new point, set it up on its tripod, determine the new HI (the height of the instrument) from a backsight (a rod reading taken along a line of known azimuth), and the coordinates of the new point must be entered. This typically can take 15-20 minutes per move.

Thus, we seek to find a minimum number of control points, from at least one of which a line of sight measurement can be taken of any given grid point. This is equivalent to finding a minimum dominating set in the line-of-sight graph for the given tract, the vertices of which correspond 1-1 with the grid points of the tract, and two vertices are adjacent if and only if the corresponding grid points permit a line-of-sight measurement. (For a general discussion on surveying, the reader is referred to *Surveying Measurements and Their Analysis* by R. Ben Buckner, Landmark Enterprises, 1983.)

1.10   Kernels

Dominating sets occur in the study of 2-person games, where they are called *kernels* (see Section 7.7). We give a simple example. The game of CHOMP (due to mathematician David Gale) is played on an $m \times n$ checkerboard, the bottom, left (s.w.) corner square of which is considered to be poisoned. Two players alternate taking turns, each of which consists of choosing any (remaining) square on the board, at a location in some row i and column j (see Figure 1.11 for an illustration of a 2×5 CHOMP board).

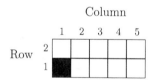

Figure 1.11: 2×5 CHOMP.

If square (i, j) is chosen, then this square and all squares north and/or east of it are erased (eaten). The player who is forced to choose the poisoned square loses the game.

For the $2 \times n$ board, one can construct a game tree of all possible plays of the game, part of which is illustrated in Figure 1.12, where the pair (r, s) indicates that remaining on the board are r squares in row 2 and s squares in row 1. Figure 1.12 shows the resulting board after all possible first moves, and some of the possible second moves.

Another description of this 2×5 game can be given by the different board in Figure 1.13, the squares of which describe the number of remaining squares

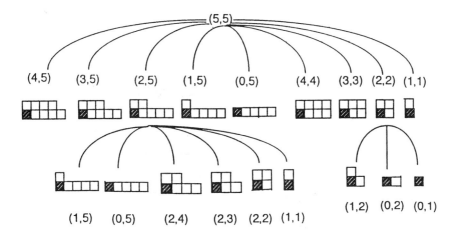

Figure 1.12: 2 × 5 CHOMP partial game tree.

on row 2 and on row 1, respectively, that are possible after a sequence of legal moves. Notice that there are, at all times, at least as many squares remaining in row 1 as there are in row 2. The top square (5,5) denotes the starting board for the 2 × 5 case, having 5 squares in both row 2 and row 1. A legal move in the 2 × 5 CHOMP game is equivalent to moving on the board in Figure 1.13 from the current square (vertex) to any square below in the same column (that is, south), to any square to the left in the same row (that is, west), or to any square below it on the major, n.e.-to-s.w. diagonal, provided that the current square is on this diagonal (that is, is one of the squares marked with an 'x' in Figure 1.13).

The first player to move has a winning strategy in the 2 × n case, if at all times the chosen move is to one of the vertices marked with a dot (see Figure 1.13). Notice that once Player 1 is on a square with a dot, no matter where Player 2 moves, Player 1 can always move again to a square with a dot. Notice also that once Player 1 has moved to a square with a dot, Player 2 cannot move to another square with a dot. Sooner or later, Player 1 will be able to move to the dotted square at position (0,1), forcing Player 2 to lose by moving to (0,0).

The set of squares marked with a dot is called a *kernel*, that is, it is a set of squares (vertices) having two properties: (i) no two vertices in this set are adjacent, and (ii) there is a move (that is, a directed edge) from every square (vertex) not in the set to a square in the set. Thus, it is an independent dominating set in the corresponding directed graph (see Figure 1.14 in which some of the edges are missing, that is, those from a vertex to all vertices to its

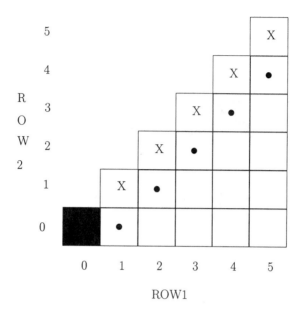

Figure 1.13: 2×5 CHOMP moves.

left). We note, in passing, that although Player 1 has a winning strategy for $2 \times n$ CHOMP boards (as indicated above), no winning strategy is known for Player 1 on $k \times n$ boards, for any $k \geq 3$, except for square $n \times n$ boards. It can also be proved that Player 1 has a winning strategy on every $m \times n$ board; the problem is that no one knows how to construct such strategies.

1.11   An Introduction to NP-Completeness

At this point we have seen examples of dominating sets of several different types (for example, dominating sets, independent dominating sets, total dominating sets, connected dominating sets, cycle dominating sets, and distance dominating sets) which arise in many different contexts and situations. Generally speaking, dominating sets arise in the study of numerous facility location problems, such as the optimal location of hospitals, fire stations, post offices, mail boxes, schools, stores, radio stations, and the like. Dominating sets also arise in situations involving sets of representatives, surveillance, personnel assignments, communications, and games. In this section we consider problems involved in computing $\gamma(G)$ and in finding γ-sets.

Given a graph $G = (V, E)$ with n vertices, the domination number clearly

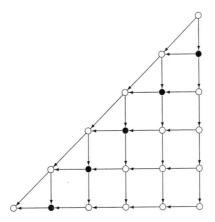

Figure 1.14: A kernel.

lies in the range $1 \leq \gamma(G) \leq n$. Thus, there are only a finite number of possible minimum cardinality dominating sets of G. Any procedure which can enumerate all 2^n subsets of $V(G)$ in nondecreasing order of cardinality and can verify, for any given subset $S \subseteq V$, whether S is a dominating set can be used to determine the value of $\gamma(G)$ simply by taking the cardinality of the first dominating set that it finds. Such an algorithm is easy to construct, but unfortunately requires $O(2^n)$ steps in the worst case, that is, has an exponential time complexity in the order of the graph G. What we would really like to know is whether there is an algorithm for determining the value of $\gamma(G)$ for an arbitrary graph G that is significantly faster than the brute-force algorithm that tries all possible subsets of vertices. To date, no one has constructed a domination algorithm that has better than exponential time complexity for arbitrary graphs. Furthermore, the theory of NP-completeness suggests that it is not likely that a polynomial algorithm can be constructed.

A brief, intuitive discussion of NP-completeness is in order. Let **P** denote the class of all problems that can be solved by a polynomial time algorithm, that is, polynomial in the length of the inputs for an instance of the problem. We can think of these algorithms as running on a relatively simple computer, for example a Turing machine, named after the British mathematician/logician Alan Turing (see [524]). Briefly, a Turing machine is a computer with (i) a two-way infinite storage tape, divided into cells, in each of which can be written one symbol chosen from a finite alphabet, and (ii) a finite-state control. The finite control can be thought of as a random access machine, or RAM. The execution time of such a RAM is usually measured by the number of operations it performs in solving an instance of a problem. Each operation can be assumed to require a constant amount of time, say C. Typical operations include addition, subtraction, multiplication, and division of two numbers, storing a number in a random access memory, and comparing two numbers.

Turing postulated the thesis that what we think of as an effective algorithm is precisely what can be done by a Turing machine or, equivalently, by a RAM with an infinite amount of auxiliary memory. Thus, we can say that a computational problem is in class **P** if there exists an algorithm for solving any instance of the problem in time $O(n^k)$ for some fixed positive integer k, where n is the length of the input for the given instance.

Typical examples of problems which can be solved in polynomial time, and are therefore in the class **P**, include:

(i) sorting n integers,

(ii) finding a shortest path between two vertices u and v in a graph G,

(iii) finding a maximum matching in a graph G,

(iv) determining whether two trees T_1 and T_2 with n vertices are isomorphic,

(v) deciding whether a given graph is bipartite or connected, and

(vi) computing the convex hull of a set of n points in the plane.

A given instance of a computational problem is represented by a set of inputs. For example, it may be a sequence of n integers (as in sorting), a set of n ordered pairs (x, y) of n points in the plane, or a set of m ordered pairs (u, v) of edges in a graph. Depending on the detail of the complexity analysis, the length of the input may be measured by the number of bits required to express the input values, for example, if any integer in the problem instance can be represented by c bits, then n integers to be sorted can be represented by an input of length $O(cn)$. The length of the input may also be the number m of edges in a graph, or the number $n + m$ of vertices plus edges in a graph.

In the theory of NP-completeness, we restrict our attention to the class of problems called *decision problems*. These are problems, every instance of which can be stated in such a way that the answer is either a 'yes' or a 'no'. Thus, for example, we do not seek an algorithm for finding the minimum cardinality of a dominating set in a graph. Instead we seek an algorithm which, given a graph G and a positive integer k, can decide whether G has a dominating set of size $\leq k$.

Let **NP** denote the class of all decision problems which can be solved in polynomial time by a nondeterministic Turing machine. Thus, **NP** stands for *Nondeterministic Polynomial* time. Again, wishing to avoid the extended discussion required to give a technical definition of a nondeterministic Turing machine, suffice it to say that such a machine has the ability to make guesses at certain points in a computation. Some of these guesses may be correct, some may be incorrect. All we seek is a guarantee that given a 'yes' instance to a problem, the nondeterministic machine can make, in some attempt to solve the problem, a sequence of 'good' guesses which produces a 'yes' answer, and the total execution time of this solution is bounded by a polynomial in the length of the input. Moreover, the machine will always answer "no" to a "no" instance.

Stated in terms of an example, suppose that we wish to decide if a given graph $G = (V, E)$ with n vertices has a dominating set S of size $\leq k$, for some given positive integer k. A nondeterministic machine for solving this problem has the ability to make a guess, for every i, $1 \leq i \leq n$, whether vertex v_i should be in the subset S, with at most k vertices. Having made these guesses, it can then verify, deterministically and in polynomial time, whether the guessed set is a dominating set. Thus, if we are given a 'yes' instance (that is, the given graph does have a dominating set of size $\leq k$), then the nondeterministic machine can find such a dominating set, by guessing, and then verify in polynomial time that it has the desired properties.

Instead of using the notion of nondeterminism, we can define the class **NP** in terms of the concept of *polynomial-time verification*. A verification algorithm is an algorithm A which takes as input an instance of a problem and a candidate solution to the problem, called a *certificate*, and verifies in polynomial time whether the certificate is a solution to the given problem instance. Thus, the class **NP** is the class of problems which can be verified in polynomial time.

The fundamental open question in computational complexity is whether the class **P** equals the class **NP**. By definition, the class **NP** contains all problems in class **P**. It is not known, however, whether all problems in **NP** can, in fact, be solved in polynomial time by deterministic Turing machines. The generally accepted belief is that $\mathbf{P} \neq \mathbf{NP}$.

In an effort to determine whether $\mathbf{P} = \mathbf{NP}$, Cook (see [524]) defined the class of NP-complete problems. We say that a problem P_1 is *polynomial-time reducible* to a problem P_2, written $P_1 \leq_p P_2$, if

(i) there exists a function f which maps any instance of P_1 to an instance of P_2 in such a way that I_1 is a 'yes' instance of P_1 if and only if $f(I_1)$ is a 'yes' instance of P_2, and

(ii) for any instance I_1, the instance $f(I_1)$ can be constructed in polynomial time.

If P_1 is polynomial-time reducible to P_2, we can say that any algorithm for solving P_2 can be used to solve P_1. Intuitively, problem P_1 is 'no harder' to solve than problem P_2. We define a problem P to be *NP-complete* if (i) $P \in \mathbf{NP}$, and (ii) for every problem $P' \in \mathbf{NP}$, $P' \leq_p P$. If a problem P can be shown to satisfy condition (ii), but not necessarily condition (i), then we say that it is *NP-hard*. Let **NPc** denote the class of NP-complete problems.

The relation $\leq_p$ is transitive. Because of this, a method frequently used in demonstrating that a given problem is NP-complete is the following:

(i) show $P \in \mathbf{NP}$, and

(ii) show there exists a problem $P' \in \mathbf{NPc}$, such that $P' \leq_p P$.

It follows from the definition of NP-completeness that if any problem in **NPc** can be solved in polynomial time, then every problem in **NPc** can be solved in polynomial time and $\mathbf{P} = \mathbf{NP}$. On the other hand, if there is some problem in **NPc** that cannot be solved in polynomial time, then no problem in **NPc** can be solved in polynomial time.

1.12   NP-Completeness of the Domination Problem

Let us proceed to the basic question: how difficult is it to compute the domination number of an arbitrary graph? We will show that the domination problem is NP-complete for arbitrary graphs.

Stated in the now accepted format, as established by Garey and Johnson in their seminal book on NP-completeness [524], the basic complexity question concerning the decision problem for the domination number takes the following form:

DOMINATING SET
INSTANCE: A graph $G = (V, E)$ and a positive integer k
QUESTION: Does G have a dominating set of size $\leq k$?

David Johnson was the first person to show that DOMINATING SET is NP-complete. We provide his proof.

Theorem 1.7 [524] *DOMINATING SET is NP-complete.*

Proof. We must do two things. First, we must show that DOMINATING SET $\in \mathbf{NP}$. This is easy to do since it is easy to verify a 'yes' instance of DOMINATING SET in polynomial time, that is, for a graph $G = (V, E)$, a positive integer k and an arbitrary set $S \subseteq V$ with $|S| \leq k$, it is easy to verify in polynomial time whether S is a dominating set.

Second, we must construct a reduction from a known NP-complete problem to DOMINATING SET. We use the well known 3-SAT problem.

3-SAT
INSTANCE: A set $U = \{u_1, u_2, \cdots, u_n\}$ of variables, and a set $C = \{C_1, C_2, \cdots, C_m\}$ of 3-element sets, called clauses, where each clause C_i contains three distinct occurrences of either a variable u_i or its complement u_i'. For example, a

clause might be $C_1 = \{u_1, u'_2, u_4\}$.

QUESTION: Does C have a satisfying truth assignment, that is, an assignment of True and False to the variables in U such that at least one variable in each clause in C is assigned the value True?

Given an instance C of 3-SAT, we construct an instance $G(C)$ of DOMINATING SET as follows. For each variable u_i, construct a triangle with vertices labelled u_i, u'_i and v_i. For each clause $C_j = \{u_i, u_k, u_l\}$ create a single vertex labelled C_j, and add edges (u_i, C_j), (u_k, C_j) and (u_l, C_j) (see Figure 1.15).

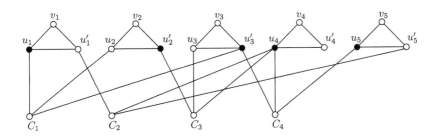

Figure 1.15: Reduction from 3-SAT to DOMINATING SET.

We must show that C has a satisfying truth assignment if and only if the graph $G(C)$ has a dominating set of cardinality $\leq k$, in this case for $k = n$, that is, C is a 'yes' instance of 3-SAT if and only if $G(C)$ is a 'yes' instance of DOMINATING SET, for $k = n$.

Suppose first that C has a satisfying truth assignment. For example in Figure 1.15, let $u_1 =$ True, $u_2 =$ False, $u_3 =$ False, $u_4 =$ True and $u_5 =$ True. Create a set S of vertices in $G(C)$ as follows: if $u_i =$ True, then put vertex u_i in S, and if $u_i =$ False, then put vertex u'_i in S. The set S will be a dominating set of $G(C)$ because (i) each triangle contains exactly one vertex in S, and therefore all vertices in a triangle are either in S or are dominated by a vertex in S, and (ii) each clause vertex C_j is dominated by at least one vertex in S, since by assumption each clause contains at least one variable whose value is assigned True, and, by construction, the corresponding vertex belongs to S. Therefore, $G(C)$ has a dominating set S of cardinality n.

Conversely, suppose that $G(C)$ has a dominating set S of cardinality $\leq n$. We must show that C has a satisfying truth assignment. Notice first that since each vertex of the form v_i must either be in S or be dominated by a vertex in S, each triangle must contain at least one vertex in S, and therefore $|S| \geq n$. In fact, each triangle must contain exactly one vertex in S. Thus, S contains no clause vertex C_j. But since S is a dominating set, each clause vertex must be dominated by at least one vertex in S. We can therefore create a satisfying truth assignment for C as follows: for each variable u_i, assign u_i the value True if $u_i \in S$, otherwise assign u_i the value False. It is straightforward to see that this is a satisfying truth assignment for C.

We must also show that the construction illustrated in Figure 1.15, for creating an instance of DOMINATING SET from an instance of 3-SAT, can be carried out in polynomial time. The length of an instance of 3-SAT is $O(3m + n)$, that is, C is specified by m sets of size three plus n variables. The graph $G(C)$ has $3n + m$ vertices and $3n + 3m$ edges. Thus, the cardinality of $G(C)$ is at most a constant times the cardinality of C, and therefore the graph $G(C)$ can be constructed from an instance of 3-SAT in polynomial time. $\square$

Theorem 1.7 is the most basic complexity result concerning domination in graphs, but it is the beginning of a vast collection of related results. It suggests that we should not expect to find soon a polynomial time algorithm for determining the domination number of an arbitrary graph. Thus, if we expect to be able to compute the domination number of a graph in polynomial time, we will have to restrict the instances to classes of graphs other than the class of arbitrary graphs. There are obviously a very large number of classes of graphs for our consideration. DOMINATING SET remains NP-complete when instances are restricted to graphs in most of these classes, while for relatively few classes we are able to compute the value of $\gamma(G)$ in polynomial time. A discussion of these complexity issues and algorithms is given in Chapter 12 of this text and in Chapters 8 and 9 of [645].

1.13   Mathematical History of Domination in Graphs

Although the mathematical study of dominating sets in graphs began around 1960, the subject has historical roots dating back to 1862 when de Jaenisch [347] studied the problem of determining the minimum number of queens which are necessary to cover (or dominate) an $n \times n$ chessboard. As reported by W. W. Rouse Ball in 1892 [72], chess enthusiasts in the late 1800s studied, among others, the following three basic types of problems:

1. Covering - what is the minimum number of chess pieces of a given type which are necessary to cover/attack/dominate every square of an $n \times n$ board? This is an example of the problem of finding a dominating set of minimum cardinality.

2. Independent Covering - what is the minimum number of mutually nonattacking chess pieces of a given type which are necessary to dominate every square of an $n \times n$ board? This is an example of the problem of finding a minimum cardinality independent dominating set.

3. Independence - what is the maximum number of chess pieces of a given type which can be placed on an $n \times n$ chessboard in such a way that no two of them attack/dominate each other? This is an example of the problem of finding the maximum cardinality of an independent set. When the chess piece is the queen, this problem is known as the *N-queens Problem*. It is known that for every positive integer $n \geq 4$, it is possible to place n nonattacking (independent) queens on an $n \times n$ board. For over a hundred years people have studied different ways of doing this.

These three problem-types were studied in detail by Yaglom and Yaglom around 1964 [1155]. These two brothers produced elegant solutions to some of these problems for the rooks, knights, kings and bishops chess pieces.

In 1958 Claude Berge [95] wrote a book on graph theory, in which he defined for the first time the concept of the domination number of a graph (although he called this number the 'coefficient of external stability'). In 1962 Oystein Ore [924] published his book on graph theory, in which he used, for the first time, the names 'dominating set' and 'domination number' (although he used the notation $d(G)$ for the domination number of a graph). In 1977 Cockayne and Hedetniemi [280] published a survey of the few results known at that time about dominating sets in graphs. In this survey paper, Cockayne and Hedetniemi were the first to use the notation $\gamma(G)$ for the domination number of a graph, which subsequently became the accepted notation.

This survey paper seems of have set in motion the modern study of domination in graphs. Some twenty years later more than 1,200 research papers have been published on this topic, and the number of papers is steadily growing. This book is inspired by the somewhat explosive growth of this field of study. It is also motivated by a desire to put some order into this huge collection of research papers, to organize the study of dominating sets in graphs into meaningful subareas, and to attempt to place the study of dominating sets in even broader mathematical and algorithmic contexts.

EXERCISES

1.1 Find dominating sets of five queens on the 8×8 chessboard for each of the following situations:

   (a) no two queens attack each other.

   (b) all queens lie on the main diagonal.

   (c) all queens lie on a common column.

1.2 List all ten minimal dominating sets of the graph in Figure 1.2.

1.3 Mimicking Theorem 1.1, characterize minimal independent dominating sets.

1.4 Prove that the total domination number of the graph in Figure 1.2 is four.

1.5 Construct a minimum length distance-1 dominating cycle for the graph in Figure 1.5.

1.6 Find a set of minimum cardinality in the hypercube Q_4 in Figure 1.7 which is

    (a) a dominating set.

    (b) an independent dominating set.

    (c) a connected dominating set.

    (d) a total dominating set.

1.7 Compute γ, i, γ_c, and γ_t for paths P_n, cycles C_n, and the graph $P_3 \times P_6$.

1.8 Show that there does not exist a vertex set $S \subseteq V(Q_4)$ in hypercube Q_4 such that $|S \cap N[v]| = 1$ for every $v \in V(Q_4)$.

1.9 Find a subset of $S \subseteq V(Q_4)$ in hypercube Q_4 such that for every $v \in V(Q_4)$, $|N[v] \cap S|$ is odd.

1.10 For positive integers j and k, find a graph $G_{j,k}$ such that the only cardinalities of minimal dominating sets are j and k.

1.11 Show that if graph G is disconnected, then $\gamma(\overline{G}) \leq 2$.

1.12 Show that for the rook graph R_n,

$$\gamma(R_n) = i(R_n) = \beta_0(R_n) = \Gamma(R_n) = n.$$

1.13 Show that for the queens graph Q_n with $n \geq 6$, $\Gamma(Q_n) \geq n + 1$. (In fact, $\Gamma(Q_n) \geq 2n - 5$ (see Weakley [1139]).)

1.14 Find locations for three radio stations so that every village is within 70 kilometers of at least one station in the graph in Figure 1.9.

1.15 Find a winning strategy for the $n \times n$ CHOMP game.

1.16 Is the proof that DOMINATING SET is NP-complete sufficient to also prove the NP-completeness of INDEPENDENT DOMINATING SET?

1.17 Find an independent set of eight queens on the chessboard, that is, no two queens attack each other.

1.18 (Cockayne, Fricke, Hedetniemi and Mynhardt [262]) Define the *boundary* of a set to be the set $B(S) = \{v : |N[v] \cap S| = 1\}$, that is, the set of vertices dominated by exactly one vertex in S. Prove that a dominating set S is a minimal dominating set if and only if $B(S)$ dominates S.

1.19 (Wolk [1150]) Prove that if a graph G does not contain either P_4 or C_4 as an induced subgraph, then $\gamma(G) = 1$, that is, G contains a vertex of degree $n - 1$.

1.20 Prove Theorem 1.4 that for any graph G, $\gamma(G) + \varepsilon_F(G) = n$.

EXERCISE for Couch Potatoes:

1.21 Determine $\gamma(K_n)$.

Chapter 2

Bounds on the Domination Number

In general, when studying subsets of a given type, we are interested in finding either a smallest or a largest such set in a graph. For instance, one considers such problems as finding the minimum cardinality of a dominating set or a cover, or finding the maximum cardinality of an independent set or a packing. Since most of these subset problems are NP-complete for arbitrary graphs, it is natural to find bounds for these numbers. In this chapter, we describe bounds for the domination number $\gamma(G)$. Additional bounds will be supplied throughout the text, particularly Chapter 3. Algorithmic problems associated with actually finding subsets with the desired properties will be discussed in Chapter 12.

2.1 Bounds in Terms of Order

An obvious upper bound on the domination number is the number of vertices in the graph. Since at least one vertex is needed to dominate a graph, we have $1 \leq \gamma(G) \leq n$ for every graph of order n. Both of these bounds are sharp. A graph obtains the lower bound if and only if it has a vertex of degree $n-1$, and it achieves the upper bound if and only if the graph $G = \overline{K}_n$, that is, G is a set of isolated vertices. Note that each isolated vertex must be in every dominating set. For graphs without isolated vertices, the upper bound is much improved in a classical result that is a direct consequence of Ore's Theorem [924] given in Chapter 1.

Theorem 2.1 (Ore) *If a graph G has no isolated vertices, then $\gamma(G) \leq n/2$.*

The *corona* of two graphs G_1 and G_2, as defined by Frucht and Harary [On the corona of two graphs. *Aequationes Math.* 4 (1970) 322–324], is the graph $G = G_1 \circ G_2$ formed from one copy of G_1 and $|V(G_1)|$ copies of G_2 where the *ith* vertex of G_1 is adjacent to every vertex in the *ith* copy of G_2. The corona $H \circ K_1$, in particular, is the graph constructed from a copy of H, where for

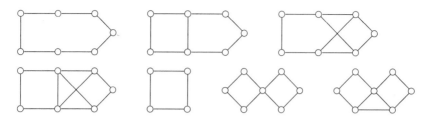

Figure 2.1: Graphs in family $\mathcal{A}$.

each vertex $v \in V(H)$, a new vertex v' and a pendant edge vv' are added. Hence, $H \circ K_1$ has even order and achieves the bound of Theorem 2.1. Graphs having no isolated vertices and domination number exactly half their order were characterized independently by Payan and Xuong and by Fink, Jacobson, Kinch, and Roberts.

Theorem 2.2 [475, 933] *For a graph G with even order n and no isolated vertices, $\gamma(G) = n/2$ if and only if the components of G are the cycle C_4 or the corona $H \circ K_1$ for any connected graph H.*

Cockayne, Haynes, and Hedetniemi [274] characterized the graphs G for which $\gamma(G) = \lfloor n/2 \rfloor$. (We note that as a result of characterizing graphs with equal covering and domination numbers, Randerath and Volkmann [963] have also characterized the graphs G for which $\gamma(G) = \lfloor n/2 \rfloor$.) We need another result before we can treat this more general situation.

Ore's theorem applies to graphs having minimum degree $\delta(G) \geq 1$. Restricting their attention to graphs G having $\delta(G) \geq 2$, McCuaig and Shepherd made another improvement on the upper bound. Let $\mathcal{A}$ be the collection of graphs in Figure 2.1.

Theorem 2.3 [883] *If G is a connected graph with $\delta(G) \geq 2$ and $G \notin \mathcal{A}$, then $\gamma(G) \leq 2n/5$.*

We note that the bound of Theorem 2.3 is sharp and is achieved by the family of graphs illustrated in Figure 2.2 (the shaded vertices form a γ-set). McCuaig and Shepherd [883] characterized the extremal (edge-minimal and edge-maximal) graphs that obtain this upper bound.

In order to present the characterization of graphs for which $\gamma(G) = \lfloor n/2 \rfloor$, we begin with two preliminary results. Let $\mathcal{B}$ be the collection of graphs in Figure 2.3.

Lemma 2.4 [274] *If G is a connected graph with $\delta(G) \geq 2$ and $\gamma(G) = \left\lfloor \frac{n}{2} \right\rfloor$, then $G \in \mathcal{A} \cup \mathcal{B}$.*

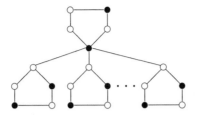

Figure 2.2: A family of graphs with $\delta \geq 2$ for which $\gamma = 2n/5$.

Proof. Let G be connected with $\delta(G) \geq 2$ and $\gamma(G) = \left\lfloor \frac{n}{2} \right\rfloor$. If $G \notin \mathcal{A}$, then by Theorem 2.3, $\gamma(G) \leq \frac{2n}{5}$. If n is even, then $\gamma(G) = \left\lfloor \frac{n}{2} \right\rfloor \not\leq \frac{2n}{5}$, a contradiction. If n is odd, then $\gamma(G) = \left\lfloor \frac{n}{2} \right\rfloor \leq \frac{2n}{5}$ implies that $n = 3$ or 5. All graphs having order three or five, $\delta(G) \geq 2$, and $\gamma(G) = \left\lfloor \frac{n}{2} \right\rfloor$ are in $\mathcal{B}$. $\square$

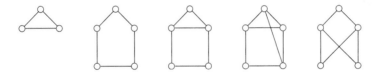

Figure 2.3: Graphs in family $\mathcal{B}$.

In view of Lemma 2.4, we only need to consider graphs with endvertices.

Lemma 2.5 [274] *If G is a connected graph and $\gamma(G) = \left\lfloor \frac{n}{2} \right\rfloor$, then there is at most one endvertex adjacent to each $v \in V$, except for possibly one vertex which may be adjacent to exactly two endvertices when n is odd.*

Proof. Let G be connected with $\gamma(G) = \left\lfloor \frac{n}{2} \right\rfloor$ and assume that X is the set of endvertices adjacent to a vertex v, where $|X| = t$. Then the induced subgraph $H = \langle V - (X \cup \{v\}) \rangle$ has no isolates and Theorem 2.1 gives

$$\gamma(H) \leq \left\lfloor \frac{n - t - 1}{2} \right\rfloor.$$

If n is even, say $n = 2k$, then

$$k = \gamma(G) \leq 1 + \left\lfloor \frac{2k - t - 1}{2} \right\rfloor,$$

which implies that $t \leq 1$ as required.

If n is odd, say $n = 2k + 1$, then

$$k = \gamma(G) \leq 1 + \left\lfloor \frac{2k + 1 - t - 1}{2} \right\rfloor$$

and we deduce that $t \leq 2$. Assume, in this case (that is, n odd), that R is the set of vertices that are adjacent to exactly two endvertices, where $|R| = r$. Note that each vertex of R is in any minimum dominating set of G. Let G' be the subgraph formed by removing R and all endvertices that are adjacent to vertices in R. The set I of isolates of G' is dominated in G by R. The graph $G' - I$ is isolate-free and has at most $2k + 1 - 3r$ vertices. Hence, by Theorem 2.1,

$$\gamma(G' - I) \leq \left\lfloor \frac{2k + 1 - 3r}{2} \right\rfloor .$$

A dominating set of $G' - I$ together with R dominates G; hence,

$$k = \gamma(G) \leq r + \left\lfloor \frac{2k + 1 - 3r}{2} \right\rfloor .$$

We deduce that $r \leq 1$, completing the proof. $\square$

In order to characterize connected graphs with $\gamma(G) = \lfloor n/2 \rfloor$, we define six classes of graphs. Let

$$\mathcal{G}_1 = \{C_4\} \cup \{G : G = H \circ K_1 \text{ where } H \text{ is connected}\}$$

and

$$\mathcal{G}_2 = \mathcal{A} \cup \mathcal{B} - \{C_4\}.$$

For any graph H, let $\mathcal{S}(H)$ denote the set of connected graphs, each of which can be formed from $H \circ K_1$ by adding a new vertex x and edges joining x to one or more vertices of H. Then define

$$\mathcal{G}_3 = \bigcup_H \mathcal{S}(H),$$

where the union is taken over all graphs H. Let y be a vertex of a copy of C_4 and, for $G \in \mathcal{G}_3$, let $\theta(G)$ be the graph obtained by joining G to C_4 with the single edge xy, where x is the new vertex added in forming G. Then define

$$\mathcal{G}_4 = \{\theta(G) : G \in \mathcal{G}_3\}.$$

Next, let u, v, w be a vertex sequence of a path P_3. For any graph H, let $\mathcal{P}(H)$ be the set of connected graphs which may be formed from $H \circ K_1$ by joining each of u and w to one or more vertices of H. Then define

$$\mathcal{G}_5 = \bigcup_H \mathcal{P}(H).$$

Let H be a graph and $X \in \mathcal{B}$. Let $\mathcal{R}(H, X)$ be the set of connected graphs which may be formed from $H \circ K_1$ by joining each vertex of $U \subseteq V(X)$ to one or more vertices of H such that no set with fewer than $\gamma(X)$ vertices of X dominates $V(X) - U$. Then define

$$\mathcal{G}_6 = \bigcup_{H,X} \mathcal{R}(H, X).$$

An example of a graph $G \in \mathcal{G}_6$ is given in Figure 2.4. (The shaded vertices form a $\gamma(G)$-set.)

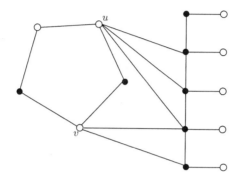

Figure 2.4: Graph $G \in \mathcal{G}_6$ with $U = \{u, v\}$.

Theorem 2.6 [274, 963] *A connected graph G satisfies $\gamma(G) = \left\lfloor \frac{n}{2} \right\rfloor$ if and only if $G \in \mathcal{G} = \bigcup_{i=1}^{6} \mathcal{G}_i$.*

Proof. It is an easy exercise to verify that all graphs G in $\mathcal{G}$ satisfy $\gamma(G) = \left\lfloor \frac{n}{2} \right\rfloor$. Conversely, assume that G is connected with $\gamma(G) = \left\lfloor \frac{n}{2} \right\rfloor$. If n is even, then by Theorem 2.2, $G \in \mathcal{G}_1$. If n is odd and $\delta(G) \geq 2$, then by Lemma 2.4, $G \in \mathcal{G}_2$. Thus we may assume that n is odd and G has a set S $(\neq \emptyset)$ of endvertices. Let T be the set of neighbors of vertices in S. If $|T| = t$, then by Lemma 2.5, $|S| = t$ or $|S| = t + 1$. There is a γ-set of G containing T. Let $G' = G - (S \cup T)$.

Case 1. $|S| = t + 1$.
In this case, G' has $n - (2t + 1)$ vertices. Suppose that G' is not empty. Since G is connected, G' has a vertex x that is adjacent to a vertex in T. Any vertex of degree 0 or 1 in G' has a neighbor in T (since vertices G' are neither isolated in G nor endvertices of G). Hence, T dominates all vertices of G except (perhaps) those vertices having degree at least two in G'. We deduce that T dominates all vertices of G which are not in Y, the set of nonisolates of $G' - x$. By Theorem

2.1, $\langle Y \rangle$ has a γ-set D' and

$$|D'| \leq \left\lfloor \frac{|Y|}{2} \right\rfloor \leq \left\lfloor \frac{n - 2t - 2}{2} \right\rfloor .$$

Since $T \cup D'$ dominates G, we have

$$\left\lfloor \frac{n}{2} \right\rfloor = \gamma(G) \leq t + \left\lfloor \frac{n - 2t - 2}{2} \right\rfloor , \tag{2.1}$$

a contradiction which shows that G' is empty and thus $G \in \mathcal{G}_3$.

Case 2. $|S| = |T| = t$ and G' has an isolated vertex y.
Then $N_G(y) \subseteq T$. If $G' - y$ is not empty, then (since G is connected) $G' - y$ has a vertex x adjacent to a vertex of T. An argument similar to that of Case 1 shows that T dominates all vertices that are not in the set Z of nonisolates of $G - \{x, y\}$. Since $|(V(G')| = n - 2t$, we deduce the inequality (2.1) for this case also. This contradiction implies that $G' - y$ is empty and since y is not an endvertex of G, $G \in \mathcal{G}_3$.

Case 3. $|S| = |T| = t$ and $\delta(G') = 1$.
Let X be the set of vertices having degree one in G'. No vertex of G' is an endvertex of G; hence, each vertex of X is adjacent to a vertex of T, that is, T dominates X. There are two subcases to consider.

Subcase (a). $G' - X$ has no isolates.
By applying Theorem 2.1 to $G' - X$, we obtain

$$\left\lfloor \frac{n}{2} \right\rfloor = \gamma(G) \leq t + \left\lfloor \frac{n - 2t - |X|}{2} \right\rfloor$$

which implies that $|X| = 1$. Let $X = \{x\}$ and let z be the unique neighbor of x in G'. By the hypothesis of this subcase, $G' - \{x\}$ is isolate-free.

If $\delta(G' - x) = 1$, then z is the unique vertex having degree one in $G' - x$ (since $|X| = 1$). Let w be the neighbor of z in $G' - x$. Since w does not have degree one in $G' - x$, there exists a vertex $y \in G' - \{x, z, w\}$ that is adjacent to w. If u is isolated in $G' - \{x, y, z, w\}$, then u is not adjacent to z and the degree of u in $G' - x$ is at least two. We deduce that u is adjacent to w and y. Therefore, $T \cup \{w\}$ dominates all vertices except (possibly) the set Y of nonisolates of $G' - \{x, y, z, w\}$. By Theorem 2.1, $\langle Y \rangle$ may be dominated with at most $\frac{|Y|}{2}$ vertices; hence,

$$\begin{aligned} \left\lfloor \tfrac{n}{2} \right\rfloor = \gamma(G) &\leq (t+1) + \tfrac{|Y|}{2} \\ &\leq (t+1) + \left\lfloor \tfrac{n-2t-4}{2} \right\rfloor , \end{aligned}$$

a contradiction.

Therefore, $\delta(G' - x) \geq 2$. Now $G' - x$ has an even number of vertices and since $\gamma(G) = \left\lfloor \frac{n}{2} \right\rfloor$, $\gamma(G' - x) = \frac{|V(G' - x)|}{2}$. Theorem 2.2 implies that $G' - x = mC_4$ where $m \geq 1$. If $m > 1$, then a vertex of one component C_4 is adjacent to a vertex of T which would imply that $\gamma(G) < \left\lfloor \frac{n}{2} \right\rfloor$. Therefore, $G' - x = C_4$ and $G \in \mathcal{G}_4$.

Subcase (b). $G' - X$ has isolated vertices.
Let I be the set of isolates of $G' - X$ where $|I| = i \geq 1$. Each $v \in I$ is not an endvertex of G' and so has at least two neighbors in X. Since $\deg_{G'}(x) = 1$ for all $x \in X$, $|X| \geq 2i$. Now T dominates $S \cup T \cup X$, I dominates itself and by Theorem 2.1, at most $\frac{|V(G') - (X \cup I)|}{2}$ vertices are required to dominate the nonisolates of $G' - X$. Hence,

$$\left\lfloor \frac{n}{2} \right\rfloor = \gamma(G) \leq t + i + \left\lfloor \frac{n - 2t - 3i}{2} \right\rfloor,$$

which implies that $i = 1$, that is, $G' - X$ has precisely one isolated vertex, say v, which has at least two neighbors u and w in X. Now $G' - (X \cup \{v\})$ has no isolates and may be dominated (Theorem 2.1) by at most

$$\left\lfloor \frac{|V(G' - (X \cup \{v\}))|}{2} \right\rfloor \leq \left\lfloor \frac{n - 2t - |N[v]|}{2} \right\rfloor$$

vertices. Also $T \cup \{v\}$ dominates $X \cup \{v\}$ hence,

$$\left\lfloor \frac{n}{2} \right\rfloor = \gamma(G) \leq t + 1 + \left\lfloor \frac{n - 2t - |N[v]|}{2} \right\rfloor. \tag{2.2}$$

This implies that $|N[v]| \leq 3$, and hence, $N(v) = \{u, w\}$. If v were dominated by T, then the right side of (2.2) can be reduced by one, giving a contradiction. Hence, in order to dominate v, one vertex from $\{u, w, v\}$ is in every γ-set of G.

Suppose that the graph $G' - \{u, w, v\}$ (which has even order) is not empty. Since $\gamma(G) = \left\lfloor \frac{n}{2} \right\rfloor$, exactly $\frac{|V(C)|}{2}$ vertices from each component C of $G' - \{u, w, v\}$ must be included in each γ-set of G. Hence, C has even order and by Theorem 2.2, $C = C_4$ or $C = H \circ K_1$ for any connected graph H. The preceding paragraphs imply that no vertex of C is adjacent to u, w, or v. If $C = C_4$, then the connectedness of G implies that a vertex of C is adjacent to a vertex of T, and it follows that there is a γ-set of G including T, v, and only one of the four vertices of C, a contradiction. Suppose that $C = H \circ K_1$. Since no vertex of C is an endvertex of G, each endvertex of C has a neighbor in T and again it is easily seen that there exists a γ-set of G containing $< \frac{|V(C)|}{2}$ vertices of C.

We conclude that $G' - \{u, w, v\}$ is empty and so $G \in \mathcal{G}_5$.

Case 4. $|S| = |T| = t$ and $\delta(G') \geq 2$.
In order that $\gamma(G) = \left\lfloor \frac{n}{2} \right\rfloor$, a γ-set of G must include $\frac{|V(C)|}{2}$ vertices from any

even order component C of G'. In this case by Theorem 2.2, $C = C_4$ and this situation is impossible (by a similar argument to that used in subcase 3(b)). If G' has more than one component of odd order, then $\gamma(G') < \left\lfloor \frac{|V(G')|}{2} \right\rfloor$, which implies that $\gamma(G) < \left\lfloor \frac{n}{2} \right\rfloor$. Hence, G' is connected and is in $\mathcal{A} \cup \mathcal{B} - \{C_4\}$ (Lemma 2.4). However, G' may be further restricted. Since G is connected, the set U of vertices G' having neighbors in T is nonempty.

It is easily checked that if $G' \in A - \{C_4\}$, then for each $U \neq \emptyset$, $V(G') - U$ may be dominated by less than $\gamma(G')$ vertices of G'. Since T dominates U, it follows that a γ-set of G contains less than $\left\lfloor \frac{|V(G')|}{2} \right\rfloor$ vertices of G', so $\gamma(G) < \left\lfloor \frac{n}{2} \right\rfloor$.

We conclude that $G' \in \mathcal{B}$ and $\gamma(G')$ vertices are necessary to dominate $V(G') - U$. Hence, $G \in \mathcal{G}_6$. □

Returning our attention to upper bounds on the domination number, we note that Reed again improved the bound by increasing the minimum degree requirement.

Theorem 2.7 [970] *If G is a connected graph with $\delta(G) \geq 3$, then $\gamma(G) \leq 3n/8$.*

The above bounds are summarized in the following table.

lower bound for δ	upper bound for γ
0	n
1	$n/2$
2, (G connected, $G \notin \mathcal{B}$)	$2n/5$
3, (G connected)	$3n/8$

An obvious conjecture seems to be that for any graph G with $\delta(G) \geq k$, $\gamma(G) \leq kn/(3k-1)$. However, for $\delta(G) \geq 7$, Caro and Roditty gave the following better bound. The question remains open for graphs G having $4 \leq \delta(G) \leq 6$.

Theorem 2.8 [182, 183] *For any graph G,*

$$\gamma(G) \leq n \left[1 - \delta \left(\frac{1}{\delta + 1} \right)^{1+1/\delta} \right].$$

Ore's theorem can also be improved for connected graphs with forbidden subgraphs. A result due to Cockayne, Ko, and Shepherd concerns graphs that do not have a claw $K_{1,3}$ or a net $K_3 \circ K_1$ as an induced subgraph.

Theorem 2.9 [287] *If a connected graph G is claw-free and net-free, then $\gamma(G) \leq \lceil n/3 \rceil$.*

Proof. Let G be a connected, claw-free, and net-free graph having $n = 3k + r$ vertices where $r \in \{0, 1, 2\}$. It is sufficient to show that V may be partitioned into $\{U_1, U_2, ..., U_k, R\}$ for $|U_i| = 3$, $|R| = r$, $\Delta(\langle U_i \rangle) \geq 2$, and $\langle R \rangle = K_r$. If no such partition exists, then we claim that there is a connected graph G with $|V| > 3$ satisfying the hypothesis such that the removal of any 3-subset $U = \{u, v, w\}$, where $\Delta(\langle U \rangle) \geq 2$, disconnects G. To see this, assume that a 3-subset U with $\Delta(\langle U \rangle) \geq 2$ can be removed from G without disconnecting the graph. Remove U to form the connected graph G'. Now $V(G')$ does not have a partition as described, for, if it did, so would $V(G)$ using U along with the partition of $V(G')$. If we keep removing 3-subsets in this manner, eventually we will have a graph for which the removal of any 3-subset disconnects the graph.

Consider a longest path in G, $P = v_1 v_2 ... v_l$ (note that $l \geq 3$). Let $S = \{v_1, v_2, ..., v_l\}$. Since P is a longest path, we have $N(v_1) \subseteq S$. If v_2 is adjacent to some $w \notin S$, then since G is claw-free and $w \notin N(v_1)$, either v_1 or w is adjacent to v_3. But if either $v_1 v_3 \in E$ or $w v_3 \in E$, we can obtain a path with length greater than l. Therefore, $N(v_2) \subseteq S$.

Since $\langle V - \{v_1, v_2, v_3\} \rangle$ is disconnected and $N(v_1) \subseteq S$ and $N(v_2) \subseteq S$, v_3 is adjacent to some $w \notin S$, it follows that $l \geq 4$ and w is neither adjacent to v_1 nor v_2. If $w \in N(v_4)$, then $v_1 v_2 v_3 w v_4 ... v_l$ is a path with length greater than l. Hence, $w v_4 \notin E$. But G is claw-free, so v_2 is adjacent to v_4, implying that $l \geq 5$.

Suppose that v_4 is adjacent to some $u \notin S$. If v_3 is also adjacent to u, then we can obtain a longer path. Again since $N(v_1) \subseteq S$ and $N(v_2) \subseteq S$, neither v_1 nor v_2 is adjacent to u. Also if v_1 is adjacent to v_4, then $v_l ... v_5 v_4 v_1 v_2 v_3 w$ is a path of length $l + 1$. On the other hand, if v_1 is adjacent to v_3, then $\langle \{v_1, v_3, w, v_4\} \rangle$ is a claw. Hence, v_1 is not adjacent to either v_3 or v_4. But this implies that the subgraph $\langle \{v_1, v_2, v_3, v_4, w, u\} \rangle$ is a net. Therefore, v_4 does not have a neighbor $u \notin S$, that is, $N(v_4) \subseteq S$.

Since a longest path has length l, the vertex v_1 is not adjacent to any of v_3, v_4, and v_5. Then $\langle \{v_1, ..., v_5, w\} \rangle$ is a net, unless v_5 is adjacent to one or both of v_2 and v_3. If $v_2 v_5 \in E$ and $v_3 v_5 \notin E$, then $\langle \{v_1, v_2, v_3, v_5\} \rangle$ is a claw. If $v_3 v_5 \in E$ and $v_2 v_5 \notin E$, then $\langle \{w, v_3, v_2, v_5\} \rangle$ is a claw. Thus $\{v_2, v_3\} \in N(v_5)$.

Suppose that there is a vertex $w' \in N(w)$ and $w' \notin S$. Then $v_l ... v_5 v_4 v_2 v_3 w w'$ is a path of length greater than l. Hence, $N(w) \subseteq S$.

Finally, if there is some vertex $w' \notin (S \cup \{w\})$ such that $w' \in N(v_3)$, then $w w' \notin E$ implies that $\langle \{v_3, w, w', v_5\} \rangle$ is a claw. But w cannot be adjacent to w' since $N(w) \subseteq S$. Therefore, $N(v_3) \subseteq S \cup \{w\}$. Hence, $\langle V - \{w, v_3, v_4\} \rangle$ is connected, a contradiction. Thus, we have shown that $V(G)$ has a partition as described and it follows that $\gamma(G) \leq \lceil n/3 \rceil$. $\square$

We conclude this section with an estimation of the domination number for almost every graph due to Weber. (In the following theorem log indicates log base 2.)

Theorem 2.10 [1141] *Let* $k = \lfloor (\log n - 2 \log \log n + \log \log e) \rfloor$. *Then for almost every graph*

$$k + 1 \le \gamma(G) \le k + 2.$$

2.2 Bounds in Terms of Order, Degree, and Packing

We begin this section with well known bounds on $\gamma(G)$ in terms of the number of vertices n and the maximum degree $\Delta(G)$. The upper bound is attributed to Berge and the lower bound to Walikar, Acharya, and Sampathkumar.

Theorem 2.11 [95, 1134] *For any graph* G,

$$\left\lceil \frac{n}{1 + \Delta(G)} \right\rceil \le \gamma(G) \le n - \Delta(G).$$

Proof. Let S be a γ-set of G. First we consider the lower bound. Each vertex can dominate at most itself and $\Delta(G)$ other vertices. Hence, $\gamma(G) \ge \lceil n/(1 + \Delta(G)) \rceil$.

For the upper bound, let v be a vertex of maximum degree $\Delta(G)$. Then v dominates $N[v]$ and the vertices in $V - N[v]$ dominate themselves. Hence, $V - N(v)$ is a dominating set of cardinality $n - \Delta(G)$, so $\gamma(G) \le n - \Delta(G)$. $\square$

Surprisingly, a statement of the following stronger lower bound for $\gamma(G)$ apparently did not appear in print until 1995.

Theorem 2.12 [1039] *If* G *has degree sequence* $(d_1, d_2, \cdots, d_n)$ *with* $d_i \ge d_{i+1}$, *then* $\gamma(G) \ge \min\{k : k + (d_1 + d_2 + \cdots + d_k) \ge n\}$.

The easy proof of the above theorem is left as an exercise. (See Exercise 2.4.) It follows from the proof of Theorem 2.11 that $\gamma(G) = n/(1 + \Delta(G))$ if and only if G has a γ-set S such that $N[u] \cap N[v] = \emptyset$ for all $u, v \in S$ and $|N(v)| = \Delta(G)$ for all $v \in S$. For example, the collection of stars $tK_{1,\Delta}$ and cycles C_{3t} have $\gamma = t = n/(1 + \Delta)$.

We note that a set S is a *2-packing* if for each pair of vertices $u, v \in S$, $N[u] \cap N[v] = \emptyset$. The *packing number* $\rho(G)$ is the cardinality of a maximum packing. Hence, the packing number provides another lower bound on $\gamma(G)$.

Theorem 2.13 *For any graph* G,

$$\rho(G) \le \gamma(G).$$

To see the sharpness of the upper bound of Theorem 2.11, consider the corona $K_p \circ K_1$ for which $n = 2p$, $\Delta = p$, and $\gamma = p$. Domke, Dunbar, and Markus characterized the trees achieving this upper bound. A *subdivision* of an edge uv is obtained by removing edge uv, adding a new vertex w, and adding edges uw and vw. A *wounded spider* is the graph formed by subdividing at most $t - 1$ of the edges of a star $K_{1,t}$ for $t \geq 0$. Examples of wounded spiders include K_1, the star $K_{1,n-1}$, the corona $K_{1,t} \circ K_1$, and the graph shown in Figure 2.5.

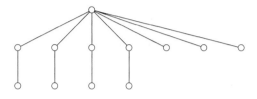

Figure 2.5: A wounded spider.

Theorem 2.14 [363] *For any tree T, $\gamma(T) = n - \Delta(T)$ if and only if T is a wounded spider.*

Proof. If T is a wounded spider, it is easy to check that $\gamma(T) + \Delta(T) = n$. Let T be a tree with $\gamma(T) + \Delta(T) = n$ and v be a vertex of maximum degree. If $T - N[v] = \emptyset$, then T is the star $K_{1,t}$ with $t \geq 0$ (a wounded spider). So assume there is at least one vertex in $T - N[v]$. Let I be any maximal independent set of $\langle T - N[v] \rangle$. Then $I \cup \{v\}$ is an independent dominating set for T. Hence, $n = \gamma(T) + \Delta(T) \leq |I| + 1 + \Delta(T) \leq n$ implying that $V - N(v)$ is an independent set. Furthermore, $N(v)$ is also an independent set since any edge in $N(v)$ creates a cycle.

The connectivity of T implies that each vertex in $V - N[v]$ must be adjacent to at least one vertex in $N(v)$. Moreover, if any vertex in $V - N[v]$ is adjacent to two or more vertices in $N(v)$, then a cycle is formed. Hence, each vertex in $V - N[v]$ is adjacent to exactly one vertex in $N(v)$. To ensure that $\Delta(T) + 1$ vertices are necessary to dominate T, there must be at least one vertex in $N(v)$ which is not adjacent to any vertex in $V - N[v]$ and each vertex in $N(v)$ has either 0 or 1 neighbors in $V - N[v]$. Therefore, T is a wounded spider. $\square$

Next we consider inequalities involving the minimum degree $\delta(G)$. Flach and Volkmann gave an upper bound in terms of the minimum and maximum degrees and the order of a graph.

Theorem 2.15 [483] *For any graph G,*

$$\gamma(G) \leq \left(n + 1 - (\delta(G) - 1)\frac{\Delta(G)}{\delta(G)} \right) / 2.$$

An immediate corollary, due to Payan, now follows.

Corollary 2.16 [930] *If a graph G has no isolated vertices, then*

$$\gamma(G) \le \frac{n + 2 - \delta(G)}{2}.$$

Marcu gave a slight improvement.

Theorem 2.17 [876] *If a graph G has no isolated vertices and $\gamma(G) \ge 3$, then $\gamma(G) \le \frac{n+1-\delta(G)}{2}$.*

The next upper bound is found in Alon and Spencer [30], Arnautov [41], and Payan [930].

Theorem 2.18 *If a graph G has no isolated vertices, then*

$$\gamma(G) \le \frac{n\left(1 + \ln(\delta(G) + 1)\right)}{\delta(G) + 1}.$$

Proof. Let $p = ln(\delta(G) + 1)/(\delta(G) + 1)$. Construct a dominating set as follows. Select a set of vertices A, where each vertex is selected independently with the probability p. The expected value of A is np. Let $B = V - N[A]$, that is, set B is not dominated by A. Clearly, $S = A \cup B$ is a dominating set. We show that the expected value of B is at most $ne^{-p(1+\delta)}$. A vertex v is in B if and only if no vertex from $N[v]$ is in A. So the probability that v is in B is $(1 - p)^{1+deg(v)}$. Since $e^{-x} \ge 1 - x$ for any nonnegative real number x, and $deg(v) \ge \delta(G)$, it follows that the probability that v is in B is at most $e^{-p(1+\delta)}$. Thus the expected value of $|B|$ is at most $ne^{-p(1+\delta)}$. Hence, the expected value of $|S|$ is at most

$$n(p + e^{-p(1+\delta)}) = n(1 + ln(\delta + 1))/(\delta + 1).$$

Since the average order of S is at most $n(1 + ln(\delta + 1))/(\delta + 1)$, there is a particular set S with at most this cardinality. $\square$

Arnautov actually proved the following bound.

Theorem 2.19 [41, 930] *If a graph G has no isolated vertices, then*

$$\gamma(G) \le \frac{n}{\delta(G) + 1} \sum_{j=1}^{\delta(G)+1} 1/j.$$

In a similar vein, Clark, Fisher, Shekhtman, and Suen [245] found bounds that are asymptotically equal to the bound in Theorem 2.18. Actually their bounds perform much better in many situations than the above result. However, the statement of their results is intricate and requires detail beyond the scope of this text. We refer the reader to [245].

2.3   Bounds in Terms of Order and Size

To aid in our development of bounds on the domination number in terms of order and size, we need a theorem from Vizing [1116] which bounds the size m of a graph having a given domination number. Since this theorem and related results due to Sanchis [1007] are interesting in their own right, we present a full discussion of them here.

Theorem 2.20 [1116] *If a graph G has $\gamma(G) \geq 2$, then*

$$m \leq \left\lfloor \frac{1}{2}(n - \gamma(G))\,(n - \gamma(G) + 2) \right\rfloor.$$

Proof. Let G be a graph with $\gamma(G) \geq 2$. We proceed by induction on the order n. If $n = 2$, then G is either K_2 or $\overline{K}_2$ and the inequality is satisfied. Assume that every graph G having order less than n and $\gamma(G) \geq 2$ satisfies the inequality.

We first consider graphs G having $\gamma(G) \geq 3$. Let v be a vertex of maximum degree $\Delta(G)$. Then, by Theorem 2.11, $|N(v)| = \Delta(G) \leq n - \gamma(G)$. That is, $|N(v)| = \Delta(G) = n - \gamma(G) - r$, where $0 \leq r \leq n - \gamma(G)$. Let $S = V - N[v]$. Then $|S| = \gamma(G) + r - 1$. If $u \in N(v)$, then the set $(S - N(u)) \cup \{u, v\}$ is a dominating set of G, and therefore $\gamma(G) \leq |S - N(u)| + 2$. Thus, $\gamma(G) \leq \gamma(G) + r - 1 - |S \cap N(u)| + 2$, and so $|N(u) \cap S| \leq r + 1$, for each vertex $u \in N(v)$. Hence, the number, say m_1, of edges between $N(v)$ and S is at most $\Delta(G) \cdot (r + 1)$.

Furthermore, if D is a γ-set of $\langle S \rangle$, then $D \cup \{v\}$ is a dominating set of G. Hence, $\gamma(G) \leq |D \cup \{v\}|$, implying that $\gamma(\langle S \rangle) \geq \gamma(G) - 1 \geq 2$. By the inductive hypothesis, the number of edges in $\langle S \rangle$, say m_2, is

$$
\begin{aligned}
m_2 \;&\leq\; \left\lfloor \tfrac{1}{2}(|S| - \gamma(\langle S \rangle))\,(|S| - \gamma(\langle S \rangle) + 2) \right\rfloor \\[4pt]
&\leq\; \left\lfloor \tfrac{1}{2}(\gamma(G) + r - 1 - (\gamma(G) - 1))\,(\gamma(G) + r - 1 - (\gamma(G) - 1) + 2) \right\rfloor \\[4pt]
&=\; \tfrac{1}{2}r(r + 2).
\end{aligned}
$$

Let $m_3 = |E(\langle N[v] \rangle)|$. Now vertex v is adjacent to $\Delta(G)$ edges and each vertex $u \in N(v)$ has degree at most $\Delta(G)$. The edges between S and $N(v)$ account for at most $r + 1$ edges incident to each $u \in N(v)$. Thus the following inequality applies for the size of G:

$$
\begin{aligned}
m &= m_1 + m_2 + m_3 \\[4pt]
&\leq \Delta(G)\,(r+1) + \tfrac{1}{2}r(r+2) + \Delta(G) + \tfrac{1}{2}\Delta(G)\,(\Delta(G) - r - 2) \\[4pt]
&= \Delta(G)\,(n - \gamma(G) - \Delta(G) + 1) + \\
&\quad\; \tfrac{1}{2}(n - \gamma(G) - \Delta(G))(n - \gamma(G) - \Delta(G) + 2) + \\
&\quad\; \Delta(G) + \tfrac{1}{2}\Delta(G)\,(2\,\Delta(G) - n + \gamma(G) - 2) \\[4pt]
&= \tfrac{1}{2}(n - \gamma(G))(n - \gamma(G) + 2) - \tfrac{1}{2}\Delta(G)\,(n - \gamma(G) - \Delta(G)) \\[4pt]
&\leq \tfrac{1}{2}(n - \gamma(G))(n - \gamma(G) + 2).
\end{aligned}
$$

Hence, the theorem holds for graphs having $\gamma(G) \geq 3$. Observe that the result also holds when $\gamma(G) \geq 2$, since adding an isolated vertex to G gives a graph with domination number $\gamma \geq 3$ and order $n + 1$, but does not increase the size or the maximum degree. $\square$

Vizing [1116] constructed the following family of graphs G which achieve the upper bound of Theorem 2.20. Let H be the graph formed from a complete graph K_t by removing the edges of a minimum edge cover (a smallest possible set of edges that contain all the vertices). For the case when $\gamma(G) = 2$, G is the join $H + \overline{K}_2$. It is easy to see that $\gamma(G) = 2$, $t = n - 2$, and

$$
m = \frac{1}{2}(n-2)(n-3) - \left\lceil \frac{1}{2}(n-2) \right\rceil + 2(n-2) = \left\lfloor \frac{1}{2}(n-2)n \right\rfloor.
$$

For graphs G with $\gamma(G) > 2$, construct G from the join $H + \overline{K}_2$ by adding a set of $\gamma(G) - 2$ isolated vertices. Again, it is a simple to see that G has domination number equal to $\gamma(G)$, $t = n - \gamma(G)$, and

$$
m = \left\lfloor \frac{1}{2}(n - \gamma(G) + 2 - 2)\,(n - \gamma(G) + 2) \right\rfloor = \left\lfloor \frac{1}{2}(n - \gamma(G))\,(n - \gamma(G) + 2) \right\rfloor.
$$

We note that Vizing's graphs also achieve the upper bound of Theorem 2.11. That is, such a graph G has $\Delta(G) = n - \gamma(G)$.

For graphs having $\Delta(G) < n - \gamma(G)$, Sanchis improved the bound of Theorem 2.20 and characterized extremal graphs achieving this better bound. This following proof to Sanchis' theorem is due to Michael Henning (personal communication).

Theorem 2.21 [1007] *If a graph G has $\gamma(G) \geq 2$ and $\Delta(G) \leq n - \gamma(G) - 1$, then*

$$
m \leq \frac{1}{2}(n - \gamma(G))\,(n - \gamma(G) + 1).
$$

Proof. From the proof to Theorem 2.20, we have

$$m \leq \left\lfloor \frac{[(n - \gamma(G))(n - \gamma(G) + 2) - \Delta(G)(n - \gamma(G) - \Delta(G))]}{2} \right\rfloor.$$

Using elementary calculus, we see that this parabolic upper bound has a minimum at $\Delta(G) = \frac{1}{2}(n - \gamma(G))$. Thus, it suffices to apply the bound for $\Delta(G) = n - \gamma(G) - 1$ and $\Delta(G) = 1$. In both cases,

$$m \leq \left\lfloor \frac{1}{2}[(n - \gamma(G))(n - \gamma(G) + 1) + 1] \right\rfloor.$$

□

Now we are ready to give bounds on $\gamma(G)$ in terms of the order and size of G. The upper bound is an immediate corollary of Vizing's theorem and the corresponding lower bound was given by Berge [95]. A *galaxy* is a forest in which each component is a star.

Theorem 2.22 *For any graph G,*

$$n - m \leq \gamma(G) \leq n + 1 - \sqrt{1 + 2m}.$$

Furthermore, $\gamma(G) = n - m$ if and only if G is a galaxy.

Proof. From Theorem 2.20, we have $2m \leq (n - \gamma(G))^2 + 2(n - \gamma(G))$. Completing the square, we have $(n - \gamma(G))^2 + 2(n - \gamma(G)) + 1 \geq 2m + 1$. Since $n - \gamma(G) \geq 0$, it follows that $n - \gamma(G) + 1 \geq \sqrt{1 + 2m}$ and hence, the upper bound on $\gamma(G)$.

Since $\gamma(G) \geq 1$, the lower bound is obvious for $m \geq n - 1$. Hence, let $m < n$. Then G has at least $n - m$ components. At least one vertex per component is required in any dominating set. Hence, $\gamma(G) \geq n - m$ with equality if and only if G has exactly $n - m$ components each with domination number equal to 1. Furthermore, G has $n - m$ components if and only if G is a forest. It follows that $\gamma(G) = n - m$ if and only if each component of G is a star. That is, G is a galaxy. □

2.4 Bounds in Terms of Degree, Diameter, and Girth

In a graph G of diameter 2, the open neighborhood of any vertex $v \in V(G)$ dominates G and the following upper bound is immediate.

Theorem 2.23 *If a graph G has $diam(G) = 2$, then $\gamma(G) \leq \delta(G)$.*

Another elementary result involving the diameter gives a lower bound for the domination number.

Theorem 2.24 *For any connected graph G,*

$$\left\lceil \frac{diam(G) + 1}{3} \right\rceil \leq \gamma(G).$$

Proof. Let S be a γ-set of a connected graph G. Consider an arbitrary path of length $diam(G)$. This diametral path includes at most two edges from the induced subgraph $\langle N[v] \rangle$ for each $v \in S$. Furthermore, since S is a γ-set, the diametral path includes at most $\gamma(G) - 1$ edges joining the neighborhoods of the vertices of S. Hence, $diam(G) \leq 2\gamma(G) + \gamma(G) - 1 = 3\gamma(G) - 1$ and the desired result follows. $\square$

Brigham, Chinn, and Dutton observed an interesting relationship between the diameter of G and domination number of its complement $\overline{G}$.

Theorem 2.25 [159] *If $\gamma(\overline{G}) \geq 3$, then $diam(G) \leq 2$.*

If G is not connected, let $diam(G) = \infty$. The contrapositive of Theorem 2.25 determines $\gamma(\overline{G})$ for any graph G having diameter at least three.

Theorem 2.26 *If a graph G has no isolated vertices and $diam(G) \geq 3$, then $\gamma(\overline{G}) = 2$.*

Proof. Let x and y be vertices of G such that $d(x, y) = diam(G) \geq 3$. Obviously, x and y dominate $\overline{G}$ since there is no vertex in G adjacent to both x and y. Hence, $\{x, y\}$ dominates $\overline{G}$ and $\gamma(\overline{G}) \leq 2$. If $\gamma(\overline{G}) = 1$, then G has an isolated vertex, contrary to the hypothesis. $\square$

Recall that the length of a shortest cycle in a graph G that contains cycles is the girth $g(G)$. A cycle with length $g(G)$ is called a *g-cycle*. Knowing the girth helps to ascertain bounds on the domination number of G. Brigham and Dutton obtained several bounds along these lines. We list a sampling of their results for graphs with sufficiently large girth.

Theorem 2.27 [163] *If a graph G has $\delta(G) \geq 2$ and $g(G) \geq 5$, then*

$$\gamma(G) \leq \left\lceil \frac{n - \lfloor g(G)/3 \rfloor}{2} \right\rceil.$$

Proof. Let G be a graph with $\delta(G) \geq 2$ and $g(G) \geq 5$. Remove a g-cycle from G to form graph G'. Suppose a vertex $v \in V(G')$ has two neighbors, say x and y on the g-cycle which was removed from G. If $d(x, y) \leq 2$, then v, x, and y are on either a C_3 or C_4 in G, contradicting the hypothesis that $g(G) \geq 5$. If

$d(x, y) \geq 3$, then replacing the path from x to y on the g-cycle with the path x, v, y reduces the girth of G, a contradiction.

Hence, no vertex in G' has two or more neighbors on the g-cycle. Since $\delta(G) \geq 2$, the graph G' has minimum degree at least $\delta(G) - 1 \geq 1$. Therefore, Ore's theorem implies that $\gamma(G') \leq \left\lfloor \frac{n-g(G)}{2} \right\rfloor$. Observe that in general, a cycle with length g can be dominated by $\left\lceil \frac{g-\lfloor g/3 \rfloor}{2} \right\rceil$ vertices. Hence,

$$\gamma(G) \leq \left\lfloor \frac{n - g(G)}{2} \right\rfloor + \left\lceil \frac{g(G) - \lfloor g(G)/3 \rfloor}{2} \right\rceil$$

and the theorem follows. $\square$

The next result establishes lower bounds on $\gamma(G)$ in terms of the minimum degree.

Theorem 2.28 [163] *For any graph G,*

(a) *if $g(G) \geq 5$, then $\gamma(G) \geq \delta(G)$.*

(b) *if $g(G) \geq 6$, then $\gamma(G) \geq 2(\delta(G) - 1)$.*

Proof. Let S be a γ-set of G. The inequalities are trivially true for $\delta(G) = 1$, so let $\delta(G) \geq 2$. By Theorem 2.11, $\gamma(G) \leq n - \Delta(G) \leq n - \delta(G) \leq n - 2$ implying that $|V - S| \geq 2$. Let $u \in V - S$. If a vertex of S dominates two or more vertices in $N(u)$, then a cycle C_3 or C_4 is formed, contradicting that $g(G) \geq 5$. Hence, each vertex of S dominates at most one vertex in $N(u)$, so $\gamma(G) \geq |N(u)| \geq \delta(G)$.

Let $g(G) \geq 6$. If there are adjacent vertices u and v in $V - S$, then each vertex of S can dominate at most one vertex in $N(u) \cup N(v) - \{u, v\}$. Thus, $\gamma(G) \geq |N(u)| + |N(v)| - 2 \geq 2(\delta(G) - 1)$. If $V - S$ is independent and $\{u, v\} \subseteq V - S$, then $N(u) \subseteq S$ and $N(v) \subseteq S$. Furthermore, since $g(G) \geq 6$, $|N(u) \cap N(v)| \leq 1$, and again $\gamma(G) \geq |N(u)| + |N(v)| - 1 \geq 2\delta(G) - 1$. $\square$

Brigham and Dutton showed that the bounds of Theorem 2.28 are sharp even for graphs with arbitrarily large maximum degree. They noted that the sharpness is interesting since $\Delta(G)$ serves as a lower bound on $\gamma(G)$ for graphs having $g(G) \geq 7$.

Theorem 2.29 [163] *If a graph G has $\delta(G) \geq 2$ and $g(G) \geq 7$, then $\gamma(G) \geq \Delta(G)$.*

Proof. Let S be a γ-set of G and v a vertex of maximum degree. If $v \notin S$, then an argument similar to the proof Theorem 2.28 shows that $\gamma(G) \geq \Delta(G)$. Let $v \in S$, $N(v) = \{v_1, v_2, ..., v_\Delta\}$, and u_i be a second neighbor of v_i, for $1 \leq i \leq \Delta(G)$. Since $g(G) \geq 7$, $N(v)$ is an independent set and no pair of vertices in $N(v)$ have a common neighbor in $V - S$. Hence, the u_i are distinct. Again since $g(G) \geq 7$, no two vertices of $\{u_1, u_2, ..., u_\Delta\}$ have a common neighbor. Thus, each u_i is either in S or has a distinct neighbor in S. It follows that $\gamma(G) \geq \Delta(G) + 1$. $\Box$

A graph is *planar* if it can be drawn in the plane with no crossing edges. MacGillivray and Seyffarth studied the interesting problem of domination numbers of planar graphs with small diameter. They determined the following sharp upper bound.

Theorem 2.30 [862] *If G is a planar graph with $diam(G) = 2$, then $\gamma(G) \leq 3$.*

We conclude this section with an upper bound for planar graphs of diameter three. It is not known if this bound is sharp.

Theorem 2.31 [862] *If G is a planar graph with $diam(G) = 3$, then $\gamma(G) \leq 10$.*

2.5   Bounds In Terms of Independence and Covering

Many bounds on the domination number will be presented throughout this text and are not contained in this chapter. However, we would be amiss if we did not at least mention a well-known inequality chain here. This chain first appeared in a paper by Cockayne, Hedetniemi, and Miller [285] and yields some very interesting bounds on γ. (The irredundance numbers, $ir(G)$ and $IR(G)$, are defined in Chapter 3.) For any graph G,

$$ir(G) \leq \gamma(G) \leq i(G) \leq \beta_0(G) \leq \Gamma(G) \leq IR(G).$$

This chain is of such significance that we devote Chapter 3 to its discussion. Here we present related results.

The *edge independence number* $\beta_1(G)$ is the maximum cardinality among the independent sets of edges of G. (This parameter $\beta_1(G)$ is the well-known *matching number* of a graph. A graph is said to have a *perfect matching* if $\beta_1(G) = n/2$.) A *vertex cover* of G is a set of vertices that covers all the edges and an *edge cover* is a set of edges that covers all the vertices. The minimum cardinality of a vertex (edge) cover is $\alpha_0(G)$ (respectively, $\alpha_1(G)$). We observe the following straightforward inequalities.

Observation 2.32 *If a graph G has no isolated vertices, then*

$$\gamma(G) \leq \alpha_0(G),$$

$$\gamma(G) \leq \beta_0(G),$$

$$\gamma(G) \leq \beta_1(G),$$
$$\gamma(G) \leq \alpha_1(G).$$

We use the following theorem due to Gallai (for the proof see Chapter 9).

Theorem 2.33 [518] *For any graph G,*

$$\alpha_0(G) + \beta_0(G) = n$$

and if G has no isolated vertices, then

$$\alpha_1(G) + \beta_1(G) = n.$$

Notice that if G has no isolated vertices, then $\gamma(G) \leq \min\{\alpha_0(G), \beta_0(G)\}$ and $\gamma(G) \leq \min\{\alpha_1(G), \beta_1(G)\}$.

A 1-*maximal* matching $M \subseteq E$ is 2-*maximal* if it is not possible to delete one edge from M and add two others to produce another matching. Let $\beta_2(G)$ denote the minimum cardinality of a 2-maximal matching of G. For example, consider the path $P_4 = v_1, v_2, v_3, v_4$. Note that the edge $v_2 v_3$ is a 1-maximal matching of P_4, but it is not a two-maximal matching since removing $v_2 v_3$ and adding $v_1 v_2$ and $v_3 v_4$ yields another matching.

A set $F \subseteq E$ is an *edge dominating set* if each edge in E is either in F or is adjacent to an edge in F and is an *independent edge dominating set* if the edges of S are independent. The *edge domination number* $\gamma'(G)$ is the smallest cardinality among all minimal edge dominating sets and $i'(G)$ denotes the *independent edge domination number*. We observe that $\gamma'(G) = i'(G) \leq \beta_2(G) \leq \beta_1(G)$.

An unpublished result due to S. T. Hedetniemi gives an improvement on the upper bound for $\gamma(G)$.

Theorem 2.34 (Hedetniemi) *If a graph G has no isolated vertices, then $\gamma(G) \leq \beta_2(G)$.*

Proof. Let $M \subseteq E$ be a 2-maximal matching with minimum cardinality in G, and let $V(M)$ denote the set of $2\beta_2(G)$ vertices incident with an edge in M. Since M is also a 1-maximal matching, $\langle V - V(M) \rangle$ is an independent set of vertices. But since G has no isolated vertices, each vertex in $V - V(M)$ is adjacent to at least one vertex in $V(M)$. We can now construct a dominating set S of order $\beta_2(G)$ as follows.

Let $\beta_2(G) = k$ and let the vertices in $V(M) = \{v_1, v_1', v_2, v_2', ..., v_k, v_k'\}$, where $v_i v_i' \in M$, and the vertices in $V - V(M) = \{u_1, u_2, ..., u_l\}$, where $2k + l = n$. Construct vertex set S as follows. Note that $N(u_i) \subseteq V(M)$ for $1 \leq i \leq l$. For each u_i, place one vertex from $N(u_i)$ in S. Now, for $1 \leq j \leq k$, if $S \cap \{v_j, v_j'\} = \emptyset$, then let $v_j \in S$.

The vertices in S now dominate the vertices in V. Also, no pair $\{v_j, v_j'\}$ are elements of S. If so, this would imply the existence of a path $u_r v_j v_j' u_s$, for $u_r \neq u_s$. But in this case, removing the edge $v_j v_j'$ from M and adding edges $u_r v_j$ and $v_j' u_s$ would produce another matching, which contradicts the assumption that M is a 2-maximal matching. Set S consists of one vertex from each pair $\{v_i, v_i'\}$, $1 \leq i \leq k$, and is a dominating set. Thus, $\gamma(G) \leq \beta_2(G)$, completing the proof. $\square$

Notice that $\gamma(G) \leq \beta_2(G)$ and $\gamma'(G) = i'(G) \leq \beta_2(G)$. However, $\gamma(G) > \gamma'(G)$ is possible. For example, consider the corona $K_3 \circ K_1$ that has $\gamma = 3$ but $\gamma' = 2$.

Corollary 2.35 *If a graph G has no isolated vertices, then*

$$\gamma(G) \leq \frac{\alpha_1(G) + \beta_2(G)}{2} \leq \frac{\alpha_1(G) + \beta_1(G)}{2} = \frac{n}{2}.$$

Corollary 2.36 *If a graph G has no isolated vertices, then*

$$\gamma(G) \leq \min\{\alpha_1(G), \beta_2(G)\}.$$

2.6   Product Graphs and Vizing's Conjecture

2.6.1   Vizing's conjecture

We now turn our attention to an upper bound on the domination number of product graphs and introduce one of the most famous open problems involving domination, Vizing's Conjecture.

For an excellent overview of different graph products and their domination parameters, the reader is referred to [605, 610, 612, 923]. In 1963 Vizing [1115] suggested the (still open) problem of determining a lower bound on the domination number of a product graph in terms of the domination numbers of its factors G and H. Later he presented the following conjecture.

Conjecture 2.37 (Vizing's Conjecture [1117]) *For any graphs G and H,*

$$\gamma(G \times H) \geq \gamma(G) \cdot \gamma(H).$$

For example, $\gamma(C_5) = 2$, $\gamma(K_2) = 1$, and $\gamma(C_5 \times K_2) = 3$ (see Figure 2.6 where the shaded vertices form a γ-set). This conjecture was listed among the top open problems in domination by E. Cockayne in his survey paper [251] and has received substantial attention in recent studies. Much of the work done on

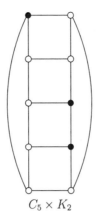

$$C_5 \times K_2$$

Figure 2.6: $\gamma(C_5 \times K_2) = 3 \geq \gamma(C_5) \cdot \gamma(K_2) = 2$.

the problem concerns showing that the conjecture holds for specific families of graphs and/or graphs satisfying a specific condition.

In 1979 Barcalkin and German proved that Vizing's conjecture is true for a large class of graphs with a given property, including some well-known families of graphs. If G is a graph with $\gamma(G) = k$ and $V(G)$ can be partitioned into k subsets $C_1, C_2, ..., C_k$ such that each of the induced subgraphs $\langle C_i \rangle$ is a complete subgraph of G, then G is said to be *decomposable*.

Theorem 2.38 [80] *If K is a spanning subgraph of a decomposable graph G where $\gamma(K) = \gamma(G)$, then for any graph H*

$$\gamma(K \times H) \geq \gamma(K) \cdot \gamma(H).$$

Since the family of graphs K in Theorem 2.38 includes trees and cycles, the next two results that we mention are actually corollaries of this theorem. Jacobson and Kinch [736] independently proved $\gamma(G \times T) \geq \gamma(G) \cdot \gamma(T)$ when T is a tree. El-Zahar and Pareek [414] showed that the conjecture holds when one of G and H is a cycle. They also showed that it is true for any graph G when $H = \overline{G}$ or when $\gamma(H) \leq 2$.

Hartnell and Rall [605, 610] constructed several infinite families of graph that achieve equality for Vizing's Conjecture. Fink et al. [475] showed that if both G and H are connected graphs of order $n \geq 4$ that obtain equality in Theorem 2.2 (that is, have domination number equal to one-half their order), then $\gamma(G \times H) = \gamma(G) \cdot \gamma(H)$. For instance, $\gamma((K_3 \circ K_1) \times P_4) = \gamma(K_3 \circ K_1) \cdot \gamma(P_4)$ (see Figure 2.7; the shaded vertices represent a γ-set). Vizing's Conjecture is discussed in detail in Chapter 7 of *Domination in Graphs: Advanced Topics* [645].

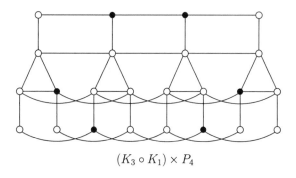

$$(K_3 \circ K_1) \times P_4$$

Figure 2.7: $\gamma((K_3 \circ K_1) \times P_4) = \gamma(K_3 \circ K_1) \cdot \gamma(P_4) = 6$.

2.6.2   Grid graphs

As we have seen, a fundamental unsolved problem concerning the bounds on the domination number of product graphs is to settle Vizing's Conjecture. Another basic problem is to find the domination number or bounds on the domination number of specific cartesian products, for example the $j \times k$ grid graph $P_j \times P_k$. This too seems to be a difficult problem. It is known that DOMINATING SET remains NP-complete when restricted to arbitrary subgraphs of grid graphs [102, 747]. However, Hare, Hare, and Hedetniemi [596] developed a linear time algorithm to solve this problem on $j \times k$ grid graphs for any fixed j.

Moreover, the domination number of $P_j \times P_k$ has been determined for small values of j. Jacobson and Kinch [735] established it for $j = 1, 2, 3$, and 4 and all k. Hare [591] developed an algorithm which she used to conjecture simple formulas for $\gamma(P_j \times P_k)$, for $1 \le j \le 10$. Chang and Clark [212] proved Hare's formulas for the domination numbers of $P_5 \times P_k$ and $P_6 \times P_k$. The domination numbers for $P_j \times P_k$, $1 \le j \le 6$, are listed below:

$$\gamma(P_1 \times P_k) = \left\lfloor \frac{k+2}{3} \right\rfloor, \ k \ge 1$$

$$\gamma(P_2 \times P_k) = \left\lfloor \frac{k+2}{2} \right\rfloor, \ k \ge 1$$

$$\gamma(P_3 \times P_k) = \left\lfloor \frac{3k+4}{4} \right\rfloor, \ k \geq 1$$

$$\gamma(P_4 \times P_k) = \begin{cases} k+1 & k = 1,\,2,\,3,\,5,\,6,\,9 \\ k & \text{otherwise for } k \geq 1 \end{cases}$$

$$\gamma(P_5 \times P_k) = \begin{cases} \left\lfloor \frac{6k+6}{5} \right\rfloor & k = 2,\,3,\,7 \\[2mm] \left\lfloor \frac{6k+8}{5} \right\rfloor & \text{otherwise for } k \geq 1 \end{cases}$$

$$\gamma(P_6 \times P_k) = \begin{cases} \left\lfloor \frac{10k+10}{7} \right\rfloor & k \geq 6 \text{ and } k \equiv 1 \bmod 7 \\[2mm] \left\lfloor \frac{10k+12}{7} \right\rfloor & \text{otherwise if } k \geq 4. \end{cases}$$

For example, the shaded circles represent a γ-set of $P_5 \times P_{13}$ shown in Figure 2.8. We next present bounds on the domination number of $P_k \times P_k$.

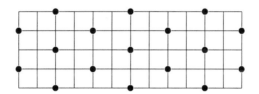

Figure 2.8: A γ-set for the grid $P_5 \times P_{13}$, $\gamma(P_5 \times P_{13}) = 17$.

Theorem 2.39 [269] *For the grid graph $P_k \times P_k$,*

$$(1/5)(k^2 + k - 3) \leq \gamma(P_k \times P_k) \leq \begin{cases} (1/5)(k^2 + 4k - 16) & k = 5t - 2 \\ (1/5)(k^2 + 4k - 17) & k = 5t - 1 \\ (1/5)(k^2 + 4k - 20) & k = 5t, \ t \geq 2 \\ (1/5)(k^2 + 4k - 20) & k = 5t + 1 \\ (1/5)(k^2 + 4k - 17) & k = 5t + 2. \end{cases}$$

The domination numbers of the cartesian products of cycles were studied in [777]. See also [213, 550, 593, 595, 596, 598].

EXERCISES

2.1 Give an infinite family of graphs that achieves the upper bound of Theorem 2.3.

2.2 Give an infinite family of graphs that achieves the upper bound of Theorem 2.7.

2.3 For $1 \leq k \leq n/2$, find a connected graph $G_{n,k}$ with order n and $\gamma(G_{n,k}) = k$.

2.4 Prove Theorem 2.12.

2.5 Give an infinite family of connected graphs G that achieves the lower bound of $\lceil n/(1 + \Delta(G)) \rceil \leq \gamma(G)$.

2.6 Verify Corollary 2.16 as a corollary of Theorem 2.15 or prove it directly.

2.7 Do 50 squats.

2.8 Find a graph G of order $n = 10$ and size $m = 24$ having $\gamma(G) = 4$.

2.9 Verify the upper bound of Theorem 2.22 for

   (a) $n = 13$, $m = 12$.
   (b) $n = 6$, $m = 8$.
   (c) $n = 10$, $m = 12$.

2.10 Show that no nontrivial tree achieves the upper bound of Theorem 2.22.

2.11 Prove Theorem 2.23 that if a graph G has $diam(G) = 2$, then $\gamma(G) \leq \delta(G)$.

2.12 Give a family of graphs with diameter two and arbitrarily large domination number.

2.13 Show that $\gamma(G) \leq n - \lfloor diam(G)/3 \rfloor$.

2.14 (a) Explain why Theorem 2.27 does not hold for graphs G with girth $g(G) = 4$.

   (b) Find graphs G with $g(G) = 4$ for which Theorem 2.27 does not hold.

   (c) What can you say about $\gamma(G)$ for graphs G with $g(G) = 4$?

2.15 Give a graph G with arbitrarily large maximum degree for which Theorem 2.28 holds.

2.16 Construct a graph G for which $\gamma(G) = 4$ and $\gamma(G) < \delta(G)$.

2.17 What is the maximum number of edges in a graph G of order $n = 10$ with $\gamma(G) = 3$?

2.18 Find a graph G of order 7 having $\gamma(G) = 3$, $\Delta(G) = 3$, and $m = 10$.

2.19 (a) Verify Observation 2.32.

(b) For each inequality in Observation 2.32, give an infinite family of graphs achieving the sharp upper bound.

2.20 Construct families of graphs G and H (in addition to the ones mentioned in the text) that obtain equality in Vizing's Conjecture.

2.21 Verify the equations given for $\gamma(P_i \times P_k)$, for $1 \leq i \leq 4$.

2.22 Determine $\gamma(P_7 \times P_7)$.

2.23 Verify that $\gamma(C_3 \times C_k) = \lceil 3k/4 \rceil$.

2.24 Verify that $\gamma(C_2 \times P_k) = \left\lceil \frac{k+1}{2} \right\rceil$.

2.25 Let $\overline{\gamma}$ denote $\gamma(\overline{G})$. Show that cycles C_n and paths P_n have $\gamma = \overline{\gamma}$ if and only if $n = 4$, 5, or 6.

2.26 Construct a cubic graph G for which $\gamma(G) = 3n/8$.

Chapter 3

Domination, Independence, and Irredundance

In Chapter 1 the concept of a dominating set and the domination number $\gamma(G)$ of a graph were introduced. Various bounds for $\gamma(G)$ in terms of a variety of other graphical parameters were presented in Chapter 2. In this chapter we will discuss the close relationships that exist among dominating sets, independent sets and irredundant sets in graphs. One result of these relationships is an inequality chain of parameters that has become one of the major focal points of the study of domination in graphs. We begin by discussing hereditary and superhereditary properties of sets of vertices in graphs. Much of this discussion is taken from a 1995 paper by Cockayne, Hattingh, Hedetniemi, Hedetniemi, and McRae [271].

3.1 Hereditary and Superhereditary Properties

Let P denote an arbitrary property of a set of vertices S in a graph $G = (V, E)$. Typical properties are expressed in terms of the type of subgraph induced by S, for example $\langle S \rangle$: is acyclic; is a tree; is planar; is 2-connected; has maximum degree $\leq k$, for some fixed positive integer k; is hamiltonian; or is a complete graph. If a set S has property P, then we say that S is a P-set; otherwise, it is a $\overline{P}$-set. Usually we are interested in considering either minimal or maximal P-sets in graphs.

A P-set is a *maximal* P-set if every proper superset $S' \supset S$ is a $\overline{P}$-set, that is, S' does not have property P. A P-set S is a *1-maximal* P-set if for every vertex $u \in V - S$, $S \cup \{u\}$ is a $\overline{P}$-set. Similarly, a P-set S is a *minimal* P-set if every proper subset $S' \subset S$ is a $\overline{P}$-set; and S is a *1-minimal* P-set if for every vertex $v \in S$, $S - \{v\}$ is a $\overline{P}$-set. Clearly, maximal P-sets are always 1-maximal P-sets, but the converse is not always true. There are properties P and graphs G in which there are sets which are 1-maximal P-sets but are not maximal P-sets.

Consider, for example, the graph G in Figure 3.1. Let P_h denote the property that the induced graph has a hamiltonian cycle (that is, a cycle containing every vertex). The set $S = \{1, 2, 3\}$ is a P_h-set, and in fact is a 1-maximal P_h-set, since none of $\langle S \cup \{4\}\rangle$, $\langle S \cup \{5\}\rangle$ or $\langle S \cup \{6\}\rangle$ is hamiltonian. However, $\langle S \cup \{4, 5\}\rangle$, $\langle S \cup \{4, 6\}\rangle$, $\langle S \cup \{5, 6\}\rangle$ and $\langle S \cup \{4, 5, 6\}\rangle$ are all hamiltonian. Thus, S is a 1-maximal P_h-set, but is not a maximal P_h-set.

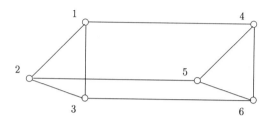

Figure 3.1: $S = \{1, 2, 3\}$ is a 1-maximal but not a maximal P_h-set.

The same situation holds for minimal and 1-minimal P-sets. Clearly minimal P-sets are always 1-minimal P-sets, but the converse is not always true. However, in some cases these definitions are equivalent.

A property P of sets of vertices is said to be *hereditary* if whenever a set S has property P, so does every proper subset $S' \subset S$. For example, if $\langle S \rangle$ is acyclic (or planar or bipartite), so is $\langle S' \rangle$ for every proper subset $S' \subset S$.

A property P is *superhereditary* if whenever a set S has property P, so does every proper superset $S' \supset S$. For example, if P denotes the property that $\langle S \rangle$ contains a chordless cycle, then P is superhereditary. Also, the property P of being a dominating set is superhereditary, since every superset of a dominating set is also a dominating set.

Proposition 3.1 *Let $G = (V, E)$ be a graph and let P be a hereditary property. Then a set S is a maximal P-set if and only if S is a 1-maximal P-set.*

Proof. By definition, every maximal P-set is 1-maximal. Conversely, let S be a 1-maximal P-set, and suppose that S is not a maximal P-set. Then there exists a superset $S'' \supset S$ which is a P-set, where $|S''| - |S| \geq 2$. But since property P is hereditary, every subset S' of a P-set S is a P-set. In particular, every subset $S' \subseteq S''$ with $|S'| = |S| + 1$ is a P-set. But this contradicts the assumption that S is a 1-maximal P-set. Thus, S must be a maximal P-set. □

A similar proposition and proof holds for superhereditary properties.

Proposition 3.2 *Let $G = (V, E)$ be a graph and let P be a superhereditary property. Then a set S is a 1-minimal P-set if and only if S is a minimal P-set.*

Proposition 3.1 asserts that in order to determine if a particular P-set S is a maximal P-set, for a hereditary property P, it suffices to determine whether $S \cup \{u\}$ is a P-set for every vertex $u \in V - S$, that is, if none of the sets $S \cup \{u\}$ for $u \in V - S$ are P-sets, then S is a maximal P-set. Similarly, Proposition 3.2 asserts that in order to determine if a particular P-set S is a minimal P-set, for some superhereditary property P, it suffices to determine whether $S - \{v\}$ is a P-set, for every vertex $v \in S$.

The importance of hereditary properties of sets of vertices is suggested by the following general theorem, due to Hedetniemi. Let $\beta(P)$ denote the maximum cardinality of a P-set in a graph G. Define a $\overline{P}$-*transversal* of G to be a set $\overline{S}$ of vertices which has the property that every $\overline{P}$-set S of G contains at least one vertex in $\overline{S}$. Let $\alpha(P)$ denote the minimum cardinality of a $\overline{P}$-transversal of G. Notice, incidentally, that the property of being a $\overline{P}$-transversal of G is superhereditary.

Theorem 3.3 [665] *For any graph G and hereditary property P,*

$$\alpha(P) + \beta(P) = n.$$

Proof. Let S be any P-set of G containing $\beta(P)$ vertices. It follows that the set $V - S$ must be a $\overline{P}$-transversal of G. For if not, then let S'' be a $\overline{P}$-set of G which does not contain any vertex in $V - S$. Thus, S'' must be a subset of S. But since the property P is hereditary, every subset of S must also be a P-set. In particular, S'' must be a P-set, which is a contradiction. Thus, $V - S$ must be a $\overline{P}$-transversal of G, and by definition, $\alpha(P) \leq |V - S| = n - \beta(P)$.

Conversely, let $\overline{S}$ be a minimum cardinality $\overline{P}$-transversal of G. It follows that the set $V - \overline{S}$ is a P-set of G, for if it were a $\overline{P}$-set, then it would have to contain a vertex in $\overline{S}$, which is a contradiction. Thus, by definition, $\beta(P) \geq |V - \overline{S}| = n - \alpha(P)$, or equivalently, $\alpha(P) \geq n - \beta(P)$. □

In the next section we will see a corollary of Theorem 3.3, involving independent sets of vertices in a graph, and we will establish a close relationship between independent sets and dominating sets in graphs.

3.2 Independent Sets

Because a set S is independent if the subgraph $\langle S \rangle$ contains no edges, or is K_2-free, one can see that the property of being an independent set of vertices is hereditary. Thus, let us define $\alpha_0(G)$ to equal the minimum cardinality of a set $\overline{S}$ of vertices with the property that every set of vertices which is not independent contains at least one vertex in $\overline{S}$. This parameter is called the vertex covering number of G, and is usually defined to be the minimum cardinality of a set of vertices S which has the property that for every edge $uv \in E$, either $u \in S$ or $v \in S$. Gallai is given credit for this now classic result, which is an immediate corollary of Theorem 3.3.

Theorem 3.4 [518] *For any graph G,*

$$\alpha_0(G) + \beta_0(G) = n.$$

We will be interested in the class of all maximal independent sets of vertices in a graph G. For example, the tree T in Figure 3.2 has maximal independent sets of three sizes: $\{1,2,3,6,7\}$, $\{1,2,3,5\}$, and $\{4,6,7\}$. Thus, $i(T) = 3$ and $\beta_0(T) = 5$.

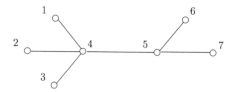

Figure 3.2: A tree with maximal independent sets of three sizes.

Consider next what makes an independent set maximal independent. Since the property of being independent is hereditary, by Proposition 3.1, we can define maximal independence in terms of 1-maximality. An independent set is maximal independent if and only if for every vertex $u \in V - S$, the set $S \cup \{u\}$ is not independent, that is, the graph $\langle S \cup \{u\}\rangle$ contains an edge. Stated in other words, an independent set S is maximal independent if and only if the following condition holds:

(1) for every vertex $u \in V - S$, there is a vertex $v \in S$ such that u is adjacent to v.

Notice that condition (1) is the definition of a dominating set; the maximality condition for an independent set is the definition of a dominating set. Thus, we can conclude the following, the necessity of which was first observed by Berge.

Proposition 3.5 [95] *An independent set S is maximal independent if and only if it is independent and dominating.*

Proof. The fact that a maximal independent set is both independent and dominating follows from the discussion above. Conversely, if a set S is both independent and dominating, we must show that it is maximal independent. Suppose that S is not maximal independent. Then, by Proposition 3.1, it is sufficient to say that there exists a vertex $u \in V - S$ for which $S \cup \{u\}$ is independent. But if $S \cup \{u\}$ is independent, then no vertex in S is adjacent to u, and hence S cannot be a dominating set. □

Therefore, every maximal independent set is a dominating set. This is why the minimum cardinality of a maximal independent set is called the independent domination number $i(G)$.

3.3 Dominating Sets

Consider next the class of all minimal dominating sets in a graph. For example, the tree T in Figure 3.2 has minimal dominating sets of several different cardinalities: $\{4,5\}$, $\{4,6,7\}$, $\{1,2,3,5\}$, and $\{1,2,3,6,7\}$. In particular, $\gamma(T) = 2$ and $\Gamma(T) = 5$.

Proposition 3.6 [95] *Every maximal independent set in a graph G is a minimal dominating set of G.*

Proof. Let S be a maximal independent set in G. Proposition 3.5 asserts that S is a dominating set. We must show that S is, in fact, a minimal dominating set. By Proposition 3.2, we can define minimal dominating sets in terms of 1-minimality, since the property of being a dominating set is superhereditary. Thus, a dominating set S is a minimal dominating set if for every vertex $v \in S$, the set $S - \{v\}$ is not a dominating set. Assume therefore that S is not a minimal dominating set. Then there exists at least one vertex $v \in S$ for which $S - \{v\}$ is a dominating set. But if $S - \{v\}$ dominates $V - (S - \{v\})$, then at least one vertex in $S - \{v\}$ must be adjacent to v. This contradicts our assumption that S is an independent set of G. Therefore, S must be a minimal dominating set. □

Corollary 3.7 *For any graph G,*

$$\gamma(G) \le i(G) \le \beta_0(G) \le \Gamma(G).$$

3.4 Irredundant Sets

Stated in other words, we can say that a dominating set S is a minimal dominating set if and only if the following condition holds:

(2) for every vertex $v \in S$, there exists a vertex $w \in V - (S - \{v\})$ which is not dominated by $S - \{v\}$.

Condition (2) can be rephrased in terms of the concept of private neighbors, as follows. In Chapter 1, we defined the private neighbor set of $v \in S$ to be $pn[v, S] = N[v] - N[S - \{v\}]$. If $pn[v, S] \neq \emptyset$ for some vertex v, then every vertex in $pn[v, S]$ is called a private neighbor of v. Consider for example, the set $S = \{4, 6, 7\}$ in the tree T in Figure 3.2. For this set S, $pn[4, S] = \{1, 2, 3, 4\}$, while $pn[6, S] = \{6\}$ (that is, vertex 6 is its own private neighbor) and $pn[7, S] = \{7\}$. We can also define the private neighbor set of a set S to equal $pn(S) = \{v : pn[v, S] \neq \emptyset\}$. Finally, we define the *private neighbor count* of a set S to equal $pnc(S) = |pn(S)|$.

It follows from the above discussion that the vertex w referred to in condition (2) must be a private neighbor of vertex v, with respect to set S (notice that it is possible that $w = v$). Thus, we can say that a dominating set S is a minimal dominating set if and only if:

(2)′ for every vertex $v \in S$, $pn[v, S] \neq \emptyset$, that is, every vertex $v \in S$ has at least one private neighbor.

We say that a set S of vertices is *irredundant* if condition (2)′ holds. Thus, the minimality condition for a dominating set is the definition of an irredundant set. From this we can easily conclude the following, the necessity of which was first observed by Cockayne, Hedetniemi and Miller.

Proposition 3.8 [285] *A dominating set S is a minimal dominating set if and only if it is dominating and irredundant.*

Proof. The fact that a minimal dominating set is both dominating and irredundant follows from the above discussion. Conversely, if a set S is both dominating and irredundant, we must show that it is minimal dominating. Suppose that S is not a minimal dominating set. Then, by Proposition 3.2 it is sufficient to say that there exists a vertex, say $v \in S$, for which $S - \{v\}$ is dominating. But since S is irredundant, we know that $pn[v, S] \neq \emptyset$. Let $w \in pn[v, S]$. By definition then, w is not adjacent to any vertex in $S - \{v\}$, that is, $S - \{v\}$ is not a dominating set, which is a contradiction. $\square$

We consider next the class of all maximal irredundant sets in a graph G. Since the property of being an irredundant set is hereditary, we can conclude by Proposition 3.1 that the property of being a maximal irredundant set can be defined in terms of 1-maximality. Thus, an irredundant set S is a maximal irredundant set if for every vertex $u \in V - S$, the set $S \cup \{u\}$ is not irredundant, which means that there exists at least one vertex $w \in S \cup \{u\}$ which does not have a private neighbor. Thus, an irredundant set S is maximal irredundant if and only if the following condition holds:

(3) for every vertex $w \in V - S$, there exists a vertex $v \in S \cup \{w\}$ for which $pn[v, S \cup \{w\}] = \emptyset$.

Notice that for every irredundant set S, $pnc(S) = |S|$, since every vertex must have at least one private neighbor. Condition (3) says that if we add any vertex in $V - S$ to a maximal irredundant set S, the private neighbor count will not increase. This gives rise to the following condition for an irredundant set to be maximal:

(3)′ for every vertex $w \in V - S$, $pnc(S \cup \{w\}) \leq pnc(S)$.

We leave it as an exercise to prove that condition (3) is equivalent to condition (3)′. The minimum cardinality of a maximal irredundant set in G is called the *irredundance number*, and is denoted $ir(G)$. The maximum cardinality of an irredundant set in G is called the *upper irredundance number*, and is denoted $IR(G)$. These terms were first defined by Cockayne, Hedetniemi, and Miller [285]. Consider for example the graph G in Figure 3.1. The sets $S = \{1, 4\}$ and $S' = \{1, 2, 3\}$ are both maximal irredundant sets. Note that for S, $pn[1, S] = \{2, 3\}$, $pn[4, S] = \{5, 6\}$, while for S', $pn[1, S'] = \{4\}$, $pn[2, S'] = \{5\}$, and $pn[3, S'] = \{6\}$. In fact, for this graph G, one can show that $ir(G) = 2$ and $IR(G) = 3$.

The following result was first observed by Bollobás and Cockayne.

Proposition 3.9 [126] *Every minimal dominating set in a graph G is a maximal irredundant set of G.*

Proof. We have already shown in Proposition 3.8 that every minimal dominating set S is irredundant. All that remains to show is that S is maximal irredundant. Suppose it is not. Then by Proposition 3.1 it suffices to consider 1-maximality, since the property of being irredundant is hereditary. Thus, if S is not maximal irredundant, there must exist a vertex $u \in V - S$ for which $S \cup \{u\}$ is irredundant. This means in particular that $pn[u, S \cup \{u\}] \neq \emptyset$, that is, there exists at least one vertex w which is a private neighbor of u with respect to $S \cup \{u\}$.

But this means that no vertex in S is adjacent to w, that is, S is not a dominating set. This contradicts the assumption that S is a dominating set. □

Since every minimal dominating set is a maximal irredundant set, we have the following inequality chain, which was first observed by Cockayne, Hedetniemi, and Miller in 1978.

Theorem 3.10 [285] *For any graph G,*

$$ir(G) \leq \gamma(G) \leq i(G) \leq \beta_0(G) \leq \Gamma(G) \leq IR(G).$$

This inequality chain has become one of the strongest focal points for research in domination theory. Many aspects of this chain will be studied in the next section. In the remainder of this section, we list some of the basic and best known results on irredundant sets in graphs.

Historically, one of the first results was the example by Slater in Figure 3.3(a) of a tree H for which $ir(H) < \gamma(G)$. A minimum cardinality, maximal irredundant set (ir-set) for H is the set $S = \{2, 3, 8, 9\}$, while a minimum cardinality dominating set (γ-set) is the set $S' = \{2, 4, 6, 8, 10\}$. A second example of a graph G for which $ir(G) < \gamma(G)$ is shown in Figure 3.3(c), and is due to Allan and Laskar [26]. This graph is the line graph $L(T)$ of the tree T in Figure 3.3(b); for this graph, called the A-L graph, $ir(L(T)) = 2$, while $\gamma(L(T)) = 3$. Both the H graph and the A-L graph played a key role in subsequent results on irredundant sets, as we will show later in this section. The first few results about the irredundance number relate it to the domination number.

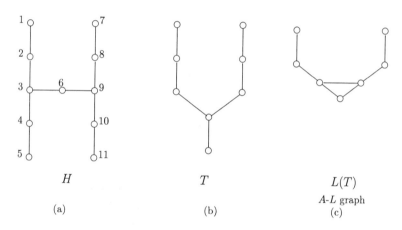

$$H \qquad\qquad T \qquad\qquad L(T)$$
A-L graph
(a) (b) (c)

Figure 3.3: Graphs illustrating irredundance.

Theorem 3.11 [26, 126] *For any graph* G,

$$\gamma(G)/2 < ir(G) \leq \gamma(G) \leq 2\,ir(G) - 1.$$

Proof. Let $ir(G) = k$ and let $S = \{v_1, v_2, ..., v_k\}$ be an ir-set of G. Since S is irredundant, $pn[v_i, S] \neq \emptyset$, for $1 \leq i \leq k$. Therefore, let $S' = \{u_1, u_2, ..., u_k\}$ where $u_i \in pn[v_i, S]$. Note that possibly $u_i = v_i$ (if v_i is its own private neighbor), but in any case $|S \cup S'| \leq 2k = 2\,ir(G)$.

We claim that the set $S'' = S \cup S'$ is a dominating set. If not, then there must exist at least one vertex $w \in V - S''$ which is not dominated by S''. This means that $w \notin N[x]$ for any vertex $x \in S''$, and therefore $pn[w, S \cup \{w\}] \neq \emptyset$.

But this also means that $u_i \notin N[w]$ for any vertex $u_i \in S'$, and therefore $pn[u_i, S \cup \{w\}] \neq \emptyset$. Thus, $S \cup \{w\}$ is an irredundant set, which contradicts the assumption that S is a maximal irredundant set. Therefore, S'' is a dominating set.

Although S'' is a dominating set it cannot be a minimal dominating set, since then, by Proposition 3.9, it would have to be a maximal irredundant set, which would again contradict the maximality of S. Therefore, $\gamma(G) \leq 2\,ir(G) - 1$ and $\gamma(G)/2 < ir(G)$. □

By making slight modifications in the above proof of Theorem 3.11, Allan, Laskar, and Hedetniemi were able to show the following, involving the total domination number $\gamma_t(G)$.

Corollary 3.12 [28] *For any connected graph G, $\gamma_t(G) \leq 2\,ir(G)$.*

Still another variant of Theorem 3.11 is due to Favaron, who settled the question of the amount by which $ir(G)$ can decrease when a vertex is deleted.

Theorem 3.13 [450] *For any graph G and $v \in V(G)$ such that $ir(G - v) \geq 2$, $ir(G) \leq 2ir(G - v) - 1$, and this bound is sharp.*

The next several results provide lower bounds for $ir(G)$ in terms of the maximum degree $\Delta(G)$.

Theorem 3.14 [128] *For any graph G,*

$$n/(2\Delta(G) - 1) \leq ir(G).$$

Theorem 3.15 [828] *A graph G has $n/(2\Delta(G)-1) = ir(G)$ if and only if every connected component of G is either a path or a cycle of length $k \equiv 0 \pmod 3$.*

A substantial improvement of this lower bound was subsequently obtained by Cockayne and Mynhardt, who also characterized the graphs for which this lower bound is achieved.

Theorem 3.16 [310] *For any graph G, $2n/(3\Delta(G)) \leq ir(G)$.*

A couple of results about $IR(G)$ are noteworthy. The proof of the following theorem by Favaron is due to Cockayne and Mynhardt [310].

Theorem 3.17 [448] *For any graph G, $IR(G) \leq n - \delta(G)$.*

Proof. Let S be an irredundant set in G and let $v \in S$. Assume that v is adjacent to k vertices in S. Since the degree of v is at least $\delta(G)$, v must be adjacent to at least $\delta(G) - k$ vertices in $V - S$. If $k = 0$, then $|V - S| \geq \delta(G)$, that is, $|S| \leq n - \delta(G)$, as required. If $k > 0$, then each neighbor of v in S must have a private neighbor in $V - S$ and these k vertices must be distinct. Hence, $|V - S| \geq (\delta(G) - k) + k = \delta(G)$, that is, $|S| \leq n - \delta(G)$. □

Cockayne and Mynhardt [310] have also characterized the class of graphs for which equality holds in Theorem 3.17.

3.5   The Domination Chain

As stated in Theorem 3.10, the domination chain is:

$$ir(G) \leq \gamma(G) \leq i(G) \leq \beta_0(G) \leq \Gamma(G) \leq IR(G). \tag{3.1}$$

Since its first publication in 1978, this chain has been the focus of more than 100 research papers, which have addressed questions such as the following:

1. Given an integer sequence, $1 \leq a \leq b \leq c \leq d \leq e \leq f$, does there exist a graph G for which $1 \leq ir(G) = a \leq \gamma(G) = b \leq i(G) = c \leq \beta_0(G) = d \leq \Gamma(G) = e \leq IR(G) = f$? If such a graph G does exist, then (a, b, c, d, e, f) is called a *domination sequence*.

2. Under what conditions are any of the parameters in (3.1) equal?

3. Are there variants of the basic independence-domination-irredundance parameters in (3.1) that satisfy a similar inequality chain?

4. Are there other graph parameters whose values are related to those in (3.1)?

In this section we will provide partial answers to these four basic questions about the domination chain.

3.5.1   Which integer sequences are domination sequences?

Although the investigation of possible domination sequences was begun by Cockayne et al. [260], these sequences were completely characterized by Cockayne and Mynhardt.

Theorem 3.18 [303] *A sequence a, b, c, d, e, f of positive integers is a domination sequence if and only if:*

(a) $a \leq b \leq c \leq d \leq e \leq f$,

(b) $a = 1$ *implies that $c = 1$,*

(c) $d = 1$ *implies that $f = 1$, and*

(d) $b \leq 2a - 1$.

The necessity of (a) follows by definition. Condition (b) implies that G contains a vertex which is adjacent to every other vertex, and hence $ir(G) = \gamma(G) = i(G) = 1$. Condition (c) implies that G must be a complete graph, and hence $\beta_0(G) = \Gamma(G) = IR(G) = 1$. Condition (d) follows from Theorem 3.11. The sufficiency follows from a detailed and impressive construction given in [303].

3.5.2 Under what conditions are any of the parameters in (3.1) equal?

Consider first, conditions under which $ir(G) = \gamma(G)$. The first two results suggest that H and the A-L graph in Figure 3.3 play a role in determining whether $ir(G) = \gamma(G)$. A graph G is *chordal* if every cycle of G of length greater than three has a chord, that is, an edge between two nonconsecutive vertices of the cycle.

Theorem 3.19 [828] *If G is a chordal graph containing no H or A-L graph as an induced subgraph, then $ir(G) = \gamma(G)$.*

Corollary 3.20 [828] *If T is a tree containing no induced H subgraph, then $ir(T) = \gamma(T)$.*

Note that Corollary 3.20 implies that the H graph is the smallest tree T with $ir(T) < \gamma(T)$.

Theorem 3.21 [828] *If a graph G has no induced $K_{1,3}$ and $ir(G) < \gamma(G)$, then G contains an A-L graph, where the only extra edges allowed are those from one of the two endvertices to a vertex of degree two or the edge between the two endvertices.*

A graph is an *interval graph* if its vertices can be associated 1-to-1 with intervals on the real line in such a way that two vertices are adjacent if and only if their corresponding intervals have a nonempty intersection. (See Section 12.4.2 for a discussion of interval graphs.)

Theorem 3.22 [830] *If G is an interval graph containing no induced A-L subgraph, then $ir(G) = \gamma(G)$.*

Corollary 3.23 *If G is a tree and an interval graph, then $ir(G) = \gamma(G)$.*

A *split graph* is a graph $G = (V, E)$ whose vertices can be partitioned into two sets V' and V'', where the vertices in V' form a complete graph and the vertices in V'' are independent.

Theorem 3.24 [830] *If a connected graph G is the complement of a bipartite graph or a split graph, then $ir(G) = \gamma(G) = \gamma_t(G) = \gamma_c(G)$.*

Theorem 3.25 [126] *If G is a graph which does not have two subgraphs isomorphic to P_4, with vertices $\{a_i, b_i, c_i, d_i\}$, $i = 1, 2$, where b_1, b_2, c_1, c_2, d_1 and d_2 are distinct and $a_1, a_2 \notin \{c_1, c_2, d_1, d_2\}$, then $ir(G) = \gamma(G)$.*

Theorem 3.26 [445] *If G is a graph containing no induced subgraph isomorphic to any of the six graphs in Figure 3.4, then $ir(G) = \gamma(G)$.*

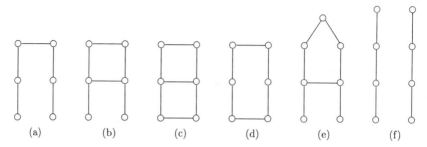

(a)   (b)   (c)   (d)   (e)   (f)

Figure 3.4: Forbidden induced subgraphs for $ir(G) = \gamma(G)$.

The conjecture in [445], that if G is ((a),(b),(c))-free, then $ir(G) = \gamma(G)$, was proved independently in [954] and [1124].

We consider next conditions under which $\gamma(G) = i(G)$, the first of which is due to Allan and Laskar, while the second is an improvement due to Favaron.

Theorem 3.27 [26] *If G is a graph containing no induced subgraph isomorphic to $K_{1,3}$, then $\gamma(G) = i(G)$.*

Theorem 3.28 [445] *If G is a graph containing no induced subgraph isomorphic to either $K_{1,3}$ or the A-L graph, then $ir(G) = \gamma(G) = i(G)$.*

Next, consider conditions under which $i(G) = \beta_0(G)$. A graph G is called *well-covered* if every maximal independent set of vertices is also maximum. For any well-covered graph G, any greedy algorithm for finding a maximal independent set always succeeds in finding a maximum independent set in G. Well-covered graphs have an extensive literature, much too large to cover in this short section; the reader is referred to an excellent 1993 survey of this literature by Plummer [946], who in 1970 [945] introduced this class of graphs. We offer a brief sample of results about well-covered graphs.

A graph is called *k-extendable* if every independent set of size k is contained in a maximum independent set. This concept generalizes that of a *B*-graph (that is, 1-extendable) introduced by Berge [98] and well-covered graphs (that is, *k*-extendable for every integer k). As we will see below, characterizations of classes of well-covered graphs can be given in terms of *k*-extendibility.

Connected, well-covered, bipartite graphs have been characterized by Ravindra [969] and Dean and Zito [350]. Let G be a bipartite graph, $e = uv$ be an edge in G, and let $G_e = \langle N(u) \cup N(v) \rangle$.

Theorem 3.29 *If G is a bipartite graph with no isolated vertices, then the following are equivalent:*

(a) *G is well-covered.*

(b) [969] *G contains a perfect matching M such that for every edge $e = uv$ in M, the graph G_e is a complete bipartite graph.*

(c) [350] *G is both 1-extendable and 2-extendable.*

Corollary 3.30 *For any tree T, the following are equivalent:*

(a) *T is well-covered.*

(b) [969] *Either T has a perfect matching consisting of pendant edges or T consists of a single vertex.*

(c) [350] *T is both 1-extendable and k-extendable, for some k between 2 and $\beta_0(T) - 1$.*

Recall that for a graph with cycles, the girth $g(G)$ is the length of a shortest cycle in G.

Theorem 3.31 [468] *If G is a graph with $g(G) \geq 8$, then G is well-covered if and only if its pendant edges form a perfect matching in G.*

Theorem 3.32 [471] *If $G \neq C_7$ is a graph with $g(G) \geq 6$, then G is well-covered if and only if its pendant edges form a perfect matching in G.*

Another structural property of well-covered graphs is mentioned by Topp and Vestergaard.

Theorem 3.33 [1092] *If G is a well-covered graph, then for each independent set I of G, the graph $G - N[I]$ is well-covered.*

A characterization of well-covered graphs was given in 1994 by Sankaranarayana. Let I_1 and I_2 be maximal independent sets of a graph $G = (V, E)$. Let $R = I_1 \cap I_2$ and let $I'_1 = I_1 - R$ and $I'_2 = I_2 - R$.

Theorem 3.34 [1008] *A graph is well-covered if and only if for every pair of maximal independent sets I_1 and I_2 of G, $\langle I'_1 \cup I'_2 \rangle$ has a perfect matching.*

We next take a brief look at graphs for which $\gamma(G) = \Gamma(G)$. A graph G is called *well-dominated* if all minimal dominating sets have the same cardinality. This concept was introduced by Finbow, Hartnell, and Nowakowski [470]. We next present a small sample of results about well-dominated graphs.

Lemma 3.35 [470] *Every well-dominated graph is well-covered.*

Proof. If a graph G is well-dominated then every minimal dominating set has the same cardinality. This implies, by Proposition 3.6, that every maximal independent set has the same cardinality. Thus, G is well-covered. □

A 5-cycle in a graph G is said to be *basic* if it does not contain two adjacent vertices of degree three or more. A graph $G = (V, E)$ is said to be in the family PC if its vertices can be partitioned into two subsets, $V = \{P, C\}$, where: P consists of the vertices incident to pendant edges of G and the set of pendant edges forms a perfect matching of P, and C consists of the vertices in the basic 5-cycles of G, and the basic 5-cycles of G form a partition of C. If C is empty, then we say that G is in the family of P.

Theorem 3.36 [470] *If G is a well-covered graph with $g(G) \geq 5$, then G is in the class PC or G is isomorphic to one of K_1, C_7, F_{10}, H_{13}, Q_{13}, or N_{14} (see Figure 3.5).*

Lemma 3.37 [470] *If a graph G is in the class PC, then G is well-dominated if and only if for every pair of basic 5-cycles either (1) no edge joins them, (2) exactly two edges join them and they are vertex disjoint, or (3) four edges join them.*

Corollary 3.38 [470] *If a graph G has $g(G) \geq 6$, then G is well-dominated if and only if G is well-covered.*

Theorem 3.39 [470] *A connected, bipartite graph G is well-dominated if and only if G is a 4-cycle or G is in the family P.*

Well-dominated and well-covered block graphs and unicyclic graphs have been studied by Topp and Volkmann. A *block* of a graph G is a maximal, 2-connected subgraph of G. A graph G is a *block graph* if and only if every block of G is a complete graph. A *unicyclic graph* is a connected graph $G = (V, E)$ with $|V| = |E|$.

Theorem 3.40 [1092] *For any block graph G or a unicyclic graph G, the following are equivalent:*

(a) $\gamma(G) = \Gamma(G)$, *that is, G is well-dominated.*

(b) $\gamma(G) = \beta_0(G)$.

(c) $i(G) = \beta_0(G)$, *that is, G is well-covered.*

(d) $i(G) = \Gamma(G)$.

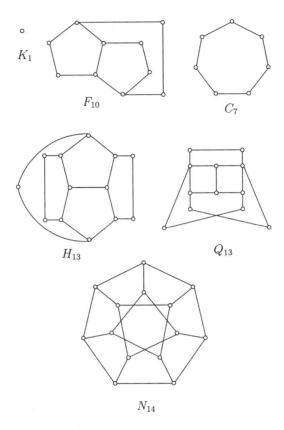

Figure 3.5: Well-covered graphs.

Finally, we consider conditions under which $\beta_0(G) = \Gamma(G) = IR(G)$. It is somewhat surprising that for a wide variety of graphs these three parameters are equal. The first two classes of graphs for which these equalities were noted are middle graphs and independence graphs. The *middle graph* of a graph $G = (V, E)$ is the graph $M(G) = (V \cup E, E')$, where $uv \in E'$ if and only if either u is a vertex of G and v is an edge of G containing u, or u and v are edges in G having a vertex in common (see Figure 3.6 for an illustration).

The *independence graph* of a graph $G = (V, E)$ is the graph $I(G) = (I, E'')$, where the vertices correspond one-to-one with independent sets of vertices in G and $uv \in E''$ if and only if the independent sets corresponding to u and v have a vertex in common.

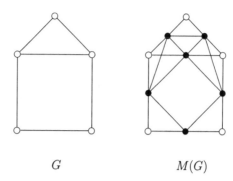

$$G \qquad\qquad M(G)$$

Figure 3.6: A graph G and its middle graph $M(G)$.

Theorem 3.41 [285] *For any graph G,*

(a) $\beta_0(M(G)) = \Gamma(M(G)) = IR(M(G)) = n,$

(b) $ir(M(G)) = \gamma(M(G)) = i(M(G)) = \alpha_1(G),$

(c) $\beta_0(I(G)) = \Gamma(I(G)) = IR(I(G)) = n,$ *and*

(d) $ir(I(G)) = \gamma(I(G)) = i(I(G)) = \chi(G).$

The next class of graphs for which these equalities hold were pointed out in 1981 by Cockayne et al.

Theorem 3.42 [260]

(a) *If a graph G has no isolated vertices and $\gamma(G) + IR(G) = n$, then $\beta_0(G) = \Gamma(G) = IR(G)$.*

(b) *If G is a bipartite graph, then $\beta_0(G) = \Gamma(G) = IR(G)$.*

In 1988, Cheston, Hare, Hedetniemi, and Laskar generalized Theorem 3.41 to upper bound graphs. A graph $G = (V, E)$ is an *upper bound graph* if there exists a partially ordered set $(P, \leq)$ such that $V = P$ and $uv \in E$ if $u \neq v$ and there exists a $w \in P$ with $u \leq w$ and $v \leq w$.

Theorem 3.43 [234] *For any upper bound graph G, $\beta_0(G) = \Gamma(G) = IR(G)$.*

The next two results along these lines were due to Jacobson and Peters.

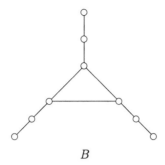

B

Figure 3.7: Forbidden subgraph of Theorem 3.44.

Theorem 3.44 [738]

(a) *If G is a chordal graph, then $\beta_0(G) = \Gamma(G) = IR(G)$.*

(b) *If G is a graph that does not contain either $K_{1,3}$, C_4, or the graph B in Figure 3.7, then $\beta_0(G) = \Gamma(G) = IR(G)$.*

In 1993, Golumbic and Laskar added circular arc graphs to this list. A graph $G = (V, E)$ is a *circular arc graph* if the vertices of G can be made to correspond one-to-one with a set of arcs of a circle in such a way that $uv \in E$ if and only if the arcs corresponding to u and v have a nonempty intersection on the circle.

Theorem 3.45 [548] *For any circular arc graph G, $\beta_0(G) = \Gamma(G) = IR(G)$.*

In 1994, Fellows, Fricke, Hedetniemi, and Jacobs added another class of graphs to the list. Given an arbitrary graph G, the *trestled graph of index k*, denoted $T_k(G)$, is the graph obtained from G by adding k copies of K_2 for each edge uv of G and joining u and v to the respective endvertices of each K_2 (see Figure 3.8 for an illustration).

Theorem 3.46 [465] *For any graph G and any positive integer $k \geq 1$, the trestled graph $T_k(G)$ satisfies: $\beta_0(T_k(G)) = \Gamma(T_k(G)) = IR(T_k(G))$.*

Finally, in 1994 Cheston and Fricke summed up all of these results and added a class which generalizes many of the above-mentioned classes. A set S of vertices is called a *stable transversal* if $|S \cap C| = 1$ for every maximal clique C of G. A graph G is called *strongly perfect* if G and each of its induced subgraphs have a stable transversal.

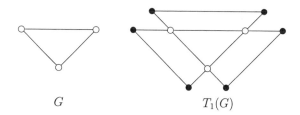

$$G \qquad\qquad T_1(G)$$

Figure 3.8: A graph G and the trestled graph of index 1, $T_1(G)$, of G.

Theorem 3.47 [232] *For any strongly perfect graph G, $\beta_0(G) = \Gamma(G) = IR(G)$.*

Proof. The general method of this proof follows that used by Jacobson and Peters in proving Theorem 3.44 for chordal graphs. Since for any graph G, $\beta_0(G) \leq \Gamma(G) \leq IR(G)$, it suffices to show that $\beta_0(G) \geq IR(G)$.

Let G be a strongly perfect graph and let I be an arbitrary IR-set of G. If I is an independent set then $\beta_0(G) \geq |I| = IR(G)$ and the proof is complete. Assume, therefore, that I is not an independent set. Let $R = \{u : u \in I$ and u is adjacent to at least one vertex v in $I\}$. Since I is an irredundant set, for each vertex $u \in R$, let u' be a private neighbor of u, that is, $u' \in pn[u, I]$ and $u' \in V - I$. Let $R' = \{u' : u \in R\}$. Consider the induced subgraph $H = \langle I \cup R' \rangle$. Notice that the vertices in $I - R$ are isolated in H. Notice also that the only edges between I and R' are in the form uu'. Now since G is strongly perfect, H is strongly perfect, and by definition, H must have a stable transversal, S. Since each edge uu' forms a maximal clique in H, exactly one of u or u' must be in S for every vertex $u \in R$. Also, each vertex of $I - R$ must be in S. Thus, $|S| = |I| = IR(G)$. But since every stable transversal must itself be an independent set, we have an independent set S in G of size $IR(G)$. Thus, $\beta_0(G) \geq |S| = IR(G)$. $\square$

Corollary 3.48 [232] *If G is a bipartite graph, a cograph, a permutation graph, a comparability graph, a chordal graph, a co-chordal graph, a peripheral graph, a parity graph, a Gallai graph, a perfectly orderable graph or a Meyniel graph, then $\beta_0(G) = \Gamma(G) = IR(G)$.*

Each of the graphs mentioned in Corollary 3.48 are classes of strongly perfect graphs; for further details the reader is referred to [232].

3.5.3 Are there variants of the basic independence-domination-irredundance parameters in (3.1) that satisfy a similar inequality chain?

Edge independence, domination, and irredundance

Let $\beta_1(G)$ and $i'(G)$ denote the maximum and minimum number of edges, respectively, in a maximal independent set of edges $F \subseteq E$, that is, no two edges in F have a vertex in common. Let $\gamma'(G)$ and $\Gamma'(G)$, denote the minimum and maximum number of edges in a minimal dominating set of edges $F \subseteq E$, where F is an edge dominating set if every edge in $E - F$ has a vertex in common with an edge in F. Finally, let $ir'(G)$ and $IR'(G)$ denote the minimum and maximum number of edges, respectively, in a maximal irredundant set of edges $F \subseteq E$, where F is an *irredundant edge set* if every edge uv in F has a private neighbor edge, that is, an edge wx which does not have a vertex in common with any edge in $F - \{uv\}$ but does have a vertex in common with uv.

For these six edge parameters, we have the following inequality chain:

$$ir'(G) \leq \gamma'(G) \leq i'(G) \leq \beta_1(G) \leq \Gamma'(G) \leq IR'(G) \tag{3.2}$$

Notice next that these edge parameters of a graph G correspond one-to-one with the vertex parameters of the line graph $L(G)$ of G. As an illustration, the subgraph induced by the shaded vertices in the graph $M(G)$ in Figure 3.6, is the line graph $L(G)$ of the graph G in Figure 3.6. It follows therefore that $ir'(G) = ir(L(G))$, $\gamma'(G) = \gamma(L(G))$, $i'(G) = i(L(G))$, etc.. Thus a re-statement of (3.2) is the following:

$$ir(L(G)) \leq \gamma(L(G)) \leq i(L(G)) \leq \beta_0(L(G)) \leq \Gamma(L(G)) \leq IR(L(G)) \tag{3.3}$$

However, two of the parameters in (3.2) are always equal:

Theorem 3.49 [26] *For any graph G, $\gamma'(G) = i'(G)$.*

Proof. This follows from (i) Theorem 3.27, (ii) the observation that $\gamma'(G) = \gamma(L(G))$, and (iii) the observation that the line graph $L(G)$ cannot contain an induced subgraph isomorphic to $K_{1,3}$. $\square$

Mixed independence, domination, and irredundance

The *total graph* $T(G)$ of a graph $G = (V, E)$ has vertices that correspond one-to-one with the elements of $V \cup E$. Two vertices are adjacent in $T(G)$ if and only if the corresponding elements are adjacent or incident in G. We say that a vertex v dominates all vertices adjacent to v and all edges incident with v. An edge uv dominates both vertices u and v and all edges having a vertex in

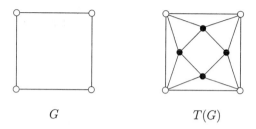

$$G \qquad\qquad T(G)$$

Figure 3.9: A graph G and its total graph $T(G)$.

common with either u or v. Figure 3.9 illustrates a graph G and its total graph $T(G)$, where the shaded vertices correspond one-to-one with the edges E of G.

A subset $W \subseteq V \cup E$ is said to be *mixed independent* if no two elements of W are either adjacent or incident. A subset $W \subseteq V \cup E$ is a *mixed dominating* set if every element $x \in (V \cup E) - W$ is either adjacent or incident to an element of W. A subset $W \subseteq V \cup E$ is a *mixed irredundant set* if every element $x \in W$ has a private element, that is, an element $y \in (V \cup E) - (W - \{x\})$ which is either equal to x or is adjacent or incident to x but is not adjacent or incident with any element of $W - \{x\}$. Given these definitions, we can define: $ir_m(G)$ and $IR_m(G)$ to equal the minimum and maximum cardinalities, respectively, of a maximal mixed irredundant set in G ; $\gamma_m(G)$ and $\Gamma_m(G)$ to equal the minimum and maximum cardinalities of a minimal mixed dominating set in G; and $i_m(G)$ and $\beta_m(G)$ to equal the minimum and maximum cardinalities of a maximal mixed independent set in G. From these definitions the following inequality chain results:

$$ir_m(G) \le \gamma_m(G) \le i_m(G) \le \beta_m(G) \le \Gamma_m(G) \le IR_m(G) \qquad (3.4)$$

But notice again that an equivalent restatement of (3.4) is the following:

$$ir(T(G)) \le \gamma(T(G)) \le i(T(G)) \le \beta_0(T(G)) \le \Gamma(T(G)) \le IR(T(G)) \quad (3.5)$$

Notice that the total graph $T(G)$ contains the graph G and the line graph $L(G)$ as disjoint subgraphs. Thus, the inequalities in Figure 3.10 hold.

For a collection of results involving the domination chain for total graphs the reader is referred to the survey paper by Hedetniemi, Hedetniemi, Laskar, McRae, and Majumdar [660].

Fractional domination and irredundance

Let $G = (V, E)$ be a graph and let $f : V \to [0, 1]$ be a function which assigns to each vertex $v \in V$ a value in the unit interval $[0, 1]$. The function f is called a

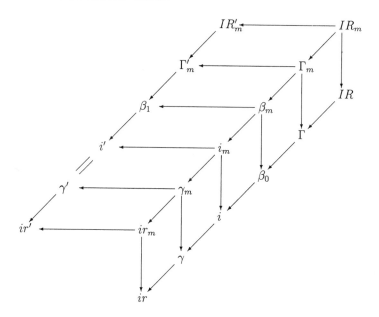

Figure 3.10: Inequalities: $\alpha \leftarrow \beta$ means $\alpha \leq \beta$.

dominating function if for every vertex $v \in V, f(N[v]) \geq 1$, that is, the sum of the values $f(w)$ for every vertex $w \in N[v]$ is at least 1. A dominating function f is a *minimal dominating function* if there does not exist a dominating function $g \neq f$, for which $g(v) \leq f(v)$ for every $v \in V$. The *weight* of a function f, denoted $w(f)$, is the sum of all values $f(v)$, for every $v \in V$. Let $\gamma_f(G)$ and $\Gamma_f(G)$ equal the minimum and maximum weights, respectively, of a minimal dominating function on G; these parameters are called the *fractional domination number* and *the upper fractional domination number*. It is worth noting that for any minimal dominating set S in a graph G, the corresponding *characteristic function* ϕ_S, defined by: $\phi_S(v) = 1$ if $v \in S$, and $\phi_S(v) = 0$ otherwise, is a minimal dominating function.

Fractional irredundance can be defined similarly. A function f is called an *irredundant function* if for every vertex $v \in V$, $f(v) > 0$ implies there exists a vertex $w \in N[v]$ for which $f(N[w]) = 1$. Let $ir_f(G)$ equal the infimum of the set $\{w(g) : g$ is a maximal irredundant function on $G\}$ and let $IR_f(G)$ equal the maximum weight of a maximal irredundant function on G. These parameters are called the *fractional irredundance numbers* of G. It can be shown that the characteristic function ϕ_S of any maximal irredundant set in G is a maximal irredundant function.

It is straightforward to show that every minimal dominating function is a maximal irredundant function. Thus, we have the following inequality chain:

$$ir_f(G) \leq \gamma_f(G) \leq \Gamma_f(G) \leq IR_f(G) \tag{3.6}$$

It is interesting to note that no acceptable definition has ever been found for fractional independence numbers of a graph, where we seek a definition of an independent function, such that every maximal independent function is also a minimal dominating function. A more detailed discussion of fractional domination is found in Chapter 10.

Iterated independence, domination, and irredundance

In a recent paper by Hedetniemi, Hedetniemi, McRae, Parks, and Telle [662], another inequality chain involving the independence, domination, and irredundance parameters is defined. Consider the process of executing the following greedy algorithm on a graph G, for any property P of sets of vertices.

Algorithm Iterated P-set Removal
Input: A graph $G = (V, E)$
Output: A partition $\{V_1, V_2, \cdots, V_k\}$ of V into P-sets for some positive integer k
Begin
    $G' = (W, E') = (V, E)$
    $k = 0$
    While $W \neq \emptyset$ **do**
        Let S be an arbitrary or random P-set of G'
        $k = k + 1$
        $V_k = S$
        $G' = G' - S$
    endwhile
End

We are interested in determining, for a given property P and a given graph G, the minimum and maximum number of iterations k that are possible before this algorithm terminates. For the property P_0 of being a maximal independent set of vertices, we define $i^*(G)$ and $\beta^*(G)$ to equal the minimum and maximum number of iterations possible for Algorithm Iterated_P_0-set_Removal. For the property P_1 of being a minimal dominating set, we define $\gamma^*(G)$ and $\Gamma^*(G)$ to equal the minimum and maximum number of iterations possible for Algorithm Iterated_P_1-set_Removal. And finally, for the property P_2 of being a maximal irredundant set of vertices, we define $ir^*(G)$ and $IR^*(G)$ to equal the minimum and maximum number of iterations possible for Algorithm Iterated_P_2-set_Removal. Notice that for properties P_0, P_1, and P_2, the corresponding Iterated_P-set Removal algorithm always terminates with a partition of V. Therefore, from the definitions given, the following inequality chain results:

$$ir^*(G) \le \gamma^*(G) \le i^*(G) \le \beta^*(G) \le \Gamma^*(G) \le IR^*(G) \qquad (3.7)$$

Although the *iterated irredundance numbers*, $ir^*(G)$ and $IR^*(G)$, and the *iterated domination numbers*, $\gamma^*(G)$ and $\Gamma^*(G)$, are new parameters, the *iterated independent domination number* $i^*(G)$ and the *iterated independence number* $\beta^*(G)$ have been previously studied under different names. For any graph G, $i^*(G) = \chi(G)$, where $\chi(G)$ equals the chromatic number of G and $\beta^*(G) = Gr(G)$, where $Gr(G)$ equals the Grundy coloring number of G.

Weakly connected domination, $\gamma_w(G)$ and $\Gamma_w(G)$

For connected graphs G, Dunbar, Grossman, Hattingh, Hedetniemi, and McRae [393] introduced the following type of domination. For a set S of vertices, the subgraph *weakly induced* by S is the graph $\langle S \rangle_w = (N[S], E_w)$ where E_w consists of the set of all edges having at least one vertex in S. A set S is a *weakly connected dominating set* of G if S is dominating and $\langle S \rangle_w$ is connected. Let $\gamma_w(G)$ and $\Gamma_w(G)$ equal the minimum and maximum cardinalities, respectively, of a minimal weakly connected dominating set in G. Note, for example, that $\gamma(P_6) = 2 < \gamma_w(P_6) = 3$.

Theorem 3.50 [393] *Every connected graph has a weakly connected independent dominating set.*

Proof. (by induction on the number n of vertices in G) If $n = 1$, then the theorem follows trivially. Assume, therefore, that every connected graph with $\le n$ vertices has a weakly connected independent dominating set, and let G be a connected graph with $n + 1 \ge 2$ vertices. Let v be a noncutvertex of G (it must always exist). Then the subgraph G' obtained by deleting v has a weakly connected independent dominating set D, by our inductive hypothesis. If v is adjacent to at least one vertex in D, then D is also a weakly connected independent dominating set for G. If not, then $D \cup \{v\}$ is a weakly connected independent dominating set for G. $\square$

Since every connected graph has a weakly connected independent dominating set, we can define the *weakly connected independent domination number* of G, $i_w(G)$, and the *upper weakly connected independent domination number* $\beta_w(G)$ to equal the minimum and maximum cardinalities, respectively, of a weakly connected independent dominating set in G.

Proposition 3.51 [393] *If G is a connected graph, then every weakly connected independent dominating set of G is a minimal weakly connected dominating set.*

Corollary 3.52 [393] *If G is a connected graph, then*

$$\gamma_w(G) \le i_w(G) \le \beta_w(G) \le \Gamma_w(G).$$

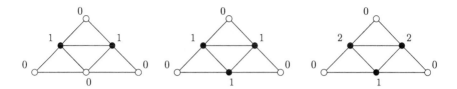

Figure 3.11: Minimum $\{k\}$-dominating functions for $k = 1,2,3$.

$\{k\}$-domination and $\{k\}$-irredundance

A function $g : V \to \{0, 1, \cdots, k\}$ is called a $\{k\}$-*dominating function* on a graph $G = (V, E)$ if for every vertex $v \in V$, $g(N[v]) \geq k$. Note that the characteristic function ϕ_S of a dominating function S of G is 1-dominating function. A $\{k\}$-dominating function g is *minimal* if for every vertex $v \in V$, if $g(v) > 0$, then there exists a vertex $w \in N[v]$ for which $g(N[w]) = k$. For a given positive integer k, the $\{k\}$-*domination number* $\gamma_{\{k\}}(G)$ and the *upper* $\{k\}$-*domination number* $\Gamma_{\{k\}}(G)$ equal the minimum and maximum weight, respectively, of a minimal $\{k\}$-dominating function of G. Figure 3.11 illustrates the $\{k\}$-domination numbers of the Hajós graph H_3, for $k = 1$, 2, and 3 $(\gamma_{\{1\}}(H_3) = 2, \gamma_{\{2\}}(H_3) = 3, \gamma_{\{3\}}(H_3) = 5)$.

In a similar way, one can define the $\{k\}$-irredundance numbers. A function $g : V \to \{0, 1, \cdots, k\}$ is called a $\{k\}$-*irredundant function* on a graph $G = (V, E)$ if for every vertex $v \in V$, if $g(v) > 0$, then there exists a vertex $w \in N[v]$ for which $g(N[w]) = k$. Let $ir_{\{k\}}(G)$ and $IR_{\{k\}}(G)$ equal the minimum and maximum weight, respectively, of a maximal $\{k\}$-irredundant function on G; these parameters are called the $\{k\}$-*irredundance numbers* of G.

The following results, due to Domke, Hedetniemi, Laskar, and Fricke, illustrate several inequalities among these parameters.

Theorem 3.53 [371] *For any graph G and any positive integer k,*

(a) $\gamma_{\{k\}}(G) \leq k\,\gamma(G) \leq k\,\Gamma(G) \leq \Gamma_{\{k\}}(G)$,

(b) $ir_{\{k\}}(G) \leq k\,ir(G) \leq k\,IR(G) \leq IR_{\{k\}}(G)$,

(c) $ir_{\{k\}}(G) \leq \gamma_{\{k\}}(G) \leq \Gamma_{\{k\}}(G) \leq IR_{\{k\}}(G)$,

(d) $ir_{\{k\}}(G)/k \leq ir(G) \leq \gamma(G) \leq i(G) \leq \beta_0(G) \leq \Gamma(G) \leq \Gamma_{\{k\}}(G)/k \leq$

$\Gamma_f(G) \leq IR(G) \leq IR_{\{k\}}(G)/k \leq IR_f(G)$, *and*

(e) $\gamma_f(G) \leq \gamma_{\{k\}}(G)/k \leq \gamma(G)$.

3.5.4 Are there graph parameters whose values are related to those in (3.1)?

Several graph theoretic parameters can be shown to be related by inequalities to various of the parameters in the domination chain in (3.1). We mention a few in this section.

The open irredundance numbers, $oir(G)$ and $OIR(G)$

A set $S \subseteq V$ is *open irredundant* if for every vertex $u \in S$, $N(u) - N[S - \{u\}] \neq \emptyset$. In an open irredundant set S, every vertex must have a private neighbor outside of S, in $V - S$; we call such a vertex an *open private neighbor*. We define the *open irredundance numbers* $oir(G)$ and $OIR(G)$ to equal the minimum and maximum cardinalities, respectively, of a maximal open irredundant set in G. Open irredundance was first studied by Farley and Shachum who observed the following.

Proposition 3.54 [438] *If a graph G has no isolated vertices, then $\gamma(G) \leq OIR(G)$.*

Later, Favaron [450] improved this result, using the following result of Bollobás and Cockayne.

Theorem 3.55 [126] *If a graph G has no isolated vertices, then G has a minimum dominating set which is open irredundant.*

Proof. Let D be a $\gamma(G)$-set for which the number of vertices in $\langle D \rangle$ having an open private neighbor is a maximum. If a vertex $u \in D$ does not have an open private neighbor, then it must be an isolate in $\langle D \rangle$, since by Proposition 3.9, D must be a maximal irredundant set. Since G has no isolated vertices, u must be adjacent to at least one vertex, say v, in $V - D$. But in this case, the set $(D - \{u\}) \cup \{v\}$ is a minimum dominating set in which vertex v has vertex u as an open private neighbor, contradicting the maximality of the number of vertices in D having an open private neighbor. Thus, D must be an open irredundant set. $\square$

Corollary 3.56 [450] *If a graph G has no isolated vertices, then*

$$oir(G) \leq \gamma(G) \leq OIR(G) \leq IR(G).$$

Proof. For a graph G, let D be a minimum dominating set which is open irredundant. Since D is dominating, it must be maximal open irredundant. Therefore, $oir(G) \leq \gamma(G) \leq OIR(G)$. The fact that $OIR(G) \leq IR(G)$ follows by definition, since every maximal open irredundant set is irredundant. $\square$

Since examples can be given (see [438] and [447]) where $OIR(G) < i(G)$, $\Gamma(G) < OIR(G)$, $ir(G) < oir(G)$, and $oir(G) < ir(G)$, the place of $oir(G)$ and $OIR(G)$ in the inequality chain in (3.1) cannot be made more precise.

Closed-open irredundance, $COIR(G)$

A set $S \subset V$ is *closed-open irredundant* if for every vertex $u \in S$, $N[u] - N(S - \{u\}) \neq \emptyset$. This means that a vertex can have a private neighbor which is either itself, another vertex in S or a vertex in $V - S$. As usual, we define $coir(G)$ and $COIR(G)$ to equal the minimum and maximum cardinalities of a maximal closed-open irredundant set in G. As originally defined by Fellows, Fricke, Hedetniemi, and Jacobs, an immediate consequence of this definition follows.

Proposition 3.57 [465] *For any graph G, $IR(G) \leq COIR(G)$.*

Weakly connected domination, $\gamma_w(G)$ and $\Gamma_w(G)$

Weakly connected domination was discussed previously in Section 3.5.3. The following result shows how these values relate to the parameters in (3.1).

Theorem 3.58 [393] *If G is a connected graph, then*

(a) $i(G) \leq i_w(G) \leq \beta_w(G) \leq \beta_0(G)$,

(b) *the parameter pairs i and γ_w, β_0 and Γ_w, and Γ and Γ_w are not comparable.*

Private domination, $\Gamma_{pvt}(G)$

As introduced by Hedetniemi, Hedetniemi, and Jacobs [657], a set S of vertices is a *private dominating set* if (i) S is a dominating set and (ii) S is an open irredundant set. Let $\gamma_{pvt}(G)$ and $\Gamma_{pvt}(G)$ denote the minimum and maximum cardinalities, respectively, of a private dominating set in a graph G. This type of domination was motivated by Theorem 3.55 of Bollobás and Cockayne [126]. An immediate corollary of Theorem 3.55 is that for any graph G, $\gamma_{pvt}(G) = \gamma(G)$.

Theorem 3.59 [657] *For any graph G,*

(a) $\gamma(G) = \gamma_{pvt}(G) \leq \Gamma_{pvt}(G) \leq \Gamma(G)$,

(b) $\Gamma_{pvt}(G) \leq OIR(G) \leq \beta_1(G)$, *and*

(c) *no strict inequality holds between either $i(G)$ or $\beta_0(G)$ and $\Gamma_{pvt}(G)$.*

Perfect neighborhood sets in graphs

In a recent paper, Fricke, Haynes, Hedetniemi, Hedetniemi, and Henning [495] introduced two new parameters closely related to those in (3.1). Let S be a set of vertices in a graph $G = (V, E)$. A vertex v is called *perfect* (with respect to S) if $|N[v] \cap S| = 1$. A vertex which is not perfect but is adjacent to a perfect vertex is called *near perfect*. A set S is defined to be a *perfect neighborhood set* of G if every vertex of G is either perfect or near perfect with respect to S. The *lower and upper perfect neighborhood numbers* of G, denoted $\theta(G)$ and $\Theta(G)$, respectively, are defined to be the minimum and maximum cardinalities of a perfect neighborhood set in G.

Proposition 3.60 [495] *For any minimal dominating set D of a graph $G = (V, E)$, there exists a perfect neighborhood set of G of cardinality $|D|$.*

Proof. Let D be a minimal dominating set of G. Let D_1 consist of all isolates in $\langle D \rangle$ and let $D_2 = D - D_1$. Since D is a minimal dominating set, every vertex in D_2 must have a private neighbor in $V - D$. For each vertex $u \in D_2$, select $u' \in pn[u, D]$ and let $D_2' = \{u' : u \in D_2\}$. Now consider the set $S = D_1 \cup D_2'$. Clearly $|N[v] \cap S| = 1$ for every vertex $v \in D$, and thus every vertex $v \in D$ is perfect with respect to S. But since D is a dominating set, every vertex in $V - D$ is adjacent to at least one vertex in D, and therefore is adjacent to a perfect vertex. Hence, every vertex of V is either perfect or near perfect, which means that S is a perfect neighborhood set of cardinality $|D|$. $\square$

Corollary 3.61 [495] *For any graph G,*

$$\theta(G) \leq \gamma(G) \leq \Gamma(G) \leq \Theta(G).$$

The next result is somewhat surprising.

Theorem 3.62 [495] *For any graph G, $\Theta(G) = \Gamma(G)$.*

Proof. From Corollary 3.61, we have $\Theta(G) \geq \Gamma(G)$. Let S be a Θ-set of G. We show that G contains a minimal dominating set of cardinality at least $|S|$. Let $S_1 = \{v \in S \mid v$ has a perfect neighbor (with respect to S) in $V - S \}$, and let $S_2 = S - S_1$. First, we show that each vertex v of S_2 is an isolate in $\langle S \rangle$. If this is not the case, then there is a $v \in S_2$ that is adjacent with some other vertex of S. Thus, $|N[v] \cap S| \geq 2$, so v is not perfect. But since S is a perfect neighborhood set of G, v must then have a perfect neighbor in $V - S$ and therefore v belongs to S_1, a contradiction. Hence, each vertex of S_2 is an isolate in $\langle S \rangle$. Thus each vertex of S_2 is perfect (with respect to S).

Let $T = N(S) \cap (V - S)$, and let $W = V - (S \cup T)$. Since $N[w] \cap S = \emptyset$ for each $w \in W$, we know that no vertex of W is perfect. Now let $T_1 = \{t \in T \mid t$ is a perfect neighbor of some vertex in $S \}$. Thus each vertex of T_1 is perfect and is adjacent with a unique vertex of S_1 and with no vertex of S_2. Let $T_2 = \{t \in T \mid t$ is adjacent with some vertex of $T_1 \cup S_2 \}$, and let $T_3 = T - (T_1 \cup T_2)$.

Each vertex of $T - T_1$ is adjacent with at least two vertices of S, so no vertex of $T - T_1$ is perfect. In particular, no vertex of T_3 is perfect. Thus each vertex of T_3 must be adjacent with a perfect vertex. Since no vertex of T_3 is adjacent with any vertex of $S_2 \cup T_1$, and since no vertex of $T_2 \cup W$ is perfect, each vertex of T_3 must have a perfect neighbor in S. Among all subsets of perfect vertices of S that dominate all the vertices of T_3, let S_1' be one of minimum cardinality. Thus, each vertex of S_1' uniquely dominates at least one vertex of T_3; that is, for each $v \in S_1'$, there exists a vertex $t \in T_3$ such that t is adjacent with v but with no other vertex of S_1'. Since each vertex of S_1' is perfect, we know that each vertex of S_1' is an isolate in $\langle S \rangle$. Furthermore, since no vertex of T_3 is adjacent with any vertex of S_2, we know that $S_1' \subseteq S_1$. Let $S_1'' = S_1 - S_1'$.

We show next that $D = S_1' \cup S_2 \cup T_1$ is a dominating set of G. By definition, each vertex of S_1'' is adjacent with some vertex of T_1 and each vertex of T_2 is adjacent with some vertex of $S_2 \cup T_1$. We also know that the set S_1' dominates T_3. Since no vertex of W is perfect, each vertex of W must have a perfect neighbor from the set T. However, the only perfect vertices of T belong to the set T_1. Hence, D is a dominating set of G. Thus there must exist a subset D^* of D that is a minimal dominating set of G.

It remains for us to show that $|D^*| \geq |S|$. Since each vertex of S_2 is an isolate in $\langle D \rangle$, $S_2 \subseteq D^*$. For each $v \in S_1'$, we know there exists a vertex $t \in T_3$ such that t is adjacent with v but with no other vertex of D. Thus each vertex of S_1' uniquely dominates some vertex of T_3, so $S_1' \subseteq D^*$. Finally, $|D^* \cap T_1| \geq |S_1''|$ since each vertex of T_1 is adjacent with at most one vertex of S_1'' while no vertex of $S_1' \cup S_2$ is adjacent with any vertex of S_1''. Hence, $|D^*| \geq |S_1'| + |S_1''| + |S_2| = |S|$. Thus D^* is a minimal dominating set of G of cardinality at least $|S|$. Consequently, $\Theta(G) = |S| \leq |D^*| \leq \Gamma(G)$. $\square$

Fricke et al. [495] conjectured that for any graph G, $\theta(G) \leq ir(G)$. Subsequently, Cockayne, Hedetniemi, Hedetniemi, and Mynhardt [275] proved that the conjecture holds for trees. However, Favaron and Puech [460] have constructed a countexample to the conjecture.

Fractional domination and irredundance numbers

The fractional domination and fractional irredundance numbers were discussed in Section 3.5.3. These parameters are related to those in (3.1) as follows:

Theorem 3.63 *For any graph G,*

(a) $ir_f(G) \leq \gamma_f(G) \leq \gamma(G) \leq i(G) \leq \beta_0(G) \leq \Gamma(G) \leq \Gamma_f(G) \leq IR_f(G)$,

(b) *the parameters $ir(G)$ and $\gamma_f(G)$ are not comparable.*

Lemma 3.64 [371] *Let f be a dominating function of a graph G. Then f is a minimal dominating function if and only if whenever $f(v) > 0$ there exists a vertex $w \in N[v]$ for which $f(N[w]) = 1$.*

Theorem 3.65 [371] *For any graph G, $\Gamma(G) \leq \Gamma_f(G) \leq IR(G)$.*

Proof. Let $g : V \to [0,1]$ be a minimal dominating function of G, where $w(g) = \Gamma_f(G)$. Let $S = \{v_1, v_2, \cdots, v_k\}$ be the set of vertices with $g(N[v_i]) = 1$ and let $P = \{v : g(v) > 0\}$. Note that since g is a minimal dominating function, by Lemma 3.64, every vertex $v \in P$ must either be in S or adjacent with a vertex in S. Therefore, S dominates the set P. Let $D \subseteq S$ be a minimal subset of S which dominates $P \cup S$. Since D is minimal, D is an irredundant set of $\langle P \cup S \rangle$, and therefore of G. Thus, $|D| \leq IR(G)$.

However,
$$
\begin{aligned}
|D| &= \sum 1, v_i \in D \\
&= \sum_{v_i \in D} g(N[v_i]) \\
&\geq \sum g(v), v \in V, \text{ since } g(v) > 0 \text{ implies } v \in P \\
&\qquad \text{and } D \text{ dominates } P \\
&= w(g) = \Gamma_f(G).
\end{aligned}
$$

Therefore, $\Gamma_f(G) \leq |D| \leq IR(G)$. $\square$

The following result is due to Fricke (personal communication).

Theorem 3.66 *For any graph G, $IR(G) = IR_f(G)$.*

k-minimal and k-maximal independence and domination

In [130], Bollobás, Cockayne, and Mynhardt generalized the concepts of minimality and maximality for sets with a given property P, and then applied this generalization to domination and independence. We briefly summarize this significant development here.

Let P be a property of vertex sets in graphs $G = (V, E)$. A subset $S \subseteq V$ is called a k-*minimal* P-*set* if S is a P-set, but for all l-subsets R of S, where $1 \leq l \leq k$, and all $(l-1)$-subsets R' of S, $(S - R) \cup R'$ is not a P-set. Given this definition, it is clear what we mean by a k-minimal dominating set. The simple example in Figure 3.12 shows a dominating set $S = \{1, 3, 5\}$ which is a 1-minimal dominating set, but is not a 2-minimal dominating set, since one can remove a set $\{1, 3\}$ of cardinality two, add back a set $\{2\}$ of cardinality one and obtain another dominating set $\{2, 5\}$.

Similarly, a subset $S \subseteq V$ is called a k-*maximal* P-*set* if S is a P-set, but for all l-subsets R of S, where $0 \leq l \leq k-1$, and all $(l+1)$-subsets R' of S, $(S - R) \cup R'$ is not a P-set. Given this definition, we will be interested in considering k-maximal independent sets. As an illustration, notice that the set $R = \{2, 5\}$ in the path in Figure 3.12 is a 1-maximal independent set, but it is not 2-maximal, since one can remove a set of cardinality one (that is, $\{2\}$)

Figure 3.12: A dominating set $S = \{1,3,5\}$ which is 1-minimal but not 2-minimal.

and add back a set of cardinality two (that is, $\{1,3\}$) and produce another independent set.

For $k \geq 2$, let $\Gamma_k(G)$ and $\beta_k(G)$ equal the maximum and minimum cardinality of a k-minimal dominating set and a k-maximal independent set, respectively. Since a k-minimal dominating set is also a $(k-1)$-minimal dominating set, and a k-maximal independent set is also a $(k-1)$-maximal independent set, the following inequalities are obvious.

Proposition 3.67 [130] *For any graph G,*

(a) $\gamma(G) \leq \cdots \leq \Gamma_k(G) \leq \cdots \leq \Gamma_2(G) \leq \Gamma(G)$,

(b) $i(G) \leq \beta_2(G) \leq \cdots \leq \beta_k(G) \leq \cdots \leq \beta_0(G)$.

Note that although $\Gamma_k(G)$ and $\beta_k(G)$ are well-defined for all positive integers k, it is clear that the sequences in (a) and (b) above contain only finitely many distinct values. It is shown in [130] that any nondecreasing sequence of integers is a k-minimal domination sequence, as in (a), for some graph G. It is much more difficult to characterize the k-maximal independent sequences.

3.6   Extensions Using Maximality and Minimality

The domination chain

$$ir(G) \leq \gamma(G) \leq i(G) \leq \beta_0(G) \leq \Gamma(G) \leq IR(G)$$

relates pairs of independence $(i(G),\ \beta_0(G))$, domination $(\gamma(G),\ \Gamma(G))$, and irredundance $(ir(G),\ IR(G))$ parameters. Each pair of parameters is defined by a property whose maximality or minimality condition serves to define the next pair of parameters. That is, $i(G)$ and $\beta_0(G)$ are the respective minimum and maximum cardinalities of a maximal independent set in G_0. What makes an independent set maximal is that it dominates. This yields $\gamma(G)$ and $\Gamma(G)$ with $\gamma(G) \leq i(G)$ and $\beta_0(G) \leq \Gamma(G)$. Similarly, minimal dominating sets are maximal irredundant sets, and this results in the relations between $\gamma(G)$ and $ir(G)$ and between $\Gamma(G)$ and $IR(G)$. In this section we consider extensions of the six parameter domination chain that maintain this maximality/minimality scheme.

3.6.1 The external redundance numbers, $er(G)$ and $ER(G)$

As defined in Section 3.4, the maximality condition for an irredundant set S is:

(3)$'$ for every vertex $w \in V - S$, $pnc(S \cup \{w\}) \leq pnc(S)$.

We say that a set S of vertices is *external redundant* if it satisfies condition (3)$'$. Stated in other words, a set S is external redundant if it is not possible to increase the private neighbor count of S by adding a new vertex. This concept is due to McRae (see Cockayne, Hattingh, Hedetniemi, Hedetniemi, and McRae [271]). Let $er(G)$ and $ER(G)$ equal the minimum and maximum cardinalities, respectively, of a minimal externally redundant set S in G; we refer to these parameters as the *external redundance numbers*.

Proposition 3.68 [271] *If $S \subseteq V$ is a maximal irredundant set in a graph G, then it is a minimal external redundant set.*

Corollary 3.69 [271] *For any graph G,*

$$er(G) \leq ir(G) \leq \gamma(G) \leq i(G) \leq \beta_0(G) \leq \Gamma(G) \leq IR(G) \leq ER(G).$$

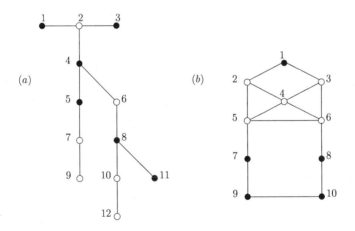

Figure 3.13: Examples of external redundant sets.

It is interesting to note that the property of being an external redundant set is not superhereditary. The set $S = \{1, 3, 4, 5, 8, 11\}$ in the tree in Figure 3.13(a) is a 1-minimal external redundant set, while the set $S - \{1, 3\}$ is also an external redundant set. This means that S is not a minimal external redundant set and therefore, in this case, 1-minimal is not equivalent to minimal. Thus, by Proposition 3.2, the property of being an external redundant set is not superhereditary. The graph G in Figure 3.13(b) provides an example of a graph for which $IR(G) = 4$ while $ER(G) = 5$ (for example, the set $S = \{1, 7, 8, 9, 10\}$ is a minimal external redundant set of G).

3.6.2   Neighborhood knockout numbers, $nk(G)$ and $NK(G)$

We next consider properties that fit the maximality/minimality scheme and yield pairs of parameters between $i(G)$ and $\beta_0(G)$. In particular, we seek a property X for which any minimal X-set is also a maximal independent set. The following discussion of "knockouts with replacement" is taken from Lampert and Slater [820]. Related results appear in [819, 821, 822, 823].

A *knockout* is a process in which one vertex is used to remove an adjacent vertex from a subset of the vertices of a graph G. A *knockout sequence* is a sequence $(S_0, S_1, S_2, ..., S_t)$ of vertex subsets, in which $S_0 = V$ and for every i, $1 \le i \le t$, set S_i is obtained from S_{i-1} by deleting a vertex $v \in S_{i-1}$ that is adjacent to another vertex $u \in S_{i-1}$. (We say that vertex u *knocks out* v in step i of the knockout process.) Further, it is required that a knockout sequence terminate with a set S_t only when no further knockouts are possible (that is, S_t is independent).

This type of process leads to the ideas of acquisition and consolidation [820]. An *acquisition sequence* is a knockout sequence in which each vertex is initially assigned a value of 1, the value of the vertex v removed from S_{i-1} is added to the adjacent vertex u used to knockout v, and, if the adjacent vertices have different values, the lower valued vertex is always the one removed. If vertex u knocks out vertex v, then u is said to *acquire* the value of v. By performing this acquisition operation until no adjacent vertices are in the set, we always find an independent set, S_t. The set S_t is said to have acquired all of the values and is called an *acquisition set*. The minimum possible cardinality of an acquisition set is the *acquisition number* $a(G)$.

Note, for example, that the four-by-four grid $G_{4,4}$ of Figure 3.14 can be acquired by a single vertex as follows. For each row i, let $u_{i,2}$ acquire $u_{i,1}$, let $u_{i,3}$ acquire $u_{i,4}$, and let $u_{i,2}$ acquire $u_{i,3}$. Then let $u_{2,2}$ acquire $u_{1,2}$ and let $u_{3,2}$ acquire $u_{4,2}$. Finally, let $u_{2,2}$ acquire $u_{3,2}$. At this point, the vertex $u_{2,2}$ has acquired all value from the grid, so $a(G_{4,4}) = 1$. Although $\{u_{2,2}\}$ is independent, it is certainly not a maximal independent set.

Because acquisitions are a form of knockouts, this shows that knockouts alone do not necessarily produce maximal independent sets. In fact, if knockouts are conducted with no additional constraints, then every independent set that has at least one vertex in each component of G can be obtained as the termination of a knockout sequence. Since we desire to have knockout sequences that produce maximal independent sets upon termination, some additional constraint or knockout method is required. One method is knockout-

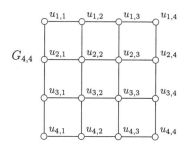

Figure 3.14: A four-by-four grid graph.

with-replacement. In *knockout-with-replacement*, each knockout of v from a set S_i results in a set $S_{i+1} = S_i - \{v\} \cup \{w \in V(G) : N[w] \cap (S_i - \{v\}) = \emptyset\}$. Thus, we add back to S_{i+1} all enclaves not in the set $S_i - \{v\}$. We will refer to any vertex w in $\{w \in V(G) : N[w] \cap (S_i - \{v\}) = \emptyset\}$ as a *replacement vertex*. For example, in the grid in Figure 3.14, if we knockout in the order given for acquisitions, then $S_8 = \{u_{1,2}, u_{1,3}, u_{2,2}, u_{2,3}, u_{3,2}, u_{3,3}, u_{4,2}, u_{4,3}\}$, which is the same set that results from the first eight acquisitions. But when $u_{1,2}$ knocks out $u_{1,3}$ in step nine, $u_{1,4}$ becomes a replacement vertex, so $S_9 = S_8 - \{u_{1,3}\} \cup \{u_{1,4}\} = \{u_{1,2}, u_{1,4}, u_{2,2}, u_{2,3}, u_{3,2}, u_{3,3}, u_{4,2}, u_{4,3}\}$. Then, $u_{2,2}$ knocks out $u_{2,3}$ with no replacement, but when $u_{3,2}$ knocks out $u_{3,3}$, there is a replacement vertex $u_{3,4}$. Possible sets in an acquisition sequence could then be: $S_{10} = \{u_{1,2}, u_{1,4}, u_{2,2}, u_{3,2}, u_{3,3}, u_{4,2}, u_{4,3}\}$, $S_{11} = \{u_{1,2}, u_{1,4}, u_{2,2}, u_{3,2}, u_{3,4}, u_{4,2}, u_{4,3}\}$, then $u_{4,2}$ knocks out $u_{4,3}$ and $S_{12} = \{u_{1,2}, u_{1,4}, u_{2,2}, u_{3,2}, u_{3,4}, u_{4,2}\}$, $u_{2,2}$ knocks out $u_{1,2}$ and $S_{13} = \{u_{1,1}, u_{1,4}, u_{2,2}, u_{3,2}, u_{3,4}, u_{4,2}\}$, $u_{3,2}$ knocks out $u_{4,2}$ and $S_{14} = \{u_{1,1}, u_{1,4}, u_{2,2}, u_{3,2}, u_{3,4}, u_{4,1}, u_{4,3}\}$, and finally $u_{2,2}$ knocks out $u_{3,2}$ and $S_{15} = \{u_{1,1}, u_{1,4}, u_{2,2}, u_{3,4}, u_{4,1}, u_{4,3}\}$. Note that since S_{14} is larger than S_{13}, this is not a pure minimization process, but S_{15} is a maximal independent set. In fact, knockout-with-replacement assures that any termination set S_t will be a maximal independent set. Because vertices knocked out from earlier S_i's might be replaced in later S_j's, it is conceivable that a knockout-with-replacement sequence could "cycle" and never terminate. However, Theorem 3.71 assures that all such sequences terminate. In the proof of Theorem 3.71, we need the following lemma.

Lemma 3.70 [820] *In any knockout-with-replacement sequence, if a vertex v is knocked out with a nonempty replacement, then vertex v can never be a replacement vertex in a subsequent knockout.*

Proof. Assume that vertex u is used to knockout vertex v in step $i + 1$ going from S_i to S_{i+1}. First, consider a vertex set T_i of a component in the subgraph

$\langle S_i \rangle$. Clearly, if v is not in T_i, then $T_i \subseteq S_{i+1}$ and no vertex in $V(G) - T_i$ that is adjacent to a vertex in T_i can be a replacement vertex in step $i+1$. Consequently, T_i is also the vertex set of a component of S_{i+1}. Next, suppose $v \in T_i$. Note that we must have $u \in T_i$. Let T_i^* be the vertex set of the component of $\langle T_i - v \rangle$ that contains u. As above, T_i^* is the vertex set of a component of S_{i+1}. That is, for any knockout in step $i + 1$ of a knockout-with-replacement sequence, no component of S_i can be eliminated or expanded.

If removal of vertex v in step $i + 1$ results in a nonempty replacement set, then all of the replaced vertices for this step are adjacent to v, and these vertices constitute the vertex set of at least one component of $\langle S_{i+1} \rangle$. Because no component can ever be completely eliminated, every S_j with $j \geq i + 1$ contains a vertex in $N(v)$, and v can not be a replacement vertex in step $j \geq i + 1$. □

Theorem 3.71 [820] *Each knockout-with-replacement sequence on a graph of order n and size m terminates in at most m steps.*

Proof. The total number of knockouts in a graph is the sum of the number of knockouts where the vertex knocked out subsequently gets replaced, and the number of knockouts where each vertex knocked out is not replaced in any later step in the sequence. We will consider an edge uv to be "responsible" for a replacement of v if the knockout of u by some other vertex w causes v to be a replacement vertex. If an edge uw is used to knockout w and w is never replaced, then we will consider uw to be "responsible" for the knockout of w. Lemma 3.70 assures that no edge can be responsible for two or more replacements, and that no edge uv responsible for a replacement will also be responsible for a knockout, since one of u or v is never replaced. Similarly, since we only hold edges to be responsible for knockouts when the vertex knocked out is never replaced, we know that no edge is held responsible for more than one knockout. As stated, the total number of knockouts is the sum of the number of replacements and the number of unreplaced vertices that are knocked out, and this procedure holds a unique distinct edge to be responsible for each. Thus the total number of knockouts in any knockout replacement sequence cannot exceed the total number m of edges. □

Theorem 3.72 [820] *Every (maximal) knockout-with-replacement sequence terminates with a maximal independent set.*

Proof. The sequence must terminate in a set S_t by Theorem 3.71. The sequence must terminate in an independent set, since otherwise it is not of maximal length. The set S_t must dominate since if a vertex v was knocked out in step t, any vertex not dominated by $S_{t-1} - \{v\}$ was replaced and is in S_t. □

From Theorem 3.72, it follows that knockout-with-replacement can provide a reasonable minimization method of finding maximal independent sets. However, with no additional constraints a knockout-with-replacement sequence can produce any maximal independent set S, as follows. In any order, each vertex $u \in S$ is used to knockout all neighbors of u that have not previously been knocked out. Thus, the problem is to find a reasonable set of additional constraints that will provide a pair of parameters interior to $i(G) \leq \beta_0(G)$.

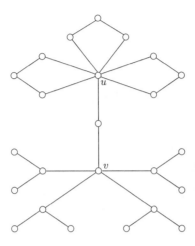

Figure 3.15: A graph G with $i(G) = 8 \leq nk(G) = 9 \leq NK(G) = 14 \leq \beta_0(G) = 15$.

One additional constraint that produces such a pair of parameters is the following. If u and v are adjacent vertices in $\langle S_i \rangle$, then we require that the vertex of lower degree in $\langle S_i \rangle$ be knocked out by the vertex of greater or equal degree in $\langle S_i \rangle$. In a *neighborhood knockout-with-replacement sequence* for u to knockout v in step $i{+}1$ going from S_i to S_{i+1}, the order of u's neighborhood in $\langle S_i \rangle$ must be at least as large as the order of v's neighborhood in $\langle S_i \rangle$. The *neighborhood knockout numbers* $nk(G)$ and $NK(G)$ denote the minimum and maximum cardinalities of a terminating set S_t in a maximal neighborhood knockout sequence, respectively. Note that $i(G) \leq nk(G) \leq NK(G) \leq \beta_0(G)$, since every terminating set is a maximal independent set. Consider the graph in Figure 3.15 for which $i(G) = 8 \leq nk(G) = 9 \leq NK(G) = 14 \leq \beta_0(G) = 15$. This demonstrates that $nk(G)$ and $NK(G)$ are parameters distinct from $i(G)$ and $\beta_0(G)$, even for bipartite graphs. To see that $i(G) < nk(G)$ for this graph, consider the vertex v of degree five. Vertex v is in no i-set and cannot be knocked out until after at least two of its neighbors are, implying that at least one pair of endvertices is present in a knockout set that does not include their common neighbor. Therefore, at least two endvertices are in any nk-set, but the endvertices are not in any i-set. To see that $NK(G) < \beta_0(G)$, consider the vertex u of degree seven, which is in no β_0-set. At least five neighbors of u must be knocked out before u can be. When five neighbors have been knocked out, at least one of the vertices at distance two from u on a four cycle is isolated, and therefore is present in the final NK-set, but no set containing such a vertex is a β_0-set. Thus, knockout-with-replacement can be used to find parameters that lie between i and β_0, and we have the following theorem.

Theorem 3.73 *For any graph* G,

$$er(G) \leq ir(G) \leq \gamma(G) \leq i(G) \leq nk(G) \leq NK(G) \leq \beta_0(G) \leq \Gamma(G) \leq IR(G) \leq ER(G).$$

We conclude this section by noting that the decision problems associated with parameters nk and NK, of deciding, for an input graph G and a positive integer k, if $nk(G) \leq k$ or if $NK(G) \geq k$, are NP-complete [820].

3.7 Summary

The main focus of this chapter has been on the relationships between dominating sets, independent sets, and irredundant sets in graphs. Particular attention has been focused on the basic domination chain:

(3.1) $ir(G) \leq \gamma(G) \leq i(G) \leq \beta_0(G) \leq \Gamma(G) \leq IR(G).$

This chain has been the motivation for many studies of these and other domination-related parameters of graphs. We have seen that numerous other parameters are closely related to those in (3.1). Thus, a number of variations of (3.1) are possible, including the following:

(a) $er(G) \leq ir(G) \leq \gamma(G) \leq i(G) \leq nk(G) \leq NK(G) \leq \beta_0(G) \leq \Gamma(G) \leq IR(G) \leq ER(G).$

(b) $ir(G) \leq \gamma(G) \leq i(G) \leq \beta_0(G) \leq \Gamma(G) = \Theta(G) \leq IR(G) \leq COIR(G).$

(c) $ir_f(G) \leq ir(G) \leq \gamma(G) \leq i(G) \leq \beta_0(G) \leq \Gamma(G) \leq \Gamma_f(G) \leq IR(G) = IR_f(G).$

(d) $ir(G) \leq \gamma(G) \leq i(G) \leq i_w(G) \leq \beta_w(G) \leq \beta_0(G) \leq \Gamma(G) \leq IR(G).$

EXERCISES

3.1 For the graph G and $S \subseteq V(G)$, which of the following properties are hereditary, superhereditary, or neither?

(a) S is a total dominating set.

(b) $\langle S \rangle$ is connected.

(c) S is a vertex cover.

(d) $\langle S \rangle$ is a tree.

(e) $\langle S \rangle$ has a maximum degree $\leq k$.

(f) $\langle S \rangle$ has a minimum degree $\geq k$.

(g) $\langle S \rangle$ has a complete graph of order $\geq k$.

3.2 Let P denote the property of a set S that $\langle S \rangle$ is a tree. If a set S has property P and is 1-maximal, is it necessarily maximal?

3.3 Prove the edge version of Theorem 3.4, namely that for any connected graph G,

$$\alpha_1(G) + \beta_1(G) = n.$$

3.4 Determine $\alpha_0(G)$ and $\beta_0(G)$ for the Petersen graph and verify Theorem 3.4 for this graph.

3.5 Verify that P_n and C_n have maximal independent sets of size k for every k, $\lceil n/3 \rceil \leq k \leq \lceil n/2 \rceil$.

3.6 Prove that condition (3) is equivalent to condition (3') for irredundant sets.

3.7 (a) Construct a graph for which $ir(G) = \gamma(G)/2 - 1$.

(b) Construct a graph for which $\gamma(G) = 2ir(G) - 1$.

3.8 Find a graph which realizes the domination sequence:

(a) (2, 3, 4, 5, 6, 7).

(b) (2, 2, 2, 5, 5, 5).

3.9 Prove Corollary 3.20, that if a tree T contains no induced H graph, then $ir(T) = \gamma(T)$.

3.10 (a) Verify that the graphs in Figure 3.5 are well-covered.

(b) Which of these graphs are well-dominated?

3.11 (a) Find an arrangement of nine queens on an 8×8 chessboard, each of which has a private neighbor, that is, each queen dominates a square that no other queen dominates.

   (b) What is the maximum number of irredundant queens that you can place on an 8×8 chessboard?

3.12 Characterize the class of graphs for which:

   (a) $ir^*(G) = 2$.

   (b) $\gamma^*(G) = 2$.

3.13 What is the weakly connected domination number of the path P_n?

3.14 Why do you think the weakly connected domination number is denoted $\gamma_W(G)$ instead of $\gamma_{WC}(G)$?

3.15 Prove Proposition 3.54, that if a graph G has no isolated vertices, then $\gamma(G) \leq OIR(G)$.

3.16 Construct a graph G_k for which

   (a) $\gamma(G) < \Gamma_k(G) < \Gamma_{k-1}(G) < \cdots < \Gamma(G)$.

   (b) $i(G) = \beta_1(G) < \beta_2(G) < \beta_3(G) < \cdots < \beta_k(G) < \beta_0(G)$.

3.17 Recalling the definition of external redundant sets and the parameters $er(G)$ and $ER(G)$, characterize the graphs that you would expect to find in an E.R.

3.18 Recall that the acquisition number $a(G)$ is the minimum number of vertices with nonzero value at the end of an acquisition sequence. As noted in Section 3.6.2, $a(G_{4,4}) = 1$. Determine $a(G_{5,5})$.

3.19 Show that $a(P_{4j} \times P_{4k}) = jk$.

3.20 (a) Is $i(T) = nk(T)$ for every tree T?

   (b) Is $NK(T) = \beta_0(T)$ for every tree T?

3.21 Find an infinite family of graphs for which $ir(G) < \gamma(G)$.

3.22 Show that the following pairs of parameters are not comparable:

   (a) $i(G)$ and $\Gamma_{pvt}(G)$.

   (b) $\beta_0(G)$ and $\Gamma_{pvt}(G)$.

   (c) $i(G)$ and $\gamma_w(G)$.

(d) $\beta_0(G)$ and $\Gamma_w(G)$.

(e) $\Gamma(G)$ and $\Gamma_w(G)$.

3.23 Prove that S is a perfect neighborhood set if and only if $B(S) = \{v : |N[v] \cap S| = 1\}$ is a dominating set.

Chapter 4

Efficiency, Redundancy, and Their Duals

For a set $S \subseteq V(G)$ to be a dominating set of G, it is required that $N[S] = V(G)$ or, equivalently, that for each $v \in V(G)$, $N[v] \cap S \neq \emptyset$. In other words, every $v \in V(G)$ must be dominated by <u>at least</u> one vertex in S (possibly itself). For the tree T_1 in Figure 4.1 with $S = \{v_3, v_6, v_9, v_{13}\}$, each of $v_1, v_2, v_4, v_5, v_7, v_8, v_9, v_{11}, v_{12}, v_{13}$, and v_{14} is dominated <u>exactly once</u>, while v_3, v_6, and v_{10} are dominated twice. The motivating idea for this chapter is to attempt to dominate every vertex exactly once. The focus shifts from the order of the (dominating, packing,...) set to the amount of domination being done.

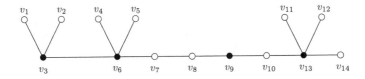

Figure 4.1: Tree T_1 with $\gamma(T_1) = 4$, $F(T_1) = 12$, and $R(T_1) = 16$.

4.1   Introduction

In this section we introduce "efficient domination" as a generalization of a "perfect code". A subset $S \subseteq V(G)$ for which $|N[v] \cap S| = 1$ for every $v \in V(G)$ is called a *perfect code*. Note that if vertex u is adjacent to endvertices v and w, then clearly u must be in every perfect code for G. Hence, if each of u_1 and u_2 has two adjacent endvertices and $u_1u_2 \in E(G)$ or the distance $d(u_1, u_2) = 2$, then G does not have a perfect code.

If $S \subseteq V(G)$ is a perfect code for G, then for every pair $u, v \in S$, we must have $d(u, v) \geq 3$. This implies that S is a *packing*. Since each vertex u dominates

$|N[u]| = 1 + deg(u)$ vertices, the following is clear.

Theorem 4.1 *The following are equivalent:*

(a) $S = \{u_1, u_2, ..., u_k\}$ *is a perfect code for G.*

(b) $\{N[u_1], N[u_2], ..., N[u_k]\}$ *is a partition of $V(G)$.*

(c) S *is a packing and $\sum_{u \in S}(1 + deg(u)) = |V(G)|$.*

For the coding theorist the important question is to decide if G has a perfect code. When G does not, how can we best approximate a perfect code? First, consider keeping the requirement that each vertex be dominated at most once, and maximizing the amount of domination done. For any subset $S \subseteq V(G)$, let $I(S) = \sum_{s \in S}(1 + deg(s))$ denote the *influence* of S, that is, the total amount of domination done by S. We say that a packing S *efficiently dominates* $\bigcup_{s \in S} N[s]$. Let $F(G)$, the *efficient domination number* of G, denote the maximum number of vertices that can be efficiently dominated by a packing. That is, $F(G) = \max\{|\bigcup_{s \in S} N[s]| : S \text{ is a packing}\} = \max\{I(S) : S \text{ is a packing}\}$. When $F(G) = n$, we say that G is *efficiently dominatable*. A perfect code S for a graph G is also called an *efficient dominating set*. In Figure 4.1, set $S_1 = \{v_1, v_4, v_8, v_{13}\}$ is a packing that shows $F(T_1) \geq 12$. Note that $S_2 = \{v_3, v_8, v_{13}\}$ also shows that $F(T_1) \geq 12$. One can easily verify that $F(T_1) = 12$, and we have two packings of different cardinalities each of which efficiently dominates $F(G) = 12$ vertices. This cannot happen when $F(G) = n$.

Theorem 4.2 [76, 77] *If G has an efficient dominating set, then the cardinality of any efficient dominating set equals the domination number $\gamma(G)$. In particular, all efficient dominating sets of G have the same cardinality.*

Proof. Assume $\{u_1, u_2, ..., u_k\}$ is an efficient dominating set, and let D be any minimum dominating set. Clearly, $k \geq |D|$ because D is a $\gamma(G)$-set. Also, if $1 \leq i < j \leq k$, we have $N[u_i] \cap N[u_j] = \emptyset$ and $D \cap N[u_i] \neq \emptyset$, and so $|D| \geq k$. □

We note that every path P_n is efficiently dominatable, and a cycle C_n is efficiently dominatable if and only if $n \equiv 0 \pmod 3$.

Alternately, one can require that every vertex is dominated at least once and attempt to minimize the amount of excess domination. Generally, a "redundance" measure is a measure of how many times vertices are dominated. The *redundance* of G (also called the *total redundance*) [555, 558] is $R(G) = \min\{\sum_{v \in V(G)} |N[v] \cap S| : S \text{ is a dominating set}\} = \min\{I(S) : S \text{ is a dominating set}\}$. As described in [749, 750], another redundance measure is the *cardinality redundance* $CR(G)$, which is the minimum number of vertices dominated more than once by a dominating set (redundantly dominated). For T_1, we have R-sets $S_3 = \{v_3, v_4, v_5, v_8, v_{13}\}$ and $S_4 = \{v_1, v_2, v_4, v_5, v_8, v_{13}\}$ achieving $R(T_1) = 16$. Note that $CR(T_1) = 1$ and S_3 is a CR-set, but S_4 is not a CR-set because it dominates each of v_3 and v_6 twice.

Theorem 4.3 [555, 749] *For any graph G,*

(a) $F(G) \leq n \leq R(G)$*, and*

(b) $F(G) = n$ *if and only if* $R(G) = n$ *if and only if* $CR(G) = 0$.

As noted, $F(P_n) = n$, so $R(P_n) = n$ and $CR(P_n) = 0$, and $F(C_{3k}) = R(C_{3k}) = 3k$. One can easily see that $R(C_{3k+1}) = R(C_{3k+2}) = 3k+3$, $CR(C_{3k+1}) = 2$, and $CR(C_{3k+2}) = 1$.

The decision problem of deciding if $F(G) = n$ was shown to be NP-complete for arbitrary graphs in [77] and a linear algorithm for computing $F(T)$ for any tree T was described. In [558] Grinstead and Slater described an extension to a linear algorithm for $F(G)$ for generalized series-parallel graphs G. Smart and Slater gave the following simplified and strengthened proof of NP-completeness for bipartite and chordal graphs.

EFFICIENT DOMINATING SET
INSTANCE: A graph $G = (V, E)$.
QUESTION: Does G have an efficient dominating set, that is, is $F(G) = n$?

Theorem 4.4 *EFFICIENT DOMINATING is NP-complete,*

(a) [466] *for planar graphs of maximum degree three,*

(b) [1049] *for bipartite graphs, and*

(c) [1049] *for chordal graphs.*

Proof. A proof of (a) is given in [466]. Clearly there is a nondeterministic polynomial time algorithm for deciding if a graph G is efficiently dominatable, that is, if $F(G) = n$. The following shows how a polynomial algorithm for deciding if $F(G) = n$ could be used to solve one-in-three 3SAT without negated literals (see [524]) in polynomial time. Assume that we are given an instance of one-in-three 3SAT, with a set $U = \{u_1, u_2, ..., u_N\}$ of variables and a set of 3-literal clauses $C = (C_1, C_2, ..., C_M)$, where each clause $C_i = u_{i1} \vee u_{i2} \vee u_{i3}$ has no negated literals, that is, each $u_{ij} \in U$. Construct the graph H in Figure 4.2, where each clause vertex C_i is made adjacent to the vertices corresponding to u_{i1}, u_{i2}, and u_{i3}. Because $|V(G)| = n = 2N + M$, H can be constructed from U and C in polynomial time.

Assume that there is a satisfying truth assignment $t : U \rightarrow \{T, F\}$ in which for each C_i, exactly one of u_{i1}, u_{i2}, and u_{i3} is assigned the value T. Define the N-set $S \subseteq V(H)$ with $u_i \in S$ if $t(u_i) = T$ and $v_i \in S$ if $t(u_i) = F$. In particular, exactly one of u_i and v_i is in S, for $1 \leq i \leq N$. If $u_i \in S$ and $u_j \in S$ with $1 \leq i < j \leq N$, then no C_h contains both u_i and u_j because only one of u_{i1}, u_{i2}, and u_{i3} is assigned the value T. Therefore, $d(u_i, u_j) > 2$. Thus, S is a packing. Clearly, $|N[u_i] \cap S| = |N[v_i] \cap S| = 1$ for $1 \le i \le N$, and $|N[C_h] \cap S| = 1$ because exactly one of u_{i1}, u_{i2}, and u_{i3} is in S. Hence, $F(H) = n$.

Conversely, let $F(H) = n$ and assume that $S \subseteq V(H)$ is an efficient dominating set. Then $|N[v_i] \cap S| = 1$ implies exactly one of u_i and v_i is in S, for $1 \le i \le N$. Because u_{h1} or v_{h1} is in S, $C_h \notin S$, for $1 \le h \le M$. And because $|N[C_k] \cap S| = 1$, exactly one of u_{h1}, u_{h2}, and u_{h3} is in S. Thus we can define $t : U \to \{T, F\}$ with $t(u_i) = T$ if $u_i \in S$ and $t(u_i) = F$ if $v_i \in S$, and t is an exact one-in-three 3SAT truth assignment.

It remains only to note that H is bipartite. And if H^* is obtained from H by adding the $M(M-1)/2$ edges making a complete subgraph on $C_1, C_2, ..., C_M$, then H^* is chordal, and the same argument works for H^*. $\square$

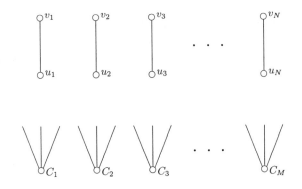

Figure 4.2: Graph H.

For the tree $T_{1,k}$ with $k \ge 2$ in Figure 4.3, we have $n = 4k+2$, and one can see that $F(T_{1,k}) = 2k+2 = 1+n/2$ (for $S = \{w\}$) and $R(T_{1,k}) = 6k+2 = 3n/2-1$ (for $S = \{w\} \cup \{x_i : 1 \le i \le 2k\}$). For $T_{2,k}$ with $k \ge 2$, we have $n = 2k^2 + 4k$, $F(T_{2,k}) = 4k = \sqrt{8(n+2)} - 4$ (for $S = \{w\}$), and $R(T_{2,k}) = 2k^2 + 6k = n + 2k$ (for $S = \{w\} \cup \{v_i : 1 \le i \le k\}$). These trees achieve the following upper and lower bounds for $R(T)$ and $F(T)$ for any tree T, respectively.

Theorem 4.5 [542] *If T is a tree of order $n \ge 2$, then $R(T) \le 3n/2 - 1$ and $F(T) \ge \sqrt{8(n+2)} - 4$.*

Proof. For a bipartite graph G with no isolated vertices, let V_1 and V_2 be disjoint independent sets with $V_1 \cup V_2 = V(G)$ and $|V_1| \le |V_2|$. Then V_1 is a dominating set and $R(G) \le I(V_1) = \sum_{v \in V_1}(1 + deg(v)) = |V_1| + |E(G)| \le n/2 + m$. For a tree T, one has $m = n - 1$, so $R(T) \le 3n/2 - 1$.

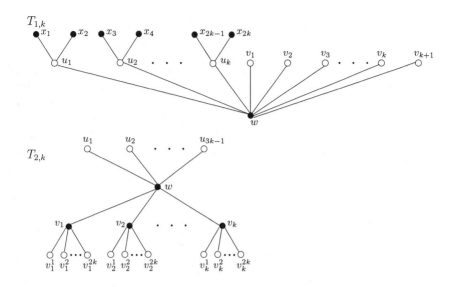

Figure 4.3: Trees $T_{1,k}$ and $T_{2,k}$.

We will prove that $F(T) \geq \sqrt{8(n+2)} - 4$ for any forest T on $n \geq 1$ vertices by induction on n. For $1 \leq n \leq 4$, we have $F(T) = n > \sqrt{8(n+2)} - 4$, so assume $n \geq 5$. Also, if every component of T is a star, then $F(T) = n > \sqrt{8(n+2)} - 4$. Let $W \subseteq V(T)$ denote the set of all vertices w such that w is adjacent to at least one endvertex and to at most one nonendvertex. For $w \in W$, let $T^* = T - N[w]$ with $deg(w) = d$. By the induction hypothesis, $F(T^*) \geq \sqrt{8(n - d + 1)} - 4$. Let S be a packing in T^* with $F(T^*) = I(S)$ in T^*, and let x be an endvertex adjacent to w. Because S contains no element of $N[w]$, $S \cup \{x\}$ is a packing in T, and $F(T) \geq I(S \cup \{x\}) \geq F(T^*) + 2 \geq \sqrt{8(n - d + 1)} - 2$. We are done if $\sqrt{8(n - d + 1)} \geq \sqrt{8(n+2)} - 2$. Consequently, we can assume that $d > \sqrt{2(n+2)} - 3/2$ for each $w \in W$.

If there are vertices w and w' in W with $d(w, w') \geq 3$, then $S = \{w, w'\}$ is a packing in T. Then $F(T) \geq I(S) > 2(\sqrt{2(n+2)} - 3/2)$, and we are finished. Consequently, we can assume T is a tree and any two elements of W are at distance at most two. In particular, T has diameter at most four. If the diameter of T is two, T is a star, and $F(T) = n > \sqrt{8(n+2)} - 4$. If the diameter of T is three, T is a double-star with adjacent vertices w_1 and w_2 and $k_i \geq 1$ endvertices adjacent to w_i and $n = k_1 + k_2 + 2$. Recalling that $n \geq 5$ and assuming $k_1 \leq k_2$, $F(T) = k_2 + 2 \geq \sqrt{8(n+2)} - 4$ (with equality only when

$n = 6$ and $k_1 = k_2 = 2$).

The final case is when T has diameter four, v is its center vertex and $N[v]$ has $k \geq 2$ elements in W and $j \geq 0$ endvertices, as in Figure 4.4.

Assume $deg(w_i) \leq deg(w_{i+1})$ for $1 \leq i \leq k - 1$, so $2 \leq deg(w_1) \leq deg(w_2) \leq$ $... \leq deg(w_k) = d$. For $S_1 = \{v\}$ and $S_2 = \{w_k, w_1', w_2', ..., w_{k-1}'\}$, $F(T) \geq$ $\max\{I(S_1), I(S_2)\} = \max\{j + k + 1, d + 2k - 1\}$. Note that $n \leq kd + j + 1$. We need to determine where the maximum of the above two bounds is minimized. It can easily be seen that this occurs when $j + k + 1 = d + 2k - 1$ and $n = kd + j + 1 = kd + d + k - 1$. Then minimizing $d + 2k - 1$, subject to $kd + d + k - 1 = n$, occurs when $k = \sqrt{(n + 2)/2} - 1$, $d = \sqrt{2n + 4} - 1$, and $j = \sqrt{9(n + 2)/2} - 4$. This minimum value $d + 2k - 1 = \sqrt{8(n + 2)} - 4$. $\square$

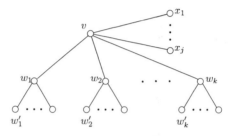

Figure 4.4: Tree T of diameter four.

If $f(n)$ is the minimum value of $F(T)$ for trees T on n vertices, then $f = O(\sqrt{n})$. For caterpillars this becomes $O(n)$. A *caterpillar* is a tree C for which the removal of all endvertices leaves a path, which is called the *spine* of C.

Theorem 4.6 [542] *For a caterpillar C of order $n \geq 2$ with k vertices on the spine, $F(C) \geq (n + 2k + 2)/3$.*

Proof. Let the spine of C be $v_1, v_2, \cdots, v_k$, and let v_0 and v_{k+1} be endvertices adjacent to v_1 and v_k, respectively. Let P_0, P_1, and P_2 be the packings of C defined by $P_i = \{v_j : j \equiv i \pmod 3$ for $0 \leq j \leq k + 1\}$, for $0 \leq i \leq 2$. Every endvertex of C is dominated in one of the three packings, with v_0 and v_{k+1} being dominated in two of them. Each vertex on the spine is dominated by all three packings. Hence, $I(P_0) + I(P_1) + I(P_2) \geq n + 2k + 2$. Thus, $F(C) \geq \max\{I(P_0), I(P_1), I(P_2)\} \geq (n + 2k + 2)/3$. $\square$

If C is a caterpillar with spine $v_1, v_2, \cdots, v_{3h}$ and each v_i is adjacent to j endvertices for $2 \leq j \leq 3h - 1$, and v_1 and v_{3h} are adjacent to $j - 1$ endvertices, then $n = 3hj + 3h - 2$, $k = 3h$, and $F(C) = h(j + 3) = (n + 2k + 2)/3$.

For arbitrary graphs, we also have the following result.

Theorem 4.7 [542] *For any graph* G,

(a) $F(G) = n$ *if* $n \leq 3$ *and* $F(G) \geq 1 + \sqrt{n-1}$ *if* $n \geq 4$.

(b) $R(G)$ *is at most* $O((nm)^{1/2})$.

4.2 Codes and Cubes

The hypercube (or simply the k-cube) Q_k has $n = 2^k$ vertices corresponding one-to-one with the 2^k binary k-tuples and two vertices are adjacent if their corresponding k-tuples differ in exactly one position. Hence, Q_k is regular of degree k, has $m = k2^{k-1}$ edges, and has diameter k. Cubes Q_3 and Q_4 are illustrated in Figure 4.5.

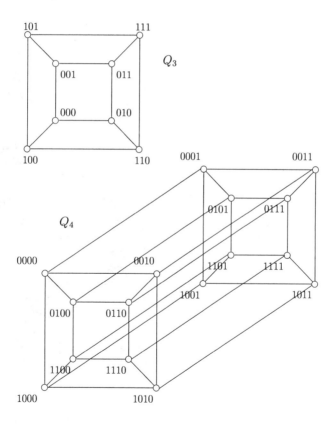

Figure 4.5: Cubes Q_3 and Q_4.

For many reasons, k-cubes have become important as architectural models for parallel computers, including the Intel iPSC and NCUBE machines. Each vertex of the k-cube represents a processor, and each edge designates a direct communication link between two processors. A k-cube structure allows for a large number of processors in a system in which each processor has relatively few direct communication links attached to it, yet the diameter is fairly small, which allows for relatively quick communication between any two processors. Another advantage of labelling processors with binary k-tuples is the application to message routing. If processor $v_1 = e_1e_2e_3...e_k$ must send a message to processor $v_2 = f_1f_2f_3...f_k$, the distance $d(v_1, v_2)$ from v_1 to v_2 is the Hamming distance, that is, the number of positions in which $e_i \neq f_i$. Moreover, it is not necessary to determine a priori the route of the message from v_1 to v_2. If a message is sent from v_1 along any communication link (perhaps the first available) that connects to a vertex whose Hamming distance to v_2 is one less, then a shortest path routing is guaranteed.

For a given parallel computer architecture, it is often necessary to distribute a limited supply of resources, such as I/O devices or certain memory units, among the processors. To effectively distribute the resources and guarantee ready access to every resource, a dominating set of processors was suggested in Chapter 1 as a good set of locations for placement of resources. As noted in [1109], various considerations, such as those involving the assignment of designated resources to processors, can make an efficient dominating set the best choice of locations for resource allocation.

For message transmission involving words of length k, each vertex in Q_k represents a possible received word. Because transmission errors could produce a change in a bit position between a word sent and the received word, a subset $S \subseteq V(Q_k)$ is agreed upon as the set of code words and only these are sent. For example, let $S = \{000, 111\} \subseteq Q_3$. Each possible received word of Q_3 is either in S or is adjacent to one element of S. In general, set $S \subseteq V(Q_k)$ is a *perfect d-code* if for each $v \in V(Q_k)$ there is exactly one element of S in $N^d[v] = \{w \in V(Q_k) : d(v, w) \leq d\}$. Equivalently, S is a perfect d-code if $\{N^d[s] : s \in S\}$ partitions $V(Q_n)$. As defined earlier, a perfect code is thus a perfect 1-code, such as $S = \{000, 111\}$ for Q_3. Note that maximum likelihood decoding (that is, decoding a received word as the nearest code word) makes a perfect d-code a d-error correcting code.

Note also that for Q_k, each $N[v]$ contains $k + 1$ vertices. Thus for Q_k to have a perfect code, 2^k must be a multiple of $k + 1$, that is, $k = 2^j - 1$ for some j. This necessary condition turns out to be sufficient. Further, the only perfect, binary, multiple error correcting code is a 3-error correcting code of order 23.

Theorem 4.8 *The k-cube Q_k has a perfect d-code if and only if*

(a) $k \leq d$ *(use any one vertex),*

(b) $d = 1$ *and* $k = 2^j - 1$, *or*

(c) $d = 3$ *and* $k = 23$.

Biggs [115] appears to have been the first person to consider perfect codes for graphs other than hypercubes. He comments, however, that "the class of all graphs is too general a setting for the perfect d-code question, since we may construct perfect codes at will by choosing suitable (but uninteresting) graphs." However, we think there are many interesting questions concerning perfect codes for arbitrary graphs. Recall, for example, Theorem 4.4 which states that deciding if a graph G has a perfect code is NP-complete. Further, specific graphs like the cube-connected cycles of Van Wieran, Livingston, and Stout [1109], almost all of which have perfect d-codes if and only if $d = 1$, appear to be quite interesting. And Dvořáková-Rulićová [410] showed that for any finite set M of positive integers and for any graph G with maximum degree $\Delta(G)$, G is an induced subgraph of a $(\Delta(G) + 1)$-regular graph containing d-perfect codes for every $d \in M$.

Several variants of efficient domination $F(G)$, redundance $R(G)$, and cardinality redundence $CR(G)$ have been studied. For example, efficient domination using open neighborhoods was studied by Gavlas and Schultz [526] and also with Slater in [527]. An *efficient open dominating set* for a graph G is a set $S \subseteq V(G)$ for which the open neighborhoods $N(v)$, $v \in S$, form a partition of $V(G)$. A graph is called *efficiently open dominatable* if it contains an efficient open dominating set. It is easily seen that a complete graph K_n with $n \geq 3$ is not efficiently open dominatable; a path P_n is efficiently open dominatable if and only if $n \not\equiv 1 \pmod 4$; and a cycle C_n is efficiently open dominatable if and only if $n \equiv 0 \pmod 4$. Recall that $\gamma_t(G)$ is the open or total domination number of G.

Theorem 4.9 [527] *If G has an efficient open dominating set S, then $|S| = \gamma_t(G)$ and all efficient open dominating sets have the same cardinality.*

Theorem 4.10 [527] *OPEN EFFICENT DOMINATING SET is NP-complete.*

Theorem 4.11 [527] *The k-cube Q_k has an efficient open dominating set if and only if $k = 2^j$.*

A dominating set S is a *perfect dominating set* if $|N(v) \cap S| = 1$ for each $v \in V - S$. Perfect domination was first studied by Weichsel (see [117]) and has been studied by Yen and Lee [1161], Cockayne, Hartnell, Hedetniemi, and Laskar [270], Livingston and Stout [852], and Fellows and Hoover [466].

Subsequently, Dunbar, Harris, Hedetniemi, Hedetniemi, McRae, and Laskar [394] defined a set S to be *nearly perfect* if for every vertex v in $V - S$, $|N(v) \cap S| \leq 1$. An interesting open problem, mentioned by Telle [1079], is that of deciding if an arbitrary graph G has two disjoint nearly perfect sets. A close variant of nearly perfect sets are *strongly stable sets*, that is, sets S such that for each vertex $v \in V$, $|N(v) \cap S| \leq 1$. These have been studied by Hochbaum and Schmoys [709].

Another variant is due to Bernhard, Hedetniemi, and Jacobs [105], who define the *efficiency* of a set $\varepsilon(S)$ to be the number of vertices $v \in V - S$ for which $|N(v) \cap S| = 1$. The efficiency of a graph G equals the maximum efficiency of any subset S of V. This concept has also been studied by Blair [117], and Telle and Proskurowski [1080].

In the next section we further investigate F and R and introduce the closed neighborhood order parameters W and P, which are their LP-duals.

4.3   Closed Neighborhoods: $F \leq W \leq n \leq P \leq R$

As stated in Theorem 4.8, the k-cube Q_k is efficiently dominatable if and only if $k = 2^j - 1$. As examples, for Q_3 we had the efficient dominating set $S = \{000, 111\}$, and for Q_7 we could use

$$S = \{0000000,\ 1101001,\ 0101010,\ 1000011,\ 1001100,\ 0100101,$$
$$1100110,\ 0011111,\ 1110000,\ 0011001,\ 1011010,\ 0110011,$$
$$0111100,\ 1010101,\ 0010110,\ 1111111\}.$$

Because every closed neighborhood has order $k + 1$, $n = 2^k$ must be a multiple of $k + 1$. More generally, the following is true. Recall that $\rho(G)$ denotes the packing number of G.

Theorem 4.12 *If an efficiently dominatable graph G is r-regular, then n is a multiple of $1 + r$, that is, $n = j(1 + r)$. For any r-regular graph G, $F(G) = (1 + r)\rho(G)$ and $R(G) = (1 + r)\gamma(G)$.*

Because Q_4 is regular of degree 4, we have $\rho(Q_4) \leq 3$ and $F(Q_4) \leq 5 \cdot 3 = 15$. The 4-cube Q_4 consists of two copies of Q_3, say Q_3' and Q_3'', joined by a matching. Note that $u, v \in V(Q_3')$ and $d(u, v) = 3$ implies that every vertex x in Q_4 is within a distance two of u or v. Hence, $\rho(Q_4) = 2$ and $F(Q_4) = 10$.

We next describe $\gamma(G)$ and $\rho(G)$ in terms of closed neighborhood subsystems of the collection C of all closed neighborhoods in G, $C = \{N[v_1], N[v_2], ..., N[v_n]\}$. The domination number $\gamma(G)$ equals the minimum order of a subcollection C' of C such that the union of the elements of C' is $V(G)$, and the packing number $\rho(G)$ is the maximum order of a subcollection of pairwise disjoint elements of C. For the closed neighborhood matrix N, the ith column gives the characteristic function of $N[v_i]$. Hence, $\gamma(G)$ can also be viewed as the minimum cardinality of a set of columns of N such that every row has a one in at least one column in the set, and $\rho(G)$ is the maximum cardinality of a set of columns such that no row has a one in two or more columns in the set. We thus have the following integer programming formulations for γ and ρ where $\vec{1}$ denotes the all-ones column n-tuple.

$$\gamma(G) = \min \sum_{i=1}^{n} X_i$$

Subject to $N \cdot X \geq \vec{1}$
$X_i \in \{0, 1\}$

$$\rho(G) = \max \sum_{i=1}^{n} X_i$$

Subject to $N \cdot X \leq \vec{1}$
$X_i \in \{0, 1\}$

Cases where the values X_i assigned to vertex v_i are in some arbitrary subset Y of the reals, rather than just $Y = \{0, 1\}$, where 1 indicates using v_i and 0 indicates not using v_i, are considered in Chapter 10 of this text and in Chapter 1 of [645]. In linear programming duality, we need to use $Y = [0, \infty)$. A function $f : V(G) \to [0, \infty)$ is a *dominating function* if $f(N[v_i]) = \sum_{w \in N[v_i]} f(w) \geq 1$, for $1 \leq i \leq n$, and f is a *packing function* if $f(N[v_i]) \leq 1$, for $1 \leq i \leq n$. The *fractional domination number* $\gamma_f(G)$ is the minimum weight of a dominating function, and the *fractional packing number* $\rho_f(G)$ is the maximum weight of a packing function, where the weight of a function f is $w(f) = f(V(G)) = \sum_{v \in V(G)} f(v)$. For example, for the Hajós graph H_3 in Figure 4.6, let $f(v_1) = f(v_4) = f(v_6) = 0$ and $f(v_2) = f(v_3) = f(v_5) = 1/2$. Then f is a dominating function and $\gamma_f(H_3) \leq 3/2$.

Letting $g(v_1) = g(v_4) = g(v_6) = 1/2$ and $g(v_2) = g(v_3) = g(v_5) = 0$, g is a packing function and $\rho_f(H_3) \geq 3/2$.

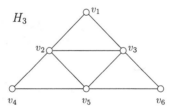

Figure 4.6: Hajós graph H_3.

Let M denote an arbitrary k-by-h real matrix, and $C = [c_1, c_2, ..., c_h]^t$ and $B = [b_1, b_2, ..., b_k]^t$ be real-valued column vectors. The following two linear programs are LP-duals of each other.

Primal $\min C^t X \quad = \sum_{i=1}^{h} c_i X_i$

   Subject to $M \cdot X \geq B$
   with $X_i \geq 0$

Dual $\max B^t X \quad = \sum_{i=1}^{k} b_i X_i$

   Subject to $M^t \cdot X \leq C$
   with $X_i \geq 0$

Theorem 4.13 (LP-duality) *If an optimum solution exists to either the primal linear program or its dual, then the other program also has an optimum solution and the two objective functions have the same optimum value.*

As a consequence of Theorem 4.13 and the existence of the functions f and g with $w(f) = w(g) = 3/2$, we have for the LP-dual functions ρ_f and γ_f (see Figure 4.7) that $\rho_f(H_3) = \gamma_f(H_3) = 3/2$. Figure 4.7 also gives the formulations for *fractional efficient domination* F_f and *fractional redundance* R_f and their LP-duals, the *fractional closed neighborhood order domination* (CLOD) and *packing* (CLOP) parameters, W_f and P_f, respectively. The efficient domination number $F(G)$ was introduced in Bange et al. [76] and then in [77, 73]. The formulation of $F_f(G)$ as a linear program suggested to Grinstead and Slater [555, 558] the formulation of fractional redundance $R_f(G)$. This was then integerized to define redundance $R(G)$. Although the (integer versions of the) following closed neighborhood parameters W_f and P_f have the following easily described applications, these parameters were first considered from their mathematical formulations as the duals of F_f and R_f, respectively, in [1037, 1040, 701] and then in [1049, 1050]. (For W_f and P_f in Figure 4.7, the n-tuple D^* has $1 + deg(v_i) = d_i^*$ as its ith entry, for $1 \leq i \leq n$. Note also that since N is symmetric, $N^t = N$.)

4.3.1   Closed neighborhood order domination $(CLOD)$: $W(G)$

Assume that the vertex set for each of the graphs G_1 and G_2 in Figure 4.8 on $n = 10$ and $n = 14$ vertices, respectively, represents a collection of towns, each of which supports its own standard-sized fire station. In the event of a major fire in one town, fire fighters from an adjacent town can be expected to respond promptly enough to help. Note that four fire stations can respond to each of the towns $u_2, u_3, u_4, u_7, u_8, u_9$, but only two can respond to u_1, u_5, u_6 or u_{10}; and eight stations can respond to town v_7 or v_8 of G_2, but only two to each other v_i. Note that if we reduce the number of fire stations in G_1 from ten to eight and

Domination

$\gamma_f(G) = \min \sum_{i=1}^{n} X_i$

subject to $N \cdot X \geq \vec{1}$

$X_i \geq 0$

Packing

$\rho_f(G) = \max \sum_{i=1}^{n} X_i$

subject to $N \cdot X \leq \vec{1}$

$X_i \geq 0$

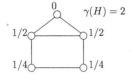
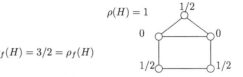

$\gamma(H) = 2$

$\gamma_f(H) = 3/2 = \rho_f(H)$

$\rho(H) = 1$

CLOD

$W_f(G) = \min \sum_{i=1}^{n} X_i$

subject to $N \cdot X \geq D^*$

$X_i \geq 0$

Efficient domination

$F_f(G) = \max \sum_{i=1}^{n} (1 + deg(v_i)) X_i$

subject to $N \cdot X \leq \vec{1}$

$X_i \geq 0$

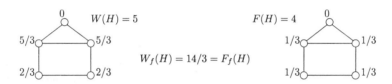

$W(H) = 5$

$W_f(H) = 14/3 = F_f(H)$

$F(H) = 4$

Redundance

$R_f(G) = \min \sum_{i=1}^{n} (1 + deg(v_i)) X_i$

subject to $N \cdot X \geq \vec{1}$

$X_i \geq 0$

CLOP

$P_f(G) = \max \sum_{i=1}^{n} X_i$

subject to $N \cdot X \leq D^*$

$X_i \geq 0$

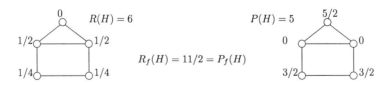

$R(H) = 6$

$R_f(H) = 11/2 = P_f(H)$

$P(H) = 5$

Figure 4.7: Linear programming formulations.

locate two of them at each of u_2, u_4, u_7, and u_9, then the same number of fire stations as before can still respond to each u_i, for $1 \leq i \leq 10$. For G_2, we can reduce the number of fire stations from 14 to eight with four at v_8 and four at v_7. There are still eight fire stations that can respond to each of v_7 and v_8. Further, the situation improves for each other v_i because now four stations can respond instead of only two.

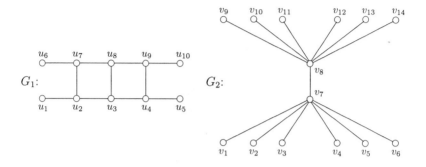

Figure 4.8: The graphs G_1 and G_2.

The parameter W for an arbitrary graph G is defined to be the minimum number of "tokens" we can place on the vertices, allowing more than one token to be placed on a vertex, so that each $N[v]$ contains at least $|N[v]|$ tokens. That is, in this sense, we seek to do at least as well as placing one token on each vertex. The LP-formulation for *closed neighborhood order domination number* W follows:

$$W(G) = \min \sum_{i=1}^{n} X_i$$
$$\text{Subject to } N \cdot X \geq D^* = [1 + deg(v_1), \cdots, 1 + deg(v_n)]^t$$
$$\text{with } X_i \in \{0, 1, 2, 3, ...\}.$$

As a feasible solution, one can always choose $f : V(G) \to \{0, 1, 2, ...\}$ with $f(v) = 1$ for every $v \in V(G)$. Hence, $W(G) \leq n$, and by duality we have the next result.

Theorem 4.14 [1037] *For any graph G,*

$$F(G) \leq F_f(G) = W_f(G) \leq W(G) \leq n.$$

In Figure 4.8, we have two examples for which $W(G) < |V(G)| = n$. We first show that $n - W(G)$ can be arbitrarily large. In fact, $W(G)/n$ can be arbitrarily small. Consider the tree $T_{j,k}$ in Figure 4.9. Vertex u has j neighbors, $N(u) = \{v_1, v_2, ..., v_j\}$. Each v_i is adjacent to k endvertices. Thus, $|V(T_{j,k})| = jk + j + 1$. The function $g : V(T_{j,k}) \rightarrow \{0, 2, k\}$ with $g(u) = k$, $g(v_i) = 2$ for $1 \le i \le j$, and $g(x) = 0$ for each endvertex x, is a $CLOD$-function with $w(g) = W(T_{j,k}) = 2j + k$. (Note that $F(T_{j,k}) = 2j + k$ also.) For a fixed k, as j increases $W(T_{j,k})/|V(T_{j,k})| = (2j + k)/(jk + j + 1)$ approaches $2/(k + 1)$. Thus we have the next theorem.

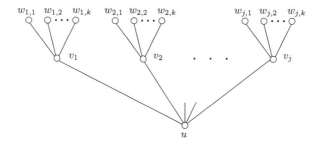

Figure 4.9: The tree $T_{i,k}$.

Theorem 4.15 [1037, 701] *The ratio $W(G)/|V(G)|$ can be arbitrarily small, even for trees.*

For the cycle C_n with $n = 3k + r$ and $0 \le r \le 2$, $F(C_n) = 3k$. But if $g(v) = 1/3$ for every $v \in V(C_n)$, then $F_f(C_n) = n$, and thus, $W_f(C_n) = W(C_n) = n$. More generally, if G is r-regular and we define $g : V(G) \rightarrow [0, \infty)$ by $g(v) = 1/(r + 1)$ for every $v \in V(G)$, we see that $F_f(G) = n$. Thus we have the following result.

Theorem 4.16 [1037] *If G is an r-regular graph, then*
$$F_f(G) = W_f(G) = W(G) = n.$$

It would seem that when G is not regular, it would be better to place higher weights on the vertices of large degree in order to better reach the amount of domination required. Specifically, each vertex v_i must be dominated by a weight of $1 + deg(v_i)$ in $N[v_i]$, so the total amount of domination required is $\sum_{i=1}^{n}(1 + deg(v_i)) = n + 2m$. However, for the graph G_{10} shown in Figure 4.10, $W(G_{10}) = 29$, and the unique $CLOD$-function g with $w(g) = 29$ is given. We note that while $g(v) = 5$, we have $g(u) = g(w) = 0$ for the vertices u and w of maximum degree $\Delta(G) = 4$. We do get the following degree sequence bound for $W(G)$.

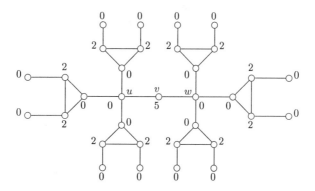

Figure 4.10: The graph G_{10}.

Theorem 4.17 [701] *Assume that the degrees d_i of vertices v_i of a graph G satisfy $d_1 \geq d_2 \geq ... \geq d_n$. If t is the largest integer for which*

$$(d_1 + 1)^2 + (d_2 + 1)^2 + ... + (d_t + 1)^2 + k(d_{t+1} + 1) \leq n + 2m,$$

where $0 \leq k \leq d_{t+1}$, and k is then maximized subject to this constraint, then $W(G) \geq W_f(G) \geq d_1 + d_2 + ... d_t + t + k$.

Proof. Let $g : V \to [0, \infty)$ be a $CLOD$-function satisfying $w(g) = W(G)$. We consider the total amount of domination N done by g, namely the sum $N = \sum \sum g(u)$, where the outer sum is over all $v \in V$ and the inner sum is over all $u \in N[v]$. Since $\sum_{u \in N[v]} g(u) = g(N[v]) \geq deg(v) + 1$ for each $v \in V$,

$$N \geq \sum_{v \in V} (deg(v) + 1) = n + 2m. \tag{4.1}$$

The sum N counts the value $g(u)$ exactly $deg(u) + 1$ times for each $u \in V$, so

$$N = \sum_{u \in V} (deg(u) + 1)g(u) = \sum_{i=1}^{t} (d_i + 1)g(v_i) + \sum_{i=t+1}^{n} (d_i + 1)g(v_i). \tag{4.2}$$

Suppose $g(v_i) \leq d_i + 1$ for all i, and assume that $W(G) < d_1 + d_2 + ... + d_t + t + k$. Then

$$\sum_{i=1}^{t} (d_i + 1) + k > W(G) = w(g) = \sum_{i=1}^{t} g(v_i) + \sum_{i=t+1}^{n} g(v_i). \tag{4.3}$$

Since each $g(v_i) \leq d_i + 1$, and $d_1 \geq d_2 \geq ... \geq d_n$, the sum in (4.2) is a maximum when $g(v_i) = d_i + 1$ for $1 \leq i \leq t$, that is, when $\sum_{i=1}^{t} g(v_i) = \sum_{i=1}^{t} (d_i + 1)$. This would imply, by (4.3), that $\sum_{i=t+1}^{n} g(v_i) < k$. Thus, by (4.2),

$$N \leq \sum_{i=1}^{t}(d_i + 1)^2 + (d_{t+1} + 1) \sum_{i=t+1}^{n} g(v_i)$$

$$< \sum_{i=1}^{t}(d_i + 1)^2 + (d_{t+1} + 1)^k$$

$$\leq n + 2m, \text{(by assumption)}$$

which contradicts equation (4.1). Hence, if each $g(v_i) \leq d_i + 1$, then the result follows.

Because $w(g) = W(G)$, if $g(v) > 0$ and one decreases $g(v)$, then we no longer have a $CLOD$-function. That is, there must be a vertex $u \in N[v]$ such that $g(N[u]) = deg(u) + 1$. Observe that if $g(v) > deg(v) + 1$, then we must have $deg(u) > deg(v)$.

Assume that we have vertices v_i with $g(v_i) > d_i + 1$. For each such v_i, select one $u_i \in N[v_i]$ such that $g(N[u_i]) = deg(u_i) + 1$. As noted, $deg(u_i) > deg(v_i)$, so u_i corresponds to a v_j where $j < i$. Also, we might have some $u_i = u_h$ with $h \neq i$. Let $g^* : V \to [0, \infty)$ be the function with $w(g^*) = w(g)$ obtained from g as follows. For each v_i with $g(v_i) > d_i + 1$, let $g^*(v_i) = 0$ and increase the function value at u_i by $g(v_i)$. We have $w(g) = w^*(g)$ and each $d^*(v_i) \leq d_i + 1$, and $\sum_{i=1}^{n}(d_i + 1)g^*(v_i) > \sum_{i=1}^{n}(d_i + 1)g(v_i)$. Finally, if $w^*(g) = w(g) = W(G) < d_1 + d_2 + ... + d_t + t + k$, then $N = \sum_{i=1}^{n}(d_i+1)g(v_i) < \sum_{i=1}^{n}(d_i+1)g^*(v_i) < n+2m$, a contradiction, completing the proof. $\square$

To illustrate Theorem 4.17, consider the graph G_1 of order $n = 10$ and size $m = 11$ shown in Figure 4.8. The graph G_1 has degree sequence $d_1, d_2, ..., d_{10}$, where $d_i = 3$ for $1 \leq i \leq 6$ and $d_i = 1$ for $7 \leq i \leq 10$. The largest integer t for which $\sum_{i=1}^{t}(d_i + 1)^2 + k(d_{t+1} + 1) \leq n + 2m = 32$, where $0 \leq k \leq d_{t+1}$, is $t = 2$ with $k = 0$. Hence, applying Theorem 4.17, we get $W(G) \geq d_1 + d_2 + 2 + 0 = 8$. As observed earlier, $W(G_1) \leq 8$, whence $W(G_1) = 8$.

4.3.2 Closed neighborhood order packing $(CLOP)$: $P(G)$

Consider a graph, such as graph G_1 in Figure 4.8, where each vertex u_i represents a community that has one "obnoxious" facility f_i, perhaps a garbage dump, or a correctional facility, where we assume that the facility affects the community having it and those immediately adjacent to it. For example, facility f_3 located at u_3 affects u_2, u_3, u_4, and u_8. In turn, community u_3 is affected by facilities f_2, f_3, f_4, and f_8. In general, u_i has $1 + deg(u_i) = |N[u_i]|$ obnoxious facilities affecting it. For G_1, consider placing two facilities at each of u_1, u_3, u_5, u_6, u_8, and u_{10} (that is, let $g(u_1) = g(u_3) = g(u_5) = g(u_6) = g(u_8) = g(u_{10}) = 2$ and $g(u_2) = g(u_4) = g(u_7) = g(u_9) = 0$). The system now supports 12 facilities (that is, $w(g) = 12$) rather than just ten with one per community, and each community u_i again has $1 + deg(u_i)$ obnoxious facilities affecting it. In general, we can seek to maximize the number of facilities in the graph where each $N[u_i]$ has at most $|N[u_i]|$ facilities. For G_2 in Figure 4.8, we can let $g(v_1) = g(v_2) = g(v_9) = g(v_{10}) = 2$, $g(v_3) = g(v_4) = g(v_5) = g(v_6) = g(v_{11}) = g(v_{12}) = g(v_{13}) = g(v_{14}) = 0$, and $g(v_7) = g(v_8) = 0$. This g is a *closed order packing function* (CLOP) achieving $P(G_2) = 16$.

$$P(G) = \max \textstyle\sum_{i=1}^{n} X_i$$

$$\text{Subject to } N \cdot X \le D^*$$
$$\text{with } X_i \in \{0, 1, 2, 3, ...\}$$

Graphs G_1 and G_2 in Figure 4.8 illustrate that we can make $P(G) > n = |V(G)|$. Thus, it is sometimes possible to pack more than n (obnoxious) facilities into a graph G while each vertex has in its closed neighborhood at most the number of facilities it would have if one facility is placed at each vertex. In fact, $P(G) - |V(G)|$ can be arbitrarily large. Consider the graph $G_t^1 = K_t \circ \overline{K}_t$ in Figure 4.11 with $|V(G_t^1)| = n = t^2 + t$ and $P(G_t^1) = 2t^2$ (obtainable by placing a weight of two on each endvertex and a weight of zero on each vertex of degree $2t - 1$). We have $P(G_t^1) - |V(G_t^1)| = t^2 - t$.

Quite interestingly, the ratio $P(G_t^1)/|V(G_t^1)| = 2t/(t+1)$, which approaches 2 as t increases. More generally, as noted in [1049], the graph G_t^j in Figure 4.11 has $n = jt + t^2$ vertices; $P(G_t^j) = (j+1)t^2$ is achieved by placing a weight of $j+1$ on each of the t^2 vertices of degree j; and so $P(G_t^j)/|V(G_t^j)| = (j+1)t/(j+t)$, which approaches $j+1$ as t increases. Note that $P(G_t^j) = t^3 + t^2$ is $O(n^{3/2})$. As a corollary to Theorem 4.7, we have the following.

Theorem 4.18 [1050] *The function* $f(n) = \max\{P(G_n) : |V(G)| = n\}$ *is* $O(n^{3/2})$.

For the tree T_k in Figure 4.12, we have $P(T_k)/|V(T_k)| = (6k + 8)/(4k + 6) < 3/2$. In contrast to the fact that we can make $P(G)/|V(G)|$ arbitrarily large, we will see that for any tree T we have $P(T)/|V(T)| < 3/2$. To achieve $P(T_k) = 6k + 8$, we place a weight of two at each endvertex. Intuitively, to maximize the weight of a packing function, it seems that one should increase as much as possible the weights on the vertices of small degree.

Theorem 4.19 [1040] *If a graph G has degree sequence $(d_1, d_2, \cdots, d_n)$ with* $deg(v_i) = d_i \le d_{i+1}$, *and if* $(d_1+1)^2 + (d_2+1)^2 + \cdots + (d_j+1)^2 + k(d_{j+1}) = n + 2m$ *where* $0 \le k \le d_{j+1}$, *then* $P(G) \le P_f(G) \le d_1 + d_2 + \cdots + d_j + j + k$.

Proof. Let $f : V(G) \to [0, \infty)$ be a P_f-function. For each $v \in V(G)$, we have

$$\sum_{x \in N[v]} f(x) \le |N[v]| = 1 + deg(v). \text{ Hence, } \sum_{i=1}^{n}(1 + d_i)f(v_i) = \sum_{i=1}^{n} \sum_{x \in N[v_i]} f(x) \le$$

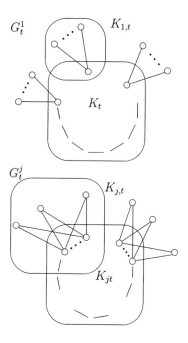

Figure 4.11: $P(G_t^j)/|V(G_t^j)| = (j+1)t^2/(jt+t^2)$.

$\sum_{i=1}^{n}(1 + d_i) = n + 2m$. That is, the weight $f(v_i)$ at v_i affects v_i and each vertex in $N(v_i)$. Therefore, $(1 + d_i)f(v_i)$ evaluates the amount of domination done by assigning a value of $f(v_i)$ at v_i, and globally, the total amount of domination done by f is bounded above by $n + 2m$, because of the closed neighborhood order constraints. Note that $f(v_i) \leq 1 + d_i = |N[v_i]|$. If we consider assigning a total weight (say $w(f)$) to the vertices in $V(G)$ with the only constraints that each v_i receives at most a weight of $1 + d_i$, then we can minimize the total amount of domination done by the assignment, by placing as much weight as is allowed on the successive vertices of smallest degree. In particular, if $w(f) = P_f(G) > (d_1 + 1) + (d_2 + 1) + ... + (d_j + 1) + k$, then the total amount of domination done would exceed $n + 2m$. Thus, $P_f(G) \leq d_1 + d_2 + ... + d_j + j + k$, completing the proof. $\square$

Corollary 4.20 *For any graph G, $P(G) \leq P_f(G) \leq (n + 2m)/(1 + \delta(G))$.*

Corollary 4.21 *If G is an r-regular graph, then $P(G) = P_f(G) = n$.*

Corollary 4.22 *For any tree T, $P(T) \leq P_f(T) \leq (3n - 2)/2 < 3n/2$.*

T_k

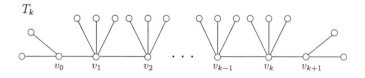

$$v_0 \qquad v_1 \qquad v_2 \qquad\qquad v_{k-1} \qquad v_k \qquad v_{k+1}$$

Figure 4.12: $P(T_k)/|V(T_k)| = (6k + 8)/(4k + 6)$.

Recall that the trees T_k illustrated by Figure 4.12 show that $P(T)/|V(T)|$ can be made arbitrarily close to $3/2$. The all-ones function on $V(G)$ shows that $n \leq P(G)$, and the Duality Theorem 4.13 yields the next result.

Theorem 4.23 [1037] *For any graph G, $n \leq P(G) \leq P_f(G) = R_f(G) \leq R(G)$.*

If G is r-regular, then $g : V(G) \to [0, \infty)$, defined by $g(v) = 1/(1 + r)$ for every $v \in V(G)$, shows that $F_f(G) = n = R_f(G)$.

Theorem 4.24 [1037] *If G is an r-regular graph, then $n = P(G) = P_f(G) = R_f(G)$.*

4.4   Computational Results

It is easy to see that $\gamma(G) \leq F(G)$ for every graph G. Combining this with Theorems 4.14 and 4.23, we get the following chain of inequalities.

Theorem 4.25 [1037] *For any graph G,*

$$\rho(G) \leq \rho_f(G) = \gamma_f(G) \leq \gamma(G) \leq F(G) \leq F_f(G) =$$

$$W_f(G) \leq W(G) \leq n \leq P(G) \leq P_f(G) = R_f(G) \leq R(G).$$

Not unexpectedly, the decision problems for $CLOD$-function W and $CLOP$-function P are NP-complete. Not only is deciding if $F(G) \geq k$, for an input constant k, an NP-complete problem, Theorem 4.4 states that simply deciding if $F(G) = n$ is NP-complete. The following results also hold for the closed neighborhood parameters.

Theorem 4.26 [1050] *The following decision problems are NP-complete, even for the class of bipartite graphs:*

(a) *Is $F(G) = n$?*

(b) *Is $W(G) < n$?*

(c) *Is $P(G) > n$?*

(d) *Is $F(G) = W(G)$?*

(e) *Is $P(G) = R(G)$?*

(f) *Is $P(G) > W(G)$?*

Given the inherent difficulty suggested by Theorem 4.26 in evaluating these parameters, the bounds provided by Theorem 4.25 can be particularly useful. As noted, if G is r-regular, then $F_f(G) = n = R_f(G)$, from which it follows that $W(G) = n = P(G)$. We note that, since fractional parameters like F_f and R_f have linear programming formulations, they can be computed in polynomial time.

Each of the parameters γ_f, ρ_f, F_f, W_f, P_f, and R_f involve n variables with n constraints. The following general Automorphism Class Theorem extends the results in [555] for γ_f, ρ_f, F_f, and W_f and in [1037] for W_f and P_f. To illustrate the result, consider the graph A in Figure 4.13. Note that the function g from the top labelling is a γ_f-function because $g(N[v_i]) \geq 1$ for each vertex v_i. Also, $\rho(G) = 4$ as illustrated by the four shaded vertices, and $w(g) = \sum_{v \in V(A)} g(v) = 4$, so $4 \leq \rho(A) \leq \rho_f(A) = \gamma_f(A) \leq 4$ which implies that $\gamma_f(A) = 4$. The γ_f-function g^* from the bottom labelling is obtained from g by letting $g^*(v)$ equal the average weight of the vertices in the automorphism class of v.

We first define general minimization and maximization parameters for column vectors $C = [c_1, c_2, ..., c_n]^t$ and $B = [b_1, b_2, ..., b_n]^t$ with the property that if v_i and v_j are in the same automorphism class then $c_i = c_j$ and $b_i = b_j$.

$$\phi(G) = \min \sum_{i=1}^n c_i X_i$$

Subject to $N \cdot X \geq B$
with $X_i \geq 0$

$$\psi(G) = \max \sum_{i=1}^n c_i X_i$$

Subject to $N \cdot X \leq B$
with $X_i \geq 0$

For $g : V(G) \to [0, \infty)$, let $C(g) = \sum_{i=1}^n c_i g(v_i)$. Then the function $g : V(G) \to [0, \infty)$ is called a ϕ_f-*function* (respectively, ψ_f-*function*) if for $g(v_i) = X_i$, we have $N \cdot X \geq B$ (respectively, $N \cdot X \leq B$) and $C(g) = \sum_{i=1}^n c_i g(v_i)$ equals $\phi_f(G)$ (respectively, equals $\psi_f(G)$).

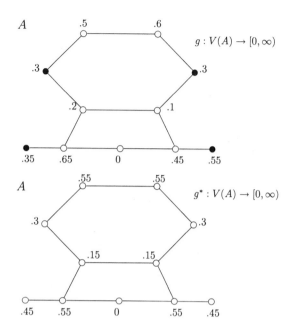

Figure 4.13: $\gamma_f(G) = 4$.

Theorem 4.27 (Automorphism Class Theorem [1044]) *For $g : V(G) \to [0, \infty)$, let $g^* : V(G) \to [0, \infty)$ be defined by $g^*(v) = \sum_{w \in [v]} g(w)/|[v]|$, where $[v]$ denotes the set of vertices in the automorphism class of v. If g is a ϕ_f-function (respectively, a ψ_f-function), then g^* is also a ϕ_f-function (respectively, a ψ_f-function) when C and B are constant on automorphism classes.*

Proof. Assume, for example, that $[v] = \{v_1, v_2, ..., v_h\}$, so $c_1 = c_2 = ... = c_h$. Then $\sum_{v_i \in [v]} c_i g(v_i) = c_1 \sum_{i=1}^{h} g(v_i) = c_1 \cdot h \cdot (g(v_1) + g(v_2) + ... + g(v_h))/h = c_1 \cdot h \cdot g^*(v) = \sum_{i=1}^{h} c_i g^*(v) = \sum_{v_i \in [v]} c_i g^*(v_i)$. It follows that $C(g^*) = C(g) = \sum_{i=1}^{n} c_i g(v_i) = \phi_f(G)$ (respectively, $\psi_f(G)$).

It remains to show that if $g(v_i) = X_i$ and $g^*(v_i) = X_i^*$ with $X = [X_1, X_2, \cdots, X_n]^t$ and $X^* = [X_1^*, X_2^*, \cdots, X_n^*]$, then (1) $N \cdot X \geq B$ implies $N \cdot X^* \geq B$ and (2) $N \cdot X \leq B$ implies $N \cdot X^* \leq B$.

Assume that $N \cdot X \geq B$. Let the automorphism classes of G be $C_1, C_2, ..., C_t$ with $v = u_1 \in C_1$ and $u_i \in C_i$ for $2 \leq i \leq t$. Let $h_j = |N[v] \cap C_j|$ for $1 \leq j \leq t$, and let $h_1 = h$ with $C_1 = \{v = u_1 = v_1, v_2, ..., v_h\}$. Note that, because each C_i is an automorphism class, for each fixed j, we have $h_j = |N[v_i] \cap C_j|$ for $1 \leq i \leq h$. And if u_i and u_i' are distinct vertices in C_i, then $|N(u_i) \cap C_1| = |N(u_i') \cap C_1|$. In particular, the number of edges connecting C_1 and C_j for $2 \leq j \leq t$ is

$h \cdot h_j = |C_j| \cdot |N(u_j) \cap C_1|$. We have the following:

$$
\begin{aligned}
\sum_{i=1}^{h} \sum_{w \in N[v_i]} g^*(w) &= h \sum_{w \in N[v_1]} g^*(w) \\
&= h \left(\sum_{w \in N[v_1] \cap C_1} g^*(w) + \sum_{w \in N[v_1] \cap C_2} g^*(w) + \right. \\
&\quad \left. \ldots + \sum_{w \in N[v_1] \cap C_t} g^*(w) \right) \\
&= h(h_1 \cdot g^*(u_1) + h_2 \cdot g^*(u_2) + \ldots + h_t \cdot g^*(u_t)) \\
&= \sum_{j=1}^{t} h \cdot h_j \cdot g^*(u_j) \\
&= \sum_{j=1}^{t} |C_j| \cdot |N[u_j] \cap C_1| \cdot g^*(u_j)
\end{aligned}
$$

$$
\begin{aligned}
&= \sum_{j=1}^{t} |N[u_j] \cap C_1| \cdot |C_j| \cdot \left(\sum_{w \in C_j} g(w) / |C_j| \right) \\
&= \sum_{j=1}^{t} |N[u_j] \cap C_1| \cdot \sum_{w \in C_j} g(w) \\
&= \sum_{j=1}^{t} \sum_{w \in C_j} |N[u_j] \cap C_1| g(w) \\
&= \sum_{i=1}^{t} \left(\sum_{w \in N[v_i]} g(w) \right) \\
&\geq b_1 + b_2 + \ldots + b_h \\
&= h \cdot b_1
\end{aligned}
$$

because the fact that B is constant on automorphism classes implies $b_1 = b_2 = \ldots = b_h$. The same argument holds for the other automorphism classes, and so, in general, $\sum_{w \in N[v_i]} g^*(w) \geq b_i$ for $1 \leq i \leq n$. That is, $N \cdot X^* \geq B$. Similarly, if $N \cdot X \leq B$, then $N \cdot X^* \leq B$, completing the proof. $\square$

Corollary 4.28 [555, 1037] *If* $g : V(G) \rightarrow [0, \infty)$ *is a* ρ_f, γ_f, F_f, W_f, P_f, *or* R_f-*function, then so is* g^*.

Thus each of these n variable LP-problems with an n-by-n constraint matrix can be reduced to a system of t variables and t constraints where t is the number of automorphism classes. For example, the Hajós graph H_3 in Figure 4.6 has only two such classes, $\{v_1, v_4, v_6\}$ and $\{v_2, v_3, v_5\}$. The Automorphism Class Theorem guarantees an optimum solution g with $g(v_1) = g(v_4) = g(v_6) = A$ and $g(v_2) = g(v_3) = g(v_5) = B$.

$$F_f(H_j) = \max \ 9A + 15B$$

$$\text{Subject to} \quad A + 2B \leq 1$$
$$2A + 3B \leq 1$$
$$A, B \geq 0$$

An optimum solution with $A = 0$ and $B = 1/3$ shows $F_f(H_3) = 5 = W_f(H_3)$. (A W_f-function has $g(v_1) = g(v_4) = g(v_6) = 0$ and $g(v_2) = g(v_3) = g(v_5) = 5/3$.) In fact, $F(H_3) = 5$ with an F-set $\{v_2\}$ and $W(H_3) = 5$, using $g(v_2) = g(v_3) = 2$, $g(v_5) = 1$, and $g(v_1) = g(v_4) = g(v_5) = 0$.

A more extensive discussion of LP-duality of graphical parameters is contained in Chapter 1 of [645].

While generalized domination and generalized efficient domination will be covered in Chapter 10, we include here a method of demonstrating that $F(G) < n$. For the tree T in Figure 4.14, if $W = [2, -1, 0, 1, 1]$, then $N \cdot W^t = \vec{1}_5$, where $\vec{1}_n$ denotes the all ones column n-tuple. In general, W represents a real n-tuple $W \in \Re^n$ and the function $f : V(G) \to \Re$ with $f(v_i) = w_i$ has weight $w(f) = \sum_{i=1}^{n} f(v_i)$. If $f(v_i) \in Y \subseteq \Re$ and $\sum_{x \in N[v]} f(x) = 1$, then f is called an *efficient Y-dominating function.* The following theorem is proved in Chapter 10.

Theorem 4.29 [74] *For a graph G, there exists a vector $W \in \Re^n$ satisfying $N \cdot W = \vec{1}_n$ if and only if $N \cdot X = N \cdot Z$ implies $\sum_{i=1}^{n} X_i = \sum_{i=1}^{n} Z_i$. In particular, for any $Y \subseteq \Re$ any two efficient Y-dominating functions have the same weight.*

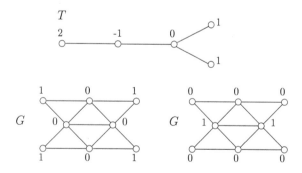

Figure 4.14: A $\{-1,0,1,2\}$-efficient labelling of T and two equivalent labellings of G of weights 4 and 2.

The two labellings of G in Figure 4.14 are equivalent in that each $N[v]$ has the same weight. By Theorem 4.29, $N \cdot W = \vec{1}_8$ has no real solution, so G is not efficiently dominatable and $F(G) < 8$. While we can not conclude it from Theorem 4.29, T is also not efficiently dominatable.

4.5   Realizability

By choosing $<$ or $=$ for each inequality in the chain $F(G) \leq W(G) \leq n \leq P(G) \leq R(G)$, there are sixteen possible inequality sequences. However, $F(G) = n$ if and only if $R(G) = n$, and this reduces the number of possible sequences to ten. All ten of these inequality sequences can be realized in the sense that, for each case, there exist infinitely many graphs G with the required inequality sequence.

Theorem 4.30 [1050] *The following ten cases are realizable:*

(a) $F(G) = W(G) = n = P(G) = R(G)$.

(b) $F(G) = W(G) < n = P(G) < R(G)$.

(c) $F(G) = W(G) < n < P(G) < R(G)$.

(d) $F(G) = W(G) < n < P(G) = R(G)$.

(e) $F(G) < W(G) = n = P(G) < R(G)$.

(f) $F(G) < W(G) = n < P(G) < R(G)$.

(g) $F(G) < W(G) = n < P(G) = R(G)$.

(h) $F(G) < W(G) < n = P(G) < R(G)$.

(i) $F(G) < W(G) < n < P(G) = R(G)$.

(j) $F(G) < W(G) < n < P(G) < R(G)$.

Note that case (a) describes the class of efficiently dominatable graphs which is trivially seen to be infinite. We present an infinite class of graphs G for case (b), where $F(G) = W(G) < n = P(G) < R(G)$, and we leave the other cases as exercises.

Let G_2 be as in Figure 4.15, where each K_s^i represents a complete subgraph on s vertices, for $1 \leq i \leq 4$, and double lines indicate that all s possible edges are included. Note that $S = \{v_2, v_5, v_9\}$ is a packing that dominates every vertex of G_2 except v_8. Thus, $F(G) \geq n - 1 = 4s + 7$. The CLOD-function $w : V(G) \rightarrow \{0, 1, 2, \cdots\}$, defined by $w(v_2) = w(v_9) = 1$, $w(v_4) = 3$, $w(v_7) = 2$, $w(v_3) = w(v_5) = w(v_8) = w(v_{10}) = 0$, and $w(v) = 1$ for each of the remaining $4s$ vertices v, shows that $W(G_2) \leq \sum_{v \in V(G_2)} w(v) = 4s + 7 = n - 1$. Because $F(G_2) \leq W(G_2)$, we have $F(G_2) = W(G_2) = n - 1 < n$. To prove that $n = P(G_2) < R(G_2)$, note that, in this particular case, graph G_2 has nine automorphism classes. They are: $C_1 = \{x : x \in V(K_s^1) \cup V(K_s^2)\}$, $C_2 = \{v_2\}$, $C_3 = \{v_3, v_{10}\}$, $C_4 = \{v_4\}$, $C_5 = \{v_5\}$, $C_6 = \{x : x \in V(K_s^3) \cup V(K_s^4)\}$, $C_7 = \{v_7\}$, $C_8 = \{v_8\}$, and $C_9 = \{v_9\}$. Therefore, by the Automorphism Class Theorem, we can formulate $P_f(G_2)$ as the following:

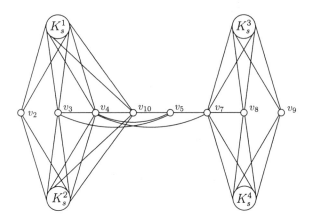

Figure 4.15: $F(G_2) = W(G_2) < n = P(G_2) < R(G_2)$.

$P_f(G_2) = \max((2s)x_1 + x_2 + 2x_3 + x_4 + x_5 + (2s)x_6 + x_7 + x_8 + x_9)$
subject to $(s)x_1 + x_2 + 2x_3 + x_4 \leq s + 4$
$(2s)x_1 + x_2 \leq 2s + 1$
$(2s)x_1 + x_3 + x_4 + x_5 \leq 2s + 3$
$(2s)x_1 + 2x_3 + x_4 + x_5 + x_7 \leq 2s + 5$
$2x_3 + x_4 + x_5 + x_7 \leq 5$
$(s)x_6 + x_7 + x_8 + x_9 \leq s + 3$
$x_4 + x_5 + (2s)x_6 + x_7 + x_8 \leq 2s + 4$
$(2s)x_6 + x_7 + x_8 \leq 2s + 2$
$(2s)x_6 + x_9 \leq 2s + 1$
$x_i \geq 0$ for $1 \leq i \leq 9$

Note that $R_f(G)$ can be formulated in a similar fashion as follows:

$R_f(G) = \min(2s(s + 4)x_1 + (2s + 1)x_2 + (4s + 6)x_3 + (2s + 5)x_4 + 5x_5$
$+2s(s + 3)x_6 + (2s + 4)x_7 + (2s + 2)x_8 + (2s + 1)x_9)$
subject to $(s)x_1 + x_2 + 2x_3 + x_4 \geq 1$
$(2s)x_1 + x_2 \geq 1$
$(2s)x_1 + x_3 + x_4 + x_5 \geq 1$
$(2s)x_1 + 2x_3 + x_4 + x_5 + x_7 \geq 1$
$2x_3 + x_4 + x_5 + x_7 \geq 1$
$(s)x_6 + x_7 + x_8 + x_9 \geq 1$
$x_4 + x_5 + (2s)x_6 + x_7 + x_8 \geq 1$
$(2s)x_6 + x_7 + x_8 \geq 1$
$(2s)x_6 + x_9 \geq 1$
$x_i \geq 0$ for $1 \leq i \leq 9$

Solving the above problem using the simplex method, we find solutions. We have $P_f(G_2) = 4s + \frac{80}{9}$, which is achieved by $x_1 = \frac{3s-1}{3s}$, $x_2 = \frac{5}{3}$, $x_3 = \frac{4}{3}$, $x_4 = 0$, $x_5 = \frac{7}{3}$, $x_6 = \frac{9s-1}{9s}$, $x_7 = 0$, $x_8 = \frac{17}{9}$, $x_9 = \frac{11}{9}$. Also, $R_f(G_2) = 4s + \frac{80}{9}$, which is achieved by $x_1 = \frac{2}{9s}$, $x_2 = \frac{5}{9}$, $x_3 = \frac{1}{9}$, $x_4 = 0$, $x_5 = \frac{4}{9}$, $x_6 = \frac{1}{3s}$, $x_7 = \frac{1}{3}$, $x_8 = 0$, $x_9 = \frac{1}{3}$. Therefore, $P(G_2) \leq 4s + 8$. But $P(G_2) \geq n = 4s + 8$, so $P(G_2) = n$. Also $R(G) \geq 4s + 9 = n + 1$. Note that the amount of domination done by $S = \{v_2, v_5, x, y\}$, where $x \in V(K_s^3)$ and $y \in V(K_s^4)$, is $4s + 12 = n + 4$. It is not difficult to see that this is best possible. Thus, $F(G_2) = W(G_2) = n - 1 < n = P(G_2) < R(G_2) = n + 4$.

EXERCISES

4.1 (a) The tree T_1 in Figure 4.1 has packings S_1 and S_2 with $F(T_1) = I(S_1) = I(S_2)$ and $|S_1| - |S_2| = 1$. For each positive integer k, find a graph G_k with packings S_1 and S_2 such that $F(G_k) = I(S_1) = I(S_2)$ and $|S_1| - |S_2| = k$.

(b) Find similar examples of graphs H_k with dominating sets S_1 and S_2 such that $R(H_k) = I(S_1) = I(S_2)$ with $|S_1| - |S_2| = k$.

4.2 Find a graph G with dominating sets S_1 and S_2 such that S_1 is an $R(G)$-set but not a $CR(G)$-set and S_2 is a $CR(G)$-set but not an $R(G)$-set (that is, $R(G) = I(S_1) < I(S_2)$, but the number of vertices dominated more than once is larger for S_1 than for S_2).

4.3 (a) For the trees in Figure 4.3, verify that $W(T_{2,k}) = 4k$ and $P(T_{1,k}) = 6k + 2$.

(b) For the caterpillar C of order $n = 3hj + 3h - 2$ described after Theorem 4.6, verify that $F(C) = W(C) = h(j + 3)$.

(c) Are the bounds in Theorem 4.5 sharp?

4.4 What is the maximum possible value of $R(C_{n,k})$ for a caterpillar $C_{n,k}$ of order n with k vertices on its spine? How about $P(C_{n,k})$?

4.5 (a) Verify that for the grid $P_3 \times P_6$ on 18 vertices, $F_f(P_3 \times P_6) = W_f(P_3 \times P_6) = 52/3$.

(b) It follows that $W(P_3 \times P_6) = 18$. What is $F(P_3 \times P_6)$?

(c) Determine $P(P_3 \times P_6)$, $R(P_3 \times P_6)$, and $P_f(P_3 \times P_6) = R_f(P_3 \times P_6)$.

4.6 The five-cycle has $F(C_5) = 1 + \sqrt{n - 1}$ as in Theorem 4.7(a). Find other such graphs.

4.7 (a) Determine $F(Q_5)$, $F(Q_6)$, $R(Q_5)$, and $R(Q_6)$ for the n-cubes Q_5 and Q_6.

   (b) Find bounds (exact values?) for $F(Q_k)$ and $R(Q_k)$.

4.8 Prove Theorem 4.9 that if S is an efficient total dominating set for G, then $|S| = \gamma_t(G)$.

4.9 (a) Apply Theorem 4.17 to the graph G_{10} in Figure 4.10.

   (b) Let $B(G) = d_1 + d_2 + ... + d_t + t + k$ as in Theorem 4.17. For graph G_{10}, $W(G_{10}) - B(G_{10}) = 5$. Find graphs G for which $W(G) - B(G)$ is arbitrarily large.

4.10 (a) Apply Theorem 4.19 to the graph G_{10} in Figure 4.10.

   (b) Determine $P(G_{10})$.

4.11 As noted in Chapter 1 of [645], we can replace the closed neighborhood matrix N with the adjacency matrix A in defining ϕ and ψ in (4.4) and (4.5). Prove the resulting open neighborhood version of the Automorphism Class Theorem.

4.12 For each of the ten cases described in Theorem 4.30, find an infinite family of graphs satisfying that particular inequality sequence.

4.13 Characterize the caterpillars that have efficient dominating sets.

4.14 (Cockayne, Hartnell, Hedetniemi, and Laskar [270]) A set $S \subseteq V(G)$ is a perfect dominating set if for every $v \in V - S$, $|N(v) \cap S| = 1$. The perfect domination number $\gamma_p(G)$ is the minimum cardinality of a perfect dominating set in G.

   (a) Construct an infinite family of graphs G for which $\gamma_p(G) = n$.

   (b) Show that if G is a connected graph with maximum degree $\Delta(G) \leq 3$, then $\gamma_p(G) < n$.

   (c) Show that for every tree T of order $n \geq 2$, $\gamma_p(T) \leq \lfloor n/2 \rfloor$ and construct an infinite family of trees achieving this bound.

Chapter 5

Changing and Unchanging Domination

An important consideration in the topological design of a network is fault tolerance, that is, the ability of the network to provide service even when it contains a faulty component or components. The behavior of a network in the presence of a fault can be analyzed by determining the effect that removing an edge (link failure) or a vertex (processor failure) from its underlying graph G has on the fault-tolerance criterion. For example, a γ-set in G represents a minimum set of processors that can communicate directly with all other processors in the system. If it is essential for file servers to have this property and that the number of processors designated as file servers be limited, then the domination number of G is the fault-tolerance criterion. In this example, it is important that $\gamma(G)$ does not increase when G is modified by removing a vertex or an edge. From another perspective, networks can be made fault-tolerant by providing redundant communication links (adding edges). Hence, we examine the effects on $\gamma(G)$ when G is modified by deleting a vertex or deleting or adding an edge.

5.1 Terminology

The semi-expository paper by Carrington, Harary, and Haynes [189] surveyed the problems of characterizing the graphs G in the following six classes. Let $G-v$ (respectively, $G-e$) denote the graph formed by removing vertex v (respectively, edge e) from G. We use acronyms to denote the following classes of graphs (C represents changing; U : unchanging; V : vertex; E : edge; R : removal; A : addition).

(CVR)	$\gamma(G - v) \neq \gamma(G)$ for all $v \in V$
(CER)	$\gamma(G - e) \neq \gamma(G)$ for all $e \in E$
(CEA)	$\gamma(G + e) \neq \gamma(G)$ for all $e \in E(\overline{G})$
(UVR)	$\gamma(G - v) = \gamma(G)$ for all $v \in V$
(UER)	$\gamma(G - e) = \gamma(G)$ for all $e \in E$
(UEA)	$\gamma(G + e) = \gamma(G)$ for all $e \in E(\overline{G})$

These six problems have been approached individually in the literature with other terminology. Here we examine them and several related problems using the above "changing and unchanging" terminology first suggested by Harary [F. Harary, Changing and unchanging invariants for graphs. *Bull. Malaysian Math. Soc.* 5 (1982) 73-78].

It is useful to partition the vertices of G into three sets according to how their removal affects $\gamma(G)$. Let $V = V^0 \cup V^+ \cup V^-$ for

$$V^0 = \{v \in V \ : \ \gamma(G - v) = \gamma(G)\}$$

$$V^+ = \{v \in V \ : \ \gamma(G - v) > \gamma(G)\}$$

$$V^- = \{v \in V \ : \ \gamma(G - v) < \gamma(G)\}.$$

Similarly, the edge set can be partitioned into

$$E^0 = \{uv \in E \ : \ \gamma(G - uv) = \gamma(G)\}$$

$$E^+ = \{uv \in E \ : \ \gamma(G - uv) > \gamma(G)\}.$$

For example, the graph in Figure 5.1 with $k \geq 3$ has $V^0 = \{u_i \ : \ 1 \leq i \leq k\} \cup \{v\}$, $V^+ = \{u\}$, $V^- = \{w\}$, $E^0 = \{uv, vw\}$, and $E^+ = \{uu_i \ : \ 1 \leq i \leq k\}$.

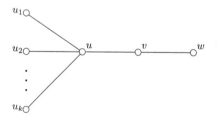

Figure 5.1: $V = V^0 \cup V^+ \cup V^-$.

5.2   Changing

We first consider the following three classes of graphs : CVR, CER, and CEA.

5.2.1   Vertex removal (CVR)

Obviously, a graph for which the domination number changes when an arbitrary vertex is removed (CVR) has $V = V^- \cup V^+$. Bauer, Harary, Nieminen, and Suffel showed that V^0 is never empty for a tree, hence, no tree is in CVR.

Theorem 5.1 [87] *For any tree T with $n \geq 2$, there exists a vertex $v \in V$, such that $\gamma(T - v) = \gamma(T)$.*

Proof. Clearly the result is true if $T = K_2$. Therefore, assume that T has at least one vertex v with $deg(v) \geq 2$ that is adjacent to at least one endvertex and at most one nonendvertex. If v is adjacent to two (or more) endvertices u_1 and u_2, then v is in every γ-set for T and $\gamma(T - u_1) = \gamma(T)$. If not, then v is adjacent to one endvertex u and $deg(v) = 2$.

Let $T' = T - v - u$. For any graph G, if $deg(u) = 1$, then $\gamma(G - u) \leq \gamma(G)$. Hence, $\gamma(T') \leq \gamma(T-u) \leq \gamma(T)$. However, $\gamma(T') \geq \gamma(T)-1$. If $\gamma(T') = \gamma(T)-1$, then $\gamma(T) = \gamma(T - v)$. Otherwise, $\gamma(T') = \gamma(T) = \gamma(T - u)$. $\square$

Note that removing a vertex can increase the domination number by more than one, but can decrease it by at most one. For examples, removing the center vertex of a star $K_{1,n-1}$ increases the domination number by $n - 2$; and removing an endvertex from the corona $G = H \circ K_1$, for any nontrivial connected graph H, decreases it by one, but removing any other vertex does not change the value of γ. The path P_{3k+1}, for $k \geq 1$, is another example of a graph for which the removal of an endvertex decreases the domination number by one. Furthermore, if S is a γ-set, then removing any vertex in $V - S$ cannot increase the domination number, so $|V^+| \leq \gamma(G)$. Obviously, every isolated vertex is in V^-. Bauer et al. characterized the vertices in V^+.

Theorem 5.2 [87] *A vertex $v \in V^+$ if and only if*

(a) *v is not an isolate and is in every γ-set of G, and*

(b) *no subset $S \subseteq V - N[v]$ with cardinality $\gamma(G)$ dominates $G - v$.*

Later, Sampathkumar and Neeralagi characterized the vertices in V^-.

Theorem 5.3 [997] *A vertex v is in V^- if and only if $pn[v, S] = \{v\}$ for some γ-set S containing v.*

Proof. Let $v \in V^-$ and D be a γ-set of $G-v$. Then $S = D \cup \{v\}$ is a γ-set of G. If D contains a vertex of $N(v)$, then D is a dominating set of G, contradicting the assumption that $v \in V^-$. Thus, $pn[v, S] = \{v\}$. Conversely, suppose that $pn[v, S] = \{v\}$ for some $\gamma(G)$-set S containing v. Obviously, $S - \{v\}$ dominates $G - v$, so $v \in V^-$. $\square$

Carrington et al. determined properties of V^+ and V^- and showed that for any graph G in CVR, $\gamma(G - v) < \gamma(G)$ for all $v \in V$, that is, $V = V^-$ and $V^+ = \emptyset$.

Theorem 5.4 [189] *For any graph G,*

(a) *if $v \in V^+$, then for every γ-set S of G, $v \in S$ and $pn[v, S]$ contains at least two nonadjacent vertices,*

(b) *if $x \in V^+$ and $y \in V^-$, then x and y are not adjacent,*

(c) *$|V^0| \geq 2|V^+|$,*

(d) *$\gamma(G) \neq \gamma(G - v)$ for all $v \in V$ if and only if $V = V^-$, and*

(e) *if $v \in V^-$ and v is not an isolate in G, then there exists a γ-set S of G such that $v \notin S$.*

Proof. (a) From Theorem 5.2, we know that each $v \in V^+$ is not an isolated vertex and is in every $\gamma(G)$-set S. If $pn[v, S] = \{v\}$, then for any $u \in N(v)$, $S - \{v\} \cup \{u\}$ is a $\gamma(G)$-set, contradicting the fact that v is in every $\gamma(G)$-set. Hence, v has a private neighbor in $V - S$. Suppose that the private neighbors of v induce a complete subgraph, contrary to the existence of two nonadjacent private neighbors of v. Then, again, $S - \{v\} \cup \{u\}$, for any $u \in pn[v, S]$, is a $\gamma(G)$-set, a contradiction.

(b) Suppose that $x \in V^+$, $y \in V^-$, and $xy \in E(G)$. Further, let S_y be a dominating set of $G - y$ with cardinality $\gamma(G) - 1$. If S_y contains x, then S_y dominates G, contradicting that $\gamma(G)$ vertices are necessary to dominate G. On the other hand, if S_y does not contain x, then $S_y \cup \{y\}$ is a $\gamma(G)$-set not containing x, violating (a).

(c) For each $v \in V^+$, (a) establishes that for every γ-set S, $v \in S$ and $pn[v, S]$ contains at least two nonadjacent vertices. That is, these private neighbors of v are in $V - S$ and hence, not in V^+. Further, since from (b), we know that v has no neighbor in V^-, these private neighbors must be in V^0.

(d) Obviously, if $V = V^-$, $\gamma(G) \neq \gamma(G - v)$ for all $v \in V$. Assume $\gamma(G) \neq \gamma(G - v)$ for all $v \in V$. Then V^+ and V^- partition V. But if $v \in V^+$, then by (c), V^0 is not empty, a contradiction. Hence, $V = V^-$.

(e) This follows directly from Theorems 5.2 and 5.3. $\square$

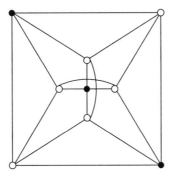

Figure 5.2: Graph $G \in CVR$ with $\gamma(G) = 3$.

Hence, any graph in CVR must have $V = V^-$, that is, $\gamma(G - v) = \gamma(G) - 1$ for all $v \in V$. The graph in Figure 5.2 is an example of a graph in CVR, where shaded vertices form a γ-set.

Corollary 5.5 *A graph $G \in CVR$ if and only if for each vertex $v \in V$, $pn[v, S] = \{v\}$ for some γ-set S containing v.*

The graphs in CVR are precisely the "vertex-critical graphs" studied by Brigham, Chinn, and Dutton [159]. They determined a sufficient condition to imply that G is not vertex-critical.

Theorem 5.6 [159] *If a graph G has a nonisolated vertex v such that the subgraph induced by $N(v)$ is complete, then $G \notin CVR$.*

Brigham et al. [159] showed that a forbidden subgraph characterization of graphs in CVR is not possible. However, they characterized those graphs in CVR having the minimum order of vertex-critical graphs $n = \gamma(G) + \Delta(G)$. They concluded their study by posing the following questions:

(1) If a graph $G \in CVR$, is $n \geq (\delta(G) + 1)(\gamma(G) - 1) + 1$?

(2) If a graph $G \in CVR$ and $n = (\Delta(G) + 1)(\gamma(G) - 1) + 1$, is G regular?

(3) If a graph $G \in CVR$, is $i(G) = \gamma(G)$?

(4) If a graph $G \in CVR$, is $diam(G) \leq 2(\gamma(G) - 1)$?

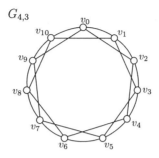

Figure 5.3: Graph $G_{4,3}$.

Fulman, Hanson, and MacGillivray [507] answered all of these questions. They defined the circulant graph G with $V = \{0, 1, ..., 16\}$, in which vertices i and j are adjacent if and only if $|i - j|$ is congruent to one of $1, 3, 5,$ or 7 modulo 17. This graph is 8-regular, vertex-critical, $n = 17 < (8 + 1)(3 - 1) + 1 = 19$, and $i(G) = 5 > \gamma(G) = 3$, which resolves questions (1) and (3). They settled question (2) as follows.

Theorem 5.7 [507] *If a graph $G \in CVR$ has order $n = (\Delta(G)+1)(\gamma(G)-1)+1$, then G is regular.*

Proof. Let $G \in CVR$ with $n = (\Delta(G)+1)(\gamma(G)-1)+1$ and let S_u denote a γ-set of $G - u$. Since $|S_u| = \gamma(G) - 1$, in order that all of the $(\Delta(G)+1)(\gamma(G)-1)$ vertices of $G - u$ are dominated, each element of S_u must dominate exactly $\Delta(G) + 1$ vertices, and therefore have degree $\Delta(G)$. This implies that no two vertices in S_u have a common neighbor.

To prove that G is regular, it therefore suffices to show that an arbitrary vertex x belongs to S_u for some u. Let $v \in S_x$. We will show $x \in S_v$. Since $G \in CVR$, $S_v \cap N[v] = \emptyset$. Each vertex of $S_x - \{v\}$ dominates a unique vertex of S_v. But the remaining vertex in S_v (if it is not x) must also be dominated by S_x, and so must be dominated by v, a contradiction. Hence, $x \in S_v$. $\square$

A family of graphs $G_{k,t}$ with order $n = (\Delta(G)+1)(\gamma(G)-1)+1$ illustrating Theorem 5.7 was described in [159]. For k even and $k, t \geq 2$, let $V(G_{k,t}) = \{v_0, v_1, ..., v_{(t-1)(k+1)}\}$ and $E(G_{k,t}) = \{v_i v_j : 1 \leq (i - j) \pmod{((t-1)(k+1) + 1))} \leq k/2\}$. Then $G_{k,t}$ is k-regular with $\gamma(G_{k,t}) = t$. Figure 5.3 shows $G_{4,3}$. The next theorem confirms question (4).

Theorem 5.8 [507] *If a graph $G \in CVR$ and $\gamma(G) \geq 2$, then $diam(G) \leq 2(\gamma(G) - 1)$.*

Fulman et al. [507] gave a general construction for graphs that achieve the bound of Theorem 5.8. We note a special case of this construction. Let G be the graph formed by replacing each edge uv of a path on k vertices by a four cycle u, u', v, v' such that u' and v' are new vertices. Then $G \in CVR$, $\gamma(G) = k$, and $diam(G) = 2(k-1)$.

Bauer et al. [87] explored a related problem of determining the minimum number of vertices whose removal changes $\gamma(G)$. Let $\mu^+(G)$ denote the minimum number of vertices whose removal increases γ and $\mu^-(G)$ the corresponding number whose removal decreases the domination number. Note that $\mu^+(G)$ may not exist, for example, $G = K_n$. Note also that $\mu^+(G) = 1$ if and only if $V^+ \neq \emptyset$, and therefore Theorem 5.2 characterizes graphs for which $\mu^+(G) = 1$.

Theorem 5.9 [87]

(a) *For any tree T, $\mu^+(T) = 2$ if and only if there are vertices u and v such that*

    (1) *every γ-set contains either u or v.*

    (2) *v is in every γ-set for $T - u$ and u is in every γ-set for $T - v$.*

    (3) *no vertex is in every γ-set.*

(b) *For any graph G, $\mu^-(G) \leq \gamma(G) + 1$.*

(c) *For any graph G, $\min\{\mu^+(G), \mu^-(G)\} \leq \delta(G) + 1$.*

(d) *If G has an endvertex, then $\mu^+(G) \geq 2$ implies $\mu^-(G) \leq 2$.*

(e) *For $n \geq 7$, $\mu^+(P_n) + \mu^-(P_n) = 4$.*

(f) *For $n \geq 8$, $\mu^+(C_n) + \mu^-(C_n) = 6$.*

5.2.2 Edge removal (CER)

Obviously, the removal of an edge from G cannot decrease γ and can increase it by at most one. Thus a graph for which the domination number changes when an arbitrary edge is removed has the property that $\gamma(G - e) = \gamma(G) + 1$, for all $e \in E$. The graphs in CER were called "γ^+-critical graphs" and independently characterized by Bauer et al. [87] and Walikar and Acharya [1133]. Recall that a galaxy is a forest in which each component is a star.

Theorem 5.10 [87, 1133] *A graph $G \in CER$ if and only if G is a galaxy.*

Proof. The sufficiency is clear. Let $G \in CER$ and S be a $\gamma(G)$-set. Note that every vertex of degree at least two must be in S. Also, note that there can be no edge between two vertices in $V - S$ or between two vertices in S, since S will still dominate $V - S$ if such an edge is removed. Therefore there can be no path in G connecting two vertices in S. Hence, G is a union of stars. $\square$

We now consider a related problem. Let the *degree of an edge uv* be $deg(u) + deg(v)$ and let $\delta'(G)$ be the smallest degree of any edge. Bauer et al. [87] defined the *bondage number $b(G)$*, which they called the "edge stability number", to be the minimum number of edges whose removal increases the domination number. In our network example, this problem corresponds to determining the smallest number of links that must fail to cause the need for at least one additional file server.

Theorem 5.11 [87] *For any graph G,*

(a) *if there is at least one vertex $v \in V^0 \cup V^+$, then $b(G) \le \Delta(G)$.*

(b) *if T is a nontrivial tree, then $b(T) \le 2$.*

(c) $b(G) \le \delta'(G) - 1$.

Theorem 5.11(b) shows that any tree has bondage number one or two. Hartnell and Rall [606] characterized the trees having bondage number two. Observe that the contrapositive of Theorem 5.11(a) establishes that a graph G having $b(G) > \Delta(G)$ is a sufficient condition for G to be in CVR. Fink, Jacobson, Kinch, and Roberts [476] studied the same concept and were the first to use the term "bondage number". They found some of the results that appeared in [87]. In addition, they determined $b(G)$ for several families of graphs including complete graphs, cycles, and paths.

Theorem 5.12 [476] *For the graphs K_n, C_n, and P_n,*

$$b(K_n) = \lceil n/2 \rceil.$$

$$b(C_n) = \begin{cases} 3 & \text{if } n \equiv 1 \ (mod \ 3) \\ 2 & \text{otherwise.} \end{cases}$$

$$b(P_n) = \begin{cases} 2 & \text{if } n \equiv 1 \ (mod \ 3) \\ 1 & \text{otherwise.} \end{cases}$$

They also established bounds on $b(G)$.

Theorem 5.13 [476] *For any graph G,*

(a) *if G is connected, then $b(G) \le n - 1$.*

(b) *if G is connected, then $b(G) \le \Delta(G) + \delta(G) - 1$.*

(c) *if $\gamma(G) \geq 2$, then $b(G) \leq (\gamma(G) - 1)\Delta(G) + 1$.*

(d) *if G is connected and $n \geq 2$, then $b(G) \leq n - \gamma(G) + 1$.*

The bounds in Theorem 5.13 are sharp as can be seen with the complete t-partite graph $K_{2,2,\ldots,2}$ achieving the bounds in (a), (c), (d) and cycles C_n for $n \equiv 1 \pmod 3$ achieving the bound in (b). Hartnell and Rall also derived several sharp upper bounds on $b(G)$ including the following. Let $\kappa_1(G)$ denote the *edge connectivity* of G, that is, the minimum number of edges whose removal disconnects G.

Theorem 5.14 [607] *If a graph G has $\kappa_1(G) = k$, then $b(G) \leq \Delta(G) + k - 1$.*

Fink et al. [476] concluded their study by conjecturing that $b(G) \leq \Delta(G) + 1$ for any nonempty graph G. Note that if $\kappa_1(G) = 2$, then Theorem 5.14 gives the bound of the conjecture. Furthermore, an upper bound of $\Delta(G)$ on the "fractional bondage number" (the linear programming relaxation of an integer linear program for finding the bondage number) due to Chvátal and Cook [238] seemed to support the conjecture. However, counterexamples to this conjecture were developed independently by Hartnell and Rall [607] and Teschner [1083]. In particular, Hartnell and Rall showed that the cartesian product $G = K_n \times K_n$ has $b(G) = 3(n - 1) = \frac{3}{2}\Delta(G)$. Teschner made use of the fact that any graph having $b(G) > \Delta(G)$ is in CVR to find his counterexample. For a more detailed discussion of these concepts, the reader is referred to Chapter 17 of *Domination in Graphs: Advanced Topics* [401, 645].

5.2.3 Edge addition (CEA)

Just as deleting an edge can increase the domination number by at most one, adding an edge can decrease it by at most one. Hence, a graph for which the domination number changes when an arbitrary edge is added, class CEA, has the property that $\gamma(G + e) = \gamma(G) - 1$ for all $e \in E(\overline{G})$. Since this concept is covered in detail in Chapter 16 of [645], we only give a brief overview here. The difficult problem of characterizing the graphs in CEA was investigated by Sumner and Blitch [1067], who called them "edge domination critical graphs". They were able to characterize these graphs only in the special cases for which $\gamma(G) = 1$ or 2.

Theorem 5.15 [1067]

(a) *A graph G with $\gamma(G) = 1$ is in CEA if and only if G is K_n.*

(b) *A graph G with $\gamma(G) = 2$ is in CEA if and only if $\overline{G}$ is a galaxy.*

Note that the graphs in CEA with $\gamma(G) = 2$ are precisely the complements of the graphs in CER characterized by Bauer et al. [87]. The problem is more difficult for graphs with $\gamma(G) \geq 3$. Sumner and Blitch [1067] gave six graphs in CEA having order $n \leq 8$ and $\gamma(G) = 3$ (see Figure 5.4(a)). Dunbar (personal communication) provided three additional examples of such graphs: the corona $K_3 \circ K_1$ and the two graphs in Figure 5.4(b).

Sumner characterized the disconnected graphs in CEA having $\gamma(G) = 3$.

Theorem 5.16 [1066] *A disconnected graph G with $\gamma(G) = 3$ is in CEA if and only if $G = A \cup B$ where either A is trivial and B is in CEA and has $\gamma(G) = 2$ or A is a complete graph and B is a complete graph minus a 1-factor.*

Although the graphs in CEA with $\gamma(G) \geq 3$ have not been characterized, many interesting properties of these graphs have been found. For example, Favaron, Sumner, and Wojcicka [461] showed that the diameter of a graph $G \in CEA$ with $\gamma(G) = k$ is at most $2k - 2$. They conjectured that the best possible bound is actually $\lfloor 3k/2 - 1 \rfloor$. This conjecture is unsolved at the present time.

Sumner and Blitch [1067] also conjectured that all graphs in CEA have equal domination and independent domination numbers. This conjecture is true for graphs in CEA which have $\gamma \leq 2$. However, Ao, Cockayne, MacGillivray, and Mynhardt [38] constructed a counterexample for graphs G in CEA with $\gamma(G) \geq 4$. The conjecture is still unsettled for graphs $G \in CEA$ with $\gamma(G) = 3$, and many people who have studied it believe it is true for this case. Wojcicka [1149] proved another conjecture by Sumner and Blitch [1067] that every connected graph in CEA with $\gamma(G) = 3$ and $n \geq 6$ has a hamiltonian path.

Relating edge addition to vertex removal, Sumner and Blitch showed that V^+ is empty for graphs in CEA.

Theorem 5.17 [1067] *If a graph $G \in CEA$, then $V = V^- \cup V^0$.*

Favaron et al. gave another property of the vertex set of a graph in CEA.

Theorem 5.18 [461] *If a graph $G \in CEA$, then the subgraph induced by V^0 is complete.*

Proof. This is obvious when $|V^0| = 1$. Assume G is in CEA and $u, v \in V^0$ are nonadjacent vertices. Then $\gamma(G + uv) < \gamma(G)$. Hence, there exists an γ-set S for $G + uv$ with cardinality $\gamma(G) - 1$ that contains exactly one of u and v, say v. Then $S \cup \{u\}$ is a γ-set for G and u is only necessary to dominate itself implying that $u \in V^-$, a contradiction. Therefore, every pair of vertices in V^0 must be adjacent. $\square$

We note that a graph in CEA may have $V = V^-$, that is, V^0 is empty. Consider C_4, for example. However, if $G \in CEA$ is not a complete graph (vacuously, $K_n \in CEA$), then V^- is not empty; in fact, the following unpublished result due to T. W. Haynes shows that $|V^-|$ is bounded below by $\gamma(G)$.

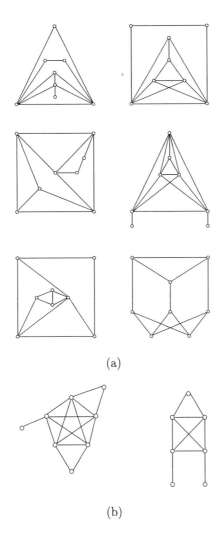

(a)

(b)

Figure 5.4: Graphs in CEA with order $n \leq 8$.

Theorem 5.19 *If a connected graph $G \in CEA$ is not complete, then $|V^-| \geq \gamma(G)$.*

Proof. Assume $G \in CEA$ is connected and not complete. By Theorem 5.17, $V = V^0 \cup V^-$. Obviously, if $V^0 = \emptyset$, the theorem holds. So suppose $|V^0| \geq 1$ and $|V^-| = k < \gamma(G)$. Since G is connected and not complete, Theorem 5.18 implies that $k \geq 1$ and at least one vertex $v \in V^-$ has a neighbor u in V^0. It also follows from Theorem 5.18 that u dominates every vertex in V^0. Hence, $\{u\} \cup (V^- - \{v\})$ is a dominating set of G with cardinality $k < \gamma(G)$, a contradiction. Therefore, $|V^-| \geq \gamma(G)$. $\square$

We now consider a related concept in which the number of edges in the graph may be increased by more than one. Kok and Mynhardt [782] defined the *reinforcement number* $r(G)$ to be the smallest number of edges which must be added to G to decrease the domination number. If $\gamma(G) = 1$, then define $r(G) = 0$. An obvious application of this theory is evident in our communications network example. Determining $r(G)$ for the network's underlying graph reveals the minimum number of communication links which must be added to the network in order to decrease the number of file servers required to service the system. Kok and Mynhardt [782] found $r(G)$ for several families of graphs and determined bounds on $r(G)$. They also used $r(G)$ to improve the bound $\gamma(G) \leq n - \Delta(G)$ (presented in Chapter 2).

Theorem 5.20 [782] *For any graph G, $\gamma(G) \leq n - \Delta(G) - r(G) + 1$.*

For a graph G with $\gamma(G) \geq 2$, let $\mu = n - \max\{|N[S]|\}$ for all $S \subset V$ having $|S| = \gamma(G) - 1$. The following result gives $r(G)$ for graphs having $\gamma(G) \geq 2$.

Theorem 5.21 [782] *If G is a graph with $\gamma(G) \geq 2$, then $r(G) = \mu(G)$.*

Proof. Let $X \subset V$ with $|X| = \gamma(G) - 1$ satisfy $|V - N[X]| = \mu(G)$. Then $\mu(G)$ vertices in G are not dominated by $|X|$. To construct a graph G' which is dominated by X, join each of these vertices to a vertex in $X \neq \emptyset$. Thus, $\gamma(G') \leq |X| < \gamma(G)$, which implies that $r(G) \leq \mu(G)$.

Now let F be a set of edges such that $|F| = r(G)$ and $\gamma(G + F) = \gamma(G) - 1$. Consider S, a γ-set for $G + F$. For every $uv \in F$, we may assume that $u \in S$ and $S \cap N_G[v] = \emptyset$, since if $\{u, v\} \cap S = \emptyset$, or if $u \in S$ and $S \cap N_G[v] \neq \emptyset$, then S dominates $G + (F - uv)$, contradicting $|F| = r(G)$. Hence, at most $r(G)$ vertices of G are not dominated by S implying that $\mu(G) \leq r(G)$. $\square$

Corollary 5.22 *If a graph G has $\gamma(G) = 2$, then $r(G) = n - \Delta(G) - 1$.*

A technique for finding $r(G)$ when $\gamma(G)$ is known is also described in [782]. For a detailed treatment of this concept, see Chapter 17 in [645].

5.3 Unchanging

We now shift our attention to the three classes of unchanging graphs UVR, UER, and UEA.

5.3.1 Vertex removal (UVR)

Clearly, if the domination number is unchanged when an arbitrary vertex is removed, class UVR, then $V = V^0$. The graphs in UVR were characterized by Carrington et al.

Theorem 5.23 [189] *A graph $G \in UVR$ if and only if G has no isolated vertices and for each vertex v, either*

(a) *there is an γ-set S' such that $v \notin S'$ and for each γ-set S such that $v \in S$, $pn[v, S]$ contains at least one vertex from $V - S$, or*

(b) *v is in every γ-set and there is a subset of $\gamma(G)$ vertices in $G - N[v]$ that dominates $G - v$.*

Proof. First consider a graph G such that $\gamma(G - v) = \gamma(G)$ for every $v \in V$. Then, obviously, there are no isolated vertices. Assume there exist γ-sets S and S' such that $v \in S$ and $v \notin S'$. Suppose $pn[v, S] = \{v\}$. Then $S - \{v\}$ dominates $G - v$, contrary to the fact that $v \in V^0$. Thus (a) holds.

Assume v is in every γ-set of G. Since $v \notin V^+$, Theorem 5.2 implies there must be a set of $\gamma(G)$ vertices of $V - N[v]$ which dominates $G - v$, proving (b).

Conversely, assume G has no isolated vertices and either (a) or (b) holds. If (a) is true for vertex v, then $\gamma(G - v) \leq \gamma(G)$ since some S' does not contain v and hence, dominates $G - v$. If $\gamma(G - v) < \gamma(G)$, then there is a γ-set S of G for which $pn[v, S] = \{v\}$, a contradiction. Thus, $\gamma(G - v) = \gamma(G)$ in this case. If (b) holds for a vertex v, then $v \notin V^-$ by Theorem 5.4(e). Thus, $\gamma(G - v) \geq \gamma(G)$. But $v \in V^+$ is excluded by Theorem 5.2, so again $\gamma(G - v) = \gamma(G)$. $\square$

For example, complete bipartite graphs $K_{r,s}$, $3 \leq r \leq s$, are in UVR.

5.3.2 Edge removal (UER)

Walikar and Acharya observed the following about graphs in UER.

Theorem 5.24 [1133] *A graph $G \in UER$ if and only if for each $e = uv \in E$, there exists a γ-set S such that one of the following conditions is satisfied:*

(a) $u, v \in S$.

(b) $u, v \in V - S$.

(c) $u \in S$ and $v \in V - S$ implies $|N(v) \cap S| \geq 2$.

For an example of graphs in UER, consider the graph G formed from $K_s \cup K_t$, $2 \leq s \leq t$, by adding an edge uv between $u \in V(K_s)$ and $v \in V(K_t)$. Hartnell and Rall [606] gave a constructive characterization for all trees in UER.

For a generic property $\mathcal{P}$, graphs with the minimum number of edges and having property $\mathcal{P}$ are called *extremal graphs*. Dutton and Brigham [409] investigated extremal graphs in UER, which they called "γ-insensitive graphs", and determined the minimum number of edges, denoted $E(n, \gamma)$, in these graphs.

Theorem 5.25 [409] *If a graph $G \in UER$ has order $n \geq 3\gamma(G)$ and $\gamma(G) \geq 2$, then*

$$E(n, \gamma) = 2n - 3\gamma(G).$$

Haynes, Brigham, and Dutton [642, 643] extended the notion of UER to (γ, k)-insensitive graphs by considering the removal of $k \geq 1$ arbitrary edges. Again we consider the example of a communications network. A network corresponding to an extremal (γ, k)-insensitive graph has minimum link cost and the fault-tolerant property that the number of file servers required to communicate directly with the other processors in the graph remains unchanged even after k links fail. A special $(\gamma, 2)$-insensitive graph, called the "G-network" was introduced in [644] as a suitable design for communications and interconnection networks.

The minimum number of edges in any (γ, k)-insensitive graph of order n is denoted $E_k(n, \gamma)$. Let F be a set of $k \geq 1$ arbitrary edges. Extremal graphs were found for the case when $\gamma(G) = 1$.

Theorem 5.26 [643, 975] *If a graph G has $\gamma(G) = 1$ and $n > 2k$ such that $\gamma(G) = \gamma(G - F)$ for any subset $F \subseteq E$, where $|F| = k$, then*

$$E_k(n, 1) = (2k + 1)(n - k - 1).$$

Proof. Let G be a $(1, k)$-insensitive graph. Suppose G has at most $2k$ vertices of degree $n - 1$. Then k edges can be selected such that each vertex of degree $n - 1$ is incident to at least one of them. No single vertex dominates the graph after these k edges are removed, contradicting that G is $(1, k)$-insensitive. Thus, G has at least $2k + 1$ vertices with degree $n - 1$, so $E_k(n, \gamma) \geq (2k+1)(n-k-1)$. On the other hand, a connected graph with exactly $2k + 1$ vertices of degree $n - 1$ and no other edges is $(1, k)$-insensitive, so equality holds. □

The value of $E_k(n, \gamma)$ and extremal graphs were determined in [642] for the case when $\gamma(G) = k = 2$.

Theorem 5.27 [642] *If G is a $(2, 2)$-insensitive graph with $n \geq 11$, then*

$$E_2(n, 2) = \lfloor (5n - 10)/2 \rfloor.$$

n even: n odd:

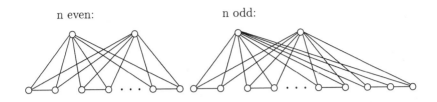

Figure 5.5: Families of extremal $(2, 2)$-insensitive graphs.

Families of extremal $(2, 2)$-insensitive graphs are shown in Figure 5.5.
Although extremal (γ, k)-insensitive graphs and the exact value of $E_k(n, \gamma)$
have not been found for the cases $\gamma(G) \geq 3, k = 2$ and $\gamma(G) \geq 2, k \geq 3$, upper
and lower bounds on $E_k(n, \gamma)$ were determined [643]. An asymptotic bound on
$E_k(n, \gamma)$ for $k \geq 2$ follows.

Theorem 5.28 [643] *For $k + 1 \leq \gamma(G) \leq 2k$, $E_k(n, \gamma)$ is asymptotically equal
to $(k + 3)n/2$ as n approaches infinity.*

More on this subject can be found in Chapter 17 of [645].

5.3.3   Edge addition (UEA)

The graphs for which the domination number is unchanged when an arbitrary
edge is added, class UEA, were characterized in terms of their vertex sets by
Carrington et al.

Theorem 5.29 [189] *A graph $G \in UEA$ if and only if V^- is empty.*

Proof. Let $G \in UEA$ and suppose that G has a vertex $x \in V^-$. Thus,
$\gamma(G - x) < \gamma(G)$. Let S be a γ-set of $G - x$. Then adding edge xy for any $y \in S$
gives $\gamma(G + xy) < \gamma(G)$ contrary to the fact that $G \in UEA$.
   To prove the converse, suppose that G has no vertices in V^- and $\gamma(G + uv) =
\gamma(G) - 1$ for some pair of nonadjacent vertices u and v. Then any minimum
dominating set S of $G + uv$ must contain exactly one of u or v, say u. Hence, S
dominates $G - v$. Thus, $v \in V^-$, a contradiction. $\square$
   For example, cycles C_n, $n \equiv 0 \pmod 3$, are in UEA.

5.4   Relationships Among Classes

There are many interesting relationships among the six classes of changing and
unchanging graphs. For example, the characterization of the graphs in UEA

relates them to the graphs in CVR. The results given in this section are due to
T. W. Haynes (unpublished). We first ascertain the vertex set for each class of
graphs. The observation follows from previous results in this chapter.

Observation 5.30

(a) A graph $G \in UVR$ if and only if $V = V^0$.

(b) If a graph $G \in UER$, then $V = V^0 \cup V^- \cup V^+$.

(c) A graph $G \in UEA$ if and only if $V = V^0 \cup V^+$ (either V^0 or V^+ may be
    empty).

(d) A graph $G \in CVR$ if and only if $V = V^-$.

(e) If a graph $G \in CER$, then $V = V^0 \cup V^- \cup V^+$. (Note that for $G \in CER$,
    $V^- = \{ v : v$ is an isolated vertex$\}$. That is, if a graph $G \in CER$ has no
    isolated vertices, then $V^- = \emptyset$.)

(f) If a graph $G \in CEA$, then $V = V^0 \cup V^-$. (It follows from (c) that $V^- \neq \emptyset$.)

Note that Theorem 5.4 implies that a graph with $V = V^0 \cup V^+$ must
have $|V^0| \geq 1$. We now consider the relationships among the classes. In order
to establish a Venn diagram representing the six classes, we do not consider
the cases that are vacuously true. For example, $K_n \in UEA$ and $K_n \in CEA$
(vacuously), so we do not allow this case.

Observation 5.31

(a) If a graph $G \in UVR$, then $G \in UEA$.

(b) A graph $G \in CER \cap UVR$ if and only if G is mK_2, $m \geq 2$.

(c) A graph $G \in (CER \cap UEA) - UVR$ if and only if G is a galaxy with no
    isolated vertices and at least one star with two or more endvertices.

(d) A graph $G \in CER - (UEA \cup CEA)$ if and only if G is a galaxy with at
    least one isolated vertex and at least two edges.

(e) A graph $G \in CER \cap CEA$ if and only if G has $n \geq 3$ vertices and exactly
    one edge.

Proof. A graph $G \in UEA$ if and only if V^- is empty. Thus, (a) follows directly
from the vertex sets for these graphs. For (b), let G be in $CER \cap UVR$. Then
$V = V^0$. Theorem 5.10 implies that the only graph in CER with $V = V^0$ is
mK_2. It is easy to see that mK_2 is in $CER \cup UVR$. Parts (c), (d), and (e)
follow from Theorem 5.10 and Observation 5.30. $\square$

The next result relates the classes CVR and UER.

Observation 5.32 *If a graph $G \in CVR$, then $G \in UER$.*

Proof. Let $G \in CVR$. Then for each $v \in V$, there exists an γ-set S such that v is in S only to dominate itself. That is, $\gamma(G) - 1$ vertices dominate $G - v$ for all $v \in V$. Removing an edge incident to any vertex v cannot increase the domination number because a set of $\gamma(G) - 1$ vertices dominates $G - v$. $\square$

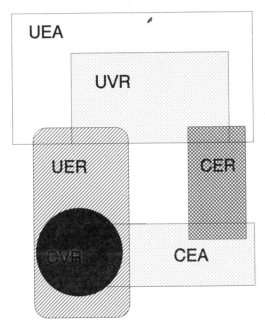

Figure 5.6: Classes of changing and unchanging graphs.

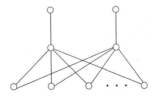

Figure 5.7: Graph $G \in (UVR \cap UEA) - (UER \cup CER)$.

The Venn diagram in Figure 5.6 illustrates these relationships among the classes. Not all graphs are in one of the six classes as can be seen by the graph in Figure 5.1 which has nonempty sets V^0, V^-, and V^+ partitioning its vertices and is not in $CER \cup UER$. Our next observation shows that no subset in the Venn diagram of Figure 5.6 is empty.

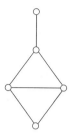

Figure 5.8: Graph $G \in UER - (CVR \cup CEA \cup UEA)$.

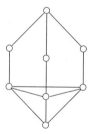

Figure 5.9: Graph $G \in (UER \cap UEA) - UVR$. Note that this graph has $V = V^0 \cup V^+$.

Observation 5.33

- The graph $mK_2 \in UEA \cap CER \cap UVR$.

- A galaxy that has at least one edge and at least one isolated vertex is in $CER - UEA$. A galaxy with at least one isolated vertex and at least two edges is in $CER - (UEA \cup CEA)$. Thus, the galaxy with $n \geq 3$ and exactly one edge is in $CER \cap CEA$.

- A galaxy with no isolated vertex and at least one star having two or more endvertices is in $(UEA \cap CER) - UVR$.

- The cycle $C_4 \in CVR \cap UER \cap CEA$.

- The graph $C_4 \cup K_2 \in CEA - (CER \cup UER)$.

- The cycles C_n for $n \equiv 0, 2 \pmod{3}$ are in $UVR \cap UER \cap UEA$.

- See Figure 5.7 for an example of $G \in (UVR \cap UEA) - (UER \cup CER)$.

- The cycles C_n for $n \equiv 1 \pmod{3}$, $n \geq 7$, are in $(CVR \cap UER) - CEA$.

- The corona $K_p \circ K_1$, $p \geq 3$, is in $(UER \cap CEA) - CVR$.

- The graph G obtained from K_p, $p \geq 2$, by adding $t \geq 1$ endvertices adjacent to exactly one of its vertices is in UEA and not in any other class, that is, $G \in UEA - (CER \cup UVR \cup UER)$.

- The graph G in Figure 5.8 is in UER and not in any other class, that is, $G \in UER - (CVR \cup CEA \cup UEA)$.

- The graph G in Figure 5.9 is in $(UER \cap UEA) - UVR$.

EXERCISES

5.1 Determine an infinite family of graphs $G = (V, E)$ with $V = V^0 \cup V^+ \cup V^-$ for nonempty sets V^0, V^+, and V^-.

5.2 Prove Theorem 5.2.

5.3 Holding this book in one hand and *Domination in Graphs : Advanced Topics* in the other, bench press 3 sets/ 8 reps. (Note that using other books is considered improper form for this exercise.)

5.4 Prove Theorem 5.6.

5.5 The family of graphs $G_{k,t}$ illustrate Theorem 5.7. Find another infinite family of graphs $G \in CVR$ with $n = (\Delta(G) + 1)(\gamma(G) - 1) + 1$.

5.6 Prove Theorem 5.13.

5.7 Determine the bondage number of the complete bipartite graph $K_{r,s}$.

5.8 Find an infinite family of graphs $G \in CVR$ other than those given in this chapter.

5.9 Prove Theorem 5.24.

5.10 What can you say about properties of (γ, k)-insensitive graphs G? For example, what can you determine about $diam(G)$?

5.11 Show that no tree T with $\gamma(T) \geq 3$ is $(\gamma, 2)$-insensitive.

5.12 Show that a (γ, k)-insensitive graph has at most γ endvertices.

5.13 (a) Show that $E_k(n, \gamma) = 2\gamma$ for $k = 2$ and $n = 2\gamma$.

(b) Determine $E_k(n, \gamma)$ for $k = 2$ and $n = 2\gamma + 1$.

5.14 Determine the reinforcement numbers of the following classes of graphs:

(a) paths P_n and cycles C_n.

(b) the complete bipartite graph $K_{r,s}$ for $2 \leq r \leq s$.

(c) the corona $G \circ K_1$.

(d) the cartesian product $G \times H$.

5.15 (a) The graph in Figure 5.7 is in $(UVR \cap UEA) - (UER \cup CER)$. Find other graphs in this class.

(b) The graph in Figure 5.8 is in $UER - (CVR \cup CEA \cup UEA)$. Find other graphs in this class.

5.16 Find graphs $G \in CEA$ with $|V^-| = \gamma(G)$.

Chapter 6

Conditions on the Dominating Set

Many domination parameters are formed by combining domination with another graph theoretical property P. In this chapter we consider the parameters defined by imposing an additional constraint on the dominating set. In Chapter 7 we will see that a condition may also be placed on the dominated set or on the method of dominating.

6.1 Introduction

In an attempt to formalize this concept, Harary and Haynes [580] defined the *conditional domination number* $\gamma(G : P)$ as the smallest cardinality of a dominating set $S \subseteq V$ such that the subgraph $\langle S \rangle$ induced by S satisfies property P. We cover just six of the many possible properties imposed on S, but mention several others.

$P1.$ $\langle S \rangle$ has no edges.

$P2.$ $\langle S \rangle$ has no isolated vertices.

$P3.$ $\langle S \rangle$ is connected.

$P4.$ $\langle S \rangle$ is a complete subgraph.

$P5.$ $\langle S \rangle$ has a perfect matching.

$P6.$ $\langle S \rangle$ has a hamiltonian cycle.

Obviously, $\gamma(G) \leq \gamma(G : P)$ for any property P. We will see that with the exception of independent domination, these conditional domination parameters do not exist for all graphs. For example, in order for a dominating set S to have property $P2$, the graph G cannot have isolated vertices and for S to have property $P3$, G must be connected.

155

The generic nature of conditional domination provides a method for defining many new parameters by considering different properties P. This open area for research seems to be rich in applications. For instance, consider a network design in which it is desirable that a smallest collection S of processors manage the system's resources and that at least one processor in S be directly accessible by each processor not in S. If the network also has specialized communication requirements for the processors in S, then conditional domination serves as a mathematical model. For example, if it is important that each pair of processors in S can communicate privately, keeping messages among the vertices in S, then in the corresponding graph G, we require that $\langle S \rangle$ is connected and $\gamma(G, P3)$ is the minimum cardinality of such a set.

6.2   Independent Dominating Sets

The cardinality of a smallest dominating set S such that $\langle S \rangle$ has no edges (property $P1$) is the *independent domination number* $\gamma(G : P1) = i(G)$. Independent dominating sets have been studied extensively and covered in detail in Chapter 3 of this book; hence, we give only a brief overview here.

We saw in Chapter 3 that $\gamma(G) \leq i(G) \leq \beta_0(G)$. These bounds are sharp as can be seen with the corona $K_{1,t} \circ K_1$, which has $\gamma = i = \beta_0 = t + 1$. On the other hand, the difference between each pair of these parameters can be made arbitrarily large. For example, the double star $S_{s,t}$ is the graph obtained by connecting the centers of two stars $K_{1,s}$ and $K_{1,t}$ with an edge. For $3 \leq s \leq t$, the double star has $\gamma = 2 < i = s + 1 < \beta_0 = s + t$.

Much work has been done in attempts to characterize the graphs G for which $\gamma(G) = i(G)$. A forbidden subgraph characterization cannot be obtained since for any graph H, the join $G = K_1 + H$ has $i(G) = \gamma(G) = 1$.

Although the characterization is still an open problem, several results give sufficient conditions for a graph G to have $\gamma(G) = i(G)$. As we saw in Chapter 3, the following such result is due to Allan and Laskar.

Theorem 6.1 [26] *If a graph G is claw-free, then $\gamma(G) = i(G)$.*

Since any line graph is claw-free, we have the following.

Corollary 6.2 *For any line graph $L(G)$ of a graph G, $\gamma(L(G)) = i(L(G))$.*

Topp and Volkmann [1097] found 16 graphs having the property that if any one of them is not an induced subgraph of a graph G, then $\gamma(G) = i(G)$. In fact, Theorem 6.1 is a corollary of their result. Another interesting corollary of this result concerns the *subdivision graph* $S(G)$ of G (that is, the graph obtained from G by subdividing each edge of G).

Corollary 6.3 [1097] *For any graph G, $\gamma(S(G)) = i(S(G))$.*

Since $\gamma(G) \leq i(G)$ for any graph G, a proof of Allan and Laskar's result can be obtained by showing that in any claw-free graph, $i(G) \leq \gamma(G)$. Bollobás and Cockayne generalized this inequality as follows.

Theorem 6.4 [126] *If a graph G does not contain the star $K_{1,k+1}, k \geq 2$, as an induced subgraph, then*

$$i(G) \leq (k-1)\gamma(G) - (k-2).$$

Proof. Let S be a γ-set of G and I a maximal independent set in $\langle S \rangle$. Let X be the set of vertices of $V - S$ not dominated by I and Y be a maximal independent set in $\langle X \rangle$. Each $v \in S - I$ is adjacent to at most $k-1$ vertices of Y, otherwise a $K_{1,k+1}$ is induced with v at the center, k vertices of Y, and a vertex of I. Therefore, $|Y| \leq (k-1)|S - I|$ and

$$
\begin{aligned}
i(G) &\leq |Y| + |I| \\
&\leq (k-1)(\gamma(G) - |I|) + |I| \\
&= (k-1)\gamma(G) - (k-2)|I| \\
&\leq (k-1)\gamma(G) - (k-2).
\end{aligned}
$$

$\square$

Harary and Livingston [586, 587] characterized the trees T for which $\gamma(T) = i(T)$. Topp and Volkmann did likewise for uncyclic graphs in [1096] and for bipartite and block graphs in [1094].

Harary and Livingston [588] applied results from coding theory to obtain information about $\gamma(G)$ and $i(G)$, when G is the hypercube Q_k of dimension k. They noted that $\gamma(Q_k) = i(Q_k)$ for infinitely many values of k and conjectured that $\gamma(Q_k)$ and $i(Q_k)$ differ only when $k = 5$, where $\gamma(Q_5) = 7$ and $i(Q_5) = 8$.

On the other hand, Barefoot, Harary, and Jones [81] constructed an infinite family of 2-connected, cubic graphs for which the difference between the domination and independent domination number may be arbitrarily large. They conjectured that a similar class exists for 1-connected, cubic graphs. Mynhardt [909] proved this conjecture and described infinite families of 1-connected and 3-connected, cubic graphs for which $i(G) - \gamma(G)$ becomes unbounded. Cockayne and Mynhardt [300], Kostochka [786], and Žerovnik and Oplerova [1207] independently found other infinite classes of 3-connected, cubic graphs for which the difference between $i(G)$ and $\gamma(G)$ may be arbitrarily large. Moreover, Seifter [1019] showed that for every triple (r, k, t), $r \geq 5$, $2 \leq k \leq r$, $t > 1$, there exist r-regular, k-connected graphs G having $i(G) - \gamma(G) > t$.

We next consider a closely related concept. A graph G is *domination perfect* if $\gamma(H) = i(H)$ for every induced subgraph H of G. Fulman [505] showed that the absence of any one of eight induced subgraphs in G is sufficient for G to be domination perfect. Three corollaries follow directly from his result.

Corollary 6.5 [505] *If a graph G is claw-free, then G is domination perfect.*

Let T_1 be the tree of order 6 with two adjacent vertices of degree three.

Corollary 6.6 [505] *A chordal graph G is domination perfect if and only if G does not contain T_1 as an induced subgraph.*

The third corollary shows that only a subset of all the subgraphs H of G need to be considered to determine if G is domination perfect.

Corollary 6.7 [505] *A graph G is domination perfect if $\gamma(H) = i(H)$ for every induced subgraph H of G with $\gamma(H) = 2$.*

Several other bounds on $i(G)$ and on the sum of $i(G)$ with other parameters can be found in [126, 260, 448, 449, 637, 884, 901, 1137, 1209]. We mention just a few. Bollobàs and Cockayne established the following sharp upper bound on $i(G)$.

Theorem 6.8 [126] *If a graph G has no isolated vertices, then*

$$i(G) \leq n - \gamma(G) + 1 - \lceil (n - \gamma(G))/\gamma(G) \rceil.$$

Favaron [449] obtained an upper bound on the independent domination number of a tree. Let $l(T)$ be the number of endvertices of tree T. Before stating the theorem, we describe a family $\mathcal{F}$ of trees, which are the extremal graphs of the theorem.

Let T_0 be the path P_4 with its endvertices labelled 0 and the vertices of degree two labelled 1. Construct a tree T_{i+1} from tree T_i by applying one of the the following two operations:

(1) Join a pendant path of length 2 to a vertex labelled 1. Assign a label of 0 to the endvertex of the pendant path and a label of 1 to its vertex of degree two.

(2) For any edge xy with at least one of x and y labelled 1, replace xy with a path $P_5 = x, u, v, w, y$ such that if x and y are labelled 1, then each of u and v is labelled 1 and w is labelled 0; if x is labelled 1 and y is labelled 0, then each of v and w are labelled 1 and u is labelled 0.

A tree T is in $\mathcal{F}$ if it can be obtained from T_0 by a finite sequence of operations (1) and/or (2). For example, the paths P_{3k-2} are in $\mathcal{F}$ for $k \geq 2$.

Theorem 6.9 [449] *For any tree T with $n \geq 2$ vertices,*

$$i(T) \leq (n + l(T))/3$$

and this bound is obtained if and only if $T \in \mathcal{F}$.

The complexity of the independent domination problem is discussed in Chapter 12. The property that a set S is independent is equivalent to the property that $\langle S \rangle$ has no P_2, that is, $\langle S \rangle$ is P_2-free. Forbidden subgraph conditions placed on $\langle S \rangle$ are open avenues for study. For example, Haynes and Henning [648] have extended the property of being P_2-free to being P_3-free. Obviously, this idea can be extended to any subgraph H.

6.3   Total (Open) Dominating Sets

A solution to the famous Five Queens Problem (see Chapter 1) inspired Cockayne, Dawes, and Hedetniemi [256] to introduce total domination. They observed that in the solution shown in Figure 6.1 not only are the squares without queens dominated by queens, but each queen is dominated by another queen.

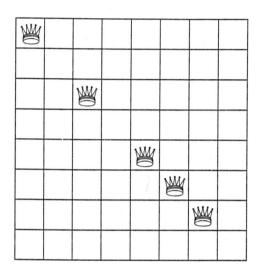

Figure 6.1: Total dominating set solution to the Five Queens Problem.

For total domination, a vertex v dominates just its open neighborhood $N(v)$ and not itself. Thus, S is a *total (open) dominating set* if $V = N(S)$ and the *total domination number* is $\gamma_t(G) = \min\{|S| : S \subseteq V(G) \text{ and } V = N(S)\}$. From this perspective, it seems we are placing a condition on the method of dominating. However, an equivalent definition conforms to the conditional domination discussed in this chapter. A dominating set S is a *total dominating set* if the subgraph induced by S has no isolated vertices, that is, S satisfies property $P2$. Any graph G with no isolated vertices has a total dominating set and hence, a total domination number.

For an application, we again consider a computer network in which a core group of fileservers has the ability to communicate directly with every computer outside the core group. In addition, each fileserver is directly linked to at least one other "backup" fileserver where duplicate information is stored. A smallest core group with this property is a γ_t-set for the graph representing the network.

Cockayne et al. [256] mentioned the example of committee selection criteria. Consider selecting a committee from a group of people such that (i) every nonmember knows at least one member of the committee, to allow direct communication to a representative, and (ii) each member of the committee has an acquaintance on the committee, to avoid feelings of isolation and thus enhance co-operation. A smallest committee with these properties is a γ_t-set of the acquaintance graph of the set of people.

For any set $S \in V$, its *boundary* $B(S) = \{v : |N[v] \cap S| = 1\}$. Similarly, the *open boundary* $OB(S) = \{v : |N(v) \cap S| = 1\}$. Hedetniemi, Jacobs, Laskar, and Pillone characterized a minimal total dominating set by its open boundary as follows.

Theorem 6.10 [668] *A dominating set S is minimal if and only if $B(S)$ dominates S. Similarly, a total dominating set is minimal if and only if $OB(S)$ dominates S.*

For a graph G with no isolated vertices, Ore's theorem gives the upper bound, $\gamma(G) \leq n/2$. However, this bound is too small for the total domination number. For instance, $\gamma_t(mK_2) = n$. We note that these are the only graphs G for which $\gamma_t(G) = n$. It is easy to see that $\gamma(G) \leq \gamma_t(G) \leq 2\gamma(G)$. The cycles C_3 and C_6 each require $2n/3$ vertices in a total dominating set. Cockayne et al. established that $2n/3$ is a best possible upper bound for $\gamma_t(G)$ for a connected graph G.

Theorem 6.11 [256] *If G is a connected graph with $n \geq 3$ vertices, then $\gamma_t(G) \leq 2n/3$.*

Proof. Let S be a γ_t-set of G. By minimality, each $v \in S$ either has a private neighbor or the induced subgraph $\langle S - \{v\}\rangle$ contains an isolated vertex.

Let $P = \{v \in S : pn[v, S] \neq \emptyset\}$. Let B be the set of isolates in $\langle P \rangle$ and $A = P - B$. Further, let C be a minimum set of vertices of $S - P$ such that each vertex of B is adjacent to some vertex of C. We note that $|C| \leq |B|$. Finally, let $D = S - (P \cup C)$. This definition of D implies that $\gamma_t(\langle D \rangle) = |D|$, and hence, $\langle D \rangle = kK_2$, $k \geq 0$.

Let $a_i b_i$, $1 \leq i \leq k$, be the distinct edges of $\langle D \rangle$. The connectivity of G implies, without loss of generality, that each a_i is adjacent to some other vertex x_i. If $x_i \in P \cup C$, then $S - \{b_i\}$ would be a smaller total dominating set than S. Hence, $x_i \in V - S$. If $x_i = x_j$ for $i \neq j$, then $S - \{b_i, b_j\} \cup \{x_i\}$ affords a similar contradiction. By the definition of D, each x_i is adjacent to at least two vertices of S and by the definition of P, there are at least $|P|$ vertices of $V - S$ that are adjacent to exactly one vertex of S. Thus,

$$|P| + k \leq |V - S|.$$

That is,

$$|A| + |B| + k \leq n - \gamma_t(G). \tag{6.1}$$

Therefore,

$$\gamma_t(G) = |A| + |B| + |C| + |D|$$
$$= (|A| + |B| + k) + (|C| + k).$$

Since $|C| \le |A| + |B|$, we have

$$\gamma_t(G) \le 2(|A| + |B| + k).$$

Hence, from (6.1)

$$\gamma_t(G) \le 2(n - \gamma_t(G)).$$

Thus, $\gamma_t(G) \le 2n/3$. $\square$

They also related the total domination number to the maximum degree.

Theorem 6.12 [256]

(a) *If a graph G has no isolated vertices, then $\gamma_t(G) \le n - \Delta(G) + 1$.*

(b) *If a graph G is connected and $\Delta(G) < n - 1$, then $\gamma_t(G) \le n - \Delta(G)$.*

Allan, Laskar, and Hedetniemi determined the following relationship between total domination numbers and irredundance numbers.

Theorem 6.13 [28] *For any graph G, $\gamma_t(G) \le 2ir(G)$.*

Allen et al. also found a bound on the sum of the independent and total domination numbers of a graph. Other such bounds on the sum of two parameters are presented in Chapter 9.

Theorem 6.14 [28] *If each component of a graph G has at least three vertices, then $i(G) + \gamma_t(G) \le n$.*

Complexity issues concerning the total domination problem are discussed in Chapter 12.

6.4 Connected Dominating Sets

Sampathkumar and Walikar [1004] defined a *connected dominating set S* to be a dominating set S whose induced subgraph $\langle S \rangle$ is connected (property $P3$). Since a dominating set must contain at least one vertex from each component of G, it follows that only connected graphs have a connected dominating set. The minimum cardinality of a connected dominating set is the *connected domination number* $\gamma(G : P3) = \gamma_c(G)$. Obviously, $\gamma(G) \le \gamma_c(G)$ and if $\gamma(G) = 1$, then $\gamma(G) = \gamma_c(G) = i(G) = 1$.

Since any nontrivial connected dominating set is also a total dominating set, $\gamma(G) \leq \gamma_t(G) \leq \gamma_c(G)$ for any connected graph G with $\Delta(G) < n - 1$. The sharpness of this inequality can be seen with the complete bipartite graph $K_{r,s}$, which has $\gamma = \gamma_t = \gamma_c = 2$. On the other hand, the inequality is strict, for example, $\gamma(C_{12k}) = 4k < \gamma_t(C_{12k}) = 6k < \gamma_c(C_{12k}) = 12k - 2 = n - 2$.

Let H be a connected spanning subgraph of a connected graph G. Then every connected dominating set of H is also a connected dominating set of G. Sampathkumar and Walikar observed the following.

Theorem 6.15 [1004] *If H is a connected spanning subgraph of G, then $\gamma_c(G) \leq \gamma_c(H)$.*

As we saw in Chapter 1, Nieminen [920] determined the domination number of a graph G in terms of ε_F (the maximum number of pendant edges in any spanning forest of G), namely, for any connected graph

$$\gamma(G) = n - \varepsilon_F(G).$$

Let ε_T be the maximum number of pendant edges in any spanning tree. Hedetniemi and Laskar gave an analogous result for the connected domination number.

Theorem 6.16 [669] *If G is a connected graph and $n \geq 3$, then*

$$\gamma_c(G) = n - \varepsilon_T(G) \leq n - 2.$$

Proof. Let T be a spanning tree of G with $\varepsilon_T(G)$ endvertices, and let L denote the set of endvertices. Then $T - L$ is a connected dominating set having $n - \varepsilon_T$ vertices, that is, $\gamma_c(G) \leq n - \varepsilon_T(G)$.

Conversely, let S be a γ_c-set. Since $\langle S \rangle$ is connected, $\langle S \rangle$ has a spanning tree T_S. A spanning tree T of G is formed by adding the remaining $n - \gamma_c(G)$ vertices of $V - S$ to T_S and adding edges of G, such that each vertex in $V - S$ is adjacent to exactly one vertex in S. Now T has at least $n - \gamma_c(G)$ endvertices. Thus, $\varepsilon_T(G) \geq n - \gamma_c(G)$ or $\gamma_c(G) \geq n - \varepsilon_T(G)$.

Hence, $\gamma_c(G) = n - \varepsilon_T(G)$, and since $\varepsilon_T(G) \geq 2$, the result follows. $\square$

The straightforward result giving the connected domination of a tree is a corollary. Let $l(T)$ denote the number of endvertices of a tree T.

Corollary 6.17 [1004] *If T is a tree with $n > 3$ vertices, then $\gamma_c(T) = n - l(T)$.*

Kleitman and West studied connected graphs that have a spanning tree with many leaves. Their results combined with Theorem 6.16 give several bounds for γ_c.

Theorem 6.18 [778] *Every connected graph G with $\delta(G) \geq k$ has at least one spanning tree with at least $n - 3\lfloor n/(k+1) \rfloor + 2$ leaves.*

Corollary 6.19 *For any connected graph G with $\delta(G) \geq k$,*

$$\gamma_c(G) \leq 3\lfloor n/(k+1)\rfloor - 2.$$

Theorem 6.20 [778] *Every connected graph G with $\delta(G) \geq 3$ has a spanning tree with at least $n/4 + 2$ leaves.*

Corollary 6.21 *For any connected graph G with $\delta(G) \geq 3$,*

$$\gamma_c(G) \leq 3n/4 - 2.$$

Theorem 6.22 [778] *Every connected graph G with $\delta(G) \geq 4$ has a spanning tree with at least $2n/5 + 2$ leaves.*

Corollary 6.23 *For any connected graph G with $\delta(G) \geq 4$,*

$$\gamma_c(G) \leq 3n/5 - 2.$$

Sampathkumar and Walikar determined the following bounds.

Theorem 6.24 [1004] *For any connected graph G,*

$$n/(\Delta(G) + 1) \leq \gamma_c(G) \leq 2m - n$$

with equality for the lower bound if and only if $\Delta(G) = n - 1$ and equality for the upper bound if and only if G is a path.

Proof. Since $n/(\Delta(G)+1) \leq \gamma(G)$ and $\gamma(G) \leq \gamma_c(G)$, the lower bound follows. It is easy to see that this lower bound is attained if and only if G has a vertex of degree $n - 1$.

From Theorem 6.16, $\gamma_c(G) \leq n - 2 = 2(n-1) - n$ and since G is connected, $m \geq n - 1$. Thus, $\gamma_c(G) \leq 2m - n$.

We now show that $\gamma_c(G) = 2m - n$ if and only if G is a path. If G is a path, then from Corollary 6.17, $\gamma_c(G) = n - 2 = 2(n-1) - n = 2m - n$. Conversely, let $\gamma_c(G) = 2m - n$. Then, by Theorem 6.16, we have $2m - n \leq n - 2$, which implies $m \leq n - 1$, so G must be a tree with $m = n - 1$. Hence, by Corollary 6.17, $\gamma_c(G) = n - l(G)$. If $l(G) \geq 3$, then $\gamma_c(G) = n - l(G) < n - 2 = 2m - n$, a contradiction to the hypothesis. Thus, $l(G) \leq 2$. But since G is a tree, $l(G) \geq 2$. Thus, $l(G) = 2$ and G is a path. $\square$

Hedetniemi and Laskar gave an inequality for connected domination number analogous to the well known result that $\gamma(G) \leq n - \Delta(G)$ from [95].

Theorem 6.25 [669] *For any connected graph G, $\gamma_c(G) \leq n - \Delta(G)$.*

Proof. Let v be a vertex in G of maximum degree. Then a spanning tree T of G can be formed in which v is adjacent to each of its neighbors in G. Hence, T has at least $\Delta(G)$ endvertices and, by Theorem 6.16, $\gamma_c(G) \leq n - \Delta(G)$. $\square$

Observing that $\gamma_c(T) = n - l(T) \leq n - \Delta(T)$ for any tree T, Hedetniemi and Laskar determined the trees which achieve the upper bound.

Corollary 6.26 [669] *For any tree T, $\gamma_c(T) = n - \Delta(T)$ if and only if T has at most one vertex of degree three or more.*

The *vertex connectivity number* $\kappa(G)$ of a graph G is the minimum number of vertices whose removal results in a disconnected or trivial graph. Since $\kappa(G) \leq \delta(G) \leq \Delta(G)$ for any graph G, we have that $n - \Delta(G) \leq n - \kappa(G)$. Joseph and Arumugam characterized the graphs G which achieve $\gamma_c(G) = n - \kappa(G)$ and hence, have $\gamma_c(G) = n - \Delta(G)$.

Theorem 6.27 [752] *Let G be a nontrivial connected graph. Then $\gamma_c(G) = n - \kappa(G)$ if and only if G is one of the following: C_n, K_n, and $K_{2k} - M$ for $k \geq 3$, where M is a perfect matching in K_{2k}.*

Proof. First we note that $\gamma_c(C_n) = n - 2$ and $\kappa(C_n) = 2$, $\gamma_c(K_n) = 1$ and $\kappa(K_n) = n - 1$, and $\gamma_c(K_{2k} - M) = 2$ and $\kappa(K_{2k} - M) = 2k - 2 = n - 2$ for $k \geq 3$. Thus, for all these graphs, $\gamma_c(G) = n - \kappa(G)$.

Conversely, let G be any connected graph for which $\gamma_c(G) = n - \kappa(G)$. From Theorems 6.16 and 6.25

$$\gamma_c(G) = n - \varepsilon_T(G) = n - \kappa(G) \leq n - \Delta(G).$$

Hence, any spanning tree has at most $\kappa(G)$ endvertices and $\Delta(G) \leq \kappa(G) \leq \delta(G)$. Thus G is regular of degree $\kappa(G)$. Since $\gamma_c(G) \leq n - 2$, $\kappa(G) \geq 2$. To obtain the desired result, it suffices to show that $\kappa(G) = 2$ or $n - 1$ or $n - 2$.

Suppose to the contrary that $3 \leq \kappa(G) \leq n - 3$. There exists a spanning tree T with a vertex u of maximum degree $\Delta(G) = \kappa(G)$ in T. From Theorems 6.15 and 6.25, $n - \Delta(G) = \gamma_c(G) \leq \gamma_c(T) \leq n - \Delta(G)$, so $\gamma_c(T) = n - \Delta(G)$. From Corollary 6.26, tree T has at most one vertex of degree three or more. Since $deg_T(u) = \Delta(G) = \kappa(G) \geq 3$, each vertex $v \neq u$ has degree at most two in T. Hence, any vertex $v \neq u$ in T is either an endvertex or else $deg_T(v) = 2$ and v is on a path from u to an endvertex.

Let $S = \{u_1, u_2, ..., u_\kappa\}$ be the set of endvertices of tree T. If $N(u) = S$, then no other vertex is in T, implying that $\langle S \rangle$ is complete in G, that is, G is the complete graph K_n and $\kappa(K_n) = n - 1$, a contradiction. Therefore there is at least one vertex, say $u_1 \in S$, that is not adjacent to u. Let x_1 be the unique neighbor of u_1. If in G, $u_1 \in S$ has another neighbor in $V - S$, say $w \neq x_1$, then $T - u_1 x_1 + w u_1$ is a spanning tree with $\kappa(G) + 1$ endvertices, a contradiction. Therefore u_1 is adjacent to the other $\kappa(G) - 1$ vertices in S. If there are at least two vertices in $S - N(u)$, say u_1 and u_2, and u is adjacent to an endvertex $u_i \in S$, then $T - x_1u_1 - x_2u_2 + u_iu_1 + u_iu_2$ is a spanning subtree with $\kappa(G) + 1$ endvertices, a contradiction. Hence, either $N(u) \cap S = \emptyset$ or there is exactly one vertex $u_1 \in S - N(u)$.

Suppose that $N(u) \cap S = \emptyset$. Then $\langle S \rangle$ is complete and each u_i has a unique neighbor x_i in T. Thus, $T - u_2x_2 - u_3x_3 + u_1u_2 + u_1u_3$ is a spanning tree with $\kappa(G) + 1$ endvertices, a contradiction.

Thus, there is exactly one vertex $u_1 \in S - N(u)$. If any vertex in $S - \{u_1\}$ is adjacent to a vertex in $(V - S) - \{u\}$, then a spanning tree with more than $\kappa(G)$ endvertices can be obtained, a contradiction. Hence, $\langle S \rangle$ is a complete subgraph of G, implying that the vertices in $(V - S) - \{u, u_1\}$ (on the path from u to u_1) must be adjacent to $\kappa(G) - 2 \geq 1$ vertices in $(V - S) - \{u, u_1\}$, in addition to its two neighbors on the path. Again we have a contradiction, since a spanning tree with more than $\kappa(G)$ vertices can be obtained. Thus, $\kappa(G) = 2$, $n - 2$, or $n - 1$. $\square$

It is straightforward that $\gamma(G) \leq 2\beta_1(G)$. Hedetniemi and Laskar extended this by showing that for any connected graph, there exists a maximum matching whose vertices constitute a connected dominating set. They also determined a lower bound on $\gamma_c(G)$ in terms of the diameter of G.

Theorem 6.28 [669] *For any connected graph G,*

$$diam(G) - 1 \leq \gamma_c(G) \leq 2\beta_1(G).$$

Proof. To prove the lower bound, we let $diam(G) = k$ and let $u, v \in V$ have $d(u, v) = k$. Let S be a connected dominating set of G and consider whether u and/or v are in S.

If $u, v \in S$, then since $\langle S \rangle$ is connected, $|S| \geq k + 1$. If $u \in S$ and $v \notin S$, then since v must have at least one neighbor in S, it follows that $|S| \geq k$. If neither u nor v is in S, then since each of u and v must have a neighbor in S, $|S| \geq k - 1$. Hence, $\gamma_c(G) \geq diam(G) - 1$.

To show the upper bound, let M be a maximum matching in G that maximizes the number of vertices of a largest component C of $\langle V(M) \rangle$. Suppose $\langle V(M) \rangle$ is not connected. If T is a spanning tree of G, there exists a component C' of $\langle V(M) \rangle$, such that C' is connected to C by a path vuv' such that $uv, uv' \in E(T)$, $v \in C$, and $v' \in C'$. We have $u \notin V(M)$. Replacing in M the edge incident to v' by the edge uv', we obtain a matching M'. But $\langle V(C) \cup \{u, v'\} \rangle$ is connected, and hence, $\langle V(M') \rangle$ has a component larger than C, a contradiction to our choice of M and C. $\square$

Duchet and Meyniel related $\gamma_c(G)$ to the vertex independence number $\beta_0(G)$ and to $\gamma(G)$.

Theorem 6.29 [388] *For any connected graph* G,

$$\gamma_c(G) \leq 2\beta_0(G) - 1 \quad and \quad \gamma_c(G) \leq 3\gamma(G) - 2.$$

Bo and Liu showed that the irredundance number can actually replace the domination number in the inequality of Theorem 6.29.

Theorem 6.30 [124] *For any connected graph* G,

$$\gamma_c(G) \leq 3ir(G) - 2.$$

To see the sharpness of the bound of Theorem 6.30, consider the graph G obtained by identifying one vertex from each of C_{3s} and C_{3t}. Then the order of G is $n = 3(s + t) - 1$, $\gamma_c(G) = 3(s + t) - 5$, and $ir(G) = \gamma(G) = s + t - 1$.

Laskar and Pfaff established the following interesting result.

Theorem 6.31 [830] *If a graph G is the complement of a bipartite graph or a split graph, then* $ir(G) = \gamma(G) = \gamma_t(G) = \gamma_c(G)$.

Nordhaus-Gaddum inequalities involving the connected domination numbers of a graph and its complement were presented in [669, 832, 1004] and will be discussed in Chapter 9. The complexity of the connected dominating set problem is presented in Chapter 12.

We conclude this section by considering related conditions that may be imposed on a connected dominating set. Among the most obvious are requirements that the subgraph induced by a dominating set be a path or a tree. None of these types of domination has been studied in depth.

The concept of connected domination was extended to *connected cutfree domination* by Joseph and Arumugam [753] who required that $\langle S \rangle$ not only be connected, but be 2-connected, that is, $\langle S \rangle$ has no cutvertex. They also imposed the even stronger condition that the set S be a block in G and called such a set a *block dominating set*. In [757] Joseph and Arumugam explored dominating sets with a required edge connectivity. A set S is a *2-edge connected dominating set* if it dominates G and the subgraph $\langle S \rangle$ is 2-edge connected. For details concerning these parameters, we refer the reader to [753, 757].

6.5   Dominating Cliques

Cozzens and Kelleher [333] required that vertices of dominating set S induce a complete graph (property $P4$) and called such a set a *dominating clique*. (We note that a *clique* is usually defined to be a maximal complete subgraph. However, here a dominating clique is not necessarily a maximal complete subgraph in G. We use the term "dominating clique" to be consistent with the literature, but "a complete dominating graph" might be a better term.) Dominating cliques have a great number of applications. In fact, as we noted in Chapter 1, the definition of this parameter stemmed from the work that Kelleher and Cozzens [772] did on applications in social networks, where a vertex represented an actor and an edge represented a relationship between two actors. Considering our computer network example, a clique dominating set in a network represents a core group of computers that not only dominates the network, but also has the ability for each member of the core group to communicate directly with each other member of the core group. Such a network has the best possible connections for rapid transfer of shared information among the members of the core group.

The minimum cardinality taken over all dominating cliques is the *clique domination number* $\gamma(G : P4) = \gamma_{cl}(G)$. Not every graph has a dominating clique; consider the cycles C_n, $n \geq 5$, for example.

Obviously, if $\gamma(G) = 1$, then $\gamma(G) = \gamma_c(G) = \gamma_{cl}(G)$. If a graph G has a dominating clique and $\gamma(G) \geq 2$, then

$$\gamma(G) \leq \gamma_t(G) \leq \gamma_c(G) \leq \gamma_{cl}(G).$$

The corona $K_p \circ K_1$ has $\gamma = \gamma_t = \gamma_c = \gamma_{cl} = p$. Strictness in the inequality chain can be seen with the following graph G. Label the vertices of K_7, u_i for $1 \leq i \leq 7$, and let $P_7 = v_1, v_2, v_3, v_4, v_5, v_6, v_7$. Construct G from $K_7 \cup P_7$ by adding edges $u_i v_i$ for $1 \leq i \leq 7$. Then $\{u_4, v_2, v_6\}$ is a γ-set, $\{v_2, v_3, u_6, v_6\}$ is a γ_t-set, $\{u_2, v_2, v_3, v_4, v_5, v_6\}$ is a γ_c-set, and $V(K_7)$ is a γ_{cl}-set.

In an attempt to determine which graphs have dominating cliques, Cozzens and Kelleher established a sufficient condition in terms of forbidden subgraphs.

Theorem 6.32 [333] *If G is a connected graph with no induced P_5 or C_5, then G has a dominating clique.*

Proof. We proceed by induction on n. The theorem is clearly true for $n = 1$. Assume that any connected graph of order n with no induced P_5 or C_5 has a dominating clique. Let G be a connected graph with $n + 1$ vertices that has no induced P_5 or C_5. Since every connected graph has a vertex that is not a cutvertex, let v be such a vertex in G and let $G' = \langle V - \{v\} \rangle$. Since G' is connected and has no induced P_5 or C_5, by the induction hypothesis, it has a dominating set S' that induces a complete graph. If v has a neighbor in S', then S' is a dominating set for G.

Suppose that $N(v) \cap S' = \emptyset$. Since G is connected, v is adjacent to a vertex $x \in (V - S')$. Let $S = \{x\} \cup (N(x) \cap S')$. We will show that S is a dominating clique of G. Clearly, S induces a complete graph. Suppose S is not a dominating set of G. Then there is a vertex $u \in (V - S') - \{v\}$ that is not adjacent to any vertex in S. However, since S' is a dominating set of G', u must be adjacent to some vertex in S'. Let $a \in S'$ be adjacent to u and let $b \in S - \{x\}$. Such a vertex exists since x is not in S' and hence, must be adjacent to some vertex in S'. If u is not adjacent to v, then v, x, b, a, u is an induced P_5, a contradiction. If u is adjacent to v, then v, x, b, a, u is an induced C_5, again a contradiction to the assumption that G has no induced P_5 or C_5. Therefore, G has a dominating clique. $\square$

The converse of the above theorem is not true. For example, the graph $K_1 + P_5$ has a dominating clique of cardinality one and an induced P_5. Bascó and Tuza determined the sufficiency.

Theorem 6.33 [67] *A connected graph G has no induced P_5 or C_5 if and only if every connected subgraph of G has a dominating clique.*

An additional forbidden subgraph allowed Cozzens and Kelleher to obtain an upper bound on the clique domination number.

Theorem 6.34 [333] *If G is a connected graph with no induced P_5, C_5, or corona $K_{k+1} \circ K_1$, then $\gamma_{cl}(G) \leq k$.*

A graph G is said to have a *dominating edge* if $uv \in E$ and $\{u, v\}$ is a dominating set of G. Hence, graph G has $\gamma_{cl}(G) = 2$ if and only if G has a dominating edge and no dominating vertex. Since almost no graphs have dominating edges [174], it follows that almost no graphs have $\gamma_{cl}(G) = 2$. We note that a tree (or a cycle) has a dominating clique if and only if it has a dominating vertex or a dominating edge. A corollary gives forbidden subgraph conditions which are sufficient for a graph to have a dominating edge.

Corollary 6.35 *If G is a nontrivial connected graph with no induced P_5, C_5, or $K_3 \circ K_1$, then G has a dominating edge.*

Another sufficient condition for a graph to have a dominating edge is due to Bascó and Tuza.

Theorem 6.36 [67] *If G is a nontrivial connected graph with no induced P_5 or C_5 and no cutset that induces a star, then G has a dominating edge.*

Bascó and Tuza supplied variations on the sufficiency condition for a graph to have a dominating clique. We list two of these results. Note that the first one also holds for connected domination.

Theorem 6.37 [67] *If G is a connected graph with no induced P_5, then G has a dominating clique or a dominating P_3.*

Theorem 6.38 [67] *If G is a connected graph with a cutvertex and no induced P_5, then G has a dominating clique.*

Kratsch, Damaschke, and Lubiw characterized the chordal graphs which have dominating cliques. We note that this result is a corollary of Theorem 6.32.

Theorem 6.39 [800] *A chordal graph G has a dominating clique if and only if G has diameter at most three.*

A chordal graph is called *strongly chordal* if every cycle of even length exceeding five has an odd chord, that is, a chord joining two nonconsecutive vertices whose distance on the cycle is odd.

Theorem 6.40 [800] *If G is a strongly chordal graph that has a dominating clique, then $\gamma(G) = \gamma_{cl}(G)$.*

Corollary 6.41 *If G is a strongly chordal graph with $diam(G) \leq 3$, then*

(a) $\gamma(G) = \gamma_{cl}(G)$.

(b) $\gamma(G) = \gamma_t(G) = \gamma_c(G) = \gamma_{cl}(G)$, *if $\gamma(G) \geq 2$.*

Cozzens and Kelleher [333] presented a polynomial time algorithm to find a dominating clique of a connected graph with no induced P_5 or C_5. Kratsch [797] discussed related complexity issues and gave polynomial time algorithms to solve the problem for strongly chordal graphs and undirected path graphs. The decision problem associated with finding a dominating clique for an arbitrary graph G is NP-complete [333, 797].

We conclude this section by mentioning a related property that may be imposed on the dominating set S. The *vertex clique cover number* $\theta_0(G)$ is the smallest number of complete subgraphs of G whose union includes all the vertices of G. Penrice [936] generalized property P4 by investigating dominating sets that have a small vertex clique cover number. That is, the condition on the dominating set S is that $\theta_0(\langle S \rangle)$ is minimized. An obvious related property, which to our knowledge has not been studied, is to minimize the edge clique cover number (smallest number of complete subgraphs whose union includes all the edges) of the graph induced by a dominating set.

6.6 Paired-Dominating Sets

A set $S \subseteq V$ is a *paired-dominating set* if S dominates and the induced subgraph $\langle S \rangle$ has a perfect matching (property P5). *Paired-domination* was introduced by Haynes and Slater [653] with the following application in mind.

If we think of each $s \in S$ as the location of a guard capable of protecting each vertex dominated by s, then for domination a guard protects itself, and for total domination each guard must be protected by another guard. For paired-domination the guards' locations must be selected as adjacent pairs of vertices so that each guard is assigned one other and they are "designated as backups" for each other. Thus, a *paired-dominating set* S *with matching* M is a dominating set $S = \{v_1, v_2, ..., v_{2t-1}, v_{2t}\}$ with independent edge set $M = \{e_1, e_2, ..., e_t\}$ where each edge e_i is incident to two vertices of S, that is, M is a perfect matching in $\langle S \rangle$.

For example, for the graph Q_3 in Figure 6.2, $S_1 = \{v_1, v_2, v_3, v_4\}$ with $M_1 = \{v_1 v_2, v_3 v_4\}$ or S_1 with $M_1' = \{v_1 v_4, v_2 v_3\}$ are paired-dominating sets with matchings. If the specific matching is not important, a set S is called a paired-dominating set if it dominates V and $\langle S \rangle$ contains at least one perfect matching. If $v_j v_k = e_i \in M$, we say that v_j and v_k are *paired* in S.

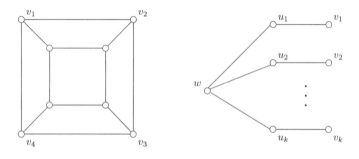

Figure 6.2: Graphs Q_3 and $S(K_{1,k})$.

The *paired-domination number* $\gamma(G : P4) = \gamma_{pr}(G)$ is the minimum cardinality of a paired-dominating set S in G. Note that $\gamma(Q_3) = 2$, while $\gamma_t(Q_3) = \gamma_{pr}(Q_3) = \gamma_c(Q_3) = 4$. For the subdivided star $S(K_{1,k})$ shown in Figure 6.2, set $S = \{u_1, u_2, ..., u_k\}$ is a γ-set, and $S \cup \{w\}$ is a γ_t-set as well as a γ_c-set. Note that if S is a paired-dominating set of $S(K_{1,k})$ and $v_i \in S$, then $u_i \in S$ with u_i and v_i paired, and if $v_i \notin S$, then $u_i \in S$ to dominate v_i and u_i and w are paired. Since w can be paired at most once, it follows that $\gamma(S(K_{1,k})) = k < \gamma_t(S(K_{1,k})) = \gamma_c(S(K_{1,k})) = k + 1 < \gamma_{pr}(S(K_{1,k})) = 2k$. Also, for the path P_{12k}, $\gamma(P_{12k}) = 4k < \gamma_t(P_{12k}) = \gamma_{pr}(P_{12k}) = 6k < \gamma_c(P_{12k}) = 12k - 2$. These examples show that $\gamma_{pr}(G)$ and $\gamma_c(G)$ are incomparable.

Both total domination and paired-domination require that there be no isolated vertices, and every paired-dominating set is a total dominating set.

Observation 6.42 *If a graph G has no isolated vertices, then $\gamma(G) \le \gamma_t(G) \le \gamma_{pr}(G)$ and $\gamma_{pr}(G)$ is even.*

For some of the standard classes of graphs, note that $\gamma_{pr}(K_n) = \gamma_{pr}(K_{r,s}) =$

$\gamma_{pr}(W_n) = 2$ and $\gamma_{pr}(P_n) = \gamma_{pr}(C_n) = 2\lceil n/4 \rceil$. For the corona, $\gamma_{pr}(G \circ K_1) = 2(|V(G)| - \beta_1(G))$.

Observation 6.43 *If a vertex u is adjacent to an endvertex of G, then u is in every paired-dominating set.*

As in previous chapters, we let $i'(G)$ denote the minimum cardinality of a maximal independent edge set.

Observation 6.44 *If a graph G has no isolated vertices, then G has a paired-dominating set, and $\gamma_{pr}(G) \leq 2i'(G) \leq 2\beta_1(G)$.*

Haynes and Slater [653] also showed that the paired-dominating set problem is NP-complete and considered bounds on $\gamma_{pr}(G)$ for arbitrary graphs. The following best possible bounds are straightforward from the definition of paired-domination.

Theorem 6.45 [653] *If a graph G has no isolated vertices, then*

$$2 \leq \gamma_{pr}(G) \leq n$$

and these bounds are sharp.

It is easy to see that any nontrivial graph with $\gamma(G) = 1$ achieves the lower bound of $\gamma_{pr}(G) = 2$. Note that $\gamma_{pr}(G) = 2$ if and only if G has a dominating edge. Recall that Corollary 6.35 gives sufficient conditions for $\gamma_{pr}(G) = 2$. A graph G has $\gamma_{pr}(G) = n$ if and only if $G = mk_2$. The following result gives the graphs which obtain the upper bound of $n-1$. Let $\mathcal{F}$ be the collection of graphs C_3, C_5, and the subdivided stars $S(K_{1,t})$.

Theorem 6.46 [653] *If G is a connected graph with $n \geq 3$, then $\gamma_{pr}(G) \leq n-1$ with equality if and only if $G \in \mathcal{F}$.*

Bollóbas and Cockayne [126] showed that every graph without isolated vertices has a minimum dominating set which is also a private dominating set (see Proposition 3.9 in Chapter 3).

Theorem 6.47 [653] *If a graph G has no isolated vertices, then*

$$\gamma_{pr}(G) \leq 2\gamma(G).$$

Proof. Let G be a graph with no isolated vertices and S be a private dominating set for G. Obviously, pairing each $v \in S$ with a private neighbor forms a paired-dominating set of cardinality $2\gamma(G)$. $\square$

The above bound is sharp as can be seen with the K_n, $K_{1,t}$, and graphs formed from a C_{3k} with vertices labelled 0 to $3k - 1$ by adding at least one endvertex adjacent to each vertex whose label is congruent to 0 modulo 3.

Consider the domination and independent domination parameters, $\gamma(G)$ and $i(G)$. Since every $i(G)$-set is also a dominating set, $\gamma(G) \leq i(G)$. Suppose that, in fact, $\gamma(G) = i(G)$ for some graph G. Then, every i-set is also a γ-set, but not every γ-set must be an i-set. For example, the path P_4 has four γ-sets, only three of which are i-sets. On the other hand, $\gamma(C_5) = i(C_5) = 2$ and each of the five γ-sets is also an i-set. We say that $\gamma(C_5)$ and $i(C_5)$ are *strongly equal*.

That is, if every minimum dominating set is independent, we say that $\gamma(G)$ "strongly equals" $i(G)$, denoted $\gamma(G) \equiv i(G)$. Haynes and Slater defined the concept of *strong equality* of parameters and showed that graphs having $\gamma_{pr}(G) = 2\gamma(G)$ also have $\gamma(G) \equiv i(G)$.

Theorem 6.48 [653] *If a graph G has $\gamma_{pr}(G) = 2\gamma(G)$, then $\gamma(G) \equiv i(G)$.*

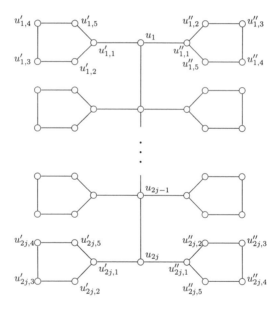

Figure 6.3: Graph G_j with $\gamma(G_j) \equiv i(G_j) = 8j$ and $\gamma_t(G_j) \equiv \gamma_{pr}(G_j) = 10j$.

The converse of Theorem 6.48 is not true as can be seen with the following graph. Let G_j denote the graph pictured in Figure 6.3 on $22j$ vertices formed by connecting to each vertex on a path P_{2j} one vertex from each of two five-cycles. Clearly every minimal dominating set has two vertices from each cycle, and

$$S = \{u'_{1,1}, u'_{1,3}, u''_{1,1}, u''_{1,3}, ..., u'_{2j,1}, u'_{2j,3}, u''_{2j,1}, u''_{2j,3}\}$$

shows $\gamma(G_j) = i(G_j) = 8j$. In fact, $\gamma(G_j) \equiv i(G_j)$, but $\gamma_{pr}(G_j) = 10j$ with a $\gamma_{pr}(G_j)$-set

$$S = (\bigcup_{i=1}^{2j} \{u'_{i,3}, u'_{i,4}, u''_{i,3}, u''_{i,4}\}) \cup (\bigcup_{k=1}^{j} \{u_{2k-1}, u_{2k}\}).$$

We note that the concept of strong equality can be applied to other parameters in an inequality, for example, one could consider graphs for which $\gamma_t(G) \equiv \gamma_{pr}(G)$.

In Section 6.3 we saw that for a connected graph G, the total domination number has a sharp upper bound of $2n/3$. We have seen that $\gamma_{pr}(mK_2) = n$ and $\gamma_{pr}(G) = n - 1$ for $G \in \mathcal{F}$. Moreover, the connected graph in Figure 6.4 has $\gamma_{pr}(G) = 4n/5$. However, if the minimum degree of connected graph G is at least two, the bound of $2n/3$ also holds for $\gamma_{pr}(G)$.

T_{5k}:

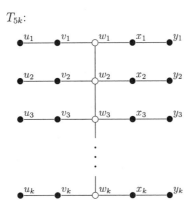

Figure 6.4: $\gamma_{pr}(T_{5k}) = 4k$. (The shaded vertices form a γ_{pr}-set.)

Theorem 6.49 [653] *If a connected graph G has $n \geq 6$ and $\delta(G) \geq 2$, then*

$$\gamma_{pr}(G) \leq \frac{2n}{3}.$$

Proof. Let G be a connected graph with $\delta(G) \geq 2$. Let S be a minimum paired-dominating set for G and let S_p be the set of paired vertices in S, that is, $S_p = \{\{u_i, v_i\} : \text{where } u_i \text{ and } v_i \text{ are paired in } S \text{ for } 1 \leq i \leq |S|/2\}$. Partition the vertices of S into:

$$A = \{u_i, v_i \in S : \text{both } u_i \text{ and } v_i \text{ have a private neighbor }\},$$

$$B = \{u_i \in S : u_i \text{ has a private neighbor and } v_i \text{ does not have a}$$
$$\text{private neighbor }\},$$

$$C = \{v_i \in S : u_i \in B\},$$

$$D = \{u_i, v_i \in S : \text{neither } u_i \text{ nor } v_i \text{ has a private neighbor }\}.$$

Let A' and B' be the private neighbors of A and B, respectively. By definition the private neighbors are in $V - S$, so $|V - S| \geq |A'| + |B'| \geq |A| + |B|$.

Consider an arbitrary pair $\{u_i, v_i\} \in D$. If both u_i and v_i have another neighbor in S, then $S - \{u_i, v_i\}$ is a paired-dominating set with order less than $\gamma_{pr}(G)$, a contradiction. Hence, without loss of generality, $|N(v_i) \cap S| = \{u_i\}$, and v_i has at least one neighbor in $V - S$ for all $\{u_i, v_i\} \in D$, because $\delta(G) \geq 2$. Let $X = N(D) \cap (V - S)$. Then $X \cap (|A'| \cup |B'|) = \emptyset$. If $|D| \geq 2$, then $|X| \geq 1$. We proceed by induction on the number of pairs in D to count the vertices in X. Assume that $|D| = 2k$ and each $v_i \in D$ has at least one unique neighbor in $|X|$, say x_i. That is, let $|X| \geq k$ implying that $|V - S| \geq |A| + |B| + k$. We want to show this is true for $k+1$ pairs, that is, for $|D| = 2k+2$. If $|D| = 2k+2$, then by the above argument v_{k+1} must have a neighbor in $V - S$. If that neighbor has not been counted in X, then we are finished. Hence, assume that v_{k+1} is adjacent to $x_j \in X$. If either u_j or u_{k+1} is adjacent to a vertex in $V - S - X - A' - B'$, then $u_j(u_{k+1})$ can be exchanged for $v_j(v_{k+1})$ in this argument and the neighbor can be added to X and counted as desired. Thus, assume that $N(u_j) \subseteq S \cup X$ and $N(u_{k+1}) \subseteq S \cup X$. Since $\delta(G) \geq 2$, we have the following cases:

Case 1. Vertex u_{k+1} has a neighbor in X, say x_i. If $x_i = x_j$, then

$$S_p - \{\{u_j, v_j\}, \{u_{k+1}, v_{k+1}\}\} \cup \{\{v_j, x_j\}\}$$

is a pairing for a paired-dominating set with order less than $\gamma_{pr}(G)$, contrary to the minimality of S. If $x_i \neq x_j$, then again we can have a paired-dominating set with order less that $\gamma_{pr}(G)$ by pairing $\{x_i, v_i\}$, $\{x_j, v_j\}$ and removing u_i, u_j, u_{k+1}, v_{k+1} from S.

Case 2. Vertex u_{k+1} is adjacent to $a \in S$.

(a) If $a \neq u_j$, then $S_p - \{\{u_j, v_j\}, \{u_{k+1}, v_{k+1}\}\} \cup \{\{v_j, x_j\}\}$ is a pairing for a paired-dominating set with order less than $\gamma_{pr}(G)$, a contradiction.

(b) Let $a = u_j$. Since $n \geq 6$, connectivity implies that one of $u_j, v_j, u_{k+1}, v_{k+1}$, and x_j must have another neighbor. We have already shown that u_{k+1} cannot have another neighbor. A similar argument holds for u_j. Let $\{u_j, u_{k+1}\}$ and $\{v_j, x_j\}$ replace $\{u_j, v_j\}$ and $\{u_{k+1}, v_{k+1}\} \in S_p$ forming a new paired-dominating set. This implies that v_j does not have another neighbor. Similar arguments show that none of the five vertices have another neighbor, contrary to the connectivity of G. Hence, we have shown for all orders of $|D|$ that $|V - S| \geq |A| + |B| + |D|/2$. That is, $n - \gamma_{pr}(G) \geq |A| + |B| + |D|/2$. Now

$$|S| = \gamma_{pr}(G) = |A| + |B| + |C| + |D| = |A| + 2|B| + |D|$$

$$\gamma_{pr}(G) \le 2(|A| + |B| + |D|/2)$$
$$\gamma_{pr}(G) \le 2(n - \gamma_{pr}(G)).$$

Thus, $\gamma_{pr} \le 2n/3$. □

The bound of Theorem 6.49 is sharp as can be seen with the cycle C_6. Although there is no known infinite family of graphs which achieves this upper bound, the family of graphs [653] shown in Figure 6.5 for which $\gamma_{pr}(G)$ approaches $2n/3$ for large n.

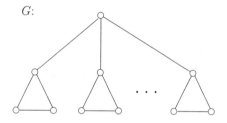

Figure 6.5: $\gamma_{pr}(G)$ approaches $2n/3$ for large n.

Returning to the relationship between paired-domination and total domination, we have seen that $\gamma(G) \le \gamma_t(G) \le \gamma_{pr}(G) \le 2\gamma(G) \le 2i(G)$. Obviously, $2 \le \gamma_t(G) \le \gamma_{pr}(G) \le 2\gamma_t(G)$.

Actually, the upper bound is even better than this.

Theorem 6.50 [653] *If a graph G has no isolated vertices, then*

$$\gamma_{pr}(G) \le 2\gamma_t(G) - 2.$$

Thus we have

$$\gamma_{pr}(G) \le 2\gamma_t(G) - 2 \qquad (6.2)$$

and

$$\gamma(G) \le \gamma_t(G) \le \gamma_{pr}(G) \le 2\gamma(G). \qquad (6.3)$$

Haynes and Slater determined for which positive integers $a \le b \le c$ satisfying (6.2) and (6.3) above, there exists a graph having $\gamma(G) = a$, $\gamma_t(G) = b$, and $\gamma_{pr}(G) = c$.

Theorem 6.51 [653] *Given positive integers $a \le b \le c$ such that c is even, $c \le 2a$, and $c \le 2b - 2$, there exists a graph G having*

$$\gamma(G) = a, \gamma_t(G) = b, \gamma_{pr}(G) = c.$$

Proof. Given positive integers a, b, and c such that $a \leq b \leq c$ and $c \leq 2b - 2$, we construct a graph having $\gamma(G) = a$, $\gamma_t(G) = b$, and $\gamma_{pr}(G) = c$.

Let $k = b - a$, $r = (2a - c)/2$, and $s = (2a - 2b + c)/2$. Begin with a $K_{r,s}$ having partite sets V_1 and V_2 with $|V_1| = r$ and $|V_2| = s$, $r \leq s$. Add two or more vertices of degree one adjacent to each vertex of $K_{r,s}$. Next select $k \geq 0$ of the edges which are incident to $y_1 \in V_2$ and an endvertex. Subdivide these k edges and label the endvertices $v_1, v_2, ..., v_k$. Finally, add two or more vertices of degree one adjacent to each v_i, $1 \leq i \leq k$. We claim that $\gamma(G) = r + s + k$, $\gamma_t(G) = r + s + 2k$, and $\gamma_{pr}(G) = 2s + 2k$. To see that $\gamma = r + s + k$, note that each vertex adjacent to two or more endvertices must be in any γ-set implying that $\gamma(G) \geq r + s + k$. The set $D = V_1 \cup V_2 \cup \{v_1, v_2, ..., v_k\}$ dominates G so $\gamma(G) \leq r + s + k$.

Note that each vertex in $V_1 \cup V_2$ has a neighbor in D and hence, is totally dominated. However, for the vertices $v_1, v_2, ..., v_k$ to be totally dominated in the most efficient way, k additional vertices are necessary. Hence, adding any neighbor for each $v_i, 1 \leq i \leq k$, forms a γ_t-set for G and $\gamma_t(G) = r + s + 2k$.

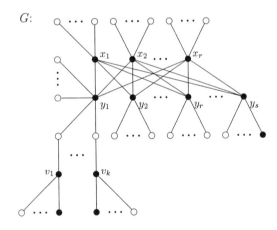

Figure 6.6: $a = \gamma(G) = r+s+k$, $b = \gamma_t(G) = r+s+2k$, and $c = \gamma_{pr}(G) = 2s+2k$.

By Observation 6.43, the vertices of $V_1 \cup V_2 \cup \{v_1, v_2, ..., v_k\}$ must be in any γ_{pr}-set. Observe that r pairs can be formed with vertices in $V_1 \cup V_2$. Hence, we need $s - r$ additional vertices to pair the remaining vertices of V_2. Also to pair $v_i, 1 \leq i \leq k$, each v_i must have a private neighbor in any γ_{pr}-set so we need at least k more vertices.

The shaded vertices in Figure 6.6 depict a γ_{pr}-set for G. Therefore, $\gamma_{pr}(G) = r + s + k + s - r + k = 2(s + k)$. Hence, $\gamma(G) = r + s + k$, $\gamma_t(G) = r + s + 2k = \gamma(G) + k$, and $\gamma_{pr}(G) = 2s + 2k = 2\gamma(G) - 2r$, for $1 \leq r \leq s$, $k \geq 0$. We note that there is a special case when $c = 2a$. It is simple to verify that the tree illustrated in Figure 6.7 has $a = \gamma(G) = j + 1$, $b = \gamma_t(G) = j + 1 + i = \gamma(G) + i$, and $c = \gamma_{pr}(G) = 2j + 2 = 2\gamma(G)$ for $i \leq j$. $\square$

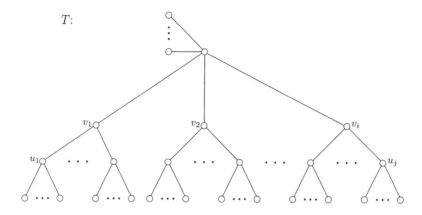

Figure 6.7: $a = \gamma(G) = j + 1$, $b = \gamma_t(G) = j + 1 + i$, and $c = \gamma_{pr}(G) = 2j + 2$.

Paired-domination is a relatively new concept which has only been studied in [603, 653, 704], and hence, there are many related problems that remain open. For instance, Studer, Haynes, and Lawson [1064] studied dominating sets S with the property that $\langle S \rangle$ induces a perfect matching, that is, $\langle S \rangle$ is $(|S|/2)K_2$. A dominating set with this property is called a *strong (or induced) paired-dominating set*. For our guard application, strongly paired-domination would pair the vertices of S with a backup guard and limit the direct communication among the guards to communication between partners only. Hence, the need for communication equipment is decreased, possible unwanted interference is minimized, and security is tightened.

6.7 Dominating Cycles

A *dominating cycle* is a cycle (not necessarily induced) in which every vertex of the graph is adjacent to at least one vertex on the cycle. Lesniak and Williamson [842] defined this concept and observed that a dominating set S such that the induced subgraph $\langle S \rangle$ is hamiltonian (property $P6$) is a dominating cycle. The smallest cardinality of a cycle dominating set is denoted by $\gamma(G : P6) = \gamma_{cy}(G)$. Obviously, not all graphs have dominating cycles (for example, trees). Bondy and Fan derived a sufficient condition for the existence of a dominating cycle.

Theorem 6.52 [132] *If G is a k-connected graph, $k \geq 2$, and any $k + 1$ independent vertices x_i $(0 \leq i \leq k)$ with $N(x_i) \cap N(x_j) = \emptyset$ $(0 \leq i \neq j \leq k)$ have* *degree-sum $\sum_{i=0}^{k} deg(x_i) \geq n - 2k$, then G has a dominating cycle.*

Proof. Suppose that G satisfies the hypothesis of the theorem, but has no dominating cycle. Since G is k-connected, G has a cycle of length at least k. Let C be a cycle in G with length at least k such that C dominates as many vertices as possible. Since C is not a dominating cycle, there is a component H of $G - C$ containing a vertex x_0 such that $N(x_0) \cap V(C) = \emptyset$. Consider all such cycles C and choose C in such a way that H has as few vertices as possible. Since G is k-connected and $|V(C)| \geq k$, there are k paths $P_1, P_2, ..., P_k$ from x_0 to C, pairwise disjoint except for x_0 and with all internal vertices in H. Let u_i for $1 \leq i \leq k$ be the endvertices of the paths P_i for $1 \leq i \leq k$, taken in order around C, and define $u_{k+1} = u_1$, $P_{k+1} = P_1$.

Let $c_1, c_2, ..., c_s$ be the vertices in order around C and define $C(c_i, c_j) = \{c_t : i < t < j\}$ and $C[c_i, c_j] = \{c_t : i \leq t \leq j\}$, where the subscripts are taken modulo s. The sets $C(c_i, c_j]$ and $C[c_i, c_j)$ are defined similarly.

If, for some i and every $v \in C(u_i, u_{i+1})$, $N(v) \cap (V - C) \cup \{v\}$ is dominated by $C - C(u_i, u_{i+1})$, then by deleting $C(u_i, u_{i+1})$ and adding the path $P_i x_0 P_{i+1}$, we obtain a new cycle which has length at least k and dominates more vertices than C, contradicting the choice of C. Thus, for each i, there is at least one vertex $y_i \in C(u_i, u_{i+1})$ such that

(a) $N(y_i) \cap (V - C) \cup \{y_i\}$ is not dominated by $C - C(u_i, u_{i+1})$.

Now subject to (a), choose y_i as close to u_i along $C(u_i, u_{i+1})$ as possible. Thus we have

(b) for each $v \in C(u_i, y_i)$, $N(v) \cap (V - C) \cup \{v\}$ is dominated by $C - C(u, u_{i+1})$.

Using (b), we will show that there is no edge joining H and $C(u_i, y_i]$ for $1 \leq i \leq k$. Suppose to the contrary, that there is an edge $e = hw$ such that $h \in V(H)$ and $w \in C(u_i, y_i]$. Choose e such that w is as close to u_i as possible. Since $h \in V(H)$ and H is connected, there is a path P from u_i to w with all internal vertices in H; by adding the path P and deleting $C(u_i, w)$ we obtain a new cycle C'. Clearly, $|V(C')| \geq k$. If C' dominates x_0, then by (b), C' dominates more vertices than C, a contradiction to the choice of C. If C' does not dominate x_0, then by (b), C' dominates at least as many vertices as C. However, in this case, the component $G - C'$ containing x_0 has fewer vertices than H, by the choice of e, and so contradicts the second criteria used in the choice of C. Hence, there is no edge joining H and $C(u_i, y_i]$ for $1 \leq i \leq k$.

Next we show that there is no path from $C(u_i, y_i]$ to $C(u_j, y_j]$ with all internal vertices in $G - C$, $1 \leq i \neq j \leq k$. Suppose that there is such a path P. Let the origin and terminus of P be $w \in C(u_i, y_i]$ and $z \in C(u_j, y_j]$, respectively. Since there is no edge joining H and $C(u_i, y_i]$ $(1 \leq i \leq k)$, $V(P) \cap V(H) = \emptyset$, and so

$$wPz...y_j...u_{j+1}...u_i P_i x_0 P_j u_j...u_{i+1}...y_i...w$$

is a cycle C'''. Choose P in such a way that w is as close to u_i as possible. By this choice, there is no path joining $C(u_i, w)$ and $C(u_j, z)$, and so, for every vertex $v \in C(u_i, w) \cup C(u_j, z)$, $N[v]$ is dominated by C''', by (b). Therefore, C''' dominates more vertices than C, a contradiction.

It follows from (a) that, for $1 \le i \le k$, there exists a vertex $x_i \in N(y_i) \cap (V - C) \cup \{y_i\}$ which is not dominated by $C - C(u_i, u_{i+1})$. But we have just shown that the x_i for $0 \le i \le k$ are $k + 1$ independent vertices.

We shall show that $N(x_i) \cap N(x_j) = \emptyset$, $0 \le i \ne j \le k$. Suppose, to the contrary, that $v \in N(x_i) \cap N(x_j)$, $i \ne j$. Since x_g $(0 \le g \le k)$ is not dominated by $C - C(u_g, u_{g+1})$, where $C(u_0, u_1) = \emptyset$, we have $v \notin C$. But then there is a path $y_i x_i v x_j y_j$ (which may be degenerate, if $y_i = x_i$ or $y_j = x_j$). That is, either there is an edge joining H and $C(u_i, y_i]$ or there is a path from $C(u_i, y_i]$ to $C(u_j, y_j]$ with all internal vertices in $G - C$. In either case, we have a contradiction.

Let $J = \{i : x_i = y_i, 1 \le i \le k\}$. Since x_i for $1 \le i \le k$ is not dominated by $C - C(u_i, u_{i+1})$, for $i \in J$,

$$|N(x_i) \cap (V - C)| \ge deg(x_i) - |C(u_i, u_{i+1})| + 1$$

and for $i \notin J$

$$|N(x_i) \cap (V - C)| \ge deg(x_i) - |C(u_i, u_{i+1})|.$$

Hence,

$$n - |V(C)| = |V(G - C)|$$

$$\ge |\{x_0\}| + |N(x_0)| + \sum_{i \notin J} |\{x_i\}| + \sum_{i \notin J} |N(x_i) \cap (V - C)| + \sum_{i \in J} |N(x_i) \cap (V - C)|$$

$$\ge 1 + k + \sum_{i=0}^{k} deg(x_i) - \sum_{i=1}^{k} |C(u_i, u_{i+1})|.$$

Since $\sum_{i=1}^{k} |C(u_i, u_{i+1})| = |V(C)| - k$,

$$n - (2k + 1) \ge \sum_{i=0}^{k} deg(x_i).$$

On the other hand, since the x_i for $1 \le i \le k$ are independent and no pair of them have a common neighbor, using the hypothesis of the theorem, we have

$$\sum_{i=0}^{k} deg(x_i) \ge n - 2k.$$

This contradiction completes the proof. $\square$

A resulting corollary proved a conjecture of Clark, Colbourne, and Erdös [241].

Corollary 6.53 [132] *If G is a k-connected graph, $k \ge 2$, having $\delta(G) \ge (n - 2k)/(k + 1)$, then G has a dominating cycle.*

As defined in Chapter 5, a graph G is edge domination critical (that is, $G \in CEA$) if $\gamma(G) = k$ and $\gamma(G + e) = k - 1$ for every edge $e \in E(\overline{G})$. Sumner and Blitch [1067] conjectured that any graph $G \in CEA$ with $\gamma(G) = 3$ and $n \geq 6$ has a hamiltonian path. In the process of proving this conjecture, Wojcicka determined the following relationship between edge domination critical graphs and graphs having a dominating cycle.

Theorem 6.54 [1149] *If a connected graph $G \in CEA$ and $\gamma(G) = 3$, then G has a dominating cycle.*

In general, the existence problem for cycle domination is NP-complete even when restricted to planar graphs [950]. However, several authors developed polynomial algorithms to solve this problem for specific families:

- Colbourn, Keil, and Stewart [316], permutation graphs

- Colbourn and Stewart [318], series-parallel graphs

- Proskurowski [949], 2-trees

- Proskurowski and Syslo [950], outerplanar graphs

- Skowrońska and Syslo [1028], Halin graphs

- Keil and Schaefer [770], circular-arc graphs.

Again we conclude the section and, hence, the chapter, with related problems. The work we have seen considered the property P6, that is, the subgraph induced by the dominating set has a hamiltonian cycle. A possible variation would be to require that $\langle S \rangle$ induces a cycle. More generally, one could require that S induces an r-regular subgraph, which for degree $r = 2$ is a collection of cycles. To our knowledge, no one has explored these topics.

EXERCISES

6.1 We have seen that if a graph G is claw-free, then $\gamma(G) = i(G)$. Demonstrate that the converse is not true, that is, give a graph G that has a claw and $\gamma(G) = i(G)$.

6.2 Show that if G is a connected bipartite graph, $i(G) \leq n/2$.

6.3 Give an infinite family of connected graphs G that obtain the sharp upper bound of Theorem 6.11, that is, $\gamma_t(G) = 2n/3$.

6.4 (a) Prove Theorem 6.28.

(b) Give an infinite family of graphs G for which $\gamma_c(G) = diam(G) - 1$.

(c) Give an infinite family of graphs G for which $\gamma_c(G) = 2\beta_1(G)$.

6.5 Find a graph G for which $\gamma_c(G) = 3ir(G) - 2$ and $ir(G) < \gamma(G)$.

6.6 Verify that $\gamma(G) \le \gamma_t(G) \le 2\gamma(G)$ and $\gamma_c(G) \le 2\beta_1(G)$.

6.7 (a) Prove Theorem 6.15, that is, for a connected spanning subgraph H of G, $\gamma_c(G) \le \gamma(H)$.

(b) For which of the following parameters does a similar result hold:

$$i(G), \gamma_t(G), \gamma_c(G), \gamma_{cl}(G), \gamma_{pr}(G), \gamma_{cy}(G).$$

6.8 Does there exist a graph $G \ne C_6$ with $n \ge 6$ and $\delta(G) \ge 2$ such that $\gamma_{pr}(G) = 2n/3$?

6.9 We have seen Gallai theorems for $\gamma(G)$, $i(G)$, $\beta_0(G)$, and $\gamma_c(G)$. Can you find corresponding Gallai theorems for other parameters?

6.10 Compute i, γ_t, γ_{pr}, γ_c, and γ_{cy} for $P_3 \times P_k$ and for $P_4 \times P_k$.

6.11 For $2 \le 2k \le n - 1$, construct a graph G of order n with $\gamma_{pr}(G) = 2k$.

6.12 Theorem 6.17 shows that for any connected graph G, $\gamma_c(G) \le 3ir - 2$. We have shown that $\gamma_c(G) \le \gamma_{cl}(G)$ and $\gamma_c(G) \le \gamma_{cy}(G)$. Is $3ir(G) - 2$ an upper bound for either $\gamma_{cl}(G)$ or $\gamma_{cy}(G)$?

6.13 Construct a graph G with $\gamma(G) = 10$, $\gamma_t(G) = 15$, and $\gamma_{pr}(G) = 16$.

6.14 Which bipartite graphs have dominating cliques?

6.15 Construct a family of graphs G for which $\gamma_t(G) = n - \Delta(G)$.

6.16 For a connected graph G, a *bridge* is an edge whose removal disconnects the graph. Find a connected *bridgeless* cubic graph that does not have a dominating cycle.

6.17 Give an infinite family of graphs G that have

$$\gamma(G) = i(G) = \gamma_c(G) = \gamma_t(G) = \gamma_{pr}(G) = \gamma_{cl}(G) = \gamma_{cy}(G).$$

6.18 Construct a 2-connected graph G with $n = 10$ and $\delta(G) = 2$, and show that G has a dominating cycle. Find as many of these graphs as you can.

Chapter 7

Varieties of Domination

In Chapter 6 we considered a variety of conditions that might be imposed on a dominating set D in a graph $G = (V, E)$. In this chapter we will consider a variety of conditions that can be imposed either on the dominated set $V - D$, or on V, or on the method by which vertices in $V - D$ are dominated. These include the following.

(i) *multiple domination*: in which we insist that each vertex in $V - D$ be dominated by at least k vertices in D for a fixed positive integer k.

(ii) *parity restrictions*: in which we insist that each vertex in V be dominated by an odd number of vertices in D, or even more generally, we specify for each vertex in V the allowable number of vertices in D which can dominate it.

(iii) *locating domination*: in which we insist that each vertex in $V - D$ has a unique set of vertices in D which dominate it.

(iv) *distance domination*: in which we insist that each vertex in $V - D$ be within distance k of at least one vertex in D for a fixed positive integer k.

(v) *strong domination*: in which we insist that each vertex v in $V - D$ be dominated by at least one vertex in D whose degree is greater than or equal to the degree of v. A similar notion of *weak domination* specifies that each vertex v in $V - D$ be dominated by at least one vertex in D whose degree is less than or equal to the degree of v.

(vi) *global domination*: in which we insist that the dominating set D also dominates the vertices $V - D$ in the complement $\overline{G}$ of G.

(vii) *directed domination in digraphs*: in which we insist that for each vertex v in $V - D$, there is a directed edge from u to v for at least one vertex u in D.

7.1   Multiple Domination

Let D be a dominating set in a graph $G = (V, E)$. If we view the dominating set as a set that either monitors or controls the vertices in $V - D$, then the removal, or failure, of an edge may result in a set which is no longer dominating. If this is an undesirable situation, then it may be necessary to increase the level of domination of each vertex so that, even if an edge fails, the set D will still be a dominating set. The idea of dominating each vertex in $V - D$ multiple times originated with Fink and Jacobson [473]. In this section we review their basic results and several recent additions about this type of domination. We begin with a simple observation by Fink and Jacobson.

Theorem 7.1 [473] *If D is a γ-set of a graph G, then at least one vertex in $V - D$ is dominated by no more than two vertices in D.*

Proof. Let D be a minimum dominating set in G and assume that every vertex in $V - D$ is dominated by three or more vertices. Let $u \in V - D$ and let v and w be two vertices in D which dominate u. It follows from our assumption that every vertex in $V - D$ is dominated by at least one vertex in $D - \{v, w\}$. Therefore, the set $D' = D - \{v, w\} \cup \{u\}$ is a dominating set. But since $|D'| < |D|$, we contradict the assumption that D is a minimum dominating set. □

This theorem shows that given any γ-set D, one can always remove two edges from G so that D is no longer a dominating set for G. If we require a greater degree of assurance that the removal of edges will not do this, then a greater degree of domination will be necessary. A vertex in $V - D$ is k-*dominated* if it is dominated by at least k vertices in D, that is, $|N(v) \cap D| \geq k$. If every vertex in $V - D$ is k-dominated, then D is called a k-*dominating set*. The minimum cardinality of an k-dominating set is called the k-*domination number* $\gamma_k(G)$. Note that if $k = 1$, then $\gamma_1(G) = \gamma(G)$, the domination number of G. Also, for $1 \leq j \leq k$, if D is a k-dominating set, it is also a j-dominating set, and therefore $\gamma_j(G) \leq \gamma_k(G)$.

It is worthwhile to mention that if $\Delta(G) \geq 3$ and $k \geq 3$, then by Theorem 7.1, no k-dominating set in G can be a minimum dominating set, since every minimum dominating set will dominate at least one vertex at most twice. Thus we have the following.

Corollary 7.2 *If G is a graph with $\Delta(G) \geq 3$ and $k \geq 3$, then $\gamma_k(G) > \gamma(G)$.*

This corollary also follows from the next lower bound for $\gamma_k(G)$.

Theorem 7.3 [473] *If G is a graph with $\Delta(G) \geq k \geq 2$, then $\gamma_k(G) \geq \gamma(G) + k - 2$.*

Proof. Let D be a minimum k-dominating set in G, let $u \in V - D$ and let $v_1, v_2, \cdots, v_k$ be distinct vertices in D which dominate u. Notice that since $\Delta(G) \geq k \geq 2$, we know that $V - D \neq \emptyset$ because there is always a k-dominating set which does not contain a vertex of degree Δ. Since D is a k-dominating set, each vertex in $V - D$ is dominated by at least one vertex in $D - \{v_2, \cdots, v_k\}$. Therefore, since u dominates each vertex in $\{v_2, \cdots, v_k\}$, we know that the set $D' = D - \{v_2, \cdots, v_k\} \cup \{u\}$ is a dominating set in G. Therefore, $\gamma(G) \leq |D'| = \gamma_k(G) - (k - 1) + 1 = \gamma_k(G) - k + 2$. $\square$

The following provides a lower bound for $\gamma_k(G)$ in terms of the maximum degree $\Delta(G)$.

Theorem 7.4 [473] *For any graph G, $\gamma_k(G) \geq kn/(\Delta(G) + k)$.*

Proof. Let D be a minimum k-dominating set and let t denote the number of edges between D and $V - D$. Since the degree of each vertex in D is at most Δ, $t \leq \Delta \gamma_k(G)$. But since each vertex in $V - D$ is adjacent with at least k vertices in D, we know $t \geq k(n - \gamma_k(G))$. Combining these two inequalities produces $\gamma_k(G) \geq kn/(\Delta(G) + k)$. $\square$

Note that for $k = 1$, this gives the lower bound on $\gamma(G)$ given in Theorem 2.11. Cockayne, Gamble, and Shepherd provided a corresponding upper bound involving the minimum degree $\delta(G)$.

Theorem 7.5 [264] *If G is a graph with $k \leq \delta(G)$, then $\gamma_k(G) \leq kn/(k + 1)$.*

Notice that for $k = 1$, this is the familiar Ore bound: $\gamma(G) \leq n/2$, (where we must also assume that G has no isolated vertices). This upper bound was later improved by Stracke and Volkmann, who used a rather complicated proof to show the following.

Theorem 7.6 [1062] *For any graph G,*

(a) $\gamma_k(G) \leq n(2k - \delta(G))/(2k - \delta(G) + 1)$, *if $k \leq \delta(G) \leq 2k - 1$.*

(b) $\gamma_k(G) \leq n/2$, *if $\delta(G) \geq 2k - 1$.*

Fink and Jacobson obtained another lower bound for γ_k involving only the number of vertices n and the number of edges m, corollaries of which yield an interesting lower bound and an extremal result for the 2-domination number of a tree.

Theorem 7.7 [473] *For any graph G, $\gamma_k(G) \geq n - (m/k)$.*

Corollary 7.8 *If T is a tree with $n \geq 2$ vertices, then $\gamma_2(T) \geq (n + 1)/2$.*

Corollary 7.9 *If T is a tree with $n \geq 2$ vertices, then $\gamma_2(T) = (n+1)/2$ if and only if T is the subdivision graph of a tree T'.*

Fink and Jacobson also provided an interesting characterization of the k-domination number in terms of spanning subgraphs. A bipartite graph G is called k-*semiregular* if every vertex in one of the two partite sets has degree k.

Theorem 7.10 [473] *If G is a graph with $\delta(G) \geq k$, then $\gamma_k(G) = \min\{\gamma_k(H)\}$ where the minimum is taken over all spanning k-semiregular bipartite subgraphs H of G.*

Proof. It is easy to see that $\gamma_k(G) \leq \gamma_k(H)$ for any spanning subgraph H of G. Therefore, $\gamma_k(G) \leq \min\{\gamma_k(H)\}$. Conversely, let D be a minimum k-dominating set in G. By definition, each vertex in $V - D$ is dominated by k or more vertices in D. Let us therefore construct a spanning subgraph H of G, as follows: add to H exactly k edges between v and D, for every $v \in V - D$. Then H will be a spanning k-semiregular bipartite subgraph of G, and D is an k-dominating set of H. Therefore, $\gamma_k(G) \geq \gamma_k(H)$ and $\gamma_k(G) \geq \min\{\gamma_k(H)\}$. □

It is interesting to point out that the classical theorem of Nieminen (see Chapter 6): for any graph G with n vertices, $\gamma(G) + \varepsilon(G) = n$, is an immediate corollary of Theorem 7.10, since $\varepsilon(G)$ equals the maximum number of pendant edges in a spanning forest of G, and this spanning forest is a 1-semiregular bipartite subgraph of G.

The following result, due to Fink and Jacobson, offers some insights into the nature of minimal k-dominating sets in graphs.

Proposition 7.11 [473] *A k-dominating set D is minimal if and only if for every vertex $v \in D$, either (1) $|N(v) \cap D| < k$ or (2) there exists a vertex $u \in V - D$ such that $|N(u) \cap D| = k$ and $u \in N(v)$.*

Proof. Let D be a minimal k-dominating set in G. Suppose there exists a vertex $v \in D$ for which $|N(v) \cap D| \geq k$ and for every vertex $u \in V - D$, either $|N(u) \cap D| > k$ or $u \notin N(v)$. Then consider the set $D' = D - \{v\}$. Since v is adjacent to at least k vertices in D', it follows that D' is an k-dominating set; contradicting the minimality of D.

Now assume that D is a k-dominating set satisfying conditions (1) and (2). Consider the set $D' = D - \{v\}$ for an arbitrary vertex $v \in D$. If condition (1) holds, then $|N(v) \cap D'| < k$, which means that D' is not an k-dominating set. If condition (2) holds, then there exists a vertex $u \in V - D$ such that $|N(u) \cap D| = k$ and $u \in N(v)$. But in this case the set D' would not k-dominate u, and hence, would not be an k-dominating set of G. Thus, in either case, D' is not an k-dominating set, and therefore D is a minimal k-dominating set. □

In 1962 Ore first observed that for any graph G, $\gamma(G) \leq \beta_0(G)$. This basic result inspired Fink and Jacobson to study a similar relationship between k-domination and a similar notion of vertex independence. For $k \geq 2$, a subset $S \subseteq V(G)$ is called a k-*dependent set* if and only if the maximum degree of a vertex in the subgraph induced by S is less than k, that is, $\Delta(\langle S \rangle) < k$. The maximum cardinality of an k-dependent set in G is the k-*dependence number* $\beta_k(G)$. Note that every k-dependent set is also an h-dependent set, for every $h > k$.

Theorem 7.12 [473] *For any graph G, $\gamma_2(G) \leq \beta_2(G)$.*

Proof. If $\Delta(G) \leq 1$, then $\gamma_2(G) = \beta_2(G) = n$. We can assume that $\Delta(G) \geq 2$. Among all maximum 2-dependent sets, let S be one for which $\langle S \rangle$ has a minimum number of edges. We claim that S is also a 2-dominating set.

Assume, to the contrary, that S is not a 2-dominating set, that is, there exists a vertex $u \in V - S$ which is not 2-dominated by S. Since $S \cup \{u\}$ is not a 2-dependent set, there must be two vertices in S, say v and w, where v is adjacent to both u and w. But since v is the only vertex in S adjacent to u, the set $S' = (S - \{v\}) \cup \{u\}$ is a maximum 2-dependent set having fewer edges in $\langle S' \rangle$ than the number of edges in $\langle S \rangle$, contradicting our choice of S. Thus, S must be a 2-dominating set and $\gamma_2(G) \leq |S| = \beta_2(G)$. $\square$

This theorem led Fink and Jacobson to conjecture that for every positive integer k, $\gamma_k(G) \leq \beta_k(G)$. This conjecture was settled by Favaron, who actually proved a stronger result. Let $m\langle S \rangle$ denote the number of edges in the subgraph induced by a set S and let $deg_S(u)$, the degree of u in S, be the number of vertices in S to which u is adjacent.

Theorem 7.13 [444] *For any graph G and any positive integer k, every k-dependent set D for which $k\,|D| - m\langle D \rangle$ is a maximum is a k-dominating set of G.*

Proof. Let D be a k-dependent set for which $k\,|D| - m\langle D \rangle$ is a maximum. Assume that D is not a k-dominating set and let $v \in V - D$ be a vertex which is not k-dominated by D. Let $C = N_D(v)$ be the set of neighbors of v in D. Notice that $0 \leq |C| < k$. Let B be the set of vertices $w \in C$ for which $deg_D(w) = k - 1$ (that is, w has degree $k - 1$ in $\langle D \rangle$). Finally, let A be a maximal independent set of vertices in $\langle B \rangle$ (see Figure 7.1).

Notice that $\emptyset \subseteq A \subseteq B \subseteq C \subseteq D$. Let $D' = D - A \cup \{v\}$. The set D' is k-dependent. This is true because of the following:

(i) $deg_{D'}(v) \leq |C| < k$;
(ii) $deg_{D'}(c) \leq deg_D(c) < k$ for every vertex c in $D - C$,
(iii) $deg_{D'}(b) \leq deg_D(b) + 1 < k$ for every vertex b in $C - B$,
(iv) $deg_{D'}(a) \leq deg_D(a) = k - 1$ for every vertex a in $B - A$,

because every vertex in $B - A$ has at least one neighbor in A, since A is a maximal independent set in $\langle B \rangle$.

Now it follows that $|D'| = |D| - |A| + 1$ and $m\langle D' \rangle = m\langle D \rangle - (k-1)|A| + |C| - |A| = m\langle D \rangle - k|A| + |C|$. Therefore, $k|D'| - m\langle D' \rangle = k|D| - m\langle D \rangle + k - |C| > k|D| - m\langle D \rangle$, which contradicts the choice of D. Thus, D is a k-dominating set. $\square$

Corollary 7.14 *For any graph G and any positive integer k, $\gamma_k(G) \le \beta_k(G)$.*

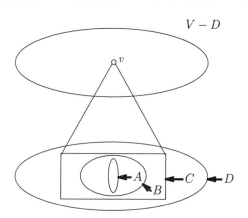

Figure 7.1: Favaron's proof.

We close this discussion of k-dominating sets by mentioning a fascinating conjecture of Fink and Jacobson [473]. Corollary 7.2 shows if G is a graph with $\Delta(G) \ge 3$ and $k \ge 3$, then $\gamma_k(G) > \gamma(G)$, so in particular, $\gamma_3(G) > \gamma_1(G)$.

This raises the question: does there exist a function $f(k)$, such that for any graph G with $\delta(G) \ge k$ and $j \ge f(k)$, we have $\gamma_k(G) < \gamma_j(G)$? Schelp (see [740]) has shown that if such a function exists, then $f(k) \ge k^2/4$. His example is the following. Consider the class of graphs defined by:

$$G_k = \overline{K}_{k+1} + (k + 1)K_k.$$

This is the graph which is the join of $k + 1$ isolated vertices and $(k + 1)$ copies of the complete graph on k vertices (see example for $k = 3$ in Figure 7.2). Observe that for Schelp's graphs, $|V(G_k)| = (k + 1)^2$ and $\delta(G_k) = 2k$. One can also see that $\gamma_{2k}(G_k) = \gamma_{k(k+1)}(G_k) = k(k + 1)$.

Chen and Jacobson proved the following result.

Theorem 7.15 [227] *If G is a graph with $\delta(G) \ge 2$, then $\gamma_2(G) < \gamma_5(G)$.*

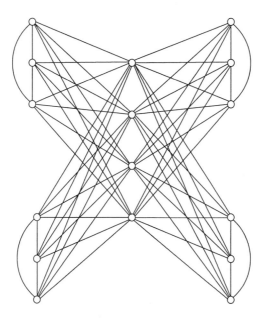

Figure 7.2: $G_3 = \overline{K}_4 + 4K_3$.

This means that the function $f(k)$ must satisfy $f(2) \leq 5$. But they point out that for the graph $K_{4,t}$, $\gamma_2(K_{4,t}) = \gamma_4(K_{4,t}) = 4$, which implies that $f(2) > 4$. Therefore, $f(2) = 5$. The question of the existence of the function $f(k)$, as originally posed in [473] remains unanswered.

Before concluding this section on multiple domination, we should mention the variation of k-domination called k-tuple domination, which was introduced by Harary and Haynes in [581]. A set D is a k-*tuple dominating set* for G if each vertex in V is dominated by at least k vertices in D. The difference between these two types of multiple domination is that with k-domination the vertices in $V - D$ are the only ones that must be multiply dominated, while with k-tuple domination every vertex in V must be multiply dominated. We note that for k-tuple domination, each vertex dominates its closed neighborhood. Hence, it is not the same as *total (open) k-domination* in which each vertex $v \in V$ must have at least k neighbors in D (see, for example, Kulli [804]).

The minimum cardinality of a k-tuple dominating set is the k-*tuple domination number* $\gamma_{\times k}(G)$. It follows from the definitions that

$$\gamma(G) \leq \gamma_k(G) \leq \gamma_{\times k}(G) \text{ and } \gamma(G) = \gamma_1(G) = \gamma_{\times 1}(G).$$

Obviously, $\gamma_{\times k}(G)$ is not defined unless $k \leq \delta(G) + 1$. Therefore, for trees, with $\delta(G) = 1$, we can only consider $\gamma_{\times 1}(G)$ and $\gamma_{\times 2}(G)$; the latter value is called the *double domination number* of G, and is denoted instead $dd(G)$. Although not much is presently known about the double domination number of a graph, the following can be observed:

(i) $dd(K_{1,n-1}) = n$;

(ii) $dd(K_n) = 2$;

(iii) $dd(C_n) = \lceil 2n/3 \rceil$;

(iv) $dd(W_n) = 1 + \lceil n/3 \rceil$;

(v) $dd(K_{r,s}) = 4$ for $r \geq 3$ and $s \geq 3$.

We conclude this section with a sharp upper bound on the double domination number.

Theorem 7.16 [581] *If G is a graph with $\delta(G) \geq 2$, then*

$$dd(G) \leq \begin{cases} \lfloor n/2 \rfloor + \gamma(G) & \text{for } n = 3 \text{ and } n = 5, \\ \lfloor n/2 \rfloor + \gamma(G) - 1 & \text{otherwise.} \end{cases}$$

To see that this bound is sharp, consider a graph G on $3+2r$ vertices for $r \geq 2$, say $\{y_1, y_2, x, u_1, u_2, \cdots, u_r, u_1', u_2', \cdots, u_r'\}$, with $E(G) = \{y_1 x, y_1 u_i, y_2 x, y_2 u_i', u_i u_i' : 1 \leq i \leq r\}$. The graph G has $\gamma(G) = 2$ and $dd(G) = r + 2 = \lfloor n/2 \rfloor + \gamma(G) - 1$.

7.2 Parity Restrictions

For a graph $G = (V, E)$ with closed neighborhood matrix N, a set $S \subseteq V$ with characteristic binary column n-tuple $X_s = [e_1, e_2, \cdots, e_n]^t$, where $e_i = 1$ if $v_i \in S$ and $e_i = 0$ if $v_i \notin S$, is:

(i) a dominating set if $N \cdot X_s \geq \vec{1}_n$,

(ii) a packing if $N \cdot X_s \leq \vec{1}_n$,

(iii) an efficient dominating set if $N \cdot X_s = \vec{1}_n$,

(iv) an enclaveless set if $N \cdot X_s \leq D$ for degree sequence D, and

(v) a k-dominating set if $N \cdot X_s \geq \overline{k}_n$ where $\overline{k}_n$ is the column n-tuple $[k, k, ..., k]^t$.

An interesting variation due to Sutner does not set upper or lower bounds for each $|N[v] \cap S|$, but rather sets a parity condition.

Theorem 7.17 [1073] *For any graph G, there exists a subset $S \subseteq V$ such that for every vertex $v \in V$, $|N[v] \cap S|$ is odd.*

Sutner's result helped to motivate the following general domination problem, defined in [33]. To each vertex v_i, assign a set R_i of nonnegative integers. Does there exist a set $S \subseteq V$ such that $|N[v_i] \cap S| \in R_i$, for $1 \leq i \leq n$? For dominating sets S, $|N[v_i] \cap S| \in R_i = \{1, 2, 3, \cdots\}$; for packings, $R_i = \{0, 1\}$; for efficient dominating sets, $R_i = \{1\}$ (a condition which is not always achievable); for enclaveless sets, $R_i = \{0, 1, \cdots, deg(v_i)\}$; for k-dominating sets, $R_i = \{k, k + 1, \cdots\}$; and for odd parity dominating sets (Sutner's problem), $R_i = \{1, 3, 5, \cdots\}$. In this section we present results for the case where each R_i is either $\{1, 3, 5, \cdots\}$ or $\{0, 2, 4, \cdots\}$, that is, each v_i has an odd or even closed neighborhood parity assignment. Note that Sutner's Theorem can be restated as follows: for every graph G, the equation $N \cdot X = \vec{1}_n$ (mod 2) has a solution with $X \in \{0, 1\}^n$.

For the graph G in Figure 7.3, we have for $S_1 = \{v_1, v_4\}$ and $S_2 = \{v_1, v_3, v_4, v_5\}$ that $N \cdot X_{S_1} = [1, 1, 1, 1, 1]^t$ and $N \cdot X_{S_2} = [1, 3, 3, 3, 3]^t = [1, 1, 1, 1, 1]^t \pmod{2}$. Also, for $S_3 = \{v_1, v_2, v_3, v_4\}$ and $S_4 = \{v_1, v_2, v_4, v_5\}$, we have $N \cdot X_{S_3} = [0, 1, 1, 0, 1]^t \pmod 2 = N \cdot X_{S_4} = X_D$ where $D = \{v_2, v_3, v_5\}$. Given a set $D \subseteq V(G)$, a set $S \subseteq V(G)$ is called a D-*parity dominating set* if $|N[v_i] \cap S|$ is odd for $v_i \in D$ and even for $v_i \notin D$, that is, $N \cdot X_S \pmod 2 = X_D$. A $V(G)$-parity set (respectively, $\emptyset$-parity set) is called an *all-odd* (respectively, *all-even*) parity set. For the graph G in Figure 7.3, the all-odd parity sets are S_1 and S_2; the all-even parity sets are the empty set, $\emptyset$, and $\{v_3, v_5\}$; and the $\{v_2, v_3, v_5\}$-parity sets are S_3 and S_4.

Sutner [1073] observed that the decision problem associated with finding the minimum cardinality of an all-odd parity set S is NP-hard; Dawes [345] provided a linear algorithm to compute this number for trees; and a linear algorithm to find the minimum cardinality of a D-parity set S for an arbitrary $D \subseteq V$ for series-parallel graphs, when such an S exists, is given in [33]. In the following, we consider how many D-parity sets there are for a given $D \subseteq V$ and how many vertex sets D have a D-parity set.

Letting $\oplus$ denote componentwise addition (modulo 2) of binary n-tuples, $(Z_2^n, \oplus)$ is a vector space over Z_2. Similarly, for subsets R and S of $V(G)$, we let $R \oplus S$ denote the symmetric difference of R and S. Let $\otimes$ denote matrix multiplication (modulo 2). We have $N \otimes X_S = X_D$ if and only if S is a D-parity set. We note that the function $L : Z_2^n \to Z_2^n$ defined by $L(X) = N \otimes X$ is a linear transformation. Its null space in Z_2^n is $\mathcal{N}(L) = \{X \in Z_2^n : N \otimes X = [0, 0, \cdots, 0]^t\}$.

Hence, letting $NL(G)$ denote the collection of vertex subsets $S \subseteq V$ such that $|N[v] \cap S|$ is even for every $v \in V$ (that is, the collection of all-even parity assignments), $NL(G) = \{S : S \subseteq V, X_S \in \mathcal{N}(L)\}$. We define the *parity dimension* of G, denoted $PD(G)$, to be the dimension of $\mathcal{N}(L)$. For a fixed

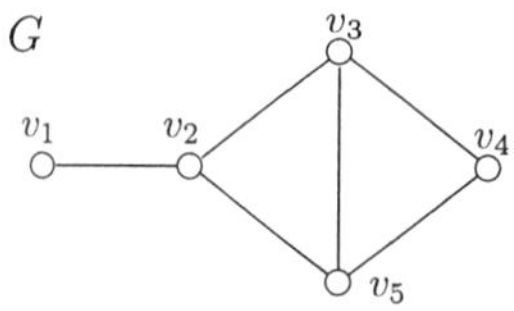

G

$$N = \begin{bmatrix} 1 & 1 & 0 & 0 & 0 \\ 1 & 1 & 1 & 0 & 1 \\ 0 & 1 & 1 & 1 & 1 \\ 0 & 0 & 1 & 1 & 1 \\ 0 & 1 & 1 & 1 & 1 \end{bmatrix}$$

Subset Vector X										Parity Assignments Realized, $B = NX$				
v_1	v_2	v_3	v_4	v_5	v_1	v_2	v_3	v_4	v_5	v_1	v_2	v_3	v_4	v_5
1	0	0	1	0	1	0	1	1	1	1	1	1	1	1
0	0	0	0	0	0	0	1	0	1	0	0	0	0	0
0	0	0	1	0	0	0	1	1	1	0	0	1	1	1
0	0	1	0	0	0	0	0	0	1	0	1	1	1	1
0	0	1	1	0	0	0	0	1	1	0	1	0	0	0
0	1	0	0	0	0	1	1	0	1	1	1	1	0	1
0	1	0	1	0	0	1	1	1	1	1	1	0	1	0
0	1	1	0	0	0	1	0	0	1	1	0	0	1	0
0	1	1	1	0	0	1	0	1	1	1	0	1	0	1
1	0	0	0	0	1	0	1	0	1	1	1	0	0	0
1	0	1	0	0	1	0	0	0	1	1	0	1	1	1
1	0	1	1	0	1	0	0	1	1	1	0	0	0	0
1	1	0	0	0	1	1	1	0	1	0	0	1	0	1
1	1	0	1	0	1	1	1	1	1	0	0	0	1	0
1	1	1	0	0	1	1	0	0	1	0	1	0	1	0
1	1	1	1	0	1	1	0	1	1	0	1	1	0	1

Figure 7.3: Parity assignments for closed neighborhoods. $PD(G) = 1$.

$D \subseteq V$, if $Y_0 \in Z_2^n$ satisfies $N \otimes Y_0 = X_D$, then the set of all solutions to $N \otimes Y = X_D$ is $\{Y_0 \oplus X : X \in \mathcal{N}(L)\}$. In particular, for $D \subseteq V$ the number of D-parity sets is either 0 or $2^{PD(G)}$, and Sutner's Theorem is that the number of $V(G)$-parity sets is $2^{PD(G)} \geq 1$. Note that there exists a (unique) D-parity set for every $D \subseteq V$ if and only if $PD(G) = 0$, so the *all-parity realizable graphs* (or *APR graphs*) defined in [33, 34] are precisely those with parity dimension zero. We have the following theorem.

Theorem 7.18 [34] *Let G be a graph with $PD(G) = k$ and let $D \subseteq V$. If G has a D-parity set, then there are exactly 2^k distinct D-parity sets.*

Every graph has a $\emptyset$-parity set, namely the empty set, and by Sutner's Theorem every graph has a $V(G)$-parity set. Thus, we have the following.

Corollary 7.19 *If G is a graph with $PD(G) = k$, then there are exactly 2^k distinct $\emptyset$-parity sets and 2^k distinct $V(G)$-parity sets.*

Corollary 7.20 *A graph G is APR if and only if $S = \emptyset$ is the only all-even parity set.*

Proposition 7.21 [34] *If a vertex u is in some all-even parity set of G, then u is in exactly $2^{PD(G)-1}$ all-even parity sets of G.*

Proof. Let n_1 and n_2 denote the number of all-even parity sets containing and not containing u, respectively. Let $u \in S \subseteq V$ with $N \otimes X_S = [0, 0, ..., 0]^t$, that is, S is an all-even parity set containing u. For each set $Y \subseteq V$ with $N \otimes X_Y = [0, 0, ..., 0]^t$ and $u \notin Y$, the set $S' = S \oplus Y$ with $u \in S'$ is an all-even parity set. Hence, $n_1 \geq n_2$.

If Z is any all-even parity set with $u \in Z$, then $u \notin S \oplus Z$, and $S \oplus Z$ is an all-even parity set. Hence, $n_2 \geq n_1$. Therefore, $n_1 = n_2$ and $n_1 + n_2 = 2^{PD(G)}$. $\square$

Lemma 7.22 [33] *If $S \subseteq V$ is an all-even parity set, then the cardinality of S is even.*

Proposition 7.23 [34] *For any graph G, $PD(G) \leq n - 1$ and equality holds if and only if $G = K_n$.*

Proof. By Lemma 7.22, a graph G has at most 2^{n-1} $\emptyset$-parity sets. And, by Corollary 7.19, G has exactly $2^{PD(G)}$ $\emptyset$-parity sets, and hence, $PD(G) \leq n - 1$. It is easily verified that $PD(K_n) = n-1$. Let G be a graph with $PD(G) = n-1$. Then, by Corollary 7.19, G has 2^{n-1} $V(G)$-parity sets. This implies that every odd subset of V is a $V(G)$-parity set. Assume that G is not a complete graph, and let $uv \notin E(G)$. Consider the odd subset $\{v\}$ of V. Then $|N[u] \cap \{v\}|$ is even; a contradiction, completing the proof. $\square$

From Corollary 7.20, we note that if every vertex in G has odd degree, then V is an all-even parity set, and the next result follows.

Corollary 7.24 *If every vertex in G has odd degree, then G is not APR, that is, $PD(G) \geq 1$.*

Theorem 7.25 [34] *If every vertex v in a tree T has odd degree, then $PD(T) = 1$.*

Proof. Since $V(T)$ and $\emptyset$ are all-even parity sets, $PD(T) \geq 1$. Assume that there is an all-even parity set $S \subseteq V(T)$ with $\emptyset \neq S \neq V(T)$. There exists an edge $u_0 v_0 \in E(T)$ with $u_0 \in S$ and $v_0 \notin S$. Then $|S \cap (N(v_0) - \{u_0\})|$ is odd, so we can choose $u_1 \in S \cap N(v_0)$ with $u_1 \neq u_0$. Then $|N[u_1]|$ is even, as is $|S \cap N[u_1]|$. Since $v_0 \in N[u_1]$ with $v_0 \notin S$, we can choose a vertex $v_1 \neq v_0$ where $v_1 \in N[u_1]$ but $v_1 \notin S$. Iterating this procedure, we could obtain an arbitrarily long path $u_0, v_0, u_1, v_1, u_2, v_2, \cdots$, in T with each $u_i \in S$ and each $v_i \notin S$. This contradicts the finiteness of $V(T)$. $\square$

Theorem 7.26 [34] *If there exists exactly one vertex of even degree in a tree T, then T is APR (that is, $PD(T) = 0$).*

Proof. Assume that some tree T is a smallest counterexample, where a vertex v has even degree and all other vertices $u \neq v$ have odd degree. Let S be a nonempty, all-even parity set. Such a set S, with $|N[w] \cap S|$ even for every vertex $w \in V(T)$, exists by Corollary 7.20.

First, suppose $v \notin S$. Let v' be a vertex in S closest to v with $v', w, \cdots, v$ being the $v' - v$ path in T (possibly having $w = v$). Then the subtree formed by the component of $T - v'w$ containing v' would be a smaller counterexample. Thus, $v \in S$. Consider T as being rooted at v. Because $v \in S$, $deg(v)$ is even and $|N[v] \cap S|$ is even, it follows that some child of v, say v_1, is not in S. Because $|N[v_1] \cap S|$ is even, some child of v_1, say v_2, is in S. The component $T - v_1 v_2$ containing v_2 would be a smaller counterexample, completing the proof. □

Assume that a path P_n has $V(P_n) = \{v_1, v_2, \cdots, v_n\}$ with $v_i v_{i+1} \in E(P_n)$ for $1 \leq i \leq n - 1$, and let S be an all-even parity set. If $v_1 \notin S$ and v_t is the first vertex on P_n with $v_t \in S$, then $|N[v_{t-1}] \cap S| = 1$, a contradiction. Hence, $v_1 \notin S$ implies $S = \emptyset$. On the other hand, by considering that $|S \cap N[v_i]|$ must be even for $i = 1, 2, 3, \cdots, n$, it is easy to verify that if $v_1 \in S$, then $v_2 \in S$, $v_3 \notin S$, $v_4 \in S$, $v_5 \in S$, $v_6 \notin S$, etc. In general, $v_i \notin S$ if and only if $i = 3k$ and we must have $n = 2 \pmod 3$.

Proposition 7.27 [34] *For the path P_n, $PD(P_{3k}) = PD(P_{3k+1}) = 0$ and $PD(P_{3k+2}) = 1$.*

Theorem 7.28 [34] *If U is the set of endvertices of a tree T, then $PD(T) \leq |U| - 1$.*

Proof. Let $U = \{u_1, u_2, \cdots, u_k\}$ be the set of endvertices of a tree T and let $U^* = U - \{u_k\}$. To see that the number of all-even parity sets is at most 2^{k-1}, we will show that if S_1 and S_2 are distinct all-even parity sets then $S_1 \cap U^* \neq S_2 \cap U^*$. Thus we root T at u_k and assume $S_1 \neq S_2$, but $S_1 \cap U^* = S_2 \cap U^*$. Then we can find an interior vertex v such that (1) if x is a descendant of v, then either $x \in S_1 \cap S_2$ or $x \notin S_1 \cup S_2$ and (2) v is in exactly one of S_1 and S_2, say $v \in S_1$ and $v \notin S_2$. But if w is a child of v, then $N[w] \cap S_1 = (N[w] \cap S_2) \cup \{v\}$, a contradiction because both sets must be even. □

Along with considering certain special classes of trees such as spiders, caterpillars, and k-ary trees, Amin and Slater presented a constructive characterization of *APR*-trees.

Theorem 7.29 [34] *If T_n is a tree on $n \geq 5$ vertices, then $PD(T_{2k}) \leq k - 2$ and $PD(T_{2k+1}) \leq k - 1$, and these bounds are sharp.*

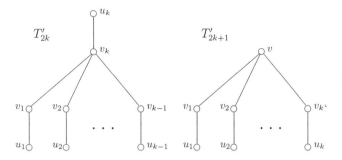

Figure 7.4: $PD(T'_{2k}) = k - 2$ and $PD(T'_{2k+1}) = k - 1$.

Proof. For the trees T'_{2k} and T'_{2k+1} in Figure 7.4, one can verify that $PD(T'_{2k}) = k - 2$ and $PD(T'_{2k+1}) = k - 1$. For T'_{2k}, the all-even parity sets are those that consist of an even subset of $\{u_1, u_2, \cdots, u_{k-1}\}$ and the corresponding v_i's. No all-even parity set contains u_k and v_k. For T'_{2k+1}, no all-even parity set contains v, and the all-even parity sets are those that consist of an even subset of $\{u_1, u_2, \cdots, u_k\}$ and the corresponding v_i's.

One can verify that each of the three trees on five vertices and six trees on six vertices (Figure 4 of the Prolegomenon) has parity dimension 0 or 1. Thus, we assume that tree T on $n \geq 7$ vertices is a smallest counterexample to the theorem. Suppose T has two endvertices with a common neighbor, say $deg(u_1) = deg(u_2) = 1$ with $N(u_1) = N(u_2) = \{v\}$. Let $T^* = T - u_1 - u_2$. We will show that $PD(T^*) = PD(T)$.

First, note that any all-even parity set S for T has $|S \cap \{v, u_1, u_2\}| = 0$ or 3. Let S_1 and S_2 be distinct all-even parity sets for T. It follows that each $S_i^* = S_i \cap V(T^*)$ is an all-even parity set for T^*. As noted, if $\{u_1, u_2\} \subseteq S_i$, then $v \in S_i^*$, and thus $S_1^* \neq S_2^*$. Hence, $PD(T^*) \geq PD(T)$.

Second, for the converse, if S_1^* and S_2^* are distinct all-even parity sets for T^*, then (1) $v \notin S_i^*$ implies $S_i = S_i^*$ is also an all-even parity set for T and (2) $v \in S_i^*$ implies $S_i = S_i^* \cup \{u_1, u_2\}$ is an all-even parity set for T. It follows that S_1 and S_2 are distinct all-even parity sets for T, and $PD(T) \geq PD(T^*)$. Thus, $PD(T) = PD(T^*)$ and T^* would be a smaller counterexample.

Suppose no two endvertices of T have a common neighbor. Then the set U of endvertices of T has $|U| \leq n/2$. If $n = 2k + 1$ then, by Theorem 7.28, $PD(T) \leq |U| - 1 \leq k - 1$, a contradiction, and we are finished. If $n = 2k$ and $|U| \leq k - 1$, then $PD(T) \leq |U| - 1 \leq k - 2$, and we are finished. The only remaining case is for $n = 2k$, $|U| = k$, and each endvertex is adjacent to a distinct vertex. That is, $T - U$ is a tree on $k \geq 4$ vertices with each $v \in T - U$ adjacent to a unique $v' \in T$. Observe that for each all-even parity set S, the elements in $S \cap (T - U)$ are paired with elements of $S \cap U$. Thus, it suffices to show that $S \cap (T - U)$ contains at most 2^{k-2} subsets of $T - U$. For $v \in T - U$, if $v \in S$, then the vertex in U adjacent to v is in S and $N(v) \cap (T - U)$ contains an even number of vertices. If $v \in T - U$ and $v \notin S$, then the vertex in U adjacent to v is also not in S, so again $|N(v) \cap (T - U)|$ is even. Letting x and y be adjacent vertices in $T - U$ and $S^* \subseteq (T - U) - \{x, y\}$, at most one of S^*, $S^* \cup \{x\}$, $S^* \cup \{y\}$, and $S^* \cup \{x, y\}$ can have an even number of elements in $N(x)$ and $N(y)$, so there are at most 2^{k-2} possibilities for $S \cap (T - U)$. Thus, $PD(T) \leq k - 2$, completing the proof. □

Theorem 7.30 [34] *For every integer k, $0 \leq k \leq \lfloor (n - 3)/2 \rfloor$, there exists a tree T with n vertices and $PD(T) = k$. (See Figure 7.5)*

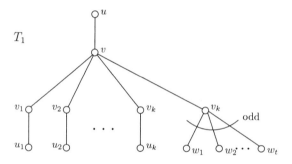

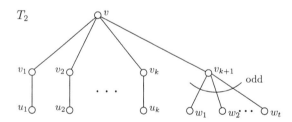

Figure 7.5: $PD(T_1) = k$ and $PD(T_2) = k$, odd.

We conclude this section with the following results.

Theorem 7.31 [34]

(a) *If $F_n = P_n + K_1$ is a fan on $n + 1$ vertices, then $PD(F_n) = 1$ if n is odd, $PD(F_n) = 0$ if $n \neq 3k + 2$ is even, and $PD(F_n) = 2$ if $n = 3k + 2$ is even.*

(b) *For a cycle* C_n, $PD(C_{3k+1}) = PD(C_{3k+2}) = 0$ *and* $PD(C_{3k}) = 2$.

(c) *For the wheel,* $W_n = C_n + K_1$, $PD(W_n) = 0$ *if* $n \neq 3k$ *is even,* $PD(W_n) = 1$ *if* $n \neq 3k$ *is odd,* $PD(W_n) = 2$ *if* $n = 3k$ *is even and* $PD(W_n) = 3$ *if* $n = 3k$ *is odd.*

(d) *Let* $K_{n_1,n_2,\cdots,n_k}$ *be a complete k-partite graph. If j denotes the number of odd values in* $\{n_1, n_2, \cdots, n_k\}$*, then* $PD(K_{n_1,n_2,\cdots,n_k})$ *is 0 if $j = 0$ and is 1 if $j \neq 0$.*

From Theorem 7.29, we see that the parity domination for a tree on n vertices is less that $n/2$. In contrast, for arbitrary graphs we have the following result.

Theorem 7.32 [34] *For every integer k, $0 \leq k < n$, there exists a graph G on n vertices with $PD(G) = k$.*

7.3 Locating-Domination

In the main, conditions on a dominating set D put restrictions either on the subgraph induced by D, or on the number of times each vertex v in $V(G)$ (or perhaps just in $V(G) - D$) must be dominated. For example, each vertex v must be dominated at least k times, at most k times, exactly k times, at least $deg(v)$ times, or, as in Section 7.2, a certain parity number of times. In this section we consider the restriction that the sets $N(v) \cap D$ must be distinct as v ranges over $V(G) - D$. We outline the merging of the concepts of locating sets and dominating sets.

When considering locating sets in graphs, as introduced in [1030], one can think of selecting a minimum set S of vertices of a graph G to achieve the function "location through triangulation", (analogous to locating sonar or Loran stations). The stations are to be established so that each vertex is uniquely determined by its distances to the sites. For a given k-tuple of vertices $(v_1, v_2, \cdots, v_k)$, assign to each vertex $v \in V(G)$ the k-tuple of its distances to these vertices, $f(v) = (d(v, v_1), d(v, v_2), \cdots, d(v, v_k))$. Letting $S = (v_1, v_2, \cdots, v_k)$, the k-tuple $f(v)$ is called the *S-location of v*. A set S is a *locating set* for G if no two vertices have the same S-location, and the *location number* $L(G)$ is the minimum cardinality of a locating set. For example, $L(G) = 1$ if and only if G is a path. The path P_n with $n \geq 2$ vertices has two $L(P_n)$-sets, each consisting of an endvertex. For the tree T in Figure 7.6, the set $\{v_1, v_2, v_3, v_4, v_5, v_6, v_7, v_8\}$ is an $L(T)$-set. For example, $f(v_2) = (2, 0, 2, 4, 3, 4, 4, 5)$, $f(x) = (1, 1, 1, 3, 2, 3, 3, 4)$, and $f(y) = (2, 2, 2, 2, 3, 4, 4, 5)$.

A simple characterization of $L(T)$-sets for trees T follows from the next theorem.

Theorem 7.33 [1030] *Let T be a tree with $n \geq 3$ vertices. Set S is a locating set if and only if for each vertex u there are vertices in S contained in at least $deg(u) - 1$ of the $deg(u)$ components of $T - u$.*

Proof. Assume that two components of $T - u$ do not contain any vertices in S, and let v_1 and v_2 be the vertices in $N(u)$ in these components adjacent to u. Then $f(v_1) = f(v_2)$ and S is not a locating set.

For the sufficiency, suppose $S = (v_1, v_2, \cdots, v_k)$ is a k-tuple of vertices from T. Assume $d(u, v_1) = d(w, v_1)$, and let P be the path from u to w in T. If $v_1 \in V(P)$, then one can assume that some v_i with $2 \leq i \leq k$ is in the branch from v_1 that contains u. Then $d(u, v_i) \leq d(u, v_1) - 1 + d(v_1, v_i) - 1 < d(w, v_1) + d(v_1, v_i) = d(w, v_i)$. Hence, $f(u) \neq f(w)$. If $v_1 \notin V(P)$, let v be the first vertex on both the u to v_1 and w to v_1 paths in T. One can assume that the branch at v containing u contains some v_i. As above, with v in place of v_1, one obtains $d(u, v_i) < d(w, v_i)$, and $f(u) \neq f(w)$, completing the proof. □

For safeguards analysis of a facility, such as a fire protection study of nuclear power plants (as in Hulme, Shiver, and Slater, A Boolean algebraic analysis of fire protection, *Annals Discrete Math.* 19 (1984) 215–228), the facility can be modelled by a graph or network. For such applications, a vertex can represent a room, hallway, stairwell, courtyard, etc. Each edge can connect two areas that are either physically adjacent or within sight or sound of each other. One primary function of a safeguards system is "detection" of some object – perhaps detection of a fire, or detection of an intruder, such as a saboteur. Suppose we have a detection device located at a vertex and the device can supply three outputs: (i) there is an object at that vertex, (ii) there is an object at one of the vertices adjacent to that vertex (but which adjacent vertex cannot be specified), or (iii) there is no object at that vertex or any adjacent vertex. It is necessary to determine a collection of vertices at which to place detection devices so that if there is an object at any vertex in the graph, it can be detected, and its position uniquely identified.

In order to detect an object which might be at any vertex in $V(G)$, it is necessary to have a dominating set D. The additional problem of uniquely identifying the location of an object requires a "locating" feature. For each $v \in V$, let $N_D(v) = N(v) \cap D$. A dominating set D is called a *locating-dominating set*, or simply an *LD-set*, if for any two vertices v and w in $V - D$, $N_D(v) \neq N_D(w)$. The *locating-domination number* $\gamma_L(G)$ is the minimum cardinality of an LD-set for G. For the tree T in Figure 7.6, $L(T) = 8$. One can verify that $\gamma_L(T) = 13$.

Theorem 7.34 [1036]

(a) *For a complete graph, $\gamma_L(K_n) = n - 1$.*

(b) *For paths and cycles, $\gamma_L(P_{5k}) = \gamma_L(C_{5k}) = 2k$,*
$\gamma_L(P_{5k+1}) = \gamma_L(C_{5k+1}) = \gamma_L(P_{5k+2}) = \gamma_L(C_{5k+2}) = 2k + 1$, *and*
$\gamma_L(P_{5k+3}) = \gamma_L(C_{5k+3}) = \gamma_L(P_{5k+4}) = \gamma_L(C_{5k+4}) = 2k + 2$.

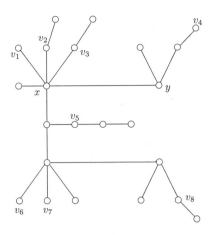

Figure 7.6: Tree T with $L(T) = 8$.

Thus, a locating-dominating set in a path requires at least 40% of the vertices. The next theorem shows that any locating-dominating set for an arbitrary tree contains more than one-third of its vertices.

Theorem 7.35 [1035] *For any tree* T, $\gamma_L(T) > n/3$.

Proof. Assume that G is a graph with n vertices and at most $n - 1$ edges, and let D be a vertex set of order $|D| \leq n/3$. If D is a $\gamma_L(G)$-set, then at most $n/3$ vertices v of $V - D$ have $|N_D(v)| = 1$. Thus, at least $n/3$ vertices of $V - D$ have $|N_D(v)| \geq 2$. But this implies that G has at least n edges, a contradiction. $\square$

Restating Theorem 7.35, if we have k detection devices available and construct a tree of order n with $\gamma_L(T) = k$, then $n \leq 3k - 1$. For arbitrary graphs, we have the following result.

Theorem 7.36 [1036] *If a graph* G *has* $\gamma_L(G) = k$, *then* $n \leq k + 2^k - 1$.

The maximum order of a graph G with $\gamma_L(G) = k$ was computed by Rall and Slater [960] for several other classes of graphs G including planar and outerplanar graphs.

Theorem 7.37 [960] *Let* G *be a graph with* $\gamma_L(G) = k$. *Then*

(a) *if* G *is planar, then* $n \leq 7k - 10$ *for* $k \geq 4$,

(b) *if* G *is outerplanar, then* $n \leq \lfloor (7k - 3)/2 \rfloor$, *and*

(c) *these bounds are sharp.*

The next result (also presented in Chapter 2) is easy to see.

Theorem 7.38 [1039] *If a graph G has a degree sequence $(d_1, d_2, \cdots, d_n)$ with $d_i \geq d_{i+1}$, then $\gamma(G) \geq \min\{k : k + (d_1 + d_2 + \cdots + d_k) \geq n\}$.*

If D is a dominating set and $v_i \in D$, then v_i dominates $1 + deg(v_i)$ vertices. For $\gamma_L(G)$, in addition, any two vertices u and w in $N(v_i) \cap (V - D)$ must satisfy $N_D(u) \neq N_D(w)$. For each $v \in D$, let

$$\gamma(v, D) = 1 + \sum_{w \in N(v) \cap (V-D)} \frac{1}{|N_D(w)|}.$$

For example, if $x \in V - D$ and $N_D(x) = \{v_1, v_2, v_3\}$, then x contributes $1/3$ to each $\gamma(v_i, D)$, for $1 \leq i \leq 3$. We can consider $\gamma(v, D)$ to be the amount of domination done by v for D. Note that $\sum_{v \in D} \gamma(v, D) = n$. Let D be an LD-set with $v \in D$. Each vertex $w \in N(v) \cap D$ does not contribute to $\gamma(v, D)$. Note that at most one vertex w in $N(v) \cap (V - D)$ can have $N_D(w) = \{v\}$ and contribute one to $\gamma(v, D)$. Any other vertex x in $N(v) \cap (V - D)$ (that is, any $x \in N(v)$ which is not a private neighbor of v) contributes $\frac{1}{|N_D(x)|} \leq 1/2$. Consequently, $\gamma(v, D) \leq 1 + 1 + |N(v) - 1|/2 = 2 + (deg(v) - 1)/2 = (3 + deg(v))/2$, and the next result follows.

Theorem 7.39 [1039]

(a) *If a graph G has degree sequence $(d_1, d_2, \cdots, d_n)$ with $d_i \geq d_{i+1}$, then $\gamma_L(G) \geq \min\{k : (3k + d_1 + d_2 + \cdots + d_k)/2 \geq n\}$.*

(b) *If G is an r-regular graph, then $\gamma_L(G) \geq 2n/(r + 3)$.*

The complexity questions for $\gamma_L(G)$ have also been considered.

LOCATING-DOMINATING SET
INSTANCE: A graph $G = (V, E)$ and a positive integer k.
QUESTION: Does G have a locating-dominating set of cardinality $\leq k$?

Theorem 7.40 [317]

(a) *LOCATING-DOMINATING SET is NP-complete.*

(b) *For trees and series-parallel graphs G, $\gamma_L(G)$ can be determined in linear time.*

Recall that a graph G is well-covered if $i(G) = \beta(G)$, that is, all maximal independent sets have the same cardinality. Note that the set $\{u, v\}$ in the graph of Figure 7.7 is dominating but not locating. We conclude this section with an amazing connection to well-covered graphs due to Finbow and Hartnell.

Theorem 7.41 [469] *If a graph G has girth $g(G) \geq 5$, then G is well-covered if and only if every independent dominating set of G is locating.*

Figure 7.7: Every independent dominating set of G is locating.

7.4 Distance Domination

A set S of vertices in a graph $G = (V, E)$ is a *distance-k dominating* set if every vertex in $V - S$ is within distance k of at least one vertex in S. Nearly all of the concepts involving dominating sets, independent sets, irredundant sets and other related types of domination can be given a distance-version, as we will see in this section. In many ways the distance versions of these concepts are more applicable to modeling real-world problems. And, indeed, much of the motivation for the study of domination in graphs stems from problems involving the placement of a minimum number of objects (hospitals, fire stations, post offices, police stations, warehouses, service centers and the like) within acceptable distances of a given population, or conversely, the placement of undesirable objects (e.g., toxic wastes, nuclear reactors, airports) at maximum distances from a given population. Slater [1031] considers the general case where the "acceptable" distance can vary from vertex to vertex. For example, elementary schools might require fire stations to be closer than a college requires.

The *open k-neighborhood* of a vertex $v \in V$, denoted $N_k(v)$, is the set $N_k(v) = \{u : u \neq v \text{ and } d(u, v) \leq k\}$. The set $N_k[v] = N_k(v) \cup \{v\}$ is called the *closed k-neighborhood of v*. Every vertex $w \in N_k[v]$ is said to be *k-adjacent* to v. Thus, we can define the *k-degree $deg_k(v)$* of a vertex as $|N_k(v)|$. The *minimum k-degree* $\delta_k(G)$ equals $\min\{deg_k(v) : v \in V\}$, while the *maximum k-degree* $\Delta_k(G)$ equals $\max\{deg_k(v) : v \in V\}$. Finally, for a set S of vertices, we define $N_k(S)$ to be the union of the open k-neighborhoods of vertices in S, while $N_k[S]$ is the union of the closed k-neighborhoods of vertices in S.

A set S is a *distance-k dominating set* if $N_k[S] = V$. The *distance-k domination number* $\gamma_{\leq k}(G)$ of G equals the minimum cardinality of a distance-k dominating set in G. Notice that $\gamma_{\leq k}(G)$ is equal to $\gamma(G^k)$, where G^k is the kth power graph. For a comprehensive survey of results on distance domination in graphs the reader is referred to Henning [686], which is the basis for this short section.

Figure 7.8 illustrates a graph G having several distance-2 dominating sets of cardinality two, one of which is indicated by the shaded vertices, another by circled vertices. For this graph, in fact, $\gamma_{\leq 2}(G) = 2$. Also, $\gamma_{\leq 1}(G) = \gamma(G) = 5$, $\gamma_{\leq 3}(G) = 2$, and $\gamma_{\leq k}(G) = 1$ for every $k \geq 4$.

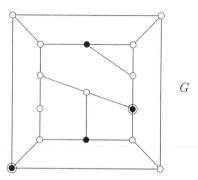

Figure 7.8: A graph with distance-2 dominating sets of cardinality 2.

We begin with a basic property of minimal distance-k dominating sets, observed by Henning, Oellermann, and Swart.

Proposition 7.42 [695] *Let S be a distance-k dominating set of a graph G for some $k \geq 1$. Then S is a minimal distance-k dominating set if and only if each vertex $u \in S$ satisfies at least one of the following conditions:*

P1: there exists a vertex $v \in V(G) - S$ for which $N_k(v) \cap S = \{u\}$;
P2: $d(u, w) > k$ for every vertex $w \in S - \{u\}$.

This proposition can be reworded, given some terminology. Define $pn_k[u, S] = N_k[u] - N_k[S - \{u\}]$ to be the *private k-neighbors* of a vertex u, with respect to a set S. If $pn_k[u, S] \neq \emptyset$, then we say that every vertex in $pn_k[u, S]$ is a *private k-neighbor* of u. Notice that it is possible that $u \in pn_k[u, S]$, in which case we say that vertex u is its own private k-neighbor, with respect to S.

Proposition 7.42 states that if S is a minimal distance-k dominating set, then every vertex in S is either its own private k-neighbor ($P2$) or has a private k-neighbor in $V - S$ ($P1$), or both. Notice that if $pn_k[u, S] \neq \emptyset$, then vertex u is the only vertex in S that distance-k dominates the vertices in $pn_k[u, S]$. Furthermore, if $u \in pn_k[u, S]$, then no vertex in $S - \{u\}$ distance-k dominates u.

A generalization of a classical theorem of Bollobás and Cockayne [126] (see Chapter 3) was proved by Henning, Oellermann, and Swart. In order to present their rather ingenious proof, we will first need some notation.

Let $S = \{u_1, u_2, \cdots, u_t\}$ be a distance-k dominating set in a graph G. For each vertex $u_i \in S$, let $d_i = d(u_i, S - \{u_i\})$. Thus, from S we can generate a sequence of distances, $(d_1, d_2, \cdots, d_t)$, representing the distances from each vertex to its closest neighbor in S. We also label the vertices of S so that the corresponding distance sequence is nondecreasing. For example, the four marked vertices in the graph in Figure 7.8 define the distance sequence (2,2,2,2), as each vertex has a minimum distance of two to another marked vertex.

Given two nondecreasing sequences of positive integers, we can define the usual lexicographic ordering relation as follows: $(d_1, d_2, \cdots, d_j) \le (e_1, e_2, \cdots, e_k)$ if either $j \le k$ and $d_i = e_i$ for $1 \le i \le j$, or there exists an i, $1 \le i \le \min\{j, k\}$, such that $d_i < e_i$ and $d_h = e_h$ for all $h < i$.

Theorem 7.43 [697] *For $k \ge 1$, if G is a connected graph with at least $k + 1$ vertices and $diam(G) \ge k$, then G has a minimum distance-k dominating set S such that every vertex $u \in S$ satisfies condition P1 and has a private k-neighbor $u' \in V - S$ for which $d(u, u') = k$.*

Proof. Among all minimum distance-k dominating sets in G, let S be one that has the lexicographically smallest distance sequence, that is, let $\gamma_{\le k}$-set $S = \{u_1, u_2, \cdots, u_t\}$ and let $(d_1, d_2, \cdots, d_t)$ be the corresponding distance sequence.

We must first show that every vertex in S has property P1. Assume that S does not have property P1 and let i be the smallest index such that u_i does not have property P1. By Proposition 7.42, vertex u_i must have property P2. Therefore, $d_i \ge k + 1$. Let u_i' be any vertex in $N_k(u_i)$. Consider the set $S' = S - \{u_i\} \cup \{u_i'\}$. It follows that S' is also a minimum distance-k dominating set in G. Also, vertex u_i' must be within distance k of some vertex in $S - \{u_i\}$, since u_i' is not a private k-neighbor of u_i. Consequently, $d_i' = d(u_i', S' - \{u_i'\}) < d_i$.

Let j be the largest integer for which $d_j < d_i$, and consider the value $d_h' = d(u_h, S' - \{u_h\})$, for each h with $1 \le h \le j$. Since $d_h < d_i$, a shortest path from u_h to a vertex of $S - \{u_i\}$ does not contain u_i. It follows, therefore, that $d_h' \le d_h$ for all h, $1 \le h \le j$. This, together with the observation that $d_i' < d_i$ implies that the distance sequence of S' is less than the distance sequence of S in lexicographic order. This contradicts the choice of S, and hence, every vertex in S has property P1.

We must show that every vertex $u \in S$ has a private k-neighbor $u' \in V - S$ for which $d(u, u') = k$. For each vertex $u_i \in S$, let w_i be a vertex of $V - S$ at maximum distance from u_i satisfying $N_k(w_i) \cap S = \{u_i\}$. We claim that $d(u_i, w_i) = k$ for all i. If this is not true, then let j be a smallest index for which $d(u_j, w_j) < k$. Note that every vertex in $V - S$ at distance greater than $k - 1$ from u_j is within distance k from some vertex of $S - \{u_j\}$. Consider a shortest path from u_j to a vertex of $S - \{u_j\}$ and let u_j^* be the vertex adjacent to u_j on this path. Now let $S^* = S - \{u_j\} \cup \{u_j^*\}$. As before, S^* must be a minimum distance-k dominating set. Next, let l be the largest integer for which $d_l < d_j$ and consider the value of $d_h^* = d(u_h, S^* - \{u_h\})$ for $1 \le h \le l$. It must be the case that $d_h^* \le d_h$ for all $1 \le h \le l$. Furthermore, $d(u_j^*, S^* - \{u_j^*\}) = d_j - 1 < d_m$ for all $m > j$. Thus, the distance sequence of S^* is less than the distance sequence for S, again contradicting the choice of S. Therefore, $d(u_i, w_i) = k$ for all i. $\square$

Henning, Oellermann, and Swart also have produced a distance version of the classical result of Jaeger and Payan (see Chapter 3).

Theorem 7.44 [695] *For $k \geq 2$ and a graph G with $n \geq k + 1$ vertices,*

(a) $2 \leq \gamma_{\leq k}(G) + \gamma_{\leq k}(\overline{G}) \leq n + 1$ *and* $1 \leq \gamma_{\leq k}(G)\gamma_{\leq k}(\overline{G}) \leq n$.

(b) *if both G and $\overline{G}$ are connected graphs then $2 \leq \gamma_{\leq k}(G) + \gamma_{\leq k}(\overline{G}) \leq n/(k + 1) + 1$ and $1 \leq \gamma_{\leq k}(G)\gamma_{\leq k}(\overline{G}) \leq n/(k + 1)$.*

(c) *each of the bounds in* (a) *and* (b) *are sharp.*

The examples are just a sampling of distance domination results. The interested reader is referred to numerous papers on such topics as the following:

(i) *well distance-k dominated graphs,* that is, graphs for which every minimal distance-k dominating set has the same cardinality;

(ii) *k-packings S in graphs,* where for every $uv \in S$, $d(u, v) > k$; k-packings bear a close relationship with distance-k dominating sets;

(iii) *distance domination critical graphs,* that is, graphs for which $\gamma_{\leq k}(G - v) > \gamma_{\leq k}(G)$ for every vertex $v \in V$;

(iv) *total distance domination number,* in which every vertex in V is within distance k from some vertex in S other than itself;

(v) *independent distance domination number,* in which in addition to a set S being distance-k dominating, every vertex in S is at least distance $k + 1$ from every other vertex in S;

(vi) *distance irredundance number,* in which every vertex in a set S has a private k-neighbor.

7.5 Strong and Weak Domination

Sampathkumar and Pushpa Latha [1003] were the first to introduce the notions of strong and weak domination in graphs. Given two adjacent vertices u and v, we say that u *strongly dominates* v if $deg(u) \geq deg(v)$. Similarly, we say that v *weakly dominates* u if $deg(v) \leq deg(u)$. A set $D \subseteq V(G)$ is a *strong-dominating set* of G if every vertex in $V - D$ is strongly dominated by at least one vertex in D. Similarly, D is a *weak-dominating set* if every vertex in $V - D$ is weakly dominated by at least one vertex in D. The *strong (weak) domination number* $\gamma_S(G)$ (respectively $\gamma_W(G)$) is the minimum cardinality of a strong (weak) dominating set of G. The graphs in Figure 7.9 illustrate these definitions.

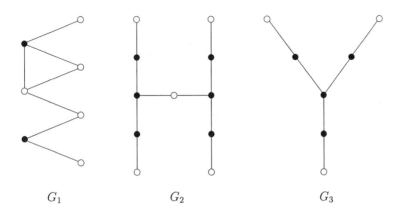

$$G_1 \qquad\qquad G_2 \qquad\qquad G_3$$

Figure 7.9: Strong dominating sets.

The shaded vertices in G_1, G_2, and G_3 in Figure 7.9 are minimum cardinality strong dominating sets, that is, $\gamma_S(G_1) = 2$, $\gamma_S(G_2) = 6$ and $\gamma_S(G_3) = 4$. One can also see that $\gamma_W(G_1) = 4$ (the four unshaded vertices at the right) and $\gamma_W(G_2) = 5$ (the unshaded vertices). Thus, it is clear that $\gamma_W(G) < \gamma_S(G)$ and $\gamma_S(G) < \gamma_W(G)$ are both possible. Notice that in G_2 the complement of the minimum strong dominating set is a minimum weak dominating set, while in G_3, the complement of the strong dominating set is not even a dominating set. The following preliminary observation gives some ideas about properties of strong and weak domination in graphs.

Theorem 7.45 [1003] *For any graph G,*

(a) $\gamma(G) \le \gamma_S(G) \le n - \Delta$ *and* $\gamma(G) \le \gamma_W(G) \le n - \delta$.

(b) $\gamma_S(\overline{G}) + \gamma_W(G) \le n + 1$.

Sampathkumar and Pushpa Latha also noted that no inequality holds between either of $\gamma_S(G)$ or $\gamma_W(G)$ and any of $i(G)$, $\gamma_c(G)$, and $\alpha_0(G)$.

We should also point out that computational complexity questions for strong and weak domination have been settled.

STRONG (WEAK) DOMINATING SET
INSTANCE: A graph $G = (V, E)$ and a positive integer k.
QUESTION: Does G have a strong (weak) dominating set of cardinality $\le k$?

It is known, but unpublished (A. McRae, private communication), that STRONG (WEAK) DOMINATING SET is NP-complete, even for bipartite and chordal graphs, and that linear algorithms exist for computing $\gamma_S(T)$ and $\gamma_W(T)$ for arbitrary trees T.

7.6   Global and Factor Domination

The notion of a dominating set D in a graph $G = (V, E)$ can be extended in a natural way to a set which is a dominating set in both G and the complement $\overline{G} = (V, \overline{E})$ of G. This concept was introduced independently by Sampathkumar [989], who used the term *global domination*, and by Brigham and Dutton [164], who used the more general term of *factor domination*. In this section, we will review the fundamental results concerning global and factor domination. All of the following results are found in [164].

A graph $H = (V, E)$ is said to have a *t-factoring* into factors $F(H) = \{G_1, G_2, \cdots, G_t\}$ if each graph $G_i = (V_i, E_i)$ has the same vertex set, $V_i = V$, and the edge sets $\{E_1, E_2, \cdots, E_t\}$ form a partition of E. Given a *t*-factoring F of H, a subset $D_f \subseteq V$ is a *factor dominating set* if D_f is a dominating set of G_i, for $1 \leq i \leq t$. The *factor domination number* $\gamma_{ft}(F(H))$ is the minimum cardinality of a factor dominating set of $F(H)$. We will write γ_{ft} when the graph H and the factoring $F(H)$ are understood, and γ_i will denote $\gamma(G_i)$. Also, for a generic parameter $\mu(G)$, we sometimes write just μ when G is understood and let $\overline{\mu}$ represent $\mu(\overline{G})$. An example of a *t*-factoring and a factor dominating set is given in Figure 7.10; note that the set of vertices $\{1,3,5\}$ in H dominates each factor G_1, G_2 and G_3, while no smaller factor dominating set exists, since G_3 has three components, at least three vertices are required to dominate G_3.

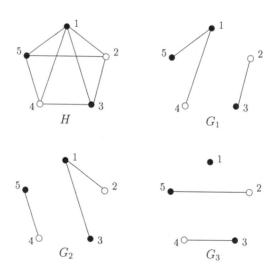

Figure 7.10: A factor dominating set.

Perhaps the first observation to be made about the factor domination number is that:

$$\max_{1 \le i \le t}\{\gamma_i\} \le \gamma_{ft} \le \sum_{i=1}^{t} \gamma_i.$$

Note that the lower bound is sharp for the graph H in Figure 7.10. The second observation is that the factor domination number is equal to the domination number of a graph H' which can be constructed from H and the graphs in the t-factoring, $F(H) = \{G_1, G_2, \cdots, G_t\}$, as follows. Construct disjoint copies of H and the graphs $G_1, G_2, \cdots, G_t$ (see Figure 7.10). Then join vertex u in $V(H)$ to the copy of u_i, in G_i and to all vertices in G_i adjacent to u_i. Do this for all vertices $u \in V(H)$ and all graphs G_i, $1 \le i \le t$. It follows from this construction that a minimum dominating set D of H' is contained in the copy of H and that D is also a factor dominating set of $F(H)$, that is, $\gamma_{ft} \le \gamma(H')$. Conversely, it follows that any factor dominating set of $F(H)$ is also a dominating set of H', that is, $\gamma(H') \le \gamma_{ft}$. Brigham and Dutton next present some upper and lower bounds for $\gamma_{ft}(F(H))$.

Theorem 7.46 [164] *For any graph,* $\gamma_{ft} \le n - \min_{1 \le i \le t}\{\delta_i\}$.

Theorem 7.47 [164] *If I is the set of vertices in $V(H)$ which are isolated in at least one G_i, then $\gamma_{ft} \le \alpha_0(H) + |I|$.*

Proof. Let S be a minimum vertex cover for H and let v be a nonisolated vertex of G_i with an incident edge e. Since e is an edge in H, at least one vertex, say u, covers e in H, and therefore this vertex u dominates v in G_i. It follows, therefore, that the set $S \cup I$ is a factor dominating set of $F(H)$. □

Theorem 7.48 [164] *For any graph H, $\gamma_{ft} \ge t$ if $t \le \Delta(H)$; $\gamma_{ft} = n$ otherwise.*

Proof. Let D_f be a minimum factor dominating set for $F(H)$. If $t \le \Delta(H)$, then any vertex v in $V(H) - D_f$ must have in H at least t edges to D_f so it can be dominated in each G_i. Hence, $|D_f| \ge t$. If $t > \Delta(H)$, no such vertex v can exist and $H - D_f$ must be empty, that is, $\gamma_{ft} = n$. □

Theorem 7.49 [164] *For any graph H, $\gamma_{ft} \le \gamma + t - 2$ if $t \le \Delta(H)$; $\gamma_{ft} = n$ otherwise.*

Proof. The proof of Theorem 7.48 shows that for $t > \Delta(H)$, $\gamma_{ft} = n$. Therefore, assume that $t \le \Delta(H)$ and let D_f be a minimum factor dominating set. If $D_f = V(H)$ then $\gamma_{ft} = n \ge \gamma + \Delta(H) \ge \gamma + t > \gamma + t - 2$. If $D_f \ne V(H)$, let $v \in V(H) - D_f$. Since v must be adjacent to at least t vertices of D_f, let $X \subseteq D_f$ be a set of $t - 1$ vertices contained in $N(v)$ in H. In H every vertex of $V(H) - (D_f \cup \{v\})$ is adjacent to at least one vertex of $D_f - X$. Thus,

$(D_f - X) \cup \{v\}$ dominates H, so $\gamma \leq |D_f - X| + 1 = \gamma_{ft} - (t - 1) + 1$, or $\gamma_{ft} \leq \gamma + t - 2$. $\square$

The lower bound $\dfrac{n}{(\Delta(G) + 1)} \leq \gamma(G)$ (presented in Chapter 2) generalizes to factor domination in a natural way.

Theorem 7.50 [164] *For any graph,* $\dfrac{nt}{(\Delta + t)} \leq \gamma_{ft}$.

Proof. Certainly, if $\gamma_{ft} = n$, the inequality is true. Assume therefore that $\gamma_{ft} < n$ and that $t \leq \Delta$. Let D_f be a minimum factor dominating set of H. Since each vertex in $V(H) - D_f$ has at least t edges to D_f, there are at least $(n - \gamma_{ft})t$ such edges and $(n - \gamma_{ft})t \leq \sum\limits_{v \in D_f} deg(v) \leq \Delta\gamma_{ft}$, from which the inequality $\dfrac{nt}{(\Delta + t)} \leq \gamma_{ft}$ follows. $\square$

Brigham and Dutton generalized Nieminen's result that $\gamma(G) + \varepsilon_F(G) = n$ (see Chapter 1) to factor domination in a straightforward way. Let $\varepsilon_{ft}(H)$ denote the maximum cardinality of a set S such that in each factor G_i, $1 \leq i \leq t$, there is a spanning forest in which S is independent and each vertex in S has degree one.

Theorem 7.51 [164] *For any graph H and any t-factoring* $F(H) = \{G_1, G_2, \cdots, G_t\}$ *of H,* $\gamma_{ft} + \varepsilon_{ft} = n$.

Proof. Certainly, if $t > \Delta(H)$, then $\gamma_{ft} = n$ and $\varepsilon_{ft} = 0$, and the equality holds. Assume, therefore, that $t \leq \Delta$ and $\gamma_{ft} < n$. Let D_f be a minimum factor dominating set. In each factor G_i, arbitrarily select for each vertex $v \in V(G_i) - D_f$ one edge between v and a vertex in D_f. The subgraph of G_i formed by these selected edges is a union of stars centered at vertices of D_f and forms a spanning forest of G_i. Notice that in each of these spanning forests, the vertices in $V(G_i) - D_f$ are endvertices and are, therefore, independent. Thus, $\varepsilon_{ft} \geq |V(G_i) - D_f| = n - \gamma_{ft}$.

Conversely, let S be a set of vertices satisfying the definition of ε_{ft}. Clearly, the vertices in $V(H) - S$ form a factor dominating set; hence, $\gamma_{ft} \leq |V(H) - S| = n - \varepsilon_{ft}$. $\square$

In the remainder of this section, we consider the simple case of a 2-factoring of the complete graph K_n. Stated in other terms, we are given a graph $G = (V, E)$ and we seek to find a minimum cardinality set D which is a dominating set of both G and the complement $\overline{G} = (V, \overline{E})$ of G. Such a set is called a *global dominating set* by Sampathkumar [989]. The minimum cardinality of a global dominating set is denoted by $\gamma_g(G)$ and is called the *global domination number*. We will use this terminology.

We suggest as an easy exercise (see Exercise 7.1) the verification of the global domination numbers of the following classes of graphs:

(i) $\gamma_g(K_n) = n$;

(ii) $\gamma_g(C_n) = 3$ if $n = 3, 5$; $\gamma_g(C_n) = \lceil n/3 \rceil$ otherwise;

(iii) $\gamma_g(W_n) = 4$ if $n = 4$; $\gamma_g(W_n) = 3$ otherwise;

(iv) $\gamma_g(K_{n_1, n_2, \cdots, n_r}) = r$.

Notice that in all of the above examples, $\gamma_g = \max\{\gamma, \overline{\gamma}\}$. This leads one to ask whether there might be other classes of graphs, or other conditions under which $\gamma_g = \max\{\gamma, \overline{\gamma}\}$.

Theorem 7.52 [164] *If either G or $\overline{G}$ is a disconnected graph, then $\gamma_g = \max\{\gamma, \overline{\gamma}\}$.*

Proof. Assume that G is disconnected, that is, G has at least two components. Then it follows that $\gamma(\overline{G}) = 2$ and any dominating set of G is automatically a dominating set of $\overline{G}$ since a dominating set of G must contain at least one vertex from each component. Therefore, $\gamma_g(G) = \gamma(G) = \max\{\gamma, \overline{\gamma}\}$. $\square$

Because of Theorem 7.52, let us assume that both G and $\overline{G}$ are connected graphs.

Theorem 7.53 [164] *If both G and $\overline{G}$ are connected graphs, then*
(a) $\gamma_g = \max\{\gamma, \overline{\gamma}\}$ *if $diam(G) + diam(\overline{G}) \geq 7$.*
(b) $\gamma_g \leq \max\{3, \gamma, \overline{\gamma}\} + 1$ *if $diam(G) + diam(\overline{G}) = 6$.*
(c) $\gamma_g \leq \max\{\gamma, \overline{\gamma}\} + 2$ *if $diam(G) + diam(\overline{G}) = 5$.*
(d) $\gamma_g \leq \min\{\delta, \overline{\delta}\} + 1$ *if $diam(G) = diam(\overline{G}) = 2$.*

Proof. We can assume that neither G nor $\overline{G}$ is a complete graph, and therefore $diam(G) \geq diam(\overline{G}) \geq 2$. Let vertices u and v be distance $diam(G)$ apart in G.

(a) It is known that if $diam(G) > 3$, then $diam(\overline{G}) < 3$. Thus, if $diam(G) + diam(\overline{G}) \geq 7$, then we can assume that $diam(G) \geq 5$ and $diam(\overline{G}) \leq 2$. In this case vertices u and v dominate $\overline{G}$, since $N_G[u] \cap N_G[v] = \emptyset$. Therefore, $\overline{\gamma} = 2$ and $\gamma_g = \gamma = \max\{\gamma, \overline{\gamma}\}$, since any minimum dominating set for G must contain at least one vertex from $N[u]$ and one from $N[v]$.

(b) In this case, either $diam(G) = diam(\overline{G}) = 3$ or $diam(G) = 4$ and $diam(\overline{G}) = 2$. But in any graph G with $diam(G) = 3$, any two vertices at distance three dominate $\overline{G}$. In this case, $\gamma = \overline{\gamma} = 2$ and therefore $\gamma_g \leq 4$. In case $diam(G) = 4$ and $diam(\overline{G}) = 2$, as in (a) above, any γ-set S of G must contain at least one vertex from $N[u]$ and one from $N[v]$. If neither u nor v are in D, then $D \cup \{u\}$ will suffice to dominate $\overline{G}$.

(c) In this case, we can assume that $diam(G) = 3$ and $diam(\overline{G}) = 2$. It follows that any two vertices at distance three in G will dominate $\overline{G}$. Therefore, if D is a γ-set of G, then $D \cup \{u, v\}$ is a global dominating set of G.

(d) In a graph of diameter two, for each $x \in V$, $N[x]$ is a global dominating set. $\square$

Theorem 7.54 [164] *If G and $\overline{G}$ are both connected graphs and* $\max\{rad(G), rad(\overline{G})\} \geq 3$, *then* $\gamma_g(G) = \max\{\gamma, \overline{\gamma}\}$.

The global domination number can be bounded from above by a variety of graphical invariants, including the minimum degree δ, maximum degree Δ, largest clique ω, matching number β_1 and vertex covering number α_0. We conclude this section by mentioning several of these results.

Theorem 7.55 [164] *If $\gamma_g > \gamma$, then $\gamma_g \leq \Delta + 1$.*

Theorem 7.56 [164] *Either $\gamma_g = \max\{\gamma, \overline{\gamma}\}$ or $\gamma_g \leq \min\{\Delta, \overline{\Delta}\} + 1$.*

Theorem 7.57 [164] *For any graph G, $\gamma_g \leq \min\{\omega + \gamma, \overline{\omega} + \overline{\gamma}\} - 1$.*

Corollary 7.58 *If G is a triangle-free graph, then $\gamma \leq \gamma_g \leq \gamma + 1$.*

Brigham and Dutton applied the previous corollary to trees and obtained the following elegant result. Let $\mathcal{T}$ denote the class of trees with $n \geq 2$ vertices and either radius one (that is, stars) or radius two having a vertex u with $deg(u) \geq 2$ and $deg(v) \geq 3$ for all $v \in N(u)$.

Theorem 7.59 [164] *If T is a tree, then either $T \in \mathcal{T}$ and $\gamma_g(T) = \gamma + 1$, or $\gamma_g(T) = \gamma$.*

Theorem 7.60 [164] *For any graph G, $\gamma_g \leq \delta + 2$ if $\delta = \overline{\delta} \leq 2$; otherwise $\gamma_g \leq \max\{\delta, \overline{\delta}\} + 1$.*

Theorem 7.61 [164] *If G and $\overline{G}$ have no isolated vertices, then*

$$\gamma_g \leq \min\{\beta_1, \overline{\beta}_1\} + 1.$$

Sampathkumar offered the following.

Theorem 7.62 [989] *If G and $\overline{G}$ have no isolated vertices, then*

$$\gamma_g \leq \min\{\alpha_0, \overline{\alpha}_0\} + 1.$$

7.7 Domination in Directed Graphs

Of more than 1,200 research papers published on the topic of domination and related concepts in graphs, approximately 10% are concerned with directed graphs, and the majority of this 10% is concerned with what are called kernels in directed graphs. In this section we present some of the basic terminology and definitions concerning domination in directed graphs and some of the fundamental results. We also raise the strong suggestion that it 'is time' for the domination research community to begin to explore the application of the current domination theory to directed graphs.

A *directed graph* (also called a *digraph*) $D = (V, A)$ consists of a finite set V of vertices and a set A of directed edges, called *arcs*, where $A \subseteq V \times V$. An arc (u, v) is said to be *directed* from u to v, in which case u is said to be a *predecessor* of v, v is a *successor* of u, and u *dominates* v. In this case, we also use the notational equivalence: $(u, v) \equiv u \to v$. For most of this section, we will assume that D contains no loops (arcs of the form (v, v)) or multiple arcs (more than one copy of an arc (u, v)).

The *outset* of a vertex u is the set $O(u) = \{v : (u, v) \in A\}$, while the *inset* is the set $I(u) = \{w : (w, u) \in A\}$. We also define $O[u] = O(u) \cup \{u\}$ and $I[u] = I(u) \cup \{u\}$. The *outdegree* of a vertex u is $od(u) = |O(u)|$ and the *indegree* of u is $in(u) = |I(u)|$. For a subset $S \subseteq V$, we also define $O(S) = \cup_{v \in S} O(v)$; $I(S)$, $O[S]$ and $I[S]$ are defined similarly.

A set $S \subseteq V$ is *independent* if no two vertices of S are joined by an arc. The *independence number* $\beta_0(D)$ is the maximum cardinality of an independent set in D. A set $S \subseteq V$ is called *absorbant* if for every vertex $v \in V - S$, there exists a vertex $u \in S$ which is a successor of v, that is, $v \to u$ is an arc in A (equivalently, S is absorbant if every vertex $v \in V - S$ dominates at least one vertex in S). A set $S \subseteq V$ is a *dominating set* of D if every vertex $v \in V - S$ is dominated by at least one vertex $u \in S$ (equivalently, every vertex in $V - S$ is a successor of at least one vertex in S, or $O[S] = V$). The *domination number* $\gamma(D)$ of a digraph D is the minimum cardinality of a dominating set in D.

A set $S \subseteq V$ is a *kernel* if it is both independent and absorbant. The existence of kernels in digraphs has been the subject of more than 80 papers; therefore it warrants some consideration here. The shaded vertices in the digraph D_1 in Figure 7.11 form a kernel of D_1; while the digraph D_2, which is obtained from D_1 by reversing the directions of three arcs, has no kernel. Thus we observe that not every digraph has a kernel. Much research, therefore, has focused on the problem of finding necessary or sufficient conditions for a digraph to have a kernel.

A set $S \subseteq V$ is called a *solution* of a digraph D if S is both independent and dominating. Kernels and solutions are related by the following simple observation. The *reversal* of a digraph $D = (V, A)$ is the digraph $D^{-1} = (V, A^{-1})$, where $(u, v) \in A$ if and only if $(v, u) \in A^{-1}$. Thus, a set S is a kernel of a

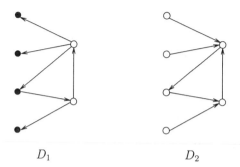

$$D_1 \qquad\qquad\qquad D_2$$

Figure 7.11: Directed graphs with and without kernels.

digraph D if and only if it is a solution of D^{-1}. A *tournament* on n vertices is a digraph $D = (V, A)$ in which for every pair u, v of vertices, either $(u, v) \in A$ or $(v, u) \in A$, but not both.

In the remainder of this section, we present some of the fundamental results about kernels in graphs and about the domination number of digraphs and tournaments. The first three results are due to Berge.

Theorem 7.63 [97] *If S is a kernel of a digraph $D = (V, A)$, then S is both maximal independent and minimal absorbant.*

Proof. Let S be a kernel of a digraph D. Assume that S is not maximal independent. Then there exists a vertex $u \in V - S$ which is not joined by an arc to any vertex in S. But this implies that S is not absorbant since there is no vertex in S which is a successor of u; this contradicts the assumption that S is a kernel. Similarly, assume that S is not a minimal absorbant set. Hence, there is a vertex $u \in S$ for which $S - \{u\}$ is absorbant. But if $S - \{u\}$ is absorbant, then there must be a vertex in $S - \{u\}$ which is a successor of u. But this implies that S is not independent, again contradicting the assumption that S is a kernel. $\square$

Theorem 7.64 [97] *A necessary and sufficient condition for a set S to be a kernel of a digraph $D = (V, A)$ is that its characteristic function ϕ_S satisfy the condition: $\phi_S(u) = 1 - \max\{\phi_S(v) : v \in O(u)\}$.*

Proof. Let S be a kernel of a digraph D and let ϕ_S be its characteristic function. Then by definition, if $u \in S$, then $\phi_S(u) = 1$. And since S is independent, for all vertices $v \in O(u)$, $v \notin S$, and $\phi_S(v) = 0$. Thus, $\max\{\phi_S(v) : v \in O(u)\} = 0$ and therefore, if $u \in S$, then $\phi_S(u) = 1 - \max\{\phi_S(v) : v \in O(u)\}$. Similarly, if $u \notin S$, then $\phi_S(u) = 0$. But since S is a kernel, S is absorbant, and therefore there must be a vertex $v \in S$ which is a successor of u, that is, $v \in O(u)$. This means that

$\phi_S(v) = 1$ and $\max\{\phi_S(v) : v \in O(u)\} = 1$. Thus, $\phi_S(u) = 0 = 1 - \max\{\phi_S(v) : v \in O(u)\} = 1 - 1 = 0$. Thus, $\phi_S(u) = 1 - \max\{\phi_S(v) : v \in O(u)\}$ holds for every vertex $u \in V$.

Conversely, assume that $\phi_S(u) = 1 - \max\{\phi_S(v) : v \in O(u)\}$ holds for some set $S \subseteq V$, its characteristic function ϕ_S, and every vertex $u \in V$. We must show that S is a kernel of D, that is, S is both independent and absorbant. Let $u \in S$. Then $\phi_S(u) = 1$ and $\phi_S(v) = 0$ for every $v \in O(u)$. But this implies that no vertex $v \in O(u)$ is in S, that is, that S is independent.

Similarly, consider any vertex $v \in V - S$. We know that $\phi_S(v) = 0$ and that $0 = 1 - \max\{\phi_S(u) : u \in O(v)\}$. But this implies that $\max\{\phi_S(u) : u \in O(v)\} = 1$, which means that for at least one vertex $u \in O(v)$, $\phi_S(u) = 1$. Therefore, $u \in S$ and there is at least one vertex in S which is a successor of v. Thus, S is absorbant. Since S is both independent and absorbant, it is a kernel. $\square$

A *symmetric digraph* is a digraph $D = (V, A)$ in which $(u, v) \in A$ implies $(v, u) \in A$. Symmetric digraphs correspond one-to-one with undirected graphs.

Theorem 7.65 [97] *If D is a symmetric digraph, then D has a kernel and $S \subseteq V$ is a kernel if and only if S is a maximal independent set.*

Proof. Note that every digraph has at least one maximal independent set. Let $S \subseteq V$ be any maximal independent set in a symmetric digraph D and let $u \in V - S$. Since S is maximal independent, there must be at least one arc between u and a vertex in S. But since D is symmetric, there must then be an arc from u to a vertex in S, that is, S is absorbant. Therefore, D has a kernel S. Conversely, let S be a kernel of D. Then by Theorem 7.63, D must be maximal independent. $\square$

A digraph $D = (V, A)$ is *transitive* if whenever (u, v) and $(v, w) \in A$ then $(u, w) \in A$.

Theorem 7.66 [97, 784] *All minimal absorbant sets of a transitive digraph $D = (V, A)$ have the same cardinality. Furthermore, a set $S \subseteq V$ is a kernel if and only if S is a minimal absorbant set.*

Corollary 7.67 *Every transitive digraph $D = (V, A)$ has a kernel and all kernels of D have the same cardinality.*

Before leaving the subject of kernels in digraphs, we should mention that several computational complexity questions have been answered. For example, Fraenkel [486] has shown that the problem of deciding whether a digraph D has a kernel is NP-complete, even if D is a cyclic planar digraph satisfying $od(u) \leq 2$, $in(u) \leq 2$ and $od(u) + in(u) \leq 3$. Fraenkel also showed that if any of these restrictions are removed, then polynomial algorithms exist for deciding whether D has a kernel.

We conclude this section on domination in digraphs by mentioning several interesting results on domination in tournaments. For tournaments T, we use the notation $u \Rightarrow S$ to mean that for every vertex $v \in S$, $u \to v$.

First of all, it is not at all obvious whether for arbitrary positive integers k, there exists a tournament T with $\gamma(T) = k$. Figure 7.12 shows tournaments with $\gamma(T) = 2$ and 3. The tournament T_3 in Figure 7.12 is called the *cyclic triple* and requires two vertices to dominate it; in fact, it is the unique smallest tournament with $\gamma(T) = 2$. The tournament T_9 consists of three cyclic triples; all of the vertices in one cyclic triple beat all of the vertices in the next cyclic triple, in cyclic order. It is a nice exercise to show that $\gamma(T_9) = 3$. The tournament T_7 is a smallest tournament with $\gamma(T) = 3$. With the vertices T_7 labelled as shown, T_7 is the quadratic residue tournament: $V(T_7) = \mathcal{Z}_7 = \{0, 1, 2, 3, 4, 5, 6\}$, where $i \to i+1$, $i+2$, $i+4$ (mod 7). This is the only 7-tournament for which $\langle O(v) \rangle$ is a cyclic triple for every vertex v (also $\langle I(v) \rangle$ is a cyclic triple).

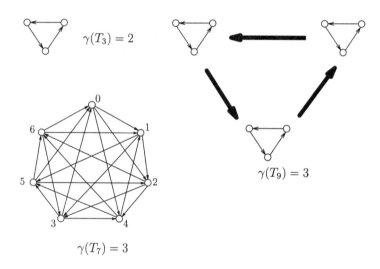

$$\gamma(T_3) = 2$$

$$\gamma(T_9) = 3$$

$$\gamma(T_7) = 3$$

Figure 7.12: Tournaments with small domination numbers.

In 1962, K. Schütte (see Erdös [420]) indirectly raised the question of whether there exist tournaments with arbitrarily large domination numbers. In fact, he raised a slightly different question: given $k > 0$, does there exist a tournament $T_{n(k)}$ in which for any set S of k vertices, there is a vertex u which dominates all vertices in S. Such a tournament is said to have property P_k.

Notice the following. If a tournament does not have property P_k, then there exists a set S' of k vertices such that for every vertex $w \notin S'$, there is a vertex $v \in S'$ for which $v \to w$. That is, S' is a dominating set in T of order k. Thus, a tournament T has property P_k if and only if $\gamma(T) > k$.

In [421], Erdös showed by probabilistic arguments that such a tournament $T_{n(k)}$ does exist, for every positive integer k. If we let $f(k)$ be the minimum value of $n(k)$ for which a $T_{n(k)}$ exists, then Erdös showed that

(1) $f(k) \le k^2 2^k (log\, 2 + \varepsilon)$, for any $\varepsilon > 0$
provided k is sufficiently large.

We can restate Erdös' theorem as follows.

Theorem 7.68 (Erdös) [421] *For every $\varepsilon > 0$, there is an integer K such that for every $k \ge K$, there exists a tournament T_k with no more than $k^2 2^k (log\, 2 + \varepsilon)$ vertices and $\gamma(T_k) > k$.*

Proof. Let T be a random tournament on n vertices, that is, for every pair of vertices u and v, choose either the $u \to v$ arc or the $v \to u$ arc with equal probability, and independently of the other arcs of T. The probability, therefore, that vertex u dominates vertex v is $1/2$. For every set S of k vertices and every vertex $u \notin S$, the probability that u dominates every vertex in S is 2^{-k}. The probability that S is a dominating set is therefore $(1 - 2^{-k})^{n-k}$. The expected number of dominating sets of cardinality k is $\binom{n}{k}(1 - 2^{-k})^{n-k}$. If n is sufficiently large, the value of this expression will be less than one, and therefore there exists a tournament T on n vertices with $\gamma(T) > k$. In fact, if $n > k^2 2^k (log\, 2 + \varepsilon)$ then $\binom{n}{k}(1 - 2^{-k})^{n-k} < 1$. $\square$

The fact that there are tournaments with $\gamma(T) > k$, for arbitrary positive integers k, is also discussed in Moon's 1968 monograph [J. W. Moon, *Topics on Tournaments.* Holt, Rinehart, and Winston, New York (1968). Exercise 5 p. 32]. Let $\gamma(n)$ be the maximum of $\gamma(T)$ over all tournaments with n vertices. For each n, there is some tournament T with n vertices with $\gamma(T) = \gamma(n)$. Moon attributes the following result to Leo Moser (without any reference):

$$\lfloor log\, n - 2log(log\, n) \rfloor \le \gamma(n) \le \lfloor log(n+1) \rfloor,$$

where $n \ge 2$ and log is to the base 2. Thus, there is a tournament T for which $\gamma(T) \ge \lfloor log\, n - 2log(log\, n) \rfloor$, that is, for every positive integer k there is a tournament T for which $\gamma(T) > k$.

Szekeres and Szekeres [1075] later established a lower bound for $f(k)$:

(2) $f(k) \ge (k+2)2^{k-1} - 1$

Still later, Graham and Spencer [549] gave an explicit construction of a tournament $T_{n(k)}$ which has property P_k, although their construction shows that $n(k)$ may be as large as $k^2 2^{2k-2}$. Their construction is as follows:

Select the smallest prime number $p > k^2 2^{2k-2}$ where $p \equiv 3 \pmod 4$. The vertices of $T_{n(k)}$ correspond to $\{0, 1, \cdots, p-1\}$. For two vertices u and v, we add the arc $u \to v$ if $u - v \equiv a \pmod p$ for some quaratic residue a of p. They pointed out, however, the following:

"The value $k^2 2^{2k-2}$ is nearly the square of the nonconstructive upper bound (1) of Erdös. Specific constructions show that much smaller values p suffice to endow T_p with property P_k. For example, T_7 has property P_2 and T_{19} has property P_3. In [1075] it is shown that $f(2) = 7$ and $f(3) = 19$, so that these tournaments are minimal. Also, it is true that T_{67} has property P_4. Since (2) gives $f(4) \geq 47$, it is possible that T_{67} is also minimal."

In 1988 Megiddo and Vishkin [891] revisited this old problem, but from a computational point of view.

TOURNAMENT DOMINATING SET

INSTANCE: A tournament $T = (V, A)$ and a positive integer k.
QUESTION: Does T have a dominating set of cardinality $\leq k$?

Theorem 7.69 [421] *If T is a tournament with $n \geq 2$ vertices, then $\gamma(T) \leq \lceil \log_2 n \rceil$.*

Proof. Clearly, $\sum_u od(u) = n(n-1)/2$. It follows that there must be at least one vertex which dominates at least $\lceil (n-1)/2 \rceil$ vertices. Select a vertex u_1 which dominates at least $\lceil (n-1)/2 \rceil$ vertices. Remove this vertex and all of the vertices it dominates. Repeat this process on the remaining tournament, which has at most $\lceil (n-1)/2 \rceil$ vertices, by selecting a second vertex u_2 which dominates at least half of the remaining vertices, and deleting u_2 and the vertices it dominates. By continuing this process, we can find a dominating set with no more than $\lceil \log_2 n \rceil$ vertices. $\square$

Corollary 7.70 [891] *A minimum dominating set in a tournament can be found in $n^{O(\log n)}$ time.*

Proof. The proof of Theorem 7.69 shows that a minimum dominating set can be found by examining all subsets of V of cardinality no greater than $\lceil \log_2 n \rceil$. There are $\sum_{i=1}^{\lceil \log_2 n \rceil} \binom{n}{i}$ such subsets. $\square$

In effect, Megiddo and Vishkin are saying that there is an algorithm for computing the domination number of a tournament which runs in subexponential, yet superpolynomial time. It remains an open problem whether it is possible to compute the domination number of a tournament in polynomial time.

For further reading on kernels and domination in directed graphs, the reader is referred to the survey chapter by Ghoshal, Laskar, and Pillone [534].

EXERCISES

7.1 Verify the global domination number of the following classes of graphs:

(a) $\gamma_g(K_n) = n$.

(b) $\gamma_g(C_n) = 3$ if $n = 3, 5$; $\gamma_g(C_n) = \lfloor n/3 \rfloor$ otherwise.

(c) $\gamma_g(W_n) = 4$ if $n = 4$; $\gamma_g(W_n) = 3$ otherwise.

(d) $\gamma_g(K_{n_1,n_2,\cdots,n_r}) = r$.

7.2 For the tournament T_9 in Figure 7.12, show that $\gamma(T_9) = 3$.

7.3 Determine the strong and weak domination numbers of the path P_n.

7.4 Find a graph G such that $\gamma_L(G) < i(G)$.

7.5 Construct an r-regular graph G for which $\gamma_L(G) = 2n/(r+3)$.

7.6 Assume that each of G and H is all-parity realizable (APR) and either G or H has an all-odd parity set of even cardinality.

(a) Show that the join $G + H$ is APR.

(b) Show that if $G + H$ is APR, then G and H are APR and either G or H has an all-odd parity set of even cardinality.

7.7 Prove: $PD(G + H) = PD(G) + PD(H) + 1$ if both G and H have odd cardinality all-odd parity sets, and $PD(G + H) = PD(G) + PD(H)$ otherwise.

7.8 Find a cubic graph G with $\gamma^-(G) < \gamma(G)$.

7.9 For each positive integer k, find a cubic graph H_k such that $\beta_0(H_k) - i(H_k) \geq k$.

7.10 Considering the proof of Theorem 7.1, can you construct a γ-set for a graph G such that every vertex except one is dominated at least three times?

7.11 Theorem 7.5 asserts that the 2-domination number of a graph G with $\delta(G) \geq 2$ is no more than $2n/3$. Can you construct a graph G for which $\gamma_2(G) = 2n/3$?

7.12 Theorem 7.12 asserts that for every graph G, $\gamma_2(G) \leq \beta_2(G)$. Construct a graph G for which $\gamma_2(G) = \beta_2(G)$.

7.13 Theorem 7.33 characterizes locating sets in trees. Can this characterization be used to design an algorithm for locating sets in trees, that is, for locating a locating set in an arbitrary tree?

7.14 (a) Is every minimum dominating set in an $m \times n$ grid locating-dominating?

(b) Is every maximal independent set in an $m \times n$ grid locating-dominating?

7.15 Give a characterization of minimal strong dominating sets that is similar to the characterization in Chapter 1 of minimal dominating sets.

7.16 In [1003] Sampathkumar and Pushpa Latha conjectured that for any tree T, $i(t) \leq \gamma_W(T)$. Can you either prove or disprove this conjecture.

7.17 [994] The *independent strong domination number* $i_S(G)$ is the minimum cardinality of a strong dominating set which is also independent.

(a) Does every graph have an independent strong dominating set?

(b) Prove that if a graph G does not have an induced $K_{1,3}$, then $\gamma_S(G) = i_S(G)$.

7.18 Consider the path P_n, for $n = 2k + 1$.

(a) What can you say about the sequence of distance-k domination numbers:
$$(\gamma_1(P_n), \gamma_2(P_n), \gamma_3(P_n), ..., \gamma_k(P_n))?$$

(b) What is the maximum possible number of consecutive values in this sequence that can be equal?

(c) What is the maximum possible number of consecutive values in this sequence that can be distinct?

7.19 Does $dd(G)$ exist for every graph G? If not, under what conditions does $dd(G)$ exist?

7.20 Determine the graphs G for which $dd(G) = \gamma(G)$.

7.21 Show that the Schelp graphs (Figure 7.2) have $\gamma_{2k} = \gamma_{k(k+1)}$.

7.22 (Richardson's Theorem [967].) Show that every digraph without odd circuits has a kernel.

Chapter 8

Multiproperty and Multiset Parameters

In each of the previous chapters in this book, the fundamental problems involve a single dominating set S which might be required to have additional properties, such as being independent, connected, or pairable. In most cases, we seek to minimize the cardinality of S subject to the stated required properties. The parameters in Chapter 4 are different in that they involve, for example, $\sum_{s \in S} |N[s]|$ rather than just $|S|$, but they still involve a single dominating set or dominating function. In this chapter we consider multiset problems and single set multiproperty problems.

8.1 Overview

The term "multiset" quite naturally suggests that we seek more than one dominating set. For example, we could be interested in finding the maximum number of disjoint dominating sets in G, which is the domatic number $d(G)$ discussed in Section 8.3. Alternately, we might have the following scenario. For a network system to be operational, we require a set of detection devices be located at the vertices in a dominating set. Normal detection devices need regular maintenance, and, while one dominating set of devices is down for repair, a backup dominating set is required. Perhaps the initial cost for establishing a normal detection device is low and the principal expense results from operating costs which are proportional to the number of devices in operation and are very high per unit. And, finally, for a large initial investment one can "harden" a normal detection device so that it never needs maintenance. If there are two disjoint minimum dominating sets, then they form an optimum solution. At the other extreme, if there is a unique minimum dominating set then the optimum solution is to harden each location in the set. In general, one seeks two minimum dominating sets with minimum possible intersection. For example, for the graph G_1 in Figure 8.1, if D_1 and D_2 are disjoint dominating sets, then one of D_1 or

219

D_2 has order at least five and $|D_1 \cup D_2| \geq 7$. In the circumstances described, it is better to use the γ-sets $\{u, v\}$ and $\{u, w\}$ and to harden u. Results on this type of problem are in [556, 557].

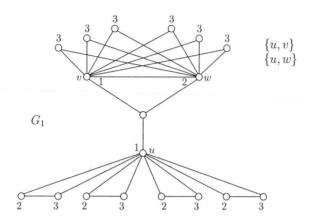

Figure 8.1: The γ-sets are $\{u, v\}$ and $\{u, w\}$.

The previously considered "multiproperty" problems in this book involved finding a single set with more than one property, such as dominating and independent, dominating and pairable, or dominating and locating. The following motivates "prioritized multiproperty" problems. Again assume we need to locate a set of detection devices at the vertices in a dominating set. Assume that each detection device is extremely expensive, so we must use a $\gamma(G)$-set. However, detection devices placed at adjacent vertices will create some tolerable but undesired interference. The optimum solution is a $\gamma(G)$-set which is as independent as possible. Insisting on having a $\gamma(G)$-set makes domination the priority property, with, in this instance, independence being a desired secondary property. Of course, if $\gamma(G) = i(G)$, then the optimum solution is to use an $i(G)$-set; but when $\gamma(G) < i(G)$, the optimum solution is a $\gamma(G)$-set and some dependence must be tolerated.

Generally, if $P1$ and $P2$ are properties of subsets of the vertex set V of a graph G and $\Omega 1(G)$ denotes the minimum or maximum cardinality of a set with property $P1$, then the prioritized $(P2, P1)$ multiproperty problem is to find a $\Omega 1(G)$-set which comes as close as possible to also having property $P2$. Note that coming as close as possible for $P2$ could lead to either maximization or minimization problems. For example, with a priority property of domination and secondary property of independence, "as independent as possible" could mean that one is seeking a $\gamma(G)$-set S with the largest possible number of isolated vertices in $\langle S \rangle$ or one is seeking the minimum possible number of edges in $\langle S \rangle$.

It could also mean minimizing $\Delta(\langle S \rangle)$. In Section 8.2 we describe prioritized multiproperty problems, and we consider sequential $(P2, P1)$ problems where one might not be required to have a $\Omega 1(G)$-set. We also consider (domination-) forcing sets, sets $S \subseteq V(G)$ that need $\gamma(G)$ vertices from $V(G)$ to dominate them.

8.2   Prioritized Multiproperty Problems and Sequential Problems

To date, relatively little work has been done on prioritized multiproperty problems, and the focus of that work has principally been on computational complexity questions. The first "single set, prioritized multiproperty" graph parameter was introduced in 1988 [1020]. There and in three subsequent papers [749, 750, 1038], the multiproperty problems were conceived as possible examples for which linear time solutions did not exist for the class of series-parallel graphs. However, linear algorithms were found to exist. In this section we outline the format for prioritized multiproperty parameters and discuss other sequential multiproperty parameters.

First we consider the pair $(P2,P1) =$ (independence, domination) with priority on the domination parameter. Thus, we seek a minimum dominating set which will be as independent as possible. To illustrate some of the possible measures of independence, consider the graph G_1 in Figure 8.2, which has 58 vertices with $V(G_1) = \{u_1, u_2, u_3, u_4, u_5, v_1, v_2, v_3, v_4, v_5\} \cup \{u_{ij}^k : 1 \leq i, j \leq 4, 1 \leq k \leq 3\}$. The edge set is, as illustrated, with each u_{ij}^k having degree two and $N(u_{ij}^k) = \{u_i, v_j\}$. Thus, each of the sixteen pairs $\{u_i, v_j\}$ has three common neighbors, each of degree two. Notice that $\gamma(G_1) = 5$, and the only γ-sets are $D_1 = \{u_1, u_2, u_3, u_4, u_5\}$ and $D_2 = \{v_1, v_2, v_3, v_4, v_5\}$. Vertex set D_1 can be considered to be more independent than D_2, because $\langle D_1 \rangle$ has two isolated vertices while $\langle D_2 \rangle$ only has one. On the other hand, D_2 can be considered to be more independent than D_1, because $\langle D_2 \rangle$ has only two edges while $\langle D_1 \rangle$ has three. Note that $\beta_0(\langle D_1 \rangle) = \beta_0(\langle D_2 \rangle) = 3$.

Using the second criterion, we define $PR(m, \gamma)(G)$ to be the minimum number of edges in a subgraph induced by a minimum dominating set of G, that is, $PR(m, \gamma)(G) = \min\{|E(\langle D \rangle)| : D$ is a $\gamma(G)$-set$\}$. In particular, $PR(m, \gamma)(G) = 0$ if and only if $\gamma(G) = i(G)$. Typical of the complexity results for problems of this type is the following.

Theorem 8.1 [1020] *Determining if $PR(m, \gamma) \leq k$ for given graph G and positive integer k is NP-hard.*

We note that the decision problem "Is $PR(m, \gamma)(G) \leq k$?" is in $\mathbf{NP}^{NP}$, but it appears to not be in $\mathbf{NP}$ because verification of a proposed solution set D

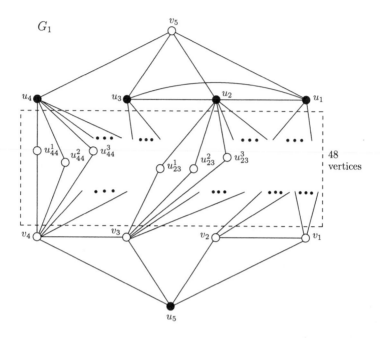

Figure 8.2: Graph G_1 with $\gamma(G_1) = 5$ and $\gamma(G_1)$-sets $\{u_1, u_2, u_3, u_4, u_5\}$ and $\{v_1, v_2, v_3, v_4, v_5\}$.

includes verifying that D is actually a minimum dominating set. Computational complexity results for problems with independence as the priority property and redundance measures as the secondary properties are presented in [749, 750]. Related to the paired-domination problem in Chapter 6 is the prioritized (maximum matching, domination) problem [653] in which one seeks a $\gamma(G)$-set S for which $\beta_1(\langle S \rangle)$ is maximized.

For the pair $(P2, P1) = $ (independence, domination), the parameter $PR(m, \gamma)$ minimized the number of edges in $\langle D \rangle$, where D is a γ-set. If one relaxes the priority domination $P1$-property and merely requires that D be an arbitrary dominating set, then one can always find an independent dominating set. However, other pairs of properties $(P2, P1)$ produce interesting problems when we sequentialize but do not prioritize (that is, we want a set D with property $P1$ but not necessarily a $\Omega1$-set). For example, Sampathkumar's d-domination number $\gamma_d(G)$ in [990] fits this format for (domination, domination). Specifically, $\gamma_d(G) = \min\{\gamma(\langle D \rangle) : D$ is a dominating set $\}$. A dominating set D is a *least dominating set* if $\gamma(\langle D \rangle) \leq \gamma(\langle S \rangle)$ for any dominating set S, and $\gamma_l(G)$ is the minimum cardinality of a least dominating set. In particular, for the path P_{10}, we have $\gamma_l(P_{10}) = 6$, and $\{v_2, v_3, v_4, v_7, v_8, v_9\}$ is a $\gamma_l(P_{10})$-set, and $\gamma_d(P_{10}) = 2$.

Proposition 8.2 [990] *For the path P_n and cycle C_n, $\gamma_l(P_n) = \gamma_l(C_n) = n - 2\lceil n/5 \rceil$ and $\gamma_d(P_n) = \gamma_d(C_n) = \lceil n/5 \rceil$.*

Sampathkumar [990] mistakenly suggested that for any graph G without isolates, $\gamma_l(G) \le \gamma_t(G)$. Note that the total domination number of the path P_{20} is $\gamma_t(P_{20}) = 10 < \gamma_l(P_{20}) = 12$. Because one can find graphs H with $\gamma_l(H) < \gamma_t(H)$, it follows that $\gamma_t(G)$ and $\gamma_l(G)$ are not comparable. Sampathkumar established the following sequence and conjectured an upper bound for $\gamma_l(G)$.

Proposition 8.3 [990] *For any graph G, $\gamma_d(G) \le \gamma(G) \le \gamma_l(G)$.*

Conjecture 8.4 [990] *If G is a connected graph on $n \ge 2$ vertices, then $\gamma_l(G) \le 3n/5$.*

For Sampathkumar's γ_d parameter, we find a dominating set S such that the cardinality of a set $D \subseteq S$ that dominates S is minimized. For the following, we consider an arbitrary $S \subseteq V$ and seek a set D of vertices that dominates S where D is not necessarily a subset of S.

For any $S \subseteq V$, an *S-dominating set* in G is a set $D \subseteq V$ such that each vertex in S is dominated by a vertex in D, thus $S \subseteq N[D]$. The order of a smallest S-dominating set in G is denoted by $\gamma(S, G)$. Because any γ-set dominates every $S \subseteq V$, we have $\gamma(S, G) \le \gamma(G)$. A question of interest is to determine how small a set $S \subseteq V$ we can find so that it requires $\gamma(G)$ vertices to dominate S. If S satisfies $\gamma(S, G) = \gamma(G)$, then S is called a *domination-forcing set* (a γ-forcing set) of G, and the cardinality of a smallest domination-forcing set of G will be denoted by $\gamma^{\#}(G)$. For example, the graph G in Figure 8.3 has $\gamma(G) = 2$ and $\{2, 5\}$ is a minimum dominating set of G. Each two vertices in a pair of distinct, nonadjacent vertices in G have a common neighbor, so $\gamma(S, G) = 1$ for $1 \le |S| \le 2$. Thus, $\gamma^{\#}(G) \ge 3$. For $S' = \{1, 4, 7\}$, we have $\gamma(S', G) = |\{2, 4\}| = 2 = \gamma(G)$, so $\gamma^{\#}(G) = 3$. Note that $\gamma(\langle S' \rangle) = 3 > \gamma(S', G) = \gamma(G) = 2$.

Theorem 8.5 [1046] *For any graph G, $\rho(G) \le \gamma(G) \le \gamma^{\#}(G)$, and if G is efficiently dominatable (that is, $F(G) = n$), then $\gamma(G) = \gamma^{\#}(G)$.*

Proof. Let packing P be a $\rho(G)$-set. Then in any dominating set D, each $v \in D$ is adjacent to at most one vertex in P, so $|D| \ge |P|$. Consequently, $\rho(G) \le \gamma(G)$. If $S \subseteq V(G)$ and $|S| < \gamma(G)$, then $\gamma(S, G) \le \gamma(\langle S \rangle) \le |S| < \gamma(G)$ and S is not a $\gamma^{\#}(G)$-set. Hence, $\gamma^{\#}(G) \ge \gamma(G)$. If D is an efficient dominating set, then D is a packing. Therefore, $\gamma(D, G) \ge |D| = \gamma(G)$. Hence, $\gamma^{\#}(G) \le |D| = \gamma(G)$ implying that $\gamma(G) = \gamma^{\#}(G)$. $\square$

Recall that a graph G is called vertex-critical if $\gamma(G - v) < \gamma(G)$ for all $v \in V$ (that is, $G \in CVR$).

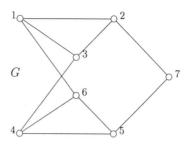

Figure 8.3: $\gamma^{\#}(G) = 3 > \gamma(\{1,4,7\}, G) = 2$.

Theorem 8.6 [1046] *For any graph G,*

(a) $\gamma^{\#}(G) = \gamma(G)$ *if and only if $\rho(G) = \gamma(G)$.*

(b) $\gamma^{\#}(G) = n$ *if and only if $G \in CVR$.*

Proof. (a) First, assume $\gamma^{\#}(G) = \gamma(G)$. Suppose there is a $\gamma^{\#}(G)$-set S containing vertices u and v with distance $d(u,v) \leq 2$. If $uv \in E(G)$, then $S - \{u\}$ dominates S. If $d(u,v) = 2$, let $w \in N(u) \cap N(v)$. Then $S - \{u,v\} \cup \{w\}$ dominates S. In either case, $\gamma(S,G) \leq |S| - 1 < |S| = \gamma^{\#}(G) = \gamma(G) = \gamma(S,G)$, a contradiction. Thus, every $\gamma^{\#}(G)$-set is a packing and $\rho(G) = \gamma(G) = \gamma^{\#}(G)$. Second, assume $\rho(G) = \gamma(G)$. Let S be a $\rho(G)$-set with $|S| = \rho(G) = \gamma(G)$. Because S is a packing, $\gamma(S,G) \geq |S| = \gamma(G)$, so $\gamma^{\#}(G) \leq |S| = \gamma(G)$. Thus $\rho(G) = \gamma(G) = \gamma^{\#}(G)$.

(b) Suppose first that $\gamma^{\#}(G) = n$. Let $v \in V(G)$, and let $S = V(G) - \{v\}$. Because $\gamma^{\#}(G) = n$ and $|S| < n$, we have $\gamma(S,G) < \gamma(G)$ and so there is a set $T \subseteq V$ with $|T| \leq \gamma(G) - 1$ such that T dominates S, but T does not dominate V. Hence, $N[v] \cap T = \emptyset$, and $\gamma(G - v) \leq |T| = \gamma(G) - 1$. Therefore $G \in CVR$. Conversely, suppose $G \in CVR$, and let $S \subseteq V$ with some $v \in V - S$. There is a subset $T \subseteq V - \{v\}$ such that $|T| < \gamma(G)$ and T dominates $G - v$. Now $S \subseteq V - \{v\}$ and T dominates S implies $\gamma(S,G) \leq |T| < \gamma(G)$. Hence, the only γ-forcing set of G is V, and $\gamma^{\#}(G) = n$. $\square$

We shall show next that, for any given positive integer $j \in N$, there exists a graph G for which $\gamma(G) = 2$, $\gamma^{\#}(G) - \gamma(G) = j$ and $n - \gamma^{\#}(G) \geq j + 1$.

Example A. For $j, t \in N$ with $t \geq j + 1$, let $m = \begin{pmatrix} t \\ j \end{pmatrix}$ and define the graph $J_{t,j}$ (as in Figure 8.4) as follows:

Let $J_1 = K_t$, $J_2 = K_m$, and $J_3 = K_1$, with $V(J_1) = \{u_1, u_2, ..., u_t\}$, $V(J_2) = \{v_1, ..., v_m\}$ and $V(J_3) = \{w\}$, and let $A_1, A_2, ..., A_m$ be the m distinct subsets of $V(J_1)$ that have cardinality j. Let $V(J_{t,j}) = V(J_1) \cup V(J_2) \cup V(J_3)$ and $E(J_{t,j}) = E(J_1) \cup E(J_2) \cup \{wv_i : i = 1, 2, ..., m\} \cup F$, where $F = \bigcup_{i=1}^{m} \{v_i u_k : u_k \in A_i\}$.

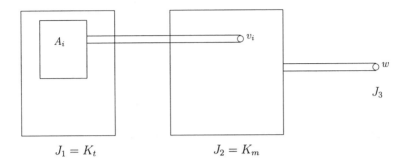

Figure 8.4: Graph $J_{t,j}$ with $\gamma(J_{t,j}) = 2$, $\gamma^{\#}(J_{t,j}) = j+2$, and $n = t + \binom{t}{j} + 1 \geq 2t + 1 \geq 2\gamma^{\#}(G) - 1$.

Proposition 8.7 [1046] *For $t, j \in N$ with $t \geq j+1$ and $G = J_{t,j}$,*

(a) $\gamma(G) = 2$,

(b) $\gamma^{\#}(G) = j + 2$, *and*

(c) $n = t + \binom{t}{j} + 1 \geq 2t + 1 \geq 2\gamma^{\#}(G) - 1$.

Proof. Let t, j and G satisfy the hypotheses of the proposition, and assume that the vertices of G are labelled as those of $J_{t,j}$ in Example A.

(a) Since $\Delta(G) < n-1$, it follows that $\gamma(G) \geq 2$; but since $\{u_t, w\}$ dominates G, $\gamma(G) = 2$.

(b) Let $B \subseteq V(G)$ such that $|B \cap V(J_1)| \leq j$. Then, there exists $k \in \{1, 2, ..., m\}$ such that $B \cap V(J_1) \subseteq A_k$; consequently, $\{v_k\}$ dominates B and $\gamma(B, G) = 1$. Hence, it follows that, if S is a $\gamma^{\#}(G)$-set (so $\gamma(S, G) = 2$), then $|S \cap V(J_1)| \geq j+1$. Furthermore, $S \not\subseteq V(J_1)$ since, otherwise, $\{u_1\}$ dominates S and $\gamma(S, G) = 1$; so $S - V(J_1) \neq \emptyset$ and $\gamma^{\#}(G) = |S| \geq (j+1) + |S - V(J_1)| \geq j+2$. To show that $\gamma^{\#}(G) \leq j + 2$, let $T = \{u_1, u_2, ..., u_{j+1}, w\}$. Then $\gamma(T, G) \geq 2$ since, otherwise, if there exists $y \in V(G)$ with $\{y\}$ dominating T, then $y \notin V(J_2) \cup V(J_3)$ (as no vertex in $V(J_2) \cup V(J_3)$ is adjacent to $j + 1$ vertices in $V(J_1)$) and so $y \in V(J_1)$, whence y does not dominate w, contradicting the fact that y dominates T. So $\gamma(G) = 2 \leq \gamma(T, G) \leq \gamma(G)$, that is, $\gamma(T, G) = \gamma(G)$ and T is a γ-forcing set of G. Hence, $\gamma^{\#}(G) \leq |T| = j+2$. Hence, $\gamma^{\#}(G) = j+2$.

(c)
$$n = t + \binom{t}{j} + 1$$

$$= t + t\frac{(t-1)(t-2)...(t-j+1)}{j(j-1)...2\cdot 1} + 1$$

$$\geq t + t + 1 = 2t + 1 \geq 2j + 3 = 2\gamma^{\#}(G) - 1,$$

completing the proof. $\square$

We remark that for $t = 2$ and $j = 1$, we obtain a graph $J_{t,j}(= J_{2,1})$ of smallest possible order (namely, $|V(J_{2,1})| = 5$), for which $\gamma^{\#}(J_{2,1}) = 3$ and $\gamma(J_{2,1}) = 2$. In this case

$$\frac{\gamma^{\#}(J_{2,1})}{n} = \frac{3}{5} > \frac{1}{2}.$$

In general, if $t = j + 1$, then

$$\frac{\gamma^{\#}(J_{j+1,j})}{n} = \frac{j+2}{2j+3} = \frac{1}{2} + \frac{1}{4j+6} \in \left(\frac{1}{2}, \frac{3}{5}\right]$$

and

$$lim_{j\to\infty}\frac{\gamma^{\#}(J_{j+1,j})}{n} = \frac{1}{2}.$$

If $t = j + 2$, then $|V(J_{t,j})| = j + \binom{j+2}{j} + 1$ and

$$lim_{j\to\infty}\frac{\gamma^{\#}(J_{j+2,j})}{n} = 0.$$

Furthermore, for any fixed $j \in N$, we see from Proposition 8.7 (b) and (c) that

$$lim_{t\to\infty}\frac{\gamma^{\#}(J_{t,j})}{n} = 0.$$

In the above example, $\gamma(J_{t,j}) = 2$. We shall show that, for prescribed $h \geq 2$, M and N, there exists a graph G for which $\gamma(G) = h$, $\gamma^{\#}(G) - \gamma(G) \geq M$ and $n - \gamma^{\#}(G) \geq N$.

Example B. For $t, j \in N$ with $h \geq 2$, $t \geq (h-1)(j+1)$, $m = \binom{t}{j}$, let

$G_1, G_2, ..., G_{h-1} = J_{t,j}$ (see Example A) and, in G_i let $V_{1i}, V_{2i}, u_{1i}, u_{2i}, ..., u_{ti}$, $v_{1i}, v_{2i}, ..., v_{mi}$ and w_i correspond to $V(J_1)$, $V(J_2)$, $u_1, u_2, ..., u_t$, $v_1, v_2, ..., v_m$ and w, respectively, in $J_{t,j}$, for $i = 1, 2, ..., h-1$. Let $J_{t,j,h}$ be the graph obtained from $G_1, G_2, ..., G_{h-1}$ by identifying the vertices $v_{i1}, v_{i2}, ..., v_{i(h-1)}$ to form a new vertex v_i^h corresponding to the vertex $v_i \in V(J_2)$ in $J_{t,j}$, for $i = 1, 2, ..., m$. Denote the resulting set $\{v_1^h, v_2^h, ..., v_m^h\}$ by V_2^h, and the subset of V_{ti} corresponding to A_k by $A_{ki}(i \in \{1, ..., h-1\}, k \in \{1, ..., m\})$. (Note that $J_{t,j,2} = J_{t,j}$.)

Proposition 8.8 [1046] *For* $t, j, h \in N$ *with* $t \geq (h-1)(j+1)$, $h \geq 2$, *and* $G = J_{t,j,h}$,

(a) $\gamma(G) = h$,

(b) $\gamma^{\#}(G) = (h-1)(j+1) + 1 = \gamma(G) + (h-1)j$, and

(c) $n = (h-1)(t+1) + \binom{t}{j} = \gamma^{\#}(G) + (h-1)(t-j) + \binom{t}{j} - 1$.

Proof. Let t, j, h, and G satisfy the hypothesis of the proposition; assume that the vertices of G are labelled as in Example B.

(a) That $\gamma(G) \leq h$ follows from the observation that $\{v_1^h, u_{11}, u_{12}, ..., u_{1(h-1)}\}$ dominates G. If there exists a dominating set D of G with $|D| \leq h-1$, then $D \not\subseteq \bigcup_{i=t}^{h-1} V_{1i}$ (otherwise D does not dominate $\{w_1, w_2, ..., w_{h-1}\}$); hence, $D \cap V_{1j} = \emptyset$ for at least one value of $j \in \{1, 2, ..., h-1\}$. Thus, V_{1j} is dominated by (at most $h-1$) vertices in $D \cap V_2^h$. However,

$$|N_G(D \cap V_2^h) \cap V_{1j}| \leq |D \cap V_2^h| \cdot j \leq (h-1)j < t = |V_{1j}|,$$

so that $D \cap V_2^h$ cannot dominate V_{1j}. Thus, any dominating set of G has cardinality at least h. Hence, $\gamma(G) = h$.

(b) Let S be a $\gamma^{\#}(G)$-set. Suppose $|S \cap \bigcup_{i=1}^{h-1} V_{1i}| < (h-1)(j+1)$. Then, for at least one $i_o \in \{1, 2, ..., h-1\}$, we have $|S \cap V_{1i_o}| \leq j$. Let $k \in \{1, 2, ..., m\}$ with $S \cap V_{1i_o} \subseteq A_{ki_o}$ and let $i_1, i_2, ..., i_l \in \{1, 2, ..., h-1\}$ be the indices i for which $S \cap V_{1i} \neq \emptyset$ and $i \neq i_o$. Then, clearly, $\{u_{1i_1}, u_{1i_2}, ..., u_{1i_l}, v_k^h\}$ dominates S (even if $S \cap (V_2^h \cup \{w_1, w_2, ..., w_{h-1}\}) \neq \emptyset$), whence $\gamma(S, G) \leq |\{u_{1i_1}, u_{1i_2}, ..., u_{1i_l}, v_k\}| \leq h-1$, a contradiction. So $|S \cap \bigcup_{i=1}^{h-1} V_{1i}| \geq (h-1)(j+1)$. Furthermore, $S \not\subseteq \bigcup_{i=1}^{h-1} V_{1i}$, since otherwise $\{u_{11}, u_{12}, ..., u_{1(h-1)}\}$ is an S-dominating set in G (contrary to $\gamma(S, G) = h$). Therefore

$$|S| > |S \cap \bigcup_{i=1}^{h-1} V_{1i}| \geq (h-1)(j+1).$$

That is, $\gamma^{\#}(G) \geq (k-1)(j+1) + 1$.

Let the set V_1 in $J_{t,j}$ be partitioned into h subsets $U_1', U_2', ..., U_h'$, where $|U_i'| = j+1$ for each $i \in \{1, 2, ..., h-1\}$ and $|U_h'| = t - (h-1)(j+1)$. Let U_i be the subset of V_{1i} corresponding to U_i' for $i = 1, 2, ..., h-1$, and let $U = \bigcup_{i=1}^{h-1} U_i$, $S = U \cup \{w_1\}$. We shall show that $\gamma(S, G) = h$. Let D be a minimum S-dominating set in G and suppose that $|D| \leq h-1$. We may assume that $D \cap \{w_1, ..., w_{h-1}\} = \emptyset$, since otherwise $D \cap \{w_1, ..., w_{h-1}\}$ may be replaced by $\{v_1\}$ in D, yielding an S-dominating set in G which is not larger than D. Let $|D \cap (\bigcup_{i=1}^{h-1} V_{1i})| = k$ and $|D \cap V_2^h| = l$. Then $k + l = |D| \leq h-1$ and $l \geq 1$ (as D dominates w_1).

Each vertex of D in $\bigcup_{i=1}^{h-1} V_{1i}$ dominates $j+1$ vertices of S (namely, those in some U_i) and each vertex of $D \cap V_2^h$ dominates w_1 and at most j vertices in $S - \{w_1\}$. Hence, the number of vertices in S dominated by D is at most

$$k(j+1) + lj + 1 \leq k(j+1) + (h-1-k) + 1.$$

However, D dominates S, so $|S| \leq k(j+1) + (h-1-k)j + 1$. That is, $(h-1)(j+1) \leq k(j+1) + (h-1-k)j = k + j(h-1)$. It follows that $k \geq h-1$ which (with $l \geq 1$) yields $|D| \geq h$, a contradiction. So $\gamma(S, G) = h$ and the desired result (b) follows.

The result in (c) is obvious. $\square$

We consider next the value of the parameter $\gamma^\#$ for cycles. (Note that the following theorem provides a nonempty graph G, namely C_{3k+1}, for which the bound $\gamma^\#(G) = n$ is attained.)

Proposition 8.9 [1046] *Let* $n \in N$ *with* $n \geq 3$. *Then*

   (a) $\gamma^\#(C_n)$ $= \gamma(C_n)$ $= n/3$ *if* $n \equiv 0 \pmod 3$

   (b) $\gamma^\#(C_n)$ $= n$ *if* $n \equiv 1 \pmod 3$

   (c) $\gamma^\#(C_n)$ $= (2n-1)/3$ *if* $n \equiv 2 \pmod 3$.

Proof. Let $n \in N$ with $n \geq 3$, and let $C_n = \langle u_0, u_1, ..., u_n(= u_0) \rangle$. Assume first that $n \equiv 0 \pmod 3$. Clearly, $D = \{u_0, u_3, u_6, ..., u_{n-3}\}$ is an efficient dominating set of C_n. Hence, by Theorem 8.5, $\gamma^\#(C_n) = \gamma(C_n) = n/3$.

Now let $n \equiv 1 \pmod 3$. Let $\emptyset \neq R \subseteq V(C_n)$ with $|R| \leq n-1$. Clearly, $\langle R \rangle \subseteq C_n - \{v\}$ for some $v \in V(C_n)$. Hence,

$$\gamma(R, C_n) \leq \gamma(R, C_n - \{v\}) \leq \gamma(C_n - \{v\}) = \frac{n-1}{3} < \left\lceil \frac{n}{3} \right\rceil = \gamma(C_n).$$

Thus $\gamma^\#(C_n) \geq |R| + 1$ for all $\emptyset \neq R \subseteq V(C_n)$, $R \neq V(C_n)$. That is, $\gamma^\#(C_n) \geq n$, so $\gamma^\#(C_n) = n$.

Finally, assume that $n \equiv 2 \pmod 3$, say $n = 3k + 2$ for some $k \in N$. Let $S = \{u_0\} \cup \{u_{3i-1}, u_{3i} : i = 1, ..., k\}$; then $|S| = 2k + 1$ and $\gamma(S, C_n) \leq \gamma(C_n) = k + 1$. Furthermore, if $T \subseteq V(C_n)$ and T dominates S, then each vertex in T dominates at most two vertices in S and so $|T| \geq \lceil |S|/2 \rceil = k + 1$. It follows that $\gamma(S, C_n) = k + 1 = \gamma(C_n)$; hence, S is a γ-forcing set of C_n and $\gamma^\#(C_n) \leq |S| = 2k + 1$.

To show that $\gamma^\#(C_n) = 2k + 1$, we assume that a γ-forcing set R of C_n exists with $|R| \leq 2k$. Let $T = V(C_n) - R$; then $t = |T| \geq k + 2$.

We observe that T is an independent set in C_n. Otherwise, if T contains two adjacent vertices, u_i and u_{i+1}, then $C_n - \{u_i, u_{i+1}\}$ is a path P of order $3k$ containing all the vertices in R and $\gamma(R, C_n) \leq \gamma(R, P) \leq \gamma(P) = k < \gamma(C_n)$, a contradiction.

It follows that $\langle R \rangle$, the subgraph of C_n induced by R, is the union of t paths, $P_1, P_2, ..., P_t$. For $i \in \{1, ..., t\}$, let l_i denote the length of P_i, and let $l = j_i$ be the smallest index of a vertex u_l in $V(P_i)$ and label the paths so that $j_1 < j_2 < ... < j_t$. Denote by m_j the number of components of $\langle R \rangle$ of order j ($j \in \{1, 2, ...\}$) and note that, as $2k \geq |R| \geq m_1 + 2(t - m_1) = 2t - m_1 \geq 2k + 4 - m_1$, it follows that $m_1 \geq 4$.

There cannot be a sequence of components of $\langle R \rangle$, namely $P_i, P_{i+1}, ..., P_{i+h}$ ($h \geq 1$) with $l_i = l_{i+h} = 1$ and $l_j = 2$ for j, $i + 1 \leq j \leq i + h - 1$ (if $h \geq 2$) may be seen as follows. Suppose that such a sequence of paths exists and let $N = N[V(P_i) \cup ... \cup V(P_{i+h})]$. Let $Q' = \langle N \rangle$ and $Q'' = \langle V(C_n) - N \rangle$; then Q' and Q'' are paths of order $3h + 2$ and $3(k - h)$, respectively. The set $R \cap V(Q')$ is dominated by the set of h vertices of T between v_{j_i} and $v_{j_{i+h}}$ in Q', hence, $\gamma(R \cap V(Q'), Q') \leq h$, while $\gamma(R \cap V(Q''), Q'') \leq \gamma(Q'') = k - h$. So $\gamma(R, C_n) \leq \gamma(R \cap V(Q'), Q') + \gamma(R \cap V(Q''), Q'') \leq h + (k - h) = k < \gamma(C_n)$, a contradiction.

We may therefore conclude that, if P_i and P_{i+h} are trivial components of $\langle R \rangle$ ($h > 0$), then there exists a component P_j of $\langle R \rangle$ of order $l_j \geq 3$ such that $i < j < i + h$. Consequently, $m_1 \leq m_3 + m_4 + ...$ and we obtain

$$2k \geq |R| \geq m_1 + 2(t - m_1 - m_3 - m_4...) + 3(m_3 + m_4 + ...)$$

$$= 2t - m_1 + (m_3 + m_4 + ...) \geq 2t \geq 2k_4,$$

from which contradiction it follows that $\gamma^\#(C_n) \geq 2k + 1$. We may therefore conclude that $\gamma^\#(C_n) = (2n - 1)/3$ if $n \equiv 2 \pmod 3$. $\square$

8.3 Domatic Number

Mimicking the definition of the chromatic number of a graph G as the minimum number of elements in a partition of $V(G)$ into independent sets, Cockayne and Hedetniemi [280] defined the *domatic number* $d(G)$ of a graph G to be the maximum number of elements in a partition of $V(G)$ into dominating sets. They and Dawes [256] subsequently defined the *total domatic number* $d_t(G)$ to be the largest number of sets in a partition of V into total dominating sets. For $d_t(G)$ we require that $\delta(G) \geq 1$. The *connected domatic number* $d_c(G)$ introduced by Hedetniemi and Laskar [669] is the maximum number of sets in a partition of V into connected dominating sets, and for this parameter to be defined, G must be connected.

Note, for example, for the cycle C_5, we have $i(C_5) = \beta_0(C_5) = 2$, so every independent dominating set has cardinality two, and $V(C_5)$ cannot be partitioned into independent dominating sets. As in Cockayne and Hedetniemi [280] and Zelinka [1175], if there exists at least one partition of V into independent dominating sets, then G is called *idomatic* and the *idomatic number* $id(G)$ equals the maximum number of sets in a partition of V into independent dominating sets. Also note that every paired-dominating set has even order, so no graph G of odd order could have its vertex set V partitioned into paired-dominating sets.

In the spirit of this chapter, we are simply interested in how many disjoint subsets we can find, each with the required properties. We define the *domatic number* $d(G)$, *total domatic number* $d_t(G)$, *connected domatic number*

$d_c(G)$, *independent domatic number* $d_i(G)$, and *paired-domatic number* $d_{pr}(G)$, respectively, to be the maximum number of disjoint dominating sets in G, total dominating sets in G, connected dominating sets in G, independent dominating sets in G, and paired-dominating sets in G, respectively.

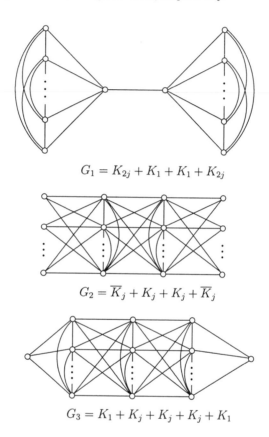

$$G_1 = K_{2j} + K_1 + K_1 + K_{2j}$$

$$G_2 = \overline{K}_j + K_j + K_j + \overline{K}_j$$

$$G_3 = K_1 + K_j + K_j + K_j + K_1$$

Figure 8.5: Demonstrating incomparability of certain pairs of parameters.

For the graphs $G_1 = K_{2j} + K_1 + K_1 + K_{2j}$, $G_2 = \overline{K}_j + K_j + K_j + \overline{K}_j$, and $G_3 = K_1 + K_j + K_j + K_j + K_1$, as illustrated in Figure 8.5, we have $d(G_1) = 2j + 1$, $d_i(G_1) = 2j + 1$, $d_t(G_1) = d_{pr}(G_1) = j + 1$, $d_c(G_1) = 1$, and $d(G_2) = j + 1$, $d_t(G_2) = d_c(G_2) = d_{pr}(G_2) = j$, $d_i(G_2) = 2$, and $d(G_3) = j + 1$, $d_t(G_3) = d_c(G_3) = d_i(G_3) = j$, and $d_{pr}(G_3) = \lfloor (3j + 2)/4 \rfloor$. The inequalities in the next theorem are fairly obvious, and the above examples demonstrate the incomparability of the other pairs of these parameters.

Theorem 8.10 *For any graph* G, $d(G) \geq d_t(G) \geq d_{pr}(G)$; $d(G) \geq d_i(G)$; *and* *if $\gamma(G) \geq 2$, then $d_t(G) \geq d_c(G)$. On the other hand, the following pairs of parameters are incomparable $\{d_t, d_i\}$, $\{d_c, d_i\}$, $\{d_c, d_{pr}\}$, and $\{d_{pr}, d_i\}$.*

For each $v \in V$, if $D \subseteq V$ is a dominating or independent dominating set, then $D \cap N[v] \neq \emptyset$, and if D is a total, connected (with $|D| \geq 2$), or paired-dominating set, then $D \cap N(v) \neq \emptyset$. Consequently, we have the next result.

Theorem 8.11 *For any graph G,*

(a) $d(G) \leq \delta(G) + 1$; $d_t(G) \leq \delta(G)$; *if* $\gamma(G) \geq 2$ *then* $d_c(G) \leq \delta(G)$; $d_i(G) \leq \delta(G) + 1$; *and* $d_{pr}(G) \leq \delta(G)$.

(b) $d(G) \leq n/\gamma(G)$; $d_t(G) \leq n/\gamma_t(G)$; $d_c(G) \leq n/\gamma_c(G)$; $d_i(G) \leq n/i(G)$; *and* $d_{pr}(G) \leq n/\gamma_{pr}(G)$.

Using the bounds from Theorem 8.11 (a), a graph for which the upper bound is achieved is called "full". Specifically, G is *domatically full* if $d(G) = 1 + \delta(G)$, *total domatically full* if $d_t(G) = \delta(G)$, etc. Note that every paired-domatically full graph is also totally domatically full.

Zelinka [1170] provided a constructive characterization of regular domatically full graphs. Our next theorem does this for paired/total domatically full graphs.

Theorem 8.12 [652] *For an r-regular graph G, the following are equivalent:*

(a) *G is paired-domatically full.*

(b) *G is totally domatically full.*

(c) *$|V(G)| = n = 2tr$; $V(G)$ contains r disjoint sets $S_1, S_2, ..., S_r$ with $|S_i| = 2t$; each subgraph induced by S_i has $\langle S_i \rangle = tK_2$; and for $1 \leq i < j \leq r$ there are $2t$ independent edges matching S_i and S_j.*

Proof. **(a)$\Leftrightarrow$(b).** If G is paired-domatically full, then there exist r disjoint paired-dominating sets, each of which is also a total dominating set, and so G is totally domatically full. Conversely, let $S_1, S_2, ..., S_r$ be disjoint total dominating sets in G. Assume some S_i, say S_1, is not paired-dominating. Then there exists a vertex $v \in S_1$ such that $|N(v) \cap S_1| \geq 2$. But the $S_i's$ are disjoint and $|N(v)| = r$ now implies that some S_j satisfies $|N[v] \cap S_j| = 0$, a contradiction.

(c)$\Rightarrow$(a). This is clear because each S_i, $1 \leq i \leq r$, is a paired-dominating set.

(a)$\Rightarrow$(c). Let $S_1, S_2, ..., S_r$ be disjoint paired-dominating sets in an r-regular graph G. For each $v \in V(G)$, we have $|N(v) \cap S_i| = 1$ for $1 \leq i \leq r$. That is, each S_i is an efficient paired-dominating set. If $|S_1| = 2t$, then because S_i efficiently totally dominates $V(G)$, we have $|V(G)| = 2tr$. Every other efficient total dominating set S_i, $2 \leq i \leq r$, must also have order $|V(G)|/r = 2t$. Efficiency implies $\langle S_i \rangle = tK_2$. Finally, if $1 \leq i < j \leq r$, then each $v \in S_j$ is adjacent to exactly one vertex in S_i, so S_i and S_j are matched. $\square$

Chang has shown that essentially all 2-dimensional grids are domatically full. It is certainly clear that the path P_n is domatically full for all $n \geq 1$. One can easily verify that $P_2 \times P_2$ and $P_4 \times P_2$ (or $P_2 \times P_4$) each have domatic number two, so they are not domatically full.

Theorem 8.13 [196] *If* $n_1 \geq n_2 \geq 2$, *then* $d(P_{n_1} \times P_{n_2}) = 3$ *with only two exceptions, namely* $P_2 \times P_2$ *and* $P_4 \times P_2$.

Proof. We first observe that for the disjoint union $G_1 \cup G_2$ of G_1 and G_2 that $d(G_1 \cup G_2) = \min\{d(G_1), d(G_2)\}$, and if H is a spanning subgraph of G then $d(H) \leq d(G)$.

Assume (n_1, n_2) is not $(2, 2)$ or $(4, 2)$. Suppose n_1 is odd, and let

$$D_1 = \{(a, b) : a \equiv 0 \pmod 2, \ 2 \leq a \leq n_1\}$$

$$D_2 = \{(a, b) : a \equiv 1 \pmod 4 \text{ and } b \equiv 1 \pmod 2, \ 1 \leq a \leq n_1, 1 \leq b \leq n_2\}$$
$$\cup \{(a, b) : a \equiv 3 \pmod 4 \text{ and } b \equiv 0 \pmod 2, \ 1 \leq a \leq n_1, 1 \leq b \leq n_2\}, \text{ and}$$

$$D_3 = \{(a, b) : a \equiv 1 \pmod 4 \text{ and } b \equiv 0 \pmod 2, \ 1 \leq a \leq n_1, 1 \leq b \leq n_2\}$$
$$\cup \{(a, b) : a \equiv 3 \pmod 4 \text{ and } b \equiv 1 \pmod 2, \ 1 \leq a \leq n_1, 1 \leq b \leq n_2\}.$$

Then $\{D_1, D_2, D_3\}$ is a domatic partition for $P_{n_1} \times P_{n_2}$. Similarly, if n_2 is odd, then we also have $d(P_{n_1} \times P_{n_2}) = 3$.

Assume n_1 and n_2 are both even. The reader can verify that $d(P_4 \times P_4) = 3$. Suppose $n_1 \geq 6$. Because $(P_3 \times P_{n_2}) \cup (P_{n_1-3} \times P_{n_2})$ is a spanning subgraph of $P_{n_1} \times P_{n_2}$ and $d(P_3 \times P_{n_2}) = d(P_{n_1-3} \times P_{n_2})$ by the above cases, we have $d(P_{n_1} \times P_{n_2}) \geq d((P_3 \times P_{n_2}) \cup (P_{n_1-3} \times P_{n_2})) \geq \min\{d(P_3 \times P_{n_2}), d(P_{n_1-3} \times P_{n_2})\} = 3$ by the initial observation. Thus, the theorem holds. $\square$

As noted, if H is a spanning subgraph of G, then $d(H) \leq d(G)$. Considering edge deleted subgraphs, Cockayne [251] defines G to be *domatically k-critical*, or *k-critical*, if $d(G) = k \geq 2$ and $d(G - e) < k$ for every edge $e \in E(G)$. The next theorem is due to Rall.

Theorem 8.14 [957] *Every domatically 2-critical or 3-critical graph is domatically full. However, for each* $k \geq 4$, *there exists a graph which is k-critical but which is not domatically full.*

Zelinka obtained lower bounds for $d(G)$ and $d_t(G)$.

Theorem 8.15 *For any graph G,*

(a) [1180]

$$d(G) \geq \left\lfloor \frac{n}{n - \delta(G)} \right\rfloor .$$

(b) [1192]

$$d_t(G) \geq \left\lfloor \frac{n}{n - \delta(G) + 1} \right\rfloor .$$

Proof. (a) Let $D \subseteq V(G)$ with $|D| \geq n - \delta(G)$.

If $v \in V(G) - D$, then $|N[v]| \geq 1 + \delta(G)$ implies $N(v) \cap D \neq \emptyset$. Thus one can take any $\lfloor n/(n - \delta(G)) \rfloor$ disjoint subsets, each of cardinality $n - \delta(G)$. Each of these subsets is a dominating set, so

$$d(G) \geq \left\lfloor \frac{n}{n - \delta(G)} \right\rfloor .$$

(b) Let $D \subseteq V(G)$ with $|D| \geq n - \delta(G) + 1$. For any vertex $v \in V(G)$, $|N(v)| \geq \delta(G)$, which implies that $N(v) \cap D \neq \emptyset$, so D is a total dominating set. Thus one can take any $\lfloor n/(n - \delta(G) + 1) \rfloor$ disjoint subsets of cardinality $n - \delta(G) + 1$ to achieve $d_t(G) \geq \lfloor n/(n - \delta(G) + 1) \rfloor$. $\square$

Cockayne and Hedetniemi [280] established the following Nordhaus-Gaddum result for the domatic number of a graph G and its complement $\overline{G}$, and with Dawes [256] presented a similar result for the total domatic number.

Theorem 8.16 [280] *For any graph G, $d(G) + d(\overline{G}) \leq n + 1$, with equality if and only if $G = K_n$ or $\overline{K}_n$.*

Theorem 8.17 [256] *For a graph G with $\delta(G) \geq 1$ and $\Delta(G) \leq n - 2$, $d_t(G) + d_t(\overline{G}) \leq n - 1$, with equality if and only if G or $\overline{G}$ is C_4.*

Proof. With $1 \leq \delta(G) \leq \Delta(G) \leq n - 2$, we have $d_t(G) \leq \delta(G)$ and $d_t(\overline{G}) \leq \delta(\overline{G}) \leq \Delta(\overline{G})$, so that $d_t(G) + d_t(\overline{G}) \leq \delta(G) + \Delta(\overline{G}) = n - 1$.

If G is not regular, then $\delta(\overline{G}) < \Delta(\overline{G})$ and so $d_t(G) + d_t(\overline{G}) < \delta(G) + \Delta(\overline{G}) = n - 1$. Hence, for equality G and $\overline{G}$ are regular, $d_t(G) = \delta(G)$, $d_t(\overline{G}) = \delta(\overline{G})$, and we can assume that $\delta(G) \leq \delta(\overline{G})$. If $\delta(\overline{G}) < (n - 1)/2$, then $\delta(G) + \delta(\overline{G}) = d_t(G) + d_t(\overline{G}) < n - 1$. Because $\delta(\overline{G}) = d_t(\overline{G}) \leq n/2$, we obtain $(n - 1)/2 \leq \delta(\overline{G}) \leq n/2$. Thus two possibilities remain: (1) n is even, $\delta(\overline{G}) = n/2$, and $\delta(G) = (n/2) - 1$, or (2) n is odd, $\delta(G) = \delta(\overline{G}) = (n - 1)/2$.

Suppose each of the $d_t(G)$ disjoint total dominating sets of G has at least three vertices. If n is odd, then $3d_t(G) = 3(n - 1)/2 \leq n$, so $n \leq 3$, but no 3-vertex graph has $1 \leq \delta(G) \leq \Delta(G) \leq 3 - 2 = 1$. If n is even, then $3d_t(G) = 3(n/2 - 1) \leq n$ implies $n \leq 6$. If $n = 6$, then $\delta(G) = \Delta(G) = 2$, and there are two disjoint total dominating sets, each with 3 vertices, say D_1 and D_2. But neither $2C_3$ nor C_6 satisfies the conditions. If $n = 4$, then $\delta(G) = \Delta(G) = 1$, so $G = 2K_2$ and $\overline{G} = C_4$, giving $d_t(G) + d_t(\overline{G}) = 1 + 2 = n - 1$.

In the remaining case, some total dominating set among the $d_t(G)$ disjoint total dominating sets has exactly two vertices, and they are adjacent. For n even, two adjacent vertices in G are adjacent to at most $2((n/2) - 2) = n - 4$ other vertices, and hence, they cannot form a dominating set. For n odd, they would dominate at most $n - 3$ others, a similar contradiction. This completes the proof. $\square$

Because $d_{pr}(G) \leq d_t(G)$, the bound for the paired-domination number follows directly.

Theorem 8.18 [652] *For a graph G with $\delta(G) \geq 1$ and $\Delta(G) \leq n-2$, $d_{pr}(G) + d_{pr}(\overline{G}) \leq n - 1$, with equality if and only if G or $\overline{G}$ is C_4.*

Theorem 8.19 *For any graph G,*

(a) [280] $\gamma(G) + d(G) \leq n + 1$, *with equality if and only if $G = K_n$ or $\overline{K}_n$.*

(b) [583] *if $\delta(G) \geq 1$, then $\gamma(G) + d(G) \leq \lfloor n/2 \rfloor + 2$, and equality requires that $\{\gamma(G), d(G)\} = \{\lfloor n/2 \rfloor, 2\}$.*

Note that the upper bound is greatly improved when G has no isolated vertices. Similar results are obtained for paired and total domination.

Theorem 8.20 [274] *If a graph G has $d_t(G) \geq 2$, then $\gamma_t(G) + d_t(G) \leq \lfloor n/2 \rfloor + 2$.*

Theorem 8.21 [652] *If a graph G has $d_{pr}(G) \geq 2$, then $\gamma_{pr}(G) + d_{pr}(G) \leq \lfloor n/2 \rfloor + 2$.*

Proof. Let G be a graph with $d_{pr}(G) \geq 2$. Then $\gamma_{pr}(G) \leq \lfloor n/2 \rfloor$. By definition $\gamma_{pr}(G) \geq 2$, so $d_{pr}(G) \leq \lfloor n/2 \rfloor$.

If either $\gamma_{pr}(G) = 2$ or $d_{pr}(G) = 2$, then the bound obviously holds. If $\gamma_{pr}(G) \geq 4$ and $d_{pr}(G) \geq 4$, then since $\gamma_{pr}(G) \cdot d_{pr}(G) \leq n$, we have

$$\gamma_{pr}(G) \leq \left\lfloor \frac{n}{d_{pr}(G)} \right\rfloor \text{ and } d_{pr}(G) \leq \left\lfloor \frac{n}{\gamma_{pr}(G)} \right\rfloor.$$

That is, $\gamma_{pr}(G) \leq \lfloor n/4 \rfloor$ and $d_{pr}(G) \leq \lfloor n/4 \rfloor$. Hence,

$$\gamma_{pr}(G) + d_{pr}(G) \leq 2 \lfloor n/4 \rfloor < \lfloor n/2 \rfloor + 2.$$

Since $\gamma_{pr}(G)$ is even, $\gamma_{pr}(G) \neq 3$, so the only case remaining is when $d_{pr}(G) = 3$. When $d_{pr}(G) = 3$,

$$\gamma_{pr}(G) + d_{pr}(G) \leq 3 + \lfloor n/3 \rfloor.$$

Since $3 = d_{pr}(G) \leq \lfloor n/2 \rfloor$, we have $n \geq 6$. For $n \geq 6$,

$$3 + \lfloor n/3 \rfloor \leq \lfloor n/2 \rfloor + 2.$$

In all three cases, $\gamma_{pr}(G) + d_{pr}(G) \leq \lfloor n/2 \rfloor + 2$. $\square$

Additional Nordhaus-Gaddum results are given in Chapter 9. We conclude this chapter by noting that Zelinka summarizes much of what is known about domatic numbers in Chapter 13 of [645].

EXERCISES

8.1 (a) For disjoint dominating sets D_1 and D_2, one can seek to minimize $|D_1 \cup D_2|$ or $\max\{|D_1|, |D_2|\}$. Let $f(G)$ denote the minimum possible value of $|D_1 \cup D_2|$, and let $g(G)$ denote the minimum possible value of $\max\{|D_1|, |D_2|\}$, where D_1 and D_2 are disjoint dominating sets. Show that f and g are incomparable (there exist graphs G and H with $f(G) < g(G)$ and $g(H) < f(H)$).

(b) Find linear algorithms to determine $f(T)$ and $g(T)$ for trees T.

8.2 In Section 8.2 several prioritized (independence, domination) multiproperty problems and parameters were described. Do the same for (domination, independence).

8.3 Do a multiset of (lat pulldowns, pullups).

8.4 Show that, even for connected graphs, total domination γ_t and least domination γ_l are incomparable.

8.5 (a) When is $\gamma(K_{n_1,n_2,...,n_t}) = i(K_{n_1,n_2,...,n_t})$?

(b) Determine $PR(m, \gamma)(K_{n_1,n_2,...,n_t})$.

(c) Determine $PR(m, \gamma)(P_j \times P_k)$ for small j and arbitrary k. (See Section 2.7.2 for $\gamma(P_j \times P_k)$.)

8.6 Verify Proposition 8.2 for $\gamma_l(P_n)$ and $\gamma_l(C_n)$.

8.7 Verify Conjecture 8.4 for trees.

8.8 For which graph (tree) G on n vertices is $i(G) - \gamma(G)$ maximized?

8.9 Use Proposition 8.8 to construct a graph G with $\gamma(G) = 3$ and $\gamma^\#(G) = 9$.

8.10 (a) Find, if possible, infinitely many triples (a, b, c) with $2 \leq a \leq b \leq c$ such that no graph G satisfies $\gamma(G) = a$, $\gamma^\#(G) = b$, and G has order $n = c$ (that is, (a, b, c) is not $(\gamma, \gamma^\#, n)$-realizable).

(b) Characterize the $(\gamma, \gamma^\#, n)$- realizable triples.

8.11 As noted, $\gamma^\#(C_{3k+1}) = 3k + 1$. For every n, find a graph G_n of order n with $\gamma^\#(G_n) = n$.

8.12 Show that $\gamma^\#(G)$ and $i(G)$ are incomparable.

8.13 Prove or disprove: for any tree T, $\gamma(T) = \gamma^\#(T)$.

8.14 For $j \le k$ determine the minimum order $n(j, k)$ of a graph G with $\gamma(G) = j$ and $\gamma^\#(G) = k$.

8.15 If H is a spanning subgraph of G, then $\gamma(G) \le \gamma(H)$. Is it always the case that $\gamma^\#(G) \le \gamma^\#(H)$?

8.16 Verify the values given for d, d_t, d_i, and d_{pr} for the graphs in Figure 8.5.

8.17 Verify Theorem 8.10 relating the domatic parameters.

8.18 Prove Theorem 8.20: If $d_t(G) \ge 2$, then $\gamma_t(G) + d_t(G) \le \lfloor n/2 \rfloor + 2$.

Chapter 9

Sums and Products of Parameters

Relationships between graphical parameters are always of interest. We consider the sums and products of two parameters, at least one of which is a domination parameter. One could consider the sum (product) of two parameters of a graph G or alternatively, the sum (product) of a parameter of G and the same parameter of its complement $\overline{G}$. We begin with the latter category, known as Nordhaus-Gaddum type results.

9.1 Nordhaus-Gaddum Type Results

In their now classical paper [922], Nordhaus and Gaddum established the following inequalities for the chromatic numbers $\chi(G)$ and $\chi(\overline{G})$:

$$2\sqrt{n} \leq \chi(G) + \chi(\overline{G}) \leq n + 1,$$

$$n \leq \chi(G)\chi(\overline{G}) \leq \frac{(n+1)^2}{4}.$$

We present similar inequalities involving domination parameters in graphs.

9.1.1 Domination number

In 1972 Jaeger and Payan published the first Nordhaus-Gaddum type results involving domination.

Theorem 9.1 [741] *For any graph G,*

$$\gamma(G) + \gamma(\overline{G}) \leq n + 1$$

and

$$\gamma(G)\gamma(\overline{G}) \leq n.$$

Proof. First we verify the upper bound on the sum. If G has an isolated vertex, then $\gamma(G) \leq n$ and $\gamma(\overline{G}) = 1$. A similar result holds when $\overline{G}$ has an isolated vertex, so in both cases, $\gamma(G) + \gamma(\overline{G}) \leq n + 1$. If neither G nor $\overline{G}$ have isolated vertices, then by Ore's theorem $\gamma(G) \leq \lfloor n/2 \rfloor$ and $\gamma(\overline{G}) \leq \lfloor n/2 \rfloor$, and so $\gamma(G) + \gamma(\overline{G}) \leq n$.

Next we show that $\gamma(G)\gamma(\overline{G}) \leq n$. For a set of vertices $X \subseteq V$, define $D_e(X)$ to be the vertices in $V - X$ which are adjacent to every vertex in X and let $d_e(X) = |D_e(X)|$. Also, define $D_i(X)$ to be the vertices in X which are adjacent to all other vertices in X and let $d_i(X) = |D_i(X)|$.

Let $S = \{x_1, x_2, \cdots, x_\gamma\}$ be a γ-set of G and partition V into γ subsets $B_1, \cdots, B_r$, such that, $x_j \in B_j$ and all the vertices in B_j are adjacent to x_j, for $1 \leq j \leq \gamma(G)$. Let P be a partition for which the sum over all integers j of $d_i(B_j)$ is a maximum.

Suppose that $d_e(B_j) \geq 1$. Then there exists an $x \in B_k$, $k \neq j$, such that x is adjacent to every vertex in B_j. If $x \in D_i(B_k)$, then $(S - \{x_j, x_k\}) \cup \{x\}$ is a dominating set of cardinality smaller than $\gamma(G)$, a contradiction. Hence, $x \notin D_i(B_k)$. But now we have a partition P' with $B_l' = B_l$ for $l \neq k$ and $l \neq j$, $B_j' = B_j \cup \{x\}$, and $B_k' = B_k - \{x\}$, for which $d_i(B_l') = d_i(B_l)$, $d_i(B_j') = d_i(B_j) + 1$, and $d_i(B_k') \geq d_i(B_k)$, contradicting our choice of P.

Hence, $d_e(B_j) = 0$ for $1 \leq j \leq \gamma(G)$. Note that any set X with $d_e(X) = 0$ dominates $\overline{G}$. Therefore, each set B_j in P dominates $\overline{G}$, so $\gamma(\overline{G}) \leq |B_j|$. Hence,

$$n = \sum_{j=1}^{\gamma(G)} |B_j| \geq \gamma(G)\gamma(\overline{G}).$$

$\square$

Notice from the proof that each set B_j is a dominating set of $\overline{G}$ and hence, P is a domatic partition of $\overline{G}$ and we have the following corollary.

Corollary 9.2 *For any graph G, $d(\overline{G}) \geq \gamma(G)$.*

Cockayne and Hedetniemi sharpened the upper bound for the sum.

Theorem 9.3 [280] *For any graph G,*

$$\gamma(G) + \gamma(\overline{G}) \leq n + 1$$

with equality if and only if $G = K_n$ or $G = \overline{K}_n$.

Laskar and Peters improved this bound for the case when both G and $\overline{G}$ are connected.

Theorem 9.4 [829] *For connected graphs G and $\overline{G}$,*

$$\gamma(G) + \gamma(\overline{G}) \leq n$$

with equality if and only if $G = P_4$.

Using a simple, yet elegant proof technique, Joseph and Arumugam obtained an upper bound on the sum, for graphs G and $\overline{G}$ with no isolated vertices, that considerably improved the bound for general graphs.

Theorem 9.5 [755] *If graphs G and $\overline{G}$ have no isolated vertices, then*

$$\gamma(G) + \gamma(\overline{G}) \leq \lfloor n/2 \rfloor + 2.$$

Proof. Let graphs G and $\overline{G}$ have no isolated vertices. Then $\gamma(G) \geq 2$ and $\gamma(\overline{G}) \geq 2$. By Ore's theorem, $\gamma(G) \leq \lfloor n/2 \rfloor$ and $\gamma(\overline{G}) \leq \lfloor n/2 \rfloor$. If $\gamma(G) = 2$ or $\gamma(\overline{G}) = 2$, the result holds. If $\gamma(G) \geq 4$ and $\gamma(\overline{G}) \geq 4$, then since $\gamma(G)\gamma(\overline{G}) \leq n$, we have

$$\gamma(G) \leq \left\lfloor \frac{n}{\gamma(\overline{G})} \right\rfloor \text{ and } \gamma(\overline{G}) \leq \left\lfloor \frac{n}{\gamma(G)} \right\rfloor.$$

That is, $\gamma(G) \leq \lfloor n/4 \rfloor$ and $\gamma(\overline{G}) \leq \lfloor n/4 \rfloor$. Hence,

$$\gamma(G) + \gamma(\overline{G}) \leq 2\lfloor n/4 \rfloor < \lfloor n/2 \rfloor + 2.$$

The only remaining possibility is that $\gamma(G) = 3$ or $\gamma(\overline{G}) = 3$. But in this case, by Theorem 9.1, $\gamma(\overline{G}) \leq \lfloor n/3 \rfloor$, and so

$$\gamma(G) + \gamma(\overline{G}) \leq 3 + \lfloor n/3 \rfloor.$$

Since $\gamma(G) = 3 \leq \lfloor n/2 \rfloor$, we have $n \geq 6$, and for $n \geq 6$, $3 + \lfloor n/3 \rfloor \leq \lfloor n/2 \rfloor + 2$. $\square$

Noting that for graphs of order $n \neq 9$, equality of the bound in Theorem 9.5 is only possible for the case when G or $\overline{G}$ has domination number equal to $\lfloor n/2 \rfloor$, Joseph and Arumugam determined a property of most graphs achieving the upper bound.

Theorem 9.6 [755] *For a graph G of order $n \neq 9$, such that G and $\overline{G}$ have no isolated vertices, $\gamma(G) + \gamma(\overline{G}) = \lfloor n/2 \rfloor + 2$ if and only if either $\gamma(G)$ or $\gamma(\overline{G}) = \lfloor n/2 \rfloor$.*

Payan and Xuong [933] proved that the only graph of order nine having $\gamma(G) = \gamma(\overline{G}) = 3$ is the self–complementary graph $K_3 \times K_3$. All other graphs G and $\overline{G}$ with no isolated vertices and order nine have $\{\gamma(G), \gamma(\overline{G})\} = \{2, 2\}$, $\{2, 3\}$ or $\{2, 4\}$. Hence, we may deduce the following from Theorem 9.6.

Corollary 9.7 *If graphs G and $\overline{G}$ have no isolated vertices, then $\gamma(G) + \gamma(\overline{G}) = \lfloor n/2 \rfloor + 2$ if and only if $G = K_3 \times K_3$ or $\{\gamma(G), \gamma(\overline{G})\} = \{\lfloor n/2 \rfloor, 2\}$.*

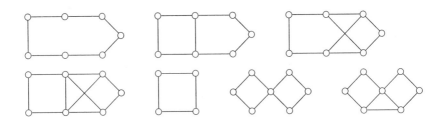

Figure 9.1: Set $\mathcal{A}$.

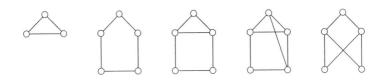

Figure 9.2: Set $\mathcal{B}$.

A characterization of the connected graphs G for which $\gamma(G) = \lfloor n/2 \rfloor$ was given in Chapter 2. Since it enables us to obtain the extremal graphs of Theorem 9.5, we repeat the relevant results here.

The sets of graphs $\mathcal{A}$ and $\mathcal{B}$ are shown in Figures 9.1 and 9.2, respectively. Let

$$\mathcal{G}_1 = C_4 \cup \{G : G = H \circ K_1 \text{ where } H \text{ is connected}\}$$

and

$$\mathcal{G}_2 = (\mathcal{A} \cup \mathcal{B}) - \{C_4\} .$$

For any graph H, let $\mathcal{S}(H)$ denote the set of connected graphs which may be formed from $H \circ K_1$ by adding a new vertex x and edges joining x to one or more vertices of H. Then define

$$\mathcal{G}_3 = \bigcup_H \mathcal{S}(H),$$

where the union is taken over all graphs H.

Let y be a vertex of a copy of C_4 and for $G \in \mathcal{G}_3$, let $\theta(G)$ be the graph obtained by connecting G to C_4 with the single edge xy, where x is the new vertex added in forming G. Then define

$$\mathcal{G}_4 = \{\theta(G) : G \in \mathcal{G}_3\} .$$

Next let u, v, w be the vertex sequence of a path P_3. For any graph H, let $\mathcal{P}(H)$ be the set of connected graphs which may be formed from $H \circ K_1$ by joining both u and w to one or more vertices of H. Then define

$$\mathcal{G}_5 = \bigcup_H \mathcal{P}(H) .$$

Let H be any graph and $X \in \mathcal{B}$. Let $\mathcal{R}(H, X)$ be the set of connected graphs which may be formed from $H \circ K_1$ by joining each vertex of $U \subseteq V(X)$ to one or more vertices of H such that $\gamma(X)$ vertices of X are required to dominate $V(X) - U$. Then define

$$\mathcal{G}_6 = \bigcup_{H,X} \mathcal{R}(H, X) .$$

Theorem 9.8 [274, 963] *A connected graph G satisfies $\gamma(G) = \lfloor n/2 \rfloor$ if and only if $G \in \mathcal{G} = \bigcup_{i=1}^{6} \mathcal{G}_i$.*

Cockayne, Haynes, and Hedetniemi characterized the set of graphs which attain the bound of Theorem 9.5.

Theorem 9.9 [274] *Let graphs G and $\overline{G}$ have no isolated vertices.*

(a) *If G is connected, then $\gamma(G) + \gamma(\overline{G}) = \lfloor n/2 \rfloor + 2$ if and only if $G = K_3 \times K_3$ or one of G, $\overline{G}$ is in $\mathcal{G}$.*

(b) *If G is disconnected with components $G_1, \ldots, G_t$ ($t \geq 2$), then $\gamma(G) + \gamma(\overline{G}) = \lfloor n/2 \rfloor + 2$ if and only if at most one G_i has an odd number of vertices and for each $i = 1, \ldots, t$, $G_i \in \mathcal{G}$.*

Proof. (a) This is immediate from Theorem 9.8 and Corollary 9.7.

(b) For $i = 1, \ldots, t$, let $|V(G_i)| = n_i$ and $\gamma_i = \gamma(G_i)$. Since G and $\overline{G}$ have no isolated vertices and $t \geq 2$, we have $\gamma(\overline{G}) = 2$ and $\gamma(G) = \sum_{i=1}^{t} \gamma_i$. Therefore (by Corollary 9.7), the bound is attained if and only if

$$\sum_{i=1}^{t} \gamma_i = \left\lfloor \frac{n}{2} \right\rfloor . \tag{9.1}$$

Suppose, without loss of generality, that $n_1 = 2k_1 + 1$ and $n_2 = 2k_2 + 1$ are odd. For each $i = 1, \ldots, t$, G_i has no isolated vertices. By Ore's Theorem, $\gamma_1 \leq k_1$, $\gamma_2 \leq k_2$ and for $i \geq 3$, $\gamma_i \leq \frac{n_i}{2}$. Hence, $\sum_{i=1}^{t} \gamma_i \leq k_1 + k_2 + \frac{1}{2} \sum_{i=3}^{t} n_i = \frac{n}{2} - 1 < \left\lfloor \frac{n}{2} \right\rfloor$, contrary to (9.1). Thus at most one n_i is odd.

Suppose, without losing generality, that $\gamma_1 < \left\lfloor \frac{n_1}{2} \right\rfloor$. Then $\sum_{i=1}^{t} \gamma_i \leq \left\lfloor \frac{n_1}{2} \right\rfloor - 1 + \sum_{i=2}^{t} \left\lfloor \frac{n_i}{2} \right\rfloor \leq \frac{n}{2} - 1 < \left\lfloor \frac{n}{2} \right\rfloor$ contrary to (9.1). Hence, each $G_i \in \mathcal{G}$.

Conversely, assume that $n_2, \ldots, n_t$ are even and for each $i = 1, \ldots t$, $\gamma_i = \left\lfloor \frac{n_i}{2} \right\rfloor$. Then $\sum_{i=1}^{t} \gamma_i = \left\lfloor \frac{n_1}{2} \right\rfloor + \sum_{i=2}^{t} (\frac{n_i}{2}) = \left\lfloor \frac{n}{2} \right\rfloor$, and the proof is complete. $\square$

Recalling the following table from Chapter 2, we consider improved bounds on the sum when restrictions are placed on the minimum degree.

lower bound for $\delta(G)$	upper bound for $\gamma(G)$
0	n
1	$n/2$
2 and $G \notin \mathcal{B}$	$2n/5$
3	$3n/8$

Noting that the upper bound from Theorem 9.1 of $n + 1$, in a sense, corresponds to the upper bound of n on $\gamma(G)$ and when neither G nor $\overline{G}$ has isolated vertices, the upper bound of $\lfloor n/2 \rfloor + 2$ of Theorem 9.5 corresponds to the upper bound of $\lfloor n/2 \rfloor$ on $\gamma(G)$, it is natural to ask the following questions:

Is $\gamma(G) + \gamma(\overline{G}) \leq \lfloor 2n/5 \rfloor + 3$ when both G and $\overline{G}$ have minimum degree 2? And similarly, is $\lfloor 3n/8 \rfloor + 4$ an upper bound on the sum when both G and $\overline{G}$ have minimum degree 3? Haynes (unpublished) gave affirmative answers to these questions. Note that this work is still in progress and there is a chance the bounds may be improved. The known results are summarized in the following table.

lower bound for $\delta(G)$, $\delta(\overline{G})$	Upper bound on sum
0	$n + 1$
1	$n/2 + 2$
2	$2n/5 + 3$
3	$3n/8 + 4$

Goddard, Henning, and Swart [541] extended Nordhaus-Gaddum type results to the general case where the complete graph is factored into more than two edge-disjoint factors. In particular, they considered factoring K_n into three factors G_1, G_2, and G_3 and determined the following bounds.

Theorem 9.10 [541] *If* $G_1 \oplus G_2 \oplus G_3 = K_n$, *then*

$$\gamma(G_1) + \gamma(G_2) + \gamma(G_3) \leq 2n + 1.$$

Theorem 9.11 [541] *If* $G_1 \oplus G_2 \oplus G_3 = K_n$, *then the maximum value of the product* $\gamma(G_1)\gamma(G_2)\gamma(G_3)$ *is* $n^3/27 + \Theta(n^2)$.

That is, there exist constants c_1 and c_2 such that the maximum triple product always lies between $n^3/27 + c_1 n^2$ and $n^3/27 + c_2 n^2$.

9.1.2 Domatic number

Several Nordhaus-Gaddum type results for domatic parameters were presented in Chapter 8. We repeat the result involving the domatic numbers of complementary graphs here.

Theorem 9.12 [280] *For any graph G,*

$$d(G) + d(\overline{G}) \leq n + 1,$$

with equality if and only if G or $\overline{G}$ is K_n.

Dunbar, Haynes, and Henning determined a corresponding upper bound for the product.

Theorem 9.13 [399] *For any graph G of order $n \geq 4$,*

$$2 \leq d(G)d(\overline{G}) \leq n^2/4,$$

and these bounds are sharp.

Proof. Since $n > 1$, at least one of G and $\overline{G}$ has no isolated vertices and therefore has domatic number at least 2. Thus, $2 \leq d(G)d(\overline{G})$. That this lower bound is sharp, may be seen by taking $G \cong K_{1,n-1}$. To verify the upper bound, we make use of Theorem 9.12. If $G \cong K_n$ or $G \cong \overline{K}_n$, then $d(G)d(\overline{G}) = n \leq n^2/4$ since $n \geq 4$. Hence, we may assume that $G \not\cong K_n$ or $G \not\cong \overline{K}_n$. Thus, by Theorem 9.12, $d(G) + d(\overline{G}) \leq n$. Since the geometric mean of two positive numbers never exceeds their arithmetic mean, we have

$$\sqrt{d(G)d(\overline{G})} \leq \frac{d(G) + d(\overline{G})}{2} \leq \frac{n}{2}.$$

Thus, $d(G)d(\overline{G}) \leq n^2/4$. That this bound is sharp may be seen by taking the complete bipartite graph $G \cong K_{s,s}$ of order $n = 2s$. Then $d(G) = s$ and $d(\overline{G}) = d(2K_s) = s$. Thus, $d(G)d(\overline{G}) = s^2 = n^2/4$. $\square$

They [399] also characterized the graphs achieving the upper bound of Theorem 9.13 that is, the graphs G for which $d(G) = d(\overline{G}) = n/2$.

Recall that the upper bound on the sum of the domination numbers of graph and its complement was reduced considerably by simply requiring that G and $\overline{G}$ have no isolated vertices. One might now ask if Theorem 9.12 can be improved in a similar manner. However, this is not possible. As pointed out in the proof of Theorem 9.13, if we take $G \cong K_{s,s}$ and $n = 2s \geq 4$, then neither G nor $\overline{G}$ has isolated vertices and $d(G) = d(\overline{G}) = s$. Thus, $d(G) + d(\overline{G}) = n$ and $d(G)d(\overline{G}) = n^2/4$. Hence, the upper bounds in Theorems 9.12 and 9.13 cannot be improved for arbitrary graphs by imposing the restriction that neither G nor $\overline{G}$ has isolated vertices. However, these bounds can be improved for trees.

Theorem 9.14 [399] *For any tree T of order $n \geq 2$,*

$$3 \leq d(T) + d(\overline{T}) \leq \lfloor n/2 \rfloor + 2 \quad and$$

$$2 \leq d(T)d(\overline{T}) \leq n,$$

and these bounds are sharp. Equality in the lower bounds occurs if and only if $T \cong K_{1,n-1}$, while equality in the upper bounds occurs if and only if $d(\overline{T}) = \lfloor n/2 \rfloor$ and, for the product, n is even.

That the upper bounds are sharp may be seen by considering the double star T obtained from two disjoint stars $K_{1,s-1}$ on $s \geq 2$ vertices by joining the two vertices of maximum degree with an edge. Then T is a tree of order $n = 2s$ satisfying $d(T) = 2$ and $d(\overline{T}) = s$. Thus $d(T) + d(\overline{T}) = n/2 + 2$ and $d(T)d(\overline{T}) = 2s$.

It is interesting to note the conditions required for equality to hold in Theorems 9.5 and 9.14, that is, $\{\gamma(G), \gamma(\overline{G})\} = \lfloor n/2 \rfloor$ and $d(\overline{T}) = \lfloor n/2 \rfloor$, respectively. Graphs having $\gamma(G) = \lfloor n/2 \rfloor$ were characterized in [274], but the characterization of graphs with $d(G) = \lfloor n/2 \rfloor$ is an open problem. However, the characterization of the extremal graphs of Theorem 9.13 determines a subclass of the graphs having $d(G) = \lfloor n/2 \rfloor$, that is, the graphs for which $d(G) = d(\overline{G}) = n/2$. And, in order to characterize the extremal graphs of Theorem 9.14, Dunbar et al. characterized those trees T for which $d(\overline{T}) = \lfloor n/2 \rfloor$. They used the following lemma.

Lemma 9.15 [399] *If T is a tree of order $n \geq 2$ with $\Delta(T) \leq \lceil n/2 \rceil$, then $d(\overline{T}) = \lfloor n/2 \rfloor$.*

Proof. We proceed by induction on the order $n \geq 2$ of the tree T with $\Delta(T) \leq \lceil n/2 \rceil$. If $2 \leq n \leq 4$, then $\Delta(T) \leq 2$. Thus, $T \cong P_n$ and $d(\overline{T}) = \lfloor n/2 \rfloor$. This establishes the base cases. Let $n \geq 5$, and assume that if T' is a tree of order n' with $2 \leq n' < n$ and with $\Delta(T') \leq \lceil n'/2 \rceil$, then $d(\overline{T'}) = \lfloor n'/2 \rfloor$. Let T be a tree of order n with $\Delta(T) \leq \lfloor n/2 \rfloor$.

If $\Delta(T) = \lceil n/2 \rceil$, then there must exist an endvertex u adjacent to a vertex of maximum degree in T. If $\Delta(T) \leq \lceil n/2 \rceil - 1$, then let u be an arbitrary endvertex of T. In both cases, let w be a vertex at maximum distance from u in T. Then w is an endvertex at distance at least three from u in T (so $\{u, w\}$ is a dominating set of $\overline{T}$). Let $T' = T - \{u, w\}$. Then by our choice of u and w, T' is a tree of order $n - 2 \geq 3$ with $\Delta(T') \leq \lceil (n-2)/2 \rceil$. Hence, an application of the inductive hypothesis gives $d(\overline{T'}) = \lfloor (n-2)/2 \rfloor$. Thus, we may partition $V(T')$ into $\lfloor n/2 \rfloor - 1$ sets, each of which is a dominating set of $\overline{T'}$. These $\lfloor n/2 \rfloor - 1$ sets, together with the set $\{u, w\}$, partition $V(T)$ into $\lfloor n/2 \rfloor$ dominating sets of $\overline{T}$. Hence, $d(\overline{T}) = \lfloor n/2 \rfloor$. $\square$

We are now in a position to give the characterization.

Theorem 9.16 [399] *Let T be a tree of order $n \geq 2$. Then $d(\overline{T}) = \lfloor n/2 \rfloor$ if and only if $\Delta(T) \leq \lceil n/2 \rceil$.*

Proof. If $d(\overline{T}) = \lfloor n/2 \rfloor$, then $\delta(\overline{T}) \geq \lfloor n/2 \rfloor - 1$. Hence, $\Delta(T) = n - 1 - \delta(\overline{T}) \leq \lceil n/2 \rceil$. This establishes the necessity. The sufficiency follows from Lemma 9.15. □

9.1.3 Other domination parameters

Cockayne and Mynhardt established Nordhaus-Gaddum type results and found extremal graphs for the upper irredundance numbers of G and $\overline{G}$.

Theorem 9.17 [297] *For any graph G,*

$$IR(G) + IR(\overline{G}) \leq n + 1$$

and K_n attains this bound.

Theorem 9.18 [297] *For any graph G,*

$$IR(G)IR(\overline{G}) \leq \left\lceil \frac{n^2 + 2n}{4} \right\rceil,$$

with equality if and only if G or $\overline{G}$ consists of

(a) *a set X of $\lfloor (n+1)/2 \rfloor$ independent vertices,*

(b) *a set Y of $\lceil (n+1)/2 \rceil$ vertices where $\langle Y \rangle$ is complete and $X \cap Y = \{x\}$, and*

(c) *any set S of edges joining vertices of $X - \{x\}$ to vertices of $Y - \{x\}$.*

The next two corollaries follow directly from the previous two theorems and the domination inequality chain,

$$ir(G) \leq \gamma(G) \leq i(G) \leq \beta_0(G) \leq \Gamma(G) \leq IR(G).$$

Corollary 9.19 *For any graph G, the sums $i(G) + i(\overline{G})$, $\beta_0(G) + \beta_0(\overline{G})$, and $\Gamma(G) + \Gamma(\overline{G})$ are all bounded above by $n + 1$.*

The complete graph K_n obtains the upper bound on the sum for all the parameters. For the sums $\beta_0(G) + \beta_0(\overline{G})$, $\Gamma(G) + \Gamma(\overline{G})$, and $IR(G) + IR(\overline{G})$, the following graph (suggested by E. J. Cockayne) shows that the upper bound of $n+1$ is also sharp for graphs G and $\overline{G}$ with no isolated vertices. Let G be the graph on $n > 4$ vertices with V partitioned into V_1 and V_2 where $|V_1| = \lfloor n/2 \rfloor$. Fix v in V_2. Let $\{v\} \cup V_1$ induce a complete graph and the only other edges be a matching from $V_2 - \{v\}$ into V_1. Then $\beta_0(G) + \beta_0(\overline{G}) = \Gamma(G) + \Gamma(\overline{G}) = IR(G) + IR(\overline{G}) = n + 1$.

Corollary 9.20 *For any graph G, the products $i(G)i(\overline{G})$, $\beta_0(G)\beta_0(\overline{G})$, and $\Gamma(G)\Gamma(\overline{G})$ are all bounded above by $\lceil (n^2 + 2n)/4 \rceil$.*

Chartrand and Schuster [240] independently obtained these results for the independence numbers of complementary graphs using Ramsey theory.

Note that the extremal graphs of Theorem 9.18 obtain the bound of Corollary 9.20 for all the products except $i(G)i(\overline{G})$. Cockayne and Mynhardt [297] demonstrated that this product can be at least $\frac{(n+4)^2}{16}$ with the following family of graphs. Let $n = 4k$, $k \geq 1$, and let $V(G)$ be partitioned into sets of equal order X, Y, where $\langle X \rangle$ is independent, $\langle Y \rangle$ is complete and the bipartite graph induced by X, Y is regular of degree $n/4$. Then $i(G) = i(\overline{G}) = n/4 + 1$, and hence, $i(G)i(\overline{G}) = \frac{(n+4)^2}{16}$.

We summarize the best possible Nordhaus-Gaddum bounds in the following table.

Parameter $\mu(G)$	$\mu(G) + \mu(\overline{G})$ Bound	$\mu(G) + \mu(\overline{G})$ Bound (isolate-free graphs)	$\mu(G)\mu(\overline{G})$ Bound
$ir(G)$	$n + 1$	$\lfloor n/2 \rfloor + 2$	n
$\gamma(G)$	$n + 1$	$\lfloor n/2 \rfloor + 2$	n
$i(G)$	$n + 1$	?	?
$\beta_0(G)$	$n + 1$	$n + 1$	$\lceil \frac{n^2+2n}{4} \rceil$
$\Gamma(G)$	$n + 1$	$n + 1$	$\lceil \frac{n^2+2n}{4} \rceil$
$IR(G)$	$n + 1$	$n + 1$	$\lceil \frac{n^2+2n}{4} \rceil$

Next, we consider Nordhaus-Gaddum type results for total, connected, and edge domination parameters. In the introductory paper on total domination, Cockayne, Dawes, and Hedetniemi proved a Nordhaus-Gaddum type result for the total domination number.

Theorem 9.21 [256] *If graphs G and $\overline{G}$ have no isolated vertices, then*

$$\gamma_t(G) + \gamma_t(\overline{G}) \leq n + 2,$$

with equality if and only if G or $\overline{G}$ is mK_2.

Arumugam and Thuraiswamy characterized the graphs G for which $\gamma_t(G) + \gamma_t(\overline{G}) = n + 1$.

Theorem 9.22 [55] *A graph G has $\gamma_t(G) + \gamma_t(\overline{G}) = n + 1$ if and only if G or $\overline{G}$ is isomorphic to C_5, $P_3 \cup rK_2$, or $C_3 \cup rK_2$, $r \geq 1$.*

The graphs G for which $\gamma_t(G) + \gamma_t(\overline{G}) = n$ are also characterized in [55]. Hedetniemi and Laskar determined the following Nordhaus-Gaddum type result for the connected domination number.

Theorem 9.23 [669] *If graphs G and $\overline{G}$ are both connected, then*

$$\gamma_c(G) + \gamma_c(\overline{G}) \leq n + 1.$$

This bound was improved by Laskar and Peters.

Theorem 9.24 [827] *If graphs G and $\overline{G}$ are both connected, then*

$$\gamma_c(G) + \gamma_c(\overline{G}) = \begin{cases} n + 1 & \text{if and only if } G = C_5, \\ n & \text{if and only if } G = C_n, n \geq 6, \ G = P_n, n \geq 4, \\ & \text{or } G \text{ is the graph of Figure 9.3.} \end{cases}$$

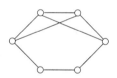

Figure 9.3: $\gamma_c(G) + \gamma_c(\overline{G}) = 6 = n$.

Schuster obtained the following Nordhaus-Gaddum type inequalities concerning the edge domination number $\gamma'(G)$. Recall that by definition $\gamma'(\overline{K}_n) = 0$.

Theorem 9.25 [1015] *For any graph G of order $n \geq 3$,*

$$\lfloor n/2 \rfloor \leq \gamma'(G) + \gamma'(\overline{G}) \leq 2\lfloor n/2 \rfloor$$

and

$$0 \leq \gamma'(G)\gamma'(\overline{G}) \leq \lfloor n/2 \rfloor^2.$$

Theorem 9.26 [1015] *For any graph G of order $n \equiv 2 \ (mod \ 4)$,*

$$n/2 \leq \gamma'(G) + \gamma'(\overline{G}) \leq n - 1$$

and

$$0 \leq \gamma'(G)\gamma'(\overline{G}) \leq n(n-2)/4.$$

9.2 Gallai Type Theorems

In 1959 Gallai presented his now classical theorem, involving the vertex cover-ing number $\alpha_0(G)$, the vertex independence number $\beta_0(G)$, the edge covering number $\alpha_1(G)$, and the edge independence number $\beta_1(G)$. (Note that Gallai theorems are also found in Chapters 2, 3, and 10.)

Theorem 9.27 (Gallai [518]) *For any graph G,*
$\alpha_0(G) + \beta_0(G) = n$, *and*
$\alpha_1(G) + \beta_1(G) = n$, *if G has no isolated vertices.*

Proof. If S is a maximal independent set, then there are no edges in $\langle S \rangle$, so every edge is incident to at least one vertex of $V - S$. Conversely, if $V - S$ covers all the edges, then there are no edges in $\langle S \rangle$. Hence, any maximum independent set is the complement of a minimum vertex cover and $\alpha_0(G) + \beta_0(G) = n$.

Let M be a maximum matching of G. We construct an edge cover of size $n - |M|$ by adding one edge incident to each unsaturated vertex. Then the total number of edges used is $n - |M|$, as desired. Hence, $\alpha_1(G) \leq n - |M| = n - \beta_1(G)$.

Conversely, let F be a minimum edge cover. If both the endvertices of some edge $uv \in F$ are incident to other edges in F, then uv is not needed in the cover. Hence, F consists of k disjoint stars, for some k. Since F has one edge for each vertex that is not a center of its stars, we have $|F| = n - k$. Then a matching M of size $k = n - |F|$ is obtained by choosing one edge arbitrarily from each star in F. Then $\beta_1(G) \geq n - k \geq n - \alpha_1(G)$. $\square$

Results of the form, $\alpha(G) + \beta(G) = n$, where $\alpha(G)$ is a minimum and $\beta(G)$ is a maximum parameter of G, are called Gallai theorems. In 1988, Cockayne, Hedetniemi, and Laskar [284] wrote a paper to unify a wide variety of Gallai theorems. One of their main results was a general theorem of this type involving hereditary set systems which can be found in Chapter 3 of this text. We list some Gallai theorems involving domination parameters, repeating a few results from previous chapters.

Theorem 9.28 [920] *For any nontrivial connected graph G, $\gamma(G) + \varepsilon_F(G) = n$.*

Theorem 9.29 [669] *For any connected graph G, $\gamma_c(G) + \varepsilon_T(G) = n$.*

The *enclaveless number* $\Psi(G)$ is the maximum cardinality of an enclaveless set of G and $\psi(G)$ is the minimum cardinality of a maximal enclaveless set.

Theorem 9.30 [1032] *For any graph G,*

$$\gamma(G) + \Psi(G) = n, \text{ and}$$

$$\psi(G) + \Gamma(G) = n.$$

Let $IR_e(G)$ denote the maximum number of vertices in a set S such that each vertex $v \in S$ has a private edge, that is, there exists an edge uv with $u \notin S$.

Theorem 9.31 [271] *For any graph G,*

$$\gamma(G) + IR_e(G) = n.$$

Let $\Lambda(G)$ equal the maximum number of vertices in a minimal vertex cover of G. McFall and Nowakowski determined the following.

Theorem 9.32 [884] *For any graph G,*

$$i(G) + \Lambda(G) = n.$$

9.3 Other Sums and Products

9.3.1 Parameters in the domination sequence

Cockayne, Favaron, Payan, and Thomason established several inequalities involving parameters from the domination sequence. First they generalized Ore's theorem.

Theorem 9.33 [260] *If a graph G has no isolated vertices and $S \subseteq V$ is an irredundant set, then $V - S$ is a dominating set.*

Corollary 9.34 [260] *If a graph G has no isolated vertices, then*

$$\gamma(G) + IR(G) \le n.$$

Proof. Let S be a maximum irredundant set. Then $V - S$ is dominating and $|V - S| \ge \gamma(G)$. Hence, $n = |S| + |V - S| \ge IR(G) + \gamma(G)$. □

They also proved that equality in any of the results of Corollary 9.34 implies equality in the three upper parameters of the domination sequence.

Theorem 9.35 [260] *If a graph G has no isolated vertices and $\gamma(G) + IR(G) = n$, then $\beta_0(G) = \Gamma(G) = IR(G)$.*

In fact, they presented a better upper bound on the sum $\gamma(G) + IR(G)$.

Theorem 9.36 [260] *If a graph G has no isolated vertices, then*

$$\gamma(G) + IR(G) \le n - \delta(G) + 2.$$

9.3.2   Domination and maximum degree

It is well known (see Chapter 2) for any graph that

$$\gamma(G) + \Delta(G) \le n.$$

It follows from the domination sequence that $ir(G) + \Delta(G) \le n$. The stronger inequality that $i(G) + \Delta(G) \le n$ is also trivial since any maximal independent set containing a vertex of maximum degree $\Delta(G)$ contains at most $n - \Delta(G)$ vertices. Sharpness for these bounds is studied in [363] and [458]. Domke, Dunbar, and Markus determined the connected bipartite graphs and trees for which equality holds for $\gamma(G)$ and $i(G)$. We give their results for trees.

Theorem 9.37 [363] *Let T be a tree with a vertex x of maximum degree. Then $i(T) + \Delta(T) = n$ if and only if $V - N[x]$ is an independent set and $|V - N[x]| \le \Delta(T) - 1$.*

Recall that a tree is a wounded spider if the tree is $K_{1,s}$, $s \ge 0$, with at most $s - 1$ of the edges subdivided.

Theorem 9.38 [363] *A tree T has $\gamma(T) + \Delta(T) = n$ if and only if T is a wounded spider.*

Favaron and Mynhardt [458] gave necessary and sufficient conditions for $\mu(G) + \Delta(G) = n$ for each of the three lower parameters, ir, γ, and i. They also showed that $\gamma(G) - ir(G)$ can be arbitrarily large, even if $\gamma(G) + \Delta(G) = n$.

9.3.3   Domination and the chromatic number

An assignment of colors to the vertices of a graph G so that adjacent vertices are assigned different colors is called a *coloring* of G. A coloring in which k-colors are used is called a *k-coloring*. As definied in the Prolegomenon, the minimum k for which a graph G has a k-coloring is the chromatic number $\chi(G)$.

Gernert showed that $\gamma(G) + \chi(G) \le n + 1$, when in fact, $\gamma(G) \le i(G) \le \beta_0(G) \le n - \chi(G) + 1$. He also considered the product of $\gamma(G)$ and $\chi(G)$.

Theorem 9.39 [531] *For any connected graph G with $n \ge 5$,*

$$\gamma(G)\chi(G) \le \frac{n^2}{4}$$

with equality if and only if G is the corona $K_{2k} \circ K_1$.

Proof. For odd cycles C_n, $n \geq 5$, we have $\gamma(C_n) = \lceil \frac{n}{3} \rceil$ and $\chi(C_n) = 3$ and the theorem holds. From Brooks theorem (see Exercise 16, Prolegomenon), if G is neither an odd cycle nor the complete graph K_n, then $\chi(G) \leq \Delta(G)$. Multiplication by $\gamma(G) \leq n - \Delta(G)$ yields

$$\gamma(G)\chi(G) \leq \Delta(G)(n - \Delta(G)).$$

The unique maximum is assumed for $\Delta(G) = n/2$ implying that $\gamma(G)\chi(G) \leq n^2/4$. Equality is reached when $\gamma(G)$ is at its maximum of $n/2$. From Theorem 2.2, G must be the corona $H \circ K_1$ or the C_4. Since $n \geq 5$, we consider the corona. In order for $\chi(G) = n/2$, H must be the complete graph on an even number of vertices and the theorem follows. $\square$

Topp and Volkmann also explored the product of $\gamma(G)$ and $\chi(G)$.

Theorem 9.40 [1100] *If G is a connected graph with $\delta(G) \geq 2$, then*

$$\gamma(G)\chi(G) \leq \frac{\delta(G)}{8(\delta(G) - 1)}(n + 1)^2.$$

For a nice summary of relations among other graph parameters, the reader is referred to Xu [1152]. Obviously, there are many other parameters $\mu(G)$ for which the sum (product) of $\mu(G)$ and a domination parameter are of interest. This area is open to future research.

9.3.4   Sums bounded by $\lfloor n/2 \rfloor + 2$

As we have seen, much work has been done to determine inequalities involving the sum of two domination related parameters. In many cases, the upper bound on the sum is either n or $n + 1$. However, sometimes a simple restriction, for example, that G have no isolated vertices, is enough to reduce the upper bound. Here we consider sums bounded above by $\lfloor n/2 \rfloor + 2$. We also examine the graphs achieving this upper bound.

Cockayne and Hedetniemi established a bound on the sum of the domination number and the domatic number of a graph.

Theorem 9.41 [280] *For any graph G, $\gamma(G) + d(G) \leq n + 1$ with equality if and only if $G = K_n$ or $\overline{K}_n$.*

Harary and Haynes, using techniques similar to Joseph and Arumugam's [755] proof technique for Theorem 9.5, improved Theorem 9.41 for graphs G and $\overline{G}$ having no isolated vertices.

Theorem 9.42 [582] *If graphs G and $\overline{G}$ have no isolated vertices, then*

$$\gamma(G) + d(G) \leq \left\lfloor \frac{n}{2} \right\rfloor + 2.$$

In fact, Theorem 9.42 is a special case of a more general result of Harary and Haynes concerning k-tuple domination.

We now consider extremal graphs of Theorem 9.42. Jaeger and Payan [953] established that $\gamma(\overline{G}) \leq d(G)$ for any graph G (Corollary 9.2). Therefore equality in

$$\gamma(G) + \gamma(\overline{G}) \leq \lfloor n/2 \rfloor + 2 \qquad (9.2)$$

implies equality in

$$\gamma(G) + d(G) \leq \lfloor n/2 \rfloor + 2 \qquad (9.3)$$

and we have the following corollary.

Corollary 9.43 [274] *All graphs G mentioned in Theorem 9.9 satisfy $\gamma(G) + d(G) = \lfloor n/2 \rfloor + 2$.*

The next two results due to Cockayne, Haynes, and Hedetniemi make some progress towards the determination of the class $\mathcal{I}$ of extremal graphs of (9.3) which are not extremal graphs of (9.2). It is clear that for each graph $F \in \mathcal{I}$, $\gamma(\overline{F})$ is strictly less than $d(F)$.

Theorem 9.44 [274] *Let graphs G and $\overline{G}$ have no isolated vertices. Then $\gamma(G) + d(G) = \lfloor n/2 \rfloor + 2$ if and only if $\{\gamma(G), d(G)\} = \{\lfloor n/2 \rfloor, 2\}$ or $n = 9$ and $\gamma(G) = d(G) = 3$.*

Proof. Let G and $\overline{G}$ have no isolated vertices, $n \neq 9$ and $\gamma(G) + d(G) = \lfloor n/2 \rfloor + 2$. Jaeger and Payan proved that $\gamma(G)d(G) \leq n$ for any graph G. Hence,

$$\gamma(G) + d(G) \leq \left\lfloor \frac{n}{d(G)} \right\rfloor + \left\lfloor \frac{n}{\gamma(G)} \right\rfloor.$$

If both $\gamma(G)$ and $d(G)$ are at least 4, then

$$\gamma(G) + d(G) \leq 2 \left\lfloor \frac{n}{4} \right\rfloor < \left\lfloor \frac{n}{2} \right\rfloor + 2,$$

a contradiction.

If $\gamma(G) = 3$ or $d(G) = 3$, then $n \geq 6$ and

$$\gamma(G) + d(G) \leq 3 + \left\lfloor \frac{n}{3} \right\rfloor \leq 2 + \left\lfloor \frac{n}{2} \right\rfloor.$$

Equality can only occur for $n = 6, 7$, and 9. If $n = 9$, equality implies $\gamma(G) = d(G) = 3$ and for equality when $n = 6$ or 7, $\gamma(G) = 3$ ($d(G) = 3$) implies $d(G) = 2$ ($\gamma(G) = 2$) as required.

The result clearly holds if $\gamma(G)$ or $d(G)$ equals 2. Since $\overline{G}$ has no isolated vertices, $\gamma(G) \geq 2$ and since G has no isolated vertices, $d(G) \geq \gamma(\overline{G}) \geq 2$. This completes the proof. $\square$

The next result follows immediately from Theorems 9.44 and 9.6.

Corollary 9.45 [274] *For graphs F such that F and $\overline{F}$ have no isolated vertices, F is in the class $\mathcal{I}$ if and only if*

(1) $n = 9$, $\gamma(F) = d(F) = 3$, and $\gamma(\overline{F}) = 2$.

   or

(2) $2 = \gamma(F) \leq \gamma(\overline{F}) < d(F) = \lfloor n/2 \rfloor$.

Graphs of Corollary 9.45 include C_9, $K_3 \times P_3$ and all subgraphs of $K_3 \times P_3$ which have C_9 as a subgraph. Graphs of Corollary 9.45 have $\lfloor n/2 \rfloor$ (respectively, $\lfloor n/2 \rfloor - 1$) disjoint minimum dominating sets (of cardinality two) if n is even (respectively, odd) and an additional dominating set of three vertices if n is odd. The possible values of $\gamma(\overline{F})$ give considerable flexibility, so the characterization of graphs having $d(G) = \lfloor n/2 \rfloor$ is an open problem.

A variety of other results give upper bounds of n or $n + 1$ for sums of two domination parameters for general graphs. The technique of Joseph and Arumugam [755] can undoubtedly be applied in some of these cases to radically improve the bound when certain fairly trivial graphs are excluded. Thus arises a large class of interesting extremal graph problems. As examples, we state here, without proof, two inequalities concerning total domination, total domatic, connected domination, and connected domatic numbers which were obtained using this proof technique.

Theorem 9.46 [274] *For any graph G with $d_t(G) \geq 2$,*

$$\gamma_t(G) + d_t(G) \leq \left\lfloor \frac{n}{2} \right\rfloor + 2,$$

with equality for $n \neq 9$ if and only if $\{\gamma_t(G), d_t(G)\} = \left\{2, \left\lfloor \frac{n}{2} \right\rfloor\right\}$.

Theorem 9.47 [274] *If a graph G has $\gamma_c(G) \geq 2$ and $d_c(G) \geq 2$, then*

$$\gamma_c(G) + d_c(G) \leq \left\lfloor \frac{n}{2} \right\rfloor + 2.$$

EXERCISES

9.1 For a path P_n, show that $d(\overline{P}_n) = \lfloor n/2 \rfloor$.

9.2 Can $IR(G) + IR(\overline{G}) = n + 1$ for G and $\overline{G}$ connected?

9.3 (a) Prove: If connected graphs G and $\overline{G}$ have $\delta(G), \delta(\overline{G}) \geq 2$, then

$$\gamma(G) + \gamma(\overline{G}) \leq 2n/5 + 3.$$

(b) Is this bound sharp?

9.4 Is it possible that $IR(G) + IR(\overline{G}) \leq n$ if both G and $\overline{G}$ are connected?

9.5 Determine sharp lower bounds for $\gamma(G) + \gamma(\overline{G})$ and for $\gamma(G)\gamma(\overline{G})$.

9.6 Determine lower and upper bounds for

(a) $\gamma(G) + i(G)$.

(b) $\gamma(G)i(G)$.

9.7 In light of Corollary 9.34 and Theorem 9.36, what can you say about $i(G) + IR(G)$?

9.8 Theorem 9.12 asserts that $d(G) + d(\overline{G}) \leq n + 1$. What can you say about $d_t(G) + d_t(\overline{G})$?

9.9 For a t-factoring $G_1, G_2, ..., G_t$ of a graph H, Haynes and Henning [647] determined the following upper bound:

$$d(G_1) + d(G_2) + ... + d(G_t) \leq n + t - 1.$$

(Note that Theorem 9.12 is a corollary of this result.)

(a) Verify this result.

(b) Give an infinite family of graphs H that obtain the upper bound.

9.10 Domke, Dunbar, and Markus [363] showed that $i(G) + \Delta(G) \leq n$ and $IR(G) + \delta(G) \leq n$.

(a) Verify these results.

(b) For each of these results, give an infinite family of graphs that achieve the bound.

Chapter 10

Dominating Functions

Part of the beauty of, and fascination with, basic subset problems, such as the problems of finding the minimum cardinality of a dominating set and the maximum cardinality of an independent set, lies in the simplicity of their statements. (This simplicity is deceptive. For example, evaluating $\gamma(G)$ is computationally difficult (that is, NP-hard) whereas in its continuous form, the corresponding linear programming problem can be solved in polynomial time.) However, as illustrated by Theorem 4.25, which states that

$$\rho(G) \leq \rho_f(G) = \gamma_f(G) \leq \gamma(G) \leq F(G) \leq F_f(G) = W_f(G) \leq W(G) \leq n$$

$$\leq P(G) \leq P_f(G) = R_f(G) \leq R(G),$$

the determination of the values of fractional parameters can be useful not only for their own applications, but also in providing information about the related integer-valued parameters. For the fractional parameters in Chapter 4, we considered functions $f : V(G) \to [0, 1]$, while for subset parameters like γ, ρ, F, and R, we can identify a subset $S \subseteq V(G)$ with its characteristic function $\phi : V(G) \to \{0, 1\}$. Recall that for parameters W and P, we consider functions $f : V(G) \to \{0, 1, 2, 3, ...\}$. This chapter is concerned with the general case of Y-valued functions $f : V(G) \to Y$ for an arbitrary subset Y of the reals, $Y \subseteq \Re$.

10.1 Introduction: Y-valued Parameters

Let $f : V(G) \to \Re$ be a real valued function definied on $V(G)$. For $S \subseteq V(G)$, we define $f(S) = \sum_{v \in S} f(v)$; and the weight of f is $w(f) = f(V(G))$. A function $f : V(G) \to \Re$ is called a *dominating function* if for each $v \in V(G)$, $f(N[v]) \geq 1$. If $f(v) \in Y$ for every $v \in V(G)$, we call f a *Y-valued function*, and the *Y-domination number* $\gamma_Y(G)$ is the minimum weight of a Y-valued dominating function. A Y-valued dominating function is called a *minimal Y-valued dominating function* if there does not exist a Y-valued dominating function g, $g \neq f$, for which $g(v) \leq f(v)$ for every $v \in V$. For $f : V(G) \to Y$, let X_f be the column vector $[f(v_1), f(v_2), ..., f(v_n)]^t$. Then f is a dominating function if and only if $N \cdot X_f \geq \vec{1}_n$ for the closed neighborhood matrix N and the all ones n-tuple $\vec{1}_n$. Hence, $\gamma_Y(G) = \min\{w(f) \mid f : V(G) \rightarrow Y \text{ and } N \cdot X_f \geq \vec{1}_n\}$. In particular, $\gamma(G) = \gamma_{\{0,1\}}(G)$ and $\gamma_f(G) = \gamma_{[0,1]}(G) = \gamma_{[0,\infty)}(G)$.

The extension of the definition of domination with $\gamma(G) = \gamma_{\{0,1\}}(G)$ to arbitrary $\gamma_Y(G)$ occurred as follows. In 1984 Farber [433] investigated the problem of determining when the corresponding linear programming formulation for γ would provide an integer solution, and he showed this to be the case for strongly chordal graphs. In 1987 Hedetniemi and Wimer [664] first formally defined fractional domination and the fractional domination number γ_f. For example, for the Hajós graph H_3 in Figure 10.1, we have $\gamma(H_3) = 2$ and $\{v_2, v_3\}$ is a γ-set, and it is easy to see that the function f illustrated in Figure 10.1(a) with $f(v_2) = f(v_3) = f(v_5) = 1/2$ and $f(v_1) = f(v_4) = f(v_6) = 0$ is a dominating function with $w(f) = \gamma_f(H_3) = 3/2$. (Earlier in 1979 Claude Berge had studied fractional transversals of hypergraphs.) Results from [369] and [555] about other fractional parameters, such as for packing ρ_f, covers α_f, independence β_f, efficient domination F_f and redundance R_f, were presented in talks by Domke and by Grinstead at the Nineteenth Southeastern Conference in 1988. Schwenk observed, in correspondence to S. T. Hedetniemi, that assignments of weights 1 to v_2, v_3 and v_5 and -1 to v_1, v_4, and v_6 as in Figure 10.1(b) maintained the property that the closed neighborhood sum of any vertex is one but the sum of the weights is reduced to zero. Allowing a negative weight of -1 motivated the definitions of minus domination with $f : V(G) \rightarrow \{-1, 0, 1\}$ in [402] and signed domination with $f : V(G) \rightarrow \{-1, 1\}$ in [404]. As noted in [77], if S is an efficient dominating set for G, then $|S| = \gamma(G)$, and, in particular, all efficient dominating sets have the same cardinality. (Recall that if $F(G) < n$, then G might have F-sets of different cardinalities.) McRae conjectured that any two efficient minus (respectively, signed) dominating functions would have the same weight. McRae's conjecture was generalized and verified by Bange, Barkauskas, Host, and Slater [74], who first defined Y-dominating functions and efficient Y-dominating functions for arbitrary $Y \subseteq \Re$. Slater [1042] defined other Y-valued parameters: upper domination Γ_Y, lower and upper independence i_Y and β_Y, lower and upper covering α_Y and Λ_Y, lower and upper enclaveless ψ_Y and Ψ_Y, edge covering α'_Y, and edge independence (matching) β'_Y. Clearly, any subset or integer parameter can be similarly so generalized.

In fact, in the companion volume [645] to this book, Slater formulates approximately 50 such parameters. Goddard and Henning [540] studied computational complexity questions for $\gamma_\Re$ and $\gamma_{\mathcal{Z}}$, where $\mathcal{Z}$ is the set of integers.

Consider the graph H_3 as in Figure 10.1(c) with $Y = \Re$. For $k \geq 1$, if $f(v_2) = f(v_3) = f(v_5) = k$ and $f(v_1) = f(v_4) = f(v_6) = (1 - 3k)/2$, then f is a dominating function with $w(f) = 3(1 - k)/2$. For the "minimum" weight of an $\Re$-valued function, we actually take the infimum and write $\gamma_\Re(H_3) = -\infty$. Again for H_3, this time with $Y = (0, 1)$ as in Figure 10.1(d) for $0 < \varepsilon < 1$, let

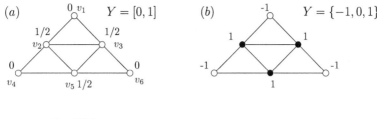

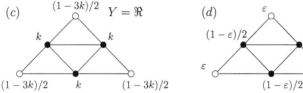

Figure 10.1: Dominating functions for the graph H_3.

$f(v_1) = f(v_4) = f(v_6) = \varepsilon$ and $f(v_2) = f(v_3) = f(v_5) = (1 - \varepsilon)/2$. Then f is a dominating function with weight $w(f) = (3 + 3\varepsilon)/2$. The minimum/infimum of these weights is $\gamma_{(0,1)}(H_3) = 3/2$. When necessary in the following definitions, "min" or "max" should be interpreted to mean infimum or supremum.

Y-domination:

$$\gamma_Y(G) = \min\{w(f) \mid f : V(G) \to Y, \ N \cdot X_f \geq \vec{1}_n\} \qquad (10.1)$$

Y-packing:

$$\rho_Y(G) = \max\{w(f) \mid f : V(G) \to Y, \ N \cdot X_f \leq \vec{1}_n\} \qquad (10.2)$$

For the following parameters, we have column n-vectors

$$D = [deg(v_1), deg(v_2), ..., deg(v_n)]^t \text{ and}$$

$$D^* = [1 + deg(v_1), 1 + deg(v_2), ..., 1 + deg(v_n)]^t.$$

Y-neighborhood domination:

$$W_Y(G) = \min\{w(f) \mid f : V(G) \to Y, \ N \cdot X_f \geq D^*\} \qquad (10.3)$$

Y-efficient domination:

$$F_Y(G) = \max\{\sum_{i=1}^{n}(1 + deg(v_i))f(v_i) \mid f : V(G) \to Y, \ N \cdot X_f \leq \vec{1}_n\} \ (10.4)$$

Y-redundance:

$$R_Y(G) = \min\{\sum_{i=1}^{n}(1+deg(v_i))f(v_i) \mid f : V(G) \to Y, \ N{\cdot}X_f \geq \vec{1}_n\} \quad (10.5)$$

Y-neighborhood packing:

$$P_Y(G) = \max\{w(f) \mid f : V(G) \to Y, \ N \cdot X_f \leq D^*\} \quad\quad (10.6)$$

Y-domination coverage:

$$\eta_Y(G) = \max\{\sum_{i=1}^{n}(deg(v_i)f(v_i) \mid f : V(G) \to Y, \ N \cdot X_f \geq \vec{1}_n\} \quad (10.7)$$

Y-enclaveless:

$$\Psi_Y(G) = \max\{w(f) \mid f : V(G) \to Y, \ N \cdot X_f \leq D\}. \quad\quad (10.8)$$

Recall that the n-by-m incidence matrix H for a graph G has $h_{i,j} = 1$ if v_i is incident with edge e_j, for $1 \leq i \leq n = |V(G)|$ and $1 \leq j \leq m = |E(G)|$. The *generalized edge covering number* α'_Y and *generalized matching number* β'_Y (respectively, the *generalized vertex covering number* α_Y and *generalized independence number* β_Y) are defined using H (respectively, its transpose H^t) as the constraint matrix, as follows.

Y-covering:

$$\alpha_Y(G) = \min\{w(f) \mid f : V(G) \to Y, \ H^t \cdot X_f \geq \vec{1}_m\} \quad\quad (10.9)$$

Y-independence:

$$\beta_Y(G) = \max\{w(f) \mid f : V(G) \to Y, \ H^t \cdot X_f \leq \vec{1}_m\} \quad\quad (10.10)$$

Y-edge covering:

$$\alpha'_Y(G) = \min\{w(f) \mid f : E(G) \to Y, \ H \cdot X_f \geq \vec{1}_n\} \quad\quad (10.11)$$

Y-matching:

$$\beta'_Y(G) = \max\{w(f) \mid f : E(G) \to Y, \ H \cdot X_f \leq \vec{1}_n\} \quad\quad (10.12)$$

Obviously, for some choices of Y the constraints cannot be satisfied. For example, if $k = 1 + deg(v)$ and $Y \subseteq (-\infty, 1/k)$ then it is impossible for a Y-valued function f to satisfy $f(N[v]) \geq 1$, as required for γ_Y, R_Y, and η_Y. However, it is easy to see when a feasible function $f : V(G) \to Y$ or $f : E(G) \to Y$ will exist. For example, for the twelve Y-valued parameters defined above we have the next theorem.

Theorem 10.1 [1042] *For any graph G,*

(a) *G has a feasible function $f : V(G) \to Y$*

    (1) *for Y-domination γ_Y, Y-redundance R_Y, and Y-domination-coverage η_Y if and only if $Y \cap [1/(1 + \delta(G)), \infty) \neq \emptyset$.*

    (2) *for Y-packing ρ_Y and Y-efficient domination if and only if $Y \cap (-\infty, 1/(1 + \Delta(G))] \neq \emptyset$,*

    (3) *for Y-closed neighborhood domination W_Y if and only if $Y \cap [1, \infty) \neq \emptyset$.*

    (4) *for Y-closed neighborhood packing P_Y if and only if $Y \cap (-\infty, 1] \neq \emptyset$, and*

    (5) *for Y-enclaveless value Ψ_Y if and only if $Y \cap (-\infty, \delta(G)/(1 + \delta(G))] \neq \emptyset$.*

    (6) *for Y-covering α_Y if and only if $Y \cap [1/2, \infty) \neq \emptyset$.*

    (7) *for Y-independence β_Y if and only if $Y \cap (-\infty, 1/2] \neq \emptyset$.*

(b) *G has a feasible function $f : E(G) \to Y$*

    (1) *for Y-edge covering α'_Y if and only if $Y \cap [1/\delta(G), \infty) \neq \emptyset$.*

    (2) *for Y-matching β'_Y if and only if $Y \cap (-\infty, 1/\Delta(G)) \neq \emptyset$.*

Henceforth, it is assumed that the set Y meets these required obvious bounds. The next theorem is also obvious.

Theorem 10.2 [1042] *If G is a graph and $Y1 \subseteq Y2 \subseteq \Re$, then*
$\gamma_{Y2}(G) \leq \gamma_{Y1}(G)$, $W_{Y2}(G) \leq W_{Y1}(G)$, $R_{Y2}(G) \leq R_{Y1}(G)$,
$\eta_{Y2}(G) \leq \eta_{Y1}(G)$, $\alpha_{Y2}(G) \leq \alpha_{Y1}(G)$, $\alpha'_{Y2}(G) \leq \alpha'_{Y1}(G)$, and
$\rho_{Y2}(G) \geq \rho_{Y1}(G)$, $F_{Y2}(G) \geq F_{Y1}(G)$, $P_{Y2}(G) \geq P_{Y1}(G)$,
$\Psi_{Y2}(G) \geq \Psi_{Y1}(G)$, $\beta_{Y2}(G) \geq \beta_{Y1}(G)$, and $\beta'_{Y2}(G) \geq \beta'_{Y1}(G)$.

We conclude this introductory section by considering efficient domination for a nested infinite family of subsets of $\Re$. Specifically, let $Y_k = \{1 - k, 2 - k, 3 - k, ..., k - 2, k - 1, k\}$ for $k \in \{1, 2, 3, ...\}$. Let $\mathcal{F}_k$ denote the family of efficiently-Y_k-dominatable graphs, that is, $\mathcal{F}_k = \{G \mid$ for some $f : V(G) \to Y_k$,

$N \cdot X_f = \vec{1}_n\} = \{G : F_{Y_k}(G) = n\}$. One can verify that for the tree T_k in Figure
10.2, the unique function $f : V(T_k) \to \Re$ with $N \cdot X_f = \vec{1}_{k+3}$ satisfies $f(x_3) = 0$,
$f(y_i) = 1$ for $1 \leq i \leq k$, $f(x_2) = 1 - k$, and $f(x_1) = k$. Hence, T_k is efficiently
Y-dominatable if and only if $\{1 - k, 0, 1, k\} \subseteq Y$. In particular, $T_k \notin \mathcal{F}_{k-1}$
and $T_k \in \mathcal{F}_k$. Thus, also using Theorem 10.2, each $\mathcal{F}_k$ is properly contained in
$\mathcal{F}_{k+1}$.

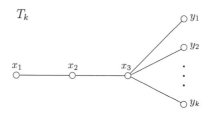

Figure 10.2: Tree $T_k \in \mathcal{F}_k$ but $T_k \notin \mathcal{F}_{k-1}$ for $k \geq 2$.

Problems like characterizing $\mathcal{F}_k$ motivate us to study Y-valued dominating
type problems for general $Y \subseteq \Re$.

10.2   Minus and Signed Domination

Let H_k with $k \geq 3$ denote the generalized Hajós graph on $n = k + \binom{k}{2}$
vertices with vertex set $V(H_k) = \{x_1, x_2, ..., x_k\} \cup \{y_{i,j} : 1 \leq i < j \leq k\}$, where
$\langle\{x_1, x_2, ..., x_k\}\rangle = K_k$, and each $y_{i,j}$ has degree two with $N(y_{i,j}) = \{x_i, x_j\}$.
Note that $deg(x_i) = 2k - 2$. The Hajós graph H_4 is illustrated in Figure 10.3.
Assume that each vertex represents a voter and each x_i votes 'yes', as indicated
by a 1, and each $y_{i,j}$ votes 'no', as indicated by a -1. The overall vote tally is
$k(1) + \binom{k}{2}(-1) = k(3 - k)/2 < 0$, for $k \geq 4$. However, viewed locally, there
is one more positive vote than negative vote in each closed neighborhood. That
is, although we might have many more negative than positive voters, one might
be able to arrange the voters with one at each vertex so that every local group
$N[v]$ would produce an overall strictly positive vote. More generally, one might
provide for abstentions or neutral responses by permitting a vote of 0.

Having a combined weight of at least one in every closed neighborhood is the
essence of domination. A *minus dominating function* is defined in [402] as a function $f : V(G) \to \{-1, 0, 1\}$ such that $f(N[v]) \geq 1$ for all $v \in V(G)$. The *minus
domination number* of graph G is $\gamma^-(G) = \min\{w(f) : f$ is a minus dominating
function $\}$. The *upper minus domination number* $\Gamma^-(G)$ is the maximum weight

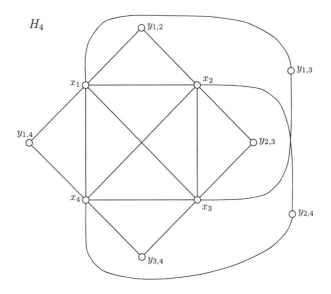

Figure 10.3: Generalized Hajós graph H_4.

of a minimal minus dominating function on G. A *signed dominating function*
is defined in [404] as a function $f : V(G) \rightarrow \{-1, 1\}$ such that $f(N[v]) \geq 1$ for
all $v \in V(G)$. The *signed domination number* $\gamma_s(G) = \min\{w(f) : f$ is a signed
dominating function on $G\}$, and the *upper signed domination number* $\Gamma_s(G)$ is
the maximum weight of a minimal signed dominating function on G. Note that
Figure 10.1(b) shows how to achieve $\gamma^-(H_3) = \gamma_s(H_3) = 0$. Other studies of mi-
nus/signed domination include [392, 403, 453, 626, 627, 702, 1205]. Specifically,
$\gamma^-(G) = \gamma_{\{-1,0,1\}}(G)$ and $\gamma_s(G) = \gamma_{\{-1,1\}}(G)$.

The next proposition follows from Theorem 10.2.

Proposition 10.3 *For any graph G, $\gamma^-(G) \leq \gamma(G)$ and $\gamma^-(G) \leq \gamma_s(G)$.*

On the other hand, γ and γ_s are incomparable. As noted, for each generalized
Hajós graph H_k with $k \geq 3$, $\gamma_s(H_k) \leq 0 < \gamma(H_k)$. Note that if $deg(v) = 1$
and $N(v) = \{w\}$, then any γ_s-function must assign a value of $+1$ to v and to
w. Therefore, for any graph G of order n, the corona $G^* = G \circ K_1$ satisfies
$\gamma(G^*) = n < 2n = \gamma_s(G^*)$.

Note that H_k is a chordal graph. We leave as an exercise the construction
of bipartite and outerplanar graphs (see [402, 404]) with γ^- and γ_s arbitrarily
small.

Proposition 10.4 [403, 404] *For any negative integer k there exists a chordal graph D_k, a bipartite graph B_k, and an outerplanar graph O_k, with $\gamma^-(D_k) \leq \gamma_s(D_k) \leq k$, $\gamma^-(B_k) \leq \gamma_s(B_k) \leq k$, and $\gamma^-(O_k) \leq \gamma_s(O_k) \leq k$.*

Lower bounds for $\gamma^-(G)$ and $\gamma_s(G)$ exist for graphs with bounded $\Delta(G)$.

Theorem 10.5 [403, 404] *If a graph G has maximum degree $\Delta(G) \leq 5$, then $0 \leq \gamma^-(G) \leq \gamma_s(G)$.*

Proof. Assume that $\Delta(G) \leq 5$ and $g : V(G) \rightarrow \{-1, 0, 1\}$ is a $\gamma^-(G)$-function. Let $P = g^{-1}(1) = \{v \in V(G) : g(v) = 1\}$ and $M = g^{-1}(-1)$. Clearly, $M = \emptyset$ implies $\gamma^-(G) = |P| \geq 1$, so assume $M \neq \emptyset$. Let k be the number of edges in $E(G)$, each of which is incident with a vertex in P and also one in M. For each $v \in M$, we have $g(N[v]) \geq 1$, so $|N(v) \cap P| \geq 2$, and $k \geq 2|M|$. For each $v \in P$, we also have $g(N[v]) \geq 1$, so $|N(v) \cap P| \geq |N(v) \cap M|$. Now $deg(v) \leq 5$ implies $|N(v) \cap M| \leq 2$. Thus, $2|P| \geq k \geq 2|M|$. Hence, $\gamma_s(G) \geq \gamma^-(G) = |P| - |M| \geq 0$. $\square$

Theorem 10.6 [404] *If a graph G has $\Delta(G) \leq 3$, then $\gamma_s(G) \geq n/3$.*

Proof. Assume $g : V(G) \rightarrow \{-1, 1\}$ is a signed dominating function with $g(V(G)) = \gamma_s(G)$. Let P, M, and k be as in the proof of Theorem 10.5. If $M = \emptyset$, then $\gamma_s(G) = n$. Otherwise, as before, $k \geq 2|M|$. But now $\Delta(G) \leq 3$ means that for each $v \in P$, we have $|N(v) \cap M| \leq 1$, so $|P| \geq k \geq 2|M|$. Because $|P| + |M| = n$, we have $|M| \leq n/3$ and so $|P| \geq 2n/3$. Hence, $\gamma_s(G) \geq 2n/3 - n/3 = n/3$. $\square$

The next result gives a lower bound on γ_s in terms of the degree sequence of the graph.

Theorem 10.7 [404] *Let G be a graph with the degree sequence $(d_1, d_2, ..., d_n)$ such that $d_1 \leq d_2 \leq ... \leq d_n$. If k is the smallest integer for which*

$$d_{n-k+1} + d_{n-k+2} + ... + d_n - (d_1 + d_2 + ... + d_{n-k}) \geq 2(n - k),$$

then $\gamma_s(G) \geq 2k - n$.

Proof. Let f be a signed dominating function satisfying $f(V) = \gamma_s(G)$. We consider the sum $S = \sum \sum f(u)$, where the outer sum is over all $v \in V$ and the inner sum is over all $u \in N[v]$. This sum counts the value $f(u)$ exactly $deg(u) + 1$ times for each $u \in V$, so $S = \sum (deg(u) + 1)f(u)$, over all $u \in V$. Since $\sum_{u \in N[v]} f(u) = f(N[v]) \geq 1$ for each $v \in V$, $S = \sum f(N[v])$ over all $v \in V$ satisfies $S \geq n$.

Hence

$$\sum_{u \in V} (deg(u) + 1)f(u) \geq n. \tag{1}$$

Let P and M be the sets of vertices of G that are assigned the values of $+1$ and -1, respectively. Then $\gamma_s(G) = f(V) = |P| - |M| = 2|P| - n$. Now

$$\sum_{u \in V}(deg(u)+1)f(u) = \sum_{u \in P}(deg(u)+1) - \sum_{u \in M}(deg(u)+1)$$

$$\leq \sum_{i=n-|P|+1}^{n}(d_i+1) - \sum_{i=1}^{n-|P|}(d_i+1)$$

$$= \sum_{i=n-|P|+1}^{n} d_i - \left(\sum_{i=1}^{n-|P|} d_i\right) + 2|P| - n. \qquad (2)$$

By (1) and (2),

$$\sum_{i=n-|P|+1}^{n} d_i - \sum_{i=1}^{n-|P|} d_i \geq 2(n - |P|).$$

By our choice of k, it follows that $|P| \geq k$, so $\gamma_s(G) = 2|P| - n \geq 2k - n$. $\square$

To illustrate Theorem 10.7, consider the graph G shown in Figure 10.4 which has degree sequence $d_1, d_2, ..., d_{20}$, where $d_i = 2$ for $1 \leq i \leq 12$ and $d_i = 6$ for $13 \leq i \leq 20$. The smallest integer k for which $d_{21-k} + d_{22-k} + ... + d_{20} - (d_1 + d_2 + ... + d_{20-k}) \geq 40 - 2k$ is $k = 8$. Hence, applying Theorem 10.7, we get $\gamma_s(G) \geq -4$. On the other hand, if g is the function on G defined by letting $g(v) = 1$ if $deg(v) = 6$ and letting $g(v) = -1$ if $deg(v) = 2$, then g is a signed dominating function on G of weight -4, so $\gamma_s(G) \leq -4$. Hence $\gamma_s(G) = -4$.

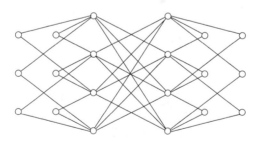

Figure 10.4: A graph for which $\gamma_s(G) = -4$.

Proposition 10.8 [404] *For any k-regular graph G, $\gamma_s(G) \geq n/(k+1)$.*

Proof. Let f be a signed dominating function satisfying $f(V) = \gamma_s(G)$. Then as in the proof of Theorem 10.7, $n \leq \sum_{u \in V}(deg(u)+1)f(u) = (k+1)\sum_{u \in V} f(u) = (k+1)f(V)$, so $f(V) \geq n/(k+1)$. $\square$

Theorem 10.9 [404] *For any graph G, $\gamma_s(G) = n$ if and only if every nonisolated vertex is either an endvertex or adjacent to an endvertex.*

Proof. We have already noted the sufficiency. The necessity follows from the observation that if we have a vertex v that is neither an endvertex nor adjacent to an endvertex, then we can assign the value -1 to v and the value $+1$ to each other vertex to produce a signed dominating function on G of weight $n - 2$. $\square$

Theorem 10.10 [403] *For any tree T, $\gamma^-(T) \geq 1$ with equality if and only if T is a star $K_{1,n-1}$.*

Proof. Let $g : V \to \{-1, 0, 1\}$ be a γ^--function on a tree $T = (V, E)$. If no vertex u satisfies $g(u) = -1$ then, clearly, g is a γ-function and $\gamma^-(T) = \gamma(T) \geq 1$. Therefore, let $g(u) = -1$ for some vertex u and let T be rooted at vertex u. Let $v_1, v_2, ..., v_k$ be the vertices adjacent to u (see Figure 10.5), which in turn are roots of subtrees $T_1, T_2, ..., T_k$, respectively. It follows that at least two neighbors of u must be assigned the value of $+1$ by g. Assume, without loss of generality, that $g(v_1) = g(v_2) = 1$. Now consider any other vertex w with $g(w) = -1$. From the given rooting of T at vertex u, it follows that every such vertex w must have at least one child w' with $g(w') = 1$. Define $M = \{u, w_1, w_2, ..., w_n\}$ to be the set of vertices w with $g(w) = -1$. It follows therefore that each vertex other than u in M has at least one child w_i, with $g(w_i') = 1$. And since vertex u has at least two children v_1 and v_2 with $g(v_1) = g(v_2) = 1$, we can conclude that $g(V) = \gamma^-(T) \geq 1$.

We need to show that $\gamma^-(T) = 1$ for a tree T if and only if T is a star $K_{1,n-1}$. But it is easy to show that $\gamma(T) = 1$ if and only if T is a star. Hence, if T is a star, $\gamma^-(T) \leq \gamma(T) = 1$. But by the above argument, $\gamma^-(T) \geq 1$, and thus $\gamma^-(T) = 1$.

It remains to show that if $\gamma^-(T) = 1$, then T is a star. Let $g : V \to \{-1, 0, 1\}$ be a γ^--function on a tree T for which $g(V) = 1$. If $g(u) \geq 0$ for every vertex $u \in V$, then g is a γ-function and $\gamma^-(T) = \gamma(T) = 1$ and T must be a star. Assume therefore that $g(u) = -1$ for at least one vertex $u \in V$. Then the number of vertices, say k, with value -1 is one less than the number of vertices (that is, $k + 1$) with value $+1$. But each vertex with value -1 must have at least two neighbors with value $+1$. This means that there must be at least $2k$ edges among these $2k + 1$ vertices. But, in addition, each vertex with value $+1$ which is adjacent to a vertex with value -1 must also have at least one neighbor with value $+1$. This implies that among these $2k + 1$ vertices there are at least $2k + 1$ edges, that is, there must be a cycle among these vertices. But this contradicts the assumption that T is a tree. Therefore, no vertex in T has the value $g(u) = -1$ if $g(V) = 1$, that is, T is a star. $\square$

Theorem 10.11 [403] *If T is a tree of order $n \geq 4$, then $\gamma(T) - \gamma^-(T) \leq (n - 4)/5$.*

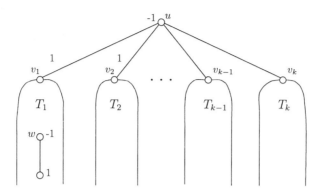

Figure 10.5: Tree T rooted at u.

Proof. We proceed by induction on the order n of a tree, where $n \geq 4$. The base case when T is a tree of order four is trivial since in this case $\gamma(T) - \gamma^-(T) = 0 = (n-4)/5$. So, assume for all trees T' of order $n' < n$, where $n > 5$, that $\gamma(T') - \gamma^-(T') \leq (n'-4)/5$. Let T be a tree of order n.

Among the minimum minus dominating functions on T, let f be one that assigns the value -1 to as few vertices as possible. If no vertex of T is assigned the value -1 under f, then $\gamma(T) - \gamma^-(T) = 0 < (n-4)/5$. So we may assume that at least one vertex of T is assigned the value -1 under f. Let P and M (standing for 'plus' and 'minus') be the sets of vertices in T that are assigned the values $+1$ and -1, respectively.

Let $T_1, T_2, ..., T_k$ be the components of $T - M$. Then each component T_i ($1 \leq i \leq k$) contains a vertex that is adjacent to some vertex of M. Thus each component T_i contains at least two vertices of P and is therefore of order at least 2. Moreover, since T is a tree, each vertex of M is adjacent to at most one vertex in each of the components T_i. We now construct a graph G with vertex set $V(G) = X \cup M'$, where $X = \{v_1, v_2, ..., v_k\}$ and $M' = \{u'_1, u'_2, ..., u'_{|M|}\}$. Two vertices v_i and u'_j are adjacent in G if and only if u_j is adjacent to some vertex in the component T_i. Furthermore, two vertices in M' are adjacent in G if and only if the corresponding vertices in M are adjacent in T. If G contains a cycle, then so too does T. Hence, G is acyclic. Furthermore, since T is connected, it follows from the way in which G is constructed that G is also connected. Thus, G is a tree.

Since each vertex of M is adjacent to at least two vertices of P, each vertex of M has degree at least two in G. Thus, the endvertices of G all belong to X. We show next that each component T_i of $T - M$ associated with an endvertex v_i of G has order at least four. Let u_j be the vertex of M adjacent in G with the endvertex v_i. Then u_j is adjacent in T with exactly one vertex of T_i. The edge joining u_j to this vertex of T_i is the only edge joining a vertex of T_i to a vertex not in T_i. Furthermore, we know that T_i contains at least two vertices of P. If T_i has order two or three, then we could reassign to some vertex of T_i that has value 1 under f the value 0 and we could reassign to u_j the value 0, to produce a new minimum minus dominating function on T that assigns the value -1 to fewer vertices than does f, contrary to assumption. Hence each component T_i of $T - M$ associated with an endvertex v_i of G has order at least four. In particular, since G has at least two endvertices and since $|M| \geq 1$, we note that T has order $n \geq 9$.

Let H be the tree obtained from G by removing all the endvertices of G. If H is the trivial tree K_1, then $|M| = 1$. Thus, $\gamma^-(T) = w(f) = |P| - |M| = |P| - 1$. However, since P is a dominating set of T, $\gamma(T) \leq |P|$. Hence, $\gamma(T) - \gamma^-(T) \leq 1 \leq (n - 4)/5$ since $n \geq 9$. So we may assume that H is a nontrivial tree, for otherwise there is nothing left to prove.

Since no two vertices of X are adjacent, all of the endvertices of H are in M. We may assume that u_1 is an endvertex of H. That is, u_1 is adjacent in G to one or more endvertices of X, say to $v_1, ..., v_l$ ($l \geq 1$), and to exactly one other vertex (which belongs to H). Let e be the edge of T that joins u_1 to a vertex of T that does not belong to any of the subtrees $T_1, ..., T_l$. Let F_1 be the component of $T - e$ that contains u_1, so F_1 consists of the subtrees $T_1, ..., T_l$ and an edge from each of these components to the vertex u_1. Let F_2 denote the component of $T - e$ different from F_1. Further, for $i = 1, 2$, let F_i be a tree of order n_i.

Since each of the components T_i ($1 \leq i \leq l$) has order at least four, it follows that $n_1 \geq 5$. We show next that $n_2 \geq 5$. Let F_5' be the tree obtained from G by removing u_1 and the l endvertices $v_1, ..., v_l$ adjacent to u_1. If F_2' contains no vertex of M, then $|M| = 1$. But then H would then be a trivial tree, contrary to our assumption. Hence F_2' contains at least one vertex u_j of M. Since u_j is adjacent in G with at least two vertices of X (which belong to F_2'), the tree F_2 consists of at least two components of $T - M$ and at least one vertex of M. Since each component of $T - M$ has order at least two, the tree F_2 therefore has order at least five, that is, $n_2 \geq 5$. We may now apply the inductive hypothesis to the tree F_2. This yields

$$\gamma(F_2) - \gamma^-(F_2) \leq \frac{n_2 - 4}{5} = \frac{n - n_1 - 4}{5} \leq \frac{n - 9}{5}.$$

Now let $f_1(f_2)$ be the restriction of f to F_1 (respectively, F_2). Then $\gamma^-(T) = w(f) = w(f_1) + w(f_2)$. Since f_2 is a minus dominating function of F_2, $\gamma^-(F_2) \leq w(f_2)$. Furthermore, the set $P \cap V(F_1)$ is a dominating set of F_1, so $\gamma(F_1) \leq |P \cap V(F_1)|$. Since u_1 is the only vertex of F_1 that is assigned the value -1 under f_1, $|P \cap V(F_1)| = w(f_1) + 1$. Hence, $\gamma(F_1) \leq w(f_1) + 1$. Thus,

$$\begin{aligned}
\gamma(T) - \gamma^-(T) &\leq \gamma(F_1) + \gamma(F_2) - w(f) \\
&\leq (w(f_1) + 1) + \gamma(F_2) - w(f_1) - w(f_2) \\
&= (\gamma(F_2) - w(f_2)) + 1 \\
&\leq (\gamma(F_2) - \gamma^-(F_2)) + 1 \\
&\leq (n-9)/5 + 1 \\
&= (n-4)/5.
\end{aligned}$$

This completes the proof of the theorem. $\square$

The bound in Theorem 10.11 is best possible. This may be seen by considering the tree T_k ($k \geq 1$) obtained from a path $v_1, v_2, ..., v_{3k+2}$ on $3k+2$ vertices by adding $2(k+1)$ new vertices $\{u_{3i+1} : 0 \leq i \leq k\} \cup \{u_{3i+2} : 0 \leq i \leq k\}$, and joining u_i to v_i with an edge for each i. (The tree T_2 is shown in Figure 10.6.) Then $\gamma(T_k) = 2(k+1)$ and $\gamma^-(T) = k+2$, and so $\gamma(T_k) - \gamma^-(T_k) = k = (|V(T_k)|-4)/5$.

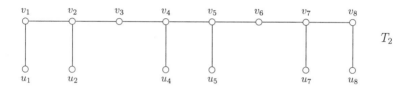

Figure 10.6: The tree T_2.

For arbitrary graphs, sharp lower bounds for the signed domination number are not known. For trees, however, there is a sharp lower bound for γ_s.

Theorem 10.12 [404] *If T is a tree of order $n \geq 2$, then $\gamma_s(T) \geq (n+4)/3$ with equality if and only if T is a path on $3k+2$ vertices for some nonnegative integer k.*

Proof. Let f be a minimum signed dominating function on T. Let $M(P)$ be the set of all vertices of T that are assigned the value -1 (respectively, $+1$) under f. Then $\gamma_s(T) = f(V) = |P| - |M|$. If $M = \emptyset$, then $\gamma_s(T) = n$. Assume that $M \neq \emptyset$ and let $v \in M$. Root the tree T at vertex v. For each vertex $x \in M$, we let P_x denote the set consisting of those vertices y satisfying (a) y is in P, (b) y is a descendant of x, and (c) each vertex of the x-y path of T, except x itself, is in P (see Figure 10.7). Necessarily, the sets P_x, $x \in M$, partition the set P.

Since $f(N[v]) \geq 1$, there are at least two children of v that belong to the set P. Furthermore, each child of v also possesses a child that is in P. It follows that $|P_v| \geq 4$. Every vertex x of $M - \{v\}$ has at least one child, and therefore at least one grandchild, that belongs to P; so $|P_x| \geq 2$. Hence $|P| = |\cup_{x \in M} P_x| = \sum_{x \in M} |P_x| \geq 2(|M| + 1)$, and so $n = |P| + |M| \geq 3|M| + 2$.

Consequently, $|M| \leq (n-2)/3$. So $\gamma_s(T) = |P| - |M| = n - 2|M| \geq n - (2(n-2))/3 = (n+4)/3$.

Moreover, if $\gamma_s(T) = (n+4)/3$, then $|P| = 2(|M|+1)$, $|P_v| = 4$ and $|P_x| = 2$ for each $x \in M - \{v\}$. We show that T is a path on $3|M| + 2$ vertices. If this is not the case, suppose that T contains a vertex w of degree at least 3. If $w = v$, then at least three children of w belong to P and so, since each child of v possesses a child that is also in P, it follows that $|P_v| \geq 6$. This contradicts the fact that $|P_v| = 4$. Assume, then, that $w \neq v$. If $w \in M$, then at least two children of w belong to P, implying that $|P_w| \geq 4$, which contradicts the fact that $|P_x| = 2$ for each $x \in M - \{v\}$. If $w \in P$, then $w \in P_x$ for some $x \in M$. If x is not the father of w, then, since w contains at least one child that belongs to P, it follows that either $|P_x| \geq 5$ if $x = v$ or $|P_x| \geq 3$ if $x \neq v$. On the other hand, if x is the father of w, then at least two children of w belong to P, and so it follows that either $|P_x| \geq 5$ if $x = v$ or $|P_x| \geq 3$ if $x \neq v$. Since all the above possibilities lead to a contradiction, we deduce that there is no vertex of T with degree exceeding two.

Conversely, if T is a path on $3k + 2$ vertices for some integer $k \geq 0$, then assigning to every third vertex on the path the value -1, and to all the remaining vertices the value $+1$, produces a signed dominating function g on T with $g(V) = k + 2 = (n + 4)/3$. This, together with the earlier observation that $\gamma_s(T) \geq (n + 4)/3$ yields $\gamma_s(T) = (n + 4)/3$. $\square$

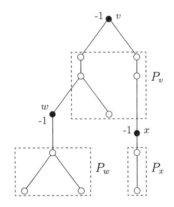

Figure 10.7: A tree rooted at v with $M = \{v, w, x\}$.

10.3 Y-dominating Parameters

In this section we consider possible extensions of results for domination γ and efficient domination F to arbitrary γ_Y and F_Y for any $Y \subseteq \Re$. In the next section, we restrict attention to complementable sets $Y \subseteq \Re$ and present a

Complementation Theorem that extends Gallai's theorem, that $\alpha_0(G)+\beta_0(G) = n = \alpha_1(G) + \beta_1(G)$, to many pairs of graphical parameters.

First, we will show that there are counterexamples to a generalized Vizing Conjecture. Recall that $F(G) = n$ implies that any two efficient dominating sets have the same cardinality. We extend this to a result even more general than the fact that any two efficient Y-dominating functions will have the same weight. These results are presented in [74].

10.3.1 Vizing's Conjecture

As presented in Chapter 2, Vizing's Conjecture states that, for any graphs G and H, the cartesian product $G \times H$ satisfies $\gamma(G \times H) \geq \gamma(G) \cdot \gamma(H)$. The following shows that Vizing's Conjecture does not hold for Y-domination γ_Y whenever $\{-1, 1\} \subseteq Y$. Recall that the generalized Hajós graph H_k on $n = k + \binom{k}{2}$ vertices has a complete graph induced on $\{x_1, x_2, ..., x_k\}$ and each pair $\{x_i, x_j\}$ is joined to a distinct vertex $y_{i,j}$ of degree two. For $k \geq 4$, if $\{-1, 1\} \subseteq Y$, define $f : V(H_k) \to Y$ by $f(x_i) = 1$ for $1 \leq i \leq k$ and $f(y_{i,j}) = -1$ for $1 \leq i < j \leq k$. Then $f(N[v]) = 1$ for every $v \in V(H_k)$, and f is an (efficient) dominating function with weight $w(f) = k - \binom{k}{2}$. Hence, $\gamma_Y(H_k) < 0$ for $k \geq 4$, and if $h \geq k \geq 4$, then $\gamma_Y(H_h) \cdot \gamma_Y(H_k) > 0$. Thus, Vizing's Conjecture will be seen not to hold for Y-domination when it is shown that $\gamma(H_k \times H_4) < 0$ for $k \geq 12$.

The graph $H_k \times H_4$ has ten copies of H_k, one corresponding to each vertex in H_4. Initially, define $f : V(H_k \times H_4) \to Y$ so that the label $f(v)$ on vertex v is $+1$ if v has degree $2k - 2$ in one of the ten copies of H_k and $f(v) = -1$ if v has degree two in one of the H_k. (Each of the ten copies of H_k is efficiently dominated as above.) Now consider one H_4 formed by the ten copies of a fixed vertex from one H_k. Either all ten of these vertices are labelled $+1$ under f or all ten are labelled -1. At each vertex which is labelled $+1$, the closed neighborhood sum in $H_k \times H_4$ is certainly positive. Consider a vertex where the label is -1, and note that the sum of its weight and the weights of its neighbors from the H_k is 1. Thus, the closed neighborhood sum in $H_k \times H_4$ is -1 if the vertex corresponds to one of degree two in the H_4 or it is -5 if the vertex corresponds to one of degree six in the H_4.

Modify the initial labelling by changing the -1 to a $+1$ at those vertices whose closed neighborhood sum under f is -5. If $f(N[v]) = -5$, then the labels on v and three of its neighbors are changed from -1 to $+1$, so the new neighborhood weight is $+3$; if $f(N[v]) = -1$, then the labels on two of its neighbors are changed from -1 to $+1$, so the new neighborhood sum is also $+3$. Now the modified labelling defines a Y-dominating function. We had $f(V(H_k \times H_4)) = 10k - 10 \binom{k}{2}$ and $4 \binom{k}{2}$ labels of -1 are changed to $+1$, so the new labelling has weight $10k - 2 \binom{k}{2}$. In summary, if $k \geq 12$ and

$\{-1, 1\} \subseteq Y$, then $\gamma_Y(H_k \times H_4) \leq 10k - 2 \binom{k}{2} < 0$, but $\gamma_Y(H_k) \cdot \gamma_Y(H_4) > 0$.

10.3.2   Efficiency

In this section we show that any two efficient Y-dominating functions for a graph G have the same weight, a generalization of the following observation made in Theorem 4.2.

Theorem 10.13 [77] *If a graph G has an efficient dominating set S, then $|S| = \gamma(G)$.*

Theorem 10.14 [74] *If a graph G has an efficient dominating set, then $\gamma(G) \leq \gamma_Y(G)$ for any weight set Y.*

Proof. Let $S = \{v_1, v_2, ..., v_k\}$ be a subset of $V(G)$ that efficiently dominates G. Then $\{N[v_1], N[v_2], ..., N[v_k]\}$ is a partition of the vertex set V. If $f : V \to Y$ is any function with $w(f) < \gamma(G)$, then at least one of these closed neighborhoods must have weight less than 1 under f, which implies that f is not a dominating function. $\square$

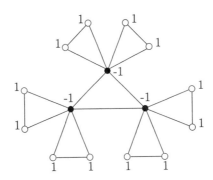

Figure 10.8: The weight of an efficient Y-dominating function is not necessarily $\gamma_Y(G)$.

In view of Theorem 10.14, it is natural to ask whether an efficient Y-dominating function will attain the minimum value $\gamma_Y(G)$ for a given set Y. In other words, is there an efficient Y-dominating function $f : V \to Y$ such that for any other Y-dominating function $g : V \to Y$, it follows that $w(f) \leq w(g)$? This is not true for the graph shown in Figure 10.8 if $Y = \{-1, 0, 1\}$. The vertex labelling shown is an efficient Y-dominating function of weight nine. However, the three vertices of degree six form a dominating set, so the graph also has a (nonefficient) Y-dominating function of weight three.

For any real-valued n-vector $\overrightarrow{B} = (b_1, b_2, ..., b_n)^t$, we can ask if there is a function $f : V(G) \to Y$ satisfying $\sum_{s \in N[v_i]} f(s) = b_i$, for $1 \le i \le n$. Clearly, this is so if and only if there is a solution $\overrightarrow{X}$ to the matrix equation $N \cdot \overrightarrow{X} = \overrightarrow{B}$. In particular, G has a real-valued efficient dominating function if and only if $N \cdot \overrightarrow{X} = \overrightarrow{1}_n$ has a real-valued solution, that is, $\overrightarrow{1}$ must be in the column space of closed neighborhood matrix N.

Theorem 10.15 [74] *Let G be a graph whose closed neighborhood matrix N is invertible. Then G has an efficient Y-dominating function for some weight set Y.*

Proof. If N is invertible, then the matrix equation $N \cdot \overrightarrow{X} = \overrightarrow{1}$ has a unique solution $\overrightarrow{X} = (x_1, x_2, ..., x_n)$ which determines the efficient Y-dominating function defined by $f(v_i) = x_i$, for $1 \le i \le n$. $\square$

As an example, it is easy to check that the determinant of the closed neighborhood matrix for the Hajós graph H_3 is nonzero. Since the labelling shown in Figure 10.1(b) is efficient, it follows that H_3 has an efficient Y-dominating function if and only if $\{-1, 1\} \subseteq Y$. Also, recall that for the tree T_k in Figure 10.2, there is a unique solution to $N \cdot X = \overrightarrow{1}_{k+3}$ using the values $0, 1, 1 - k$, and k, and $T_k \in \mathcal{F}_k$ but $T_k \notin \mathcal{F}_{k-1}$.

Two vertex labelling functions $f : V(G) \to \Re$ and $g : V(G) \to \Re$ will be called *equivalent* if $\sum_{s \in N[v_i]} f(s) = \sum_{s \in N[v_i]} g(s)$, for $1 \le i \le n$. Note that if G is a graph for which N is not invertible then, whenever $N \cdot X = B$ is consistent, the infinitely many solutions will generate equivalent vertex labelling functions for the vertices of G. However, there do exist graphs that have no efficient $\Re$-dominating function. The graph in Figure 10.9 is such a graph. Note that the vertex labellings in Figures 10.9(b) and (c) are equivalent. For this graph, we have the following matrix equation $N \cdot X = \overrightarrow{1}_8$.

$$
\begin{array}{rrrrrrrr}
x_1 & +x_2 & & +x_4 & & & & = 1 \\
x_1 & +x_2 & +x_3 & +x_4 & +x_5 & & & = 1 \\
 & x_2 & +x_3 & & +x_5 & & & = 1 \\
x_1 & +x_2 & & +x_4 & +x_5 & +x_6 & +x_7 & = 1 \\
 & x_2 & +x_3 & +x_4 & +x_5 & & +x_7 & +x_8 & = 1 \\
 & & & x_4 & & +x_6 & +x_7 & & = 1 \\
 & & & x_4 & +x_5 & +x_6 & +x_7 & +x_8 & = 1 \\
 & & & & x_5 & & +x_7 & +x_8 & = 1.
\end{array}
$$

It is easy to verify that the coefficient matrix of this system is not invertible and that the system is inconsistent, so there can be no efficient Y-dominating function for this graph.

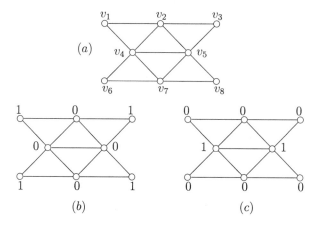

Figure 10.9: A graph with no $\Re$-valued efficient dominating function.

For a graph G to have distinct equivalent dominating functions, it is necessary that its closed neighborhood matrix be noninvertible. Under what conditions must all equivalent dominating functions have the same weight? The next
theorem characterizes such graphs.

Theorem 10.16 [74] *If G is a graph with closed neighborhood matrix N, then
there exists a vector $\vec{W}$ satisfying $N \cdot \vec{W} = \vec{1}$ if and only if $N \cdot \vec{X} = N \cdot \vec{Z} = \vec{B}$
implies that $\sum_i x_i = \sum_i z_i$.*

Proof. First, assume that there is a solution to $N \cdot \vec{W} = \vec{1}$. We claim that if $\vec{0}$ is
the 0-vector and $N \cdot \vec{X} = \vec{0}$, then $\sum_i x_i = 0$. Since $\vec{W}$ is a solution to $N \cdot \vec{W} = \vec{1}$,
the vector $\vec{1}$ is in the column space of N. But, since N is symmetric, $\vec{1}$ is also
in the row space of N. Thus, $\vec{1}$ is orthogonal to the null space of N; that is, if
$N \cdot \vec{X} = \vec{0}$, then $\vec{X} \cdot \vec{1} = 0$ or $\sum_i x_i = 0$. It then follows that if $N \cdot \vec{X} = N \cdot \vec{Z}$, we
have $N \cdot (\vec{X} - \vec{Z}) = \vec{0}$, so $\sum_i (x_i - z_i) = 0$.

Next, assume that there is no solution to $N \cdot \vec{W} = \vec{1}$, so that $\vec{1}$ is not in the
column space, and hence not in the row space of N. Since the row space is the
orthogonal complement of the null space of N and the row space is not all of $\Re^n$,
there is a nonzero vector $\vec{U}$ in the null space of N for which $\vec{U} \cdot \vec{1} = \sum_i u_i \neq 0$.
But since $N \cdot \vec{U} = N \cdot \vec{0} = \vec{0}$, it is not the case that $N \cdot \vec{X} = N \cdot \vec{Z} = \vec{B}$ implies
$\sum_i x_i = \sum_i z_i$. $\square$

Thus, all equivalent dominating functions of a graph G have the same weight
if and only if G has an efficient real-valued dominating function, and Theorem
10.16 has the following corollary.

Corollary 10.17 [74] *If f_1 and f_2 are efficient Y-dominating functions for a graph G, then $w(f_1) = w(f_2)$.*

Proof. Since $\sum_{s \in N[v]} f_1(s) = \sum_{s \in N[v]} f_2(s) = 1$ for every $v \in V$, both f_1 and f_2 are Y-dominating functions yielding the vector $\vec{1}$. Theorem 10.16 then implies that $w(f_1) = w(f_2)$. □

This corollary together with the labellings of Figures 10.9 (b) and (c) provide an alternate proof that the graph in Figure 10.9 has no efficient dominating function.

10.4 Complementarity

Gallai's Theorem (restated as Theorem 10.18 below) states, in part, that $\alpha_{\{0,1\}}(G) + \beta_{\{0,1\}}(G) = |V(G)|$. This does not generalize to arbitrary subsets $Y \subseteq \Re$. For example, for signed domination, $Y = \{-1, 1\}$, and for signed covering and independence, we have $\alpha_{\{-1,1\}}(K_5) + \beta_{\{-1,1\}}(K_5) = 5 + (-3) = 2$. However, as will be shown, when Y is "complementable", for example when $Y = Y_k = \{1 - k, 2 - k, 3 - k, ..., k - 1, k\}$, then $\alpha_{Y_k}(G) + \beta_{Y_k}(G) = |V(G)|$. For the unicyclic graph K in Figure 10.10, we have a $\beta_{Y_{10}}$-function and an $\alpha_{Y_{10}}$-function with $\beta_{Y_{10}}(K) = 107$ and $\alpha_{Y_{10}}(K) = -82$. Note that $\alpha_{Y_{10}}(K) + \beta_{Y_{10}}(K) = |V(K)|$. Recall that for the Hajós graph H_3, $\gamma_\Re(H_3) = -\infty$. For the star $K_{1,t}$, we have $\alpha_\Re(K_{1,t}) = -\infty$ and $\beta_\Re(K_{1,t}) = \infty$. While there are many interesting questions for Y-valued functions $f : V(G) \to Y$ or $f : E(G) \to Y$ when Y is unbounded, we will mainly be concerned here with the case when Y is bounded. Results presented here were presented in Slater [645, 1042].

Recall from Chapter 9, the following Gallai-type theorems.

Theorem 10.18 (Gallai [518]) *For any graph G,*

(a) $\alpha_0(G) + \beta_0(G) = n$, *and*

(b) $\alpha_1(G) + \beta_1(G) = n$ *if G has no isolated vertices.*

The restriction of Theorem 2 in Slater [1032] from hypergraphs to graphs is the next domination/enclaveless theorem.

Theorem 10.19 [1032] *For any graph G, $\gamma(G) + \Psi(G) = n = \Gamma(G) + \psi(G)$.*

The following upper-covering/lower independence theorem is due to McFall and Nowakowski.

Theorem 10.20 [884] *For any graph G, $i(G) + \Lambda(G) = n$.*

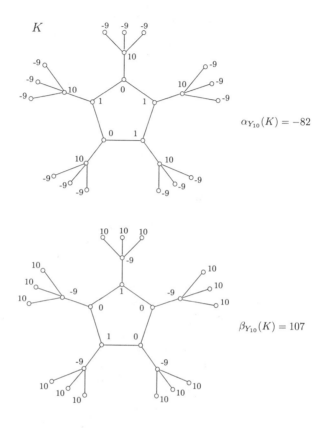

Figure 10.10: $\gamma_{Y_{10}}(K) + \beta_{Y_{10}}(K) = |V(K)|$.

In [284] Cockayne, Hedetniemi and Laskar present a wide variety of Gallai-type theorems. One of their principal results involves hereditary set systems. For a finite set S on n elements, a collection P of subsets of S is a hereditary set system if $X^* \subseteq X \in P$ implies $X^* \in P$. A subset X in P is called a P-set, and if X is a P-set then $S - X$ is called a $\overline{P}$-set. Thus every subset of a P-set is also a P-set, and every superset of a $\overline{P}$-set is a $\overline{P}$-set.

More general than the idea of hereditary set systems is the idea of complement related families of subsets of V (as used in [1032] to prove the domination/enclaveless theorem).

Families $\mathcal{F}_1$ and $\mathcal{F}_2$ of subsets of V are *complement related* when $X \in \mathcal{F}_1$ if and only if $V - X \in \mathcal{F}_2$. (Specifically, for example, $\mathcal{F}_1$ could be the P-sets and $\mathcal{F}_2$ the $\overline{P}$-sets of a hereditary set system.) If $\mathcal{F}$ is any family of subsets of V, let $M(V, \mathcal{F}) = \max\{|X| : X \in \mathcal{F}\}$ and $m(V, \mathcal{F}) = \min\{|X| : X \in \mathcal{F}\}$. Further, let

$\mathcal{F}^+$ denote the family of those members of $\mathcal{F}$ that are set-theoretically maximal with respect to membership, and $\mathcal{F}^-$ those that are minimal.

Theorem 10.21 [1032] *If $\mathcal{F}_1$ and $\mathcal{F}_2$ are complement related families of subsets of V, then $M(V, \mathcal{F}_1) + m(V, \mathcal{F}_2) = |V| = m(V, \mathcal{F}_1^+) + M(V, \mathcal{F}_2^-)$.*

Gallai's Theorem 10.18(a), Theorem 10.19, and Theorem 10.20 are corollaries of Theorem 10.21. The theorem will now be extended to generalized Y-valued functions. First, we say that $Y \subseteq \Re$ is *complementable* if $x \in Y$ implies $1 - x \in Y$. For examples, the signed domination and minus domination sets $\{-1, 1\}$ and $\{-1, 0, 1\}$ are not complementable, but each $Y_k = \{1 - k, 2 - k, 3 - k, ..., k - 1, k\}$ is a complementable set of order $2k$. For a complementable set Y and a function $f : V \to Y$, let $f^C : V \to Y$ be defined by $f^C(v) = 1 - f(v)$ for each $v \in V$. The set of all functions $f : V \to Y$ is denoted by $\mathcal{F}(V, Y)$. Subsets $\mathcal{F}_1$ and $\mathcal{F}_2$ of $\mathcal{F}(V, Y)$ are said to be *complement related families of functions* when $f \in \mathcal{F}_1$ if and only if $f^C \in \mathcal{F}_2$, in which case we write $\mathcal{F}_1^C = \mathcal{F}_2$. A function $f \in \mathcal{F}_1$ is said to be *maximal in $\mathcal{F}_1$* (respectively, *minimal in $\mathcal{F}_1$*) if $f < g$ (respectively, $g < f$) implies $g \notin \mathcal{F}_1$. Let $\mathcal{F}_1^+$ and $\mathcal{F}_1^-$ denote the family of elements of $\mathcal{F}_1$ which are maximal and minimal, respectively. If $\mathcal{F} \subseteq \mathcal{F}(V, Y)$, let $M(\mathcal{F}) = \max\{w(f) : f \in \mathcal{F}\}$ and $m(\mathcal{F}) = \min\{w(f) : f \in \mathcal{F}\}$ where we write $M(\mathcal{F}) = \infty$ if $\{w(f) : f \in \mathcal{F}\}$ is not bounded from above and $m(\mathcal{F}) = -\infty$ if $\{w(f) : f \in \mathcal{F}\}$ is not bounded from below.

Theorem 10.22 [1032] *If subsets $\mathcal{F}_1$ and $\mathcal{F}_2$ of $\mathcal{F}(V, Y)$ are complement related, and there exists $f_1 \in \mathcal{F}_1$ and $f_2 \in \mathcal{F}_2$ with $M(\mathcal{F}_1) = w(f_1)$ and $m(\mathcal{F}_2) = w(f_2)$, then $M(\mathcal{F}_1) + m(\mathcal{F}_2) = |V(G)| = m(\mathcal{F}_1^+) + M(\mathcal{F}_2^-)$.*

Proof. If $f_1 \in \mathcal{F}_1$ with $M(\mathcal{F}_1) = w(f_1)$, then $f_1^C \in \mathcal{F}_2$, so $m(\mathcal{F}_2) \leq w(f_1^C) = \sum_{v \in V} f_1^C(v) = \sum_{v \in V}(1 - f_1(v)) = |V| - w(f_1)$. Thus, $M(\mathcal{F}_1) + m(\mathcal{F}_2) \leq |V|$. And if $f_2 \in \mathcal{F}_2$ with $m(\mathcal{F}_2) = w(f_2)$, then $f_2^C \in \mathcal{F}_1$, so $M(\mathcal{F}_1) \geq w(f_2^C) = |V| - w(f_2)$. Thus, $M(\mathcal{F}_1) + m(\mathcal{F}_2) \geq |V|$. Consequently, $M(\mathcal{F}_1) + m(\mathcal{F}_2) = |V|$.

That $m(\mathcal{F}_1^+) + M(\mathcal{F}_2^-) = |V|$ follows similarly using the fact that if $\mathcal{F}_1$ and $\mathcal{F}_2$ are complement related, then so are $\mathcal{F}_1^+$ and $\mathcal{F}_2^-$. $\square$

Corollary 10.23 [1042] *If $Y \subseteq \Re$ is any finite complementable set, then for every graph G*

(a) $\alpha_Y(G) + \beta_Y(G) = |V|$ *(covering/independent)*

and $\Lambda_Y(G) + i_Y(G) = |V|$.

(b) $\gamma_Y(G) + \Psi_Y(G) = |V|$ *(dominating/enclaveless)*

and $\Gamma_Y(G) + \psi_Y(G) = |V|$.

Obviously a "complementation theorem" for covering and independence functions on $E(G)$ would produce a result of the form $\alpha_Y^1(G) + \beta_Y^1(G) = |E(G)|$ rather than the desired $|V(G)|$. Nevertheless, a generalization of Gallai's Theorem 10.18(b) is possible.

Theorem 10.24 [1042] *Let G be a graph with no isolated vertices. If Y is a set of consecutive integers $Y = \{a, a+1, ..., b\}$ with $a \le 0$ and $b \ge 1$, or if Y is a closed interval $[a, b]$ with $a \le 1/\Delta(G) \le 1/\delta(G) \le b$, then $\alpha_Y^1(G) + \beta_Y^1(G) = n$.*

Proof. For each function $h : E(G) \to Y$ the weight of h is $w(h) = \sum_{vu \in E} h(vu)$, and for $v \in V(G)$ let $w_h(v) = \sum\{h(vu) : u \in N(v)\}$. Clearly, $\sum_{v \in V} w_h(v) = 2\,w(h)$.

First, let $h : E(G) \to Y$ be a minimum Y-edge covering function, that is, an α_Y^1-function. Note that if $h(uv) > a$, then the minimality of h implies that either $w_h(u) = 1$ or $w_h(v) = 1$. Let $S_h = \{v \in V(G) : w_h(v) > 1\}$. As noted, if u and v are in S_h with $uv \in E(G)$, then $h(uv) = a$. For each $v \in S_h$, let $L(v) = \{vw \in E(G) : h(vw) > a\}$. Then $L(v_1) \cap L(v_2) = \emptyset$ for $v_1, v_2 \in S_h$. Further, because $a \le 1/\Delta(G)$, for each $v \in S_h$, we have $\sum\{h(vw) - a : vw \in L(v)\} \ge w_h(v) - 1$. Now define $f : E(G) \to Y$ by reducing the values under h on those edges e in each $L(v)$ with $v \in S_h$ and $h(e) > a$, so that $w_f(v) = 1$ for each $v \in S_h$. For each other edge xy, let $f(xy) = h(xy)$. Because $v \in V - S_h$ implies $w_h(v) = 1$, we have $w(h) - w(f) = \sum_{v \in S_h}[w_h(v) - 1] = \sum_{v \in V}[w_h(v) - 1] = 2\,w(h) - n$. Thus, $w(f) = n - w(h)$ and $w_f(v) \le 1$ for every $v \in V(G)$. Consequently, $\beta_Y^1(G) \ge w(f) = n - w(h) = n - \alpha_Y^1(G)$, so $\alpha_Y^1(G) + \beta_Y^1(G) \ge n$.

Second, let $h : E(G) \to Y$ be a $\beta_Y^1(G)$-function. Note that for $uv \in E(G)$, if $h(uv) < b$, then $w_h(u) = 1$ or $w_h(v) = 1$. Let $R_h = \{v \in V(G) : w_h(v) < 1\}$. If $\{x, y\} \subseteq R_h$ and $xy \in E(G)$, then $h(xy) = b$. For each $x \in R_h$, there is a set $S_x \subseteq N(x) \cap (V - R_h)$ such that $\sum_{v \in S_x}[b - h(xv)] \ge 1 - w_h(x)$. Define $g : E(G) \to Y$ as follows. For each $x \in R_h$, modify h by increasing g's values on the edges connecting x to S_x such that $w_g(x) = 1$, and let h and g agree on all other edges. Now $w_g(u) \ge 1$ for every $u \in V(G)$, and $w(g) - w(h) = \sum_{x \in R_h}[1 - w_h(x)] = \sum_{x \in V}[1 - w_h(x)] = n - \sum_{x \in V} w_h(x) = n - 2\,w(h)$. Thus $\alpha_Y^1(G) \le w(g) = n - w(h) = n - \beta_Y^1(G)$, and $\alpha_Y^1(G) + \beta_Y^1(G) \le n$. Consequently, $\alpha_Y^1(G) + \beta_Y^1(G) = n$. $\square$

As noted, Theorem 10.24 does not show that α_Y^1 and β_Y^1 are complement related in the sense of Theorem 10.22. Each of α_Y^1 and β_Y^1 does, however, have a complement. We complete this chapter with the Matrix Complementation Theorem in which matrix M is arbitrary but could be taken to be N for parameters like γ, or the incidence matrix H for matching β_1, or H^t for independence. Using the notation of [645], we have the following definition.

Definition. For any k-by-h matrix M, h-tuple $C = (c_1, c_2, ..., c_h)$, column k-tuple B, and $Y \in \mathfrak{R}$,

$$(M, MIN, C, B, Y) = \text{INF } \{\sum_{i=1}^{h} c_i x_i : M \cdot X \geq B, x_i \in Y\}, \text{ and}$$

$$(M, MAX, C, B, Y) = \text{SUP } \{\sum_{i=1}^{h} c_i x_i : M \cdot X \leq B, x_i \in Y\}.$$

Theorem 10.25 (Matrix Complementation Theorem) [645] *For any k-by-h matrix M, h-tuple C, k-tuple B, and complementable set $Y \in \Re$, let $L = [r_1, r_2, ..., r_k]^t$ be the row sum vector of M with $r_j = \sum_{i=1}^{h} m_{j,i}$. Then either*

$$(M, MIN, C, B, Y) = -\infty \text{ and } (M, MAX, C, L - B, Y) = \infty$$

or else

$$(M, MIN, C, B, Y) + (M, MAX, C, L - B, Y) = \sum_{i=1}^{h} c_i.$$

Proof. If $(M, MIN, C, B, Y) = K > -\infty$, then given $\varepsilon > 0$ there exists an h-tuple X in Y^h with $M \cdot X \geq B$ and $\sum_{i=1}^{h} c_i x_i \leq K + \varepsilon$.

Now $M \cdot X \geq B$

if and only if, for $1 \leq j \leq k$, we have $m_{j,1} x_1 + m_{j,2} x_2 + ... + m_{j,h} x_h \geq b_j$

if and only if, for $1 \leq j \leq k$, we have $r_j - (m_{j,1} x_1 + ... + m_{j,h} x_h) \leq r_j - b_j$

if and only if, for $1 \leq j \leq k$, we have $m_j(1 - x_1) + ... + m_{j,h}(1 - x_h) \leq r_j - b_j$

if and only if, for $M(\vec{1}_h - X) \leq L - B$.

By assumption, $\vec{1}_h - X \in Y^h$. Thus

$$(M, MAX, C, L - B, Y) \geq \sum_{i=1}^{h} c_i(1 - x_i) = \sum_{i=1}^{h} c_i - \sum_{i=1}^{h} c_i x_i$$

$$\geq \sum_{i=1}^{h} c_i - (K + \varepsilon).$$

Hence,

$$(M, MAX, C, B, Y) + (M, MAX, C, L - B, Y) \geq \sum_{i=1}^{h} c_i - \varepsilon.$$

And if $(M, MAX, C, L - B, Y) = J < \infty$, then given $\varepsilon > 0$ there exists an h-tuple X with $M \cdot X \leq L - B$ and $\sum_{i=1}^{h} c_i x_i \geq J - \varepsilon$. It follows that $M \cdot (\vec{1}_h - X) \geq B$, and

$$(M, MIN, C, B, Y) \leq \sum_{i=1}^{h} c_i(1 - x_i) = \sum_{i=1}^{h} c_i - \sum_{i=1}^{h} c_i x_i \leq \sum_{i=1}^{h} c_i - J + \varepsilon.$$

Hence,

$$(M, MIN, C, B, Y) + (M, MAX, C, L - B, Y) \leq \sum_{i=1}^{h} c_i - \varepsilon.$$

Consequently, $(M, MIN, C, B, Y) + (M, MAX, C, L - B, Y) = \sum_{i=1}^{h} c_i$.

If $(M, MIN, C, B, Y) = -\infty$ and $-\infty < K$, then there exists an h-tuple X in Y^h with $M \cdot X \geq B$ and $\sum_{i=1}^{h} c_i x_i < K$. As above, $M \cdot (\vec{1}_h - X) \leq L - B$ with $\vec{1}_h - X \in Y^h$ by complementarity. Thus

$$(M, MAX, C, L - B, Y) \geq \sum_{i=1}^{h} c_i - \sum_{i=1}^{h} c_i x_i \geq \sum_{i=1}^{h} c_i - K.$$

So $(M, MAX, C, L - B, Y) = \infty$. Similarly, $(M, MAX, C, L - B, Y) = \infty$ implies $(M, MIN, C, B, Y) = -\infty$, completing the proof. $\square$

Numerous corollaries of this theorem apply to the weighted and unweighted (that is, $Y = \{0, 1\}$) versions of complementary parameters. In particular, about 25 such pairs of parameters, including essentially all of the well-studied graph-theoretical parameters, are presented in [645].

EXERCISES

10.1 Verify Theorems 10.1 and 10.2.

10.2 For each k, find an infinite family of graphs G such that $G \in \mathcal{F}_k$ and $G \notin \mathcal{F}_{k-1}$.

10.3 Show that γ and γ_s are incomparable for graphs with arbitrarily high minimum degree.

10.4 (a) Find bipartite and outerplanar graphs with γ^- and γ_s arbitrarily small.

(b) Construct a graph G with $\Delta(G) = 5$ and $\gamma_s(G) = 0$.

(c) Show that $\gamma_s(H_k) = k(3 - k)/2$ for the Hajós graph H_k.

10.5 (a) How large can $\gamma(G) - \gamma^-(G)$ and $\gamma_s(G) - \gamma^-(G)$ be for a graph G of order n?

(b) Can you construct a graph G with $\Delta(G) \leq 3$ for which $\gamma_s(G) = n/3$?

(c) Prove that if G is cubic, then $\gamma_s(G) \leq n/2$.

(d) Also, for a cubic graph G, $\gamma_s(G) = n/2$ if and only if there exists an efficient dominating set.

10.6 Find (sharp?) lower bounds for $\gamma_s(G)$.

10.7 As explained, if $\{-1, 1\} \subseteq Y$, then $\gamma_Y(H_{12} \times H_4) < \gamma_Y(H_{12}) \cdot \gamma_Y(H_4)$. Note that $H_{12} \times H_4$ has 88 vertices. How small a graph $G \times H$ can be found with $\gamma_Y(G \times H) < \gamma_Y(G) \cdot \gamma_Y(H)$ for some Y?

10.8 Verify algebraically that $N \cdot X = \vec{1}_8$ is inconsistent for the graph in Figure 10.9.

10.9 Find the smallest graph G for which

(a) G does not have an efficient Y- dominating function for some $Y \subseteq \Re$.

(b) G does not have an efficient $\Re$-dominating function.

10.10 Interpret Theorem 10.25 for

(a) $(N, MIN, \vec{1}_n, \vec{1}_n, Y)$.

(b) $(N, MAX, D^*, \vec{1}_n, Y)$.

(c) $(A, MIN, \vec{1}_n, \vec{1}_n, Y)$.

(d) $(H^t, MAX, \vec{1}_n, \vec{1}_m, Y)$.

10.11 Prove that if f and g are minimal fractional dominating functions for G and $0 \le \alpha \le 1$, then the function $\alpha f + (1 - \alpha)g$ is a dominating function. Give an example where $\alpha f + (1 - \alpha)g$ is, in fact not a minimal dominating function. (For more results on convexity of extremal domination-related functions, see the excellent survey chapter by Cockayne and Mynhardt [307].)

Chapter 11

Frameworks for Domination

One of the major reasons why so much research has been done on domination in graphs is that the domination number appears in so many different mathematical contexts, or what we call frameworks. These include the contexts of: hypergraphs, or matrices of 0s and 1s; the algebraic solution to matrix equations of the form $N \cdot X \geq \vec{1}$; the generation of inequality chains of parameters; conditions defined on functions of the form $f : V \to \{0, 1\}$, or more generally functions of the form $f : V \to Y$, for Y a subset of real numbers; conditions defined on subgraphs induced by subsets of vertices $\langle S \rangle$; conditions defined between a set S and its complement $V - S$; and even the mathematical study of the game of chess and related games. In this chapter we discuss each of these mathematical frameworks for the domination number of a graph, several of which are discussed in detail in other chapters and will only be reviewed here. In order to be self-contained, some definitions and results will be repeated.

11.1 Hypergraphs, Matrices of 0s and 1s

A *hypergraph* $H = (X, C)$ is simply a set $X = \{x_1, x_2, \cdots, x_n\}$, elements of which are called *hypervertices*, together with a family of subsets of X, $C = \{C_1, C_2, \cdots, C_m\}$, elements of which are called *hyperedges*. We further stipulate that $\cup_{i=1}^{n} C_i = X$.

The study of hypergraphs is a field unto itself, on which books have been written. We refer the reader to the classic text of Berge, *Graphs and Hypergraphs* [96]. We will not even attempt a review of this large field of study; instead, we will consider the following parameters that are naturally associated with an arbitrary hypergraph. Given a hypergraph $H = (X, C)$:

(i) A set of hypervertices $S \subseteq X$ is a *transversal* of H if for every hyperedge C_i of C, $S \cap C_i \neq \emptyset$. The *transversal number* $\tau(H)$ is the minimum cardinality of a transversal S of X.

(ii) A set of hyperedges $D \subseteq C$ is a *cover* of H if $\cup_{C_i \in D} C_i = X$. The *covering* number $\rho(H)$ is the minimum cardinality of a cover of H.

(iii) A set of hyperedges $D \subseteq C$ is a *matching* if the hyperedges in D are pairwise disjoint. The *matching number* $\nu(H)$ is the maximum cardinality of a matching of H.

(iv) A set of hypervertices $S \subseteq X$ is *strongly stable* if for every hyperedge $C_i \in C$, $|S \cap C_i| \leq 1$. The *strong stability number* $\alpha(H)$ is the maximum cardinality of a strongly stable set in H.

With any hypergraph $H = (X, C)$, we can naturally associate a (0,1)-incidence matrix $M = \{m_{ij}\}$, which is an $|X|$ by $|C|$ matrix defined as follows:

$$m_{ij} = \left\{ \begin{array}{ll} 1 & \text{if } x_i \in C_j \\ 0 & \text{otherwise} \end{array} \right\}.$$

Thus we can associate with each hypervertex x_i an m-tuple of 0's and 1's,

$$x_i = (x_{i1}, x_{i2}, \cdots, x_{im}),$$
where $x_{ij} = 1$ if and only if $x_i \in C_j$.

We can also associate with each hyperedge C_j an n-tuple of 0's and 1's,

$$C_j = (c_{1j}, c_{2j}, \cdots, c_{nj}),$$
where $c_{lj} = 1$ if and only if $x_l \in C_j$.

With this in mind, we can say the following:

(i) the transversal number $\tau(H)$ equals the minimum number of rows of M which intersect all of the columns of M;

(ii) the covering number $\rho(H)$ equals the minimum number of columns of M which intersect all of the rows of M;

(iii) the matching number $\nu(H)$ equals the maximum number of columns in a set, no two of which have a one in a common entry;

(iv) the strong stability number $\alpha(H)$ equals the maximum number of rows in a set, no two of which have a one in a common entry.

Given a graph $G = (V, E)$, we can define a variety of hypergraphs $H = (X, C)$, as follows:

(i) Let $X = V$ and let C consist of the set of closed neighborhoods $N[v_i]$ of vertices $v_i \in V$; this is called the *neighborhood hypergraph of G*. In this case the transversal number of H satisfies $\tau(H) = \gamma(G)$, the domination number of G, and the matching number of H satisfies $\nu(H) = \rho(G)$, the packing number of G, that is, the maximum cardinality of a set S of vertices such that for every vertex $v \in V$, $|S \cap N[v]| \leq 1$. Note, also, that since the associated matrix M is symmetric (it is the neighborhood matrix of G), we have $\tau(H) = \gamma(G) = \rho(H)$ and $\nu(H) = \alpha(H) = \rho(G)$.

(ii) Let $X = V$ and let C consist of the set of open neighborhoods $N(v_i)$ of vertices $v_i \in V$; this is called the *open neighborhood hypergraph of G*. In this case the transversal number of H satisfies $\tau(H) = \gamma_t(G)$, the total domination number of G, and the matching number of H satisfies $\nu(H) = \rho^0(G)$, the *open packing number of G*, that is, the maximum cardinality of a set S of vertices such that for every vertex $v \in V$, $|S \cap N(v)| \leq 1$.

(iii) Let $X = E$ and let C consist of the set of closed neighborhoods of edges in V, that is, for each edge $e \in E$, $N[e]$ consists of the edge e together with all edges having a vertex in common with e; this is called the *edge-neighborhood hypergraph of G*. In this case the transversal number of H satisfies $\tau(H) = \gamma'(G)$, the edge domination number of G, and the matching number of H satisfies $\nu(H) = \beta^*(G)$, the *strong matching number of G*, that is, the maximum cardinality of a set F of independent edges, having the additional property that no two vertices on different edges of F are adjacent.

(iv) Let $X = E$ and let C consist of the sets of edges in the subgraphs induced by the closed neighborhoods of vertices in V; this is called the *edge-vertex neighborhood hypergraph of G*. In this case the covering number of H satisfies $\rho(H) = \rho_n(G)$, the *neighborhood number of G*, that is, the minimum cardinality of a set of vertices S for which $\cup_{v \in S}\langle N[v]\rangle = G$. This number was first studied by Sampathkumar and Neeralagi [995] who pointed out that for any connected graph G, $\gamma(G) \leq \rho_n(G) \leq \alpha_0(G)$, where $\alpha_0(G)$ is the vertex covering number of G, and if G is triangle-free then $\rho_n(G) = \alpha_0(G)$.

11.2 Linear and Integer Programming

Many interesting graph theoretic parameters can be defined in linear programming (LP) and integer and $\{0,1\}$-programming formats. Additional parameters can be defined using the concepts of LP-duality and of complementarity. For a complete discussion of this, the reader is referred to [1045]. In this section we illustrate just a few of the possibilities for defining domination parameters within the framework of linear and integer programming. Recall that A and N are the adjacency and closed neighborhood matrices, respectively.

Three particular n-tuples are of interest for a graph G: (i) the column n-vector $D = [d_i]$, which corresponds to the degree sequence of G, where d_i is the degree of vertex v_i, which equals the number of 1's in the ith row of the adjacency matrix A; (ii) $D^* = [d_i^*]$, where $d_i^* = 1 + deg(v_i)$, which equals the number of 1's in the ith row of the neighborhood matrix N; and (iii) $\vec{1}_n$ which denotes the column n-vector of all 1's.

The characteristic column vector $X_S = [x_i]$ satisfies $x_i = \phi_S(v_i)$ for $1 \le i \le n$. Given this notation, we can say that a set S is a dominating set if and only if $N \cdot X_S \ge \vec{1}_n$. This leads to the following integer programming formulation for the domination number $\gamma(G)$:

(1) $$\gamma(G) = \min \sum_{i=1}^{n} x_i$$
$$\text{subject to } N \cdot X \ge \vec{1}_n$$
$$\text{with } x_i \in \{0, 1\}$$

By allowing x_i to assume values in the unit interval [0,1], we obtain the fractional domination number $\gamma_f(G)$:

(2) $$\gamma_f(G) = \min \sum_{i=1}^{n} x_i$$
$$\text{subject to } N \cdot X \ge \vec{1}_n$$
$$\text{with } x_i \in [0, 1]$$

By changing the neighborhood matrix N to the adjacency matrix A in (1) above, we obtain the integer programming formulation of the total domination number $\gamma_t(G)$:

(3) $$\gamma_t(G) = \min \sum_{i=1}^{n} x_i$$
$$\text{subject to } A \cdot X \ge \vec{1}_n$$
$$\text{with } x_i \in \{0, 1\}$$

And by again permitting $x_i \in [0, 1]$ in (3), we obtain the open fractional domination number $\gamma_f^o(G)$:

$$(4) \qquad \gamma_f^o(G) = \min \sum_{i=1}^{n} x_i$$

$$\text{subject to } A \cdot X \geq \vec{1}_n$$
$$\text{with } x_i \in [0,1]$$

The packing number of a graph $\rho(G)$ has the following integer programming formulation:

$$(5) \qquad \rho(G) = \max \sum_{i=1}^{n} x_i$$

$$\text{subject to } N \cdot X \leq \vec{1}_n$$
$$\text{with } x_i \in \{0,1\}$$

The fractional equivalent of packing is obtained by permitting $x_i \in [0,1]$ in (5). The open packing number is defined by substituting $A \cdot X \leq \vec{1}$ for $N \cdot X \leq \vec{1}$ in (5).

The redundance number $R(G)$ of a graph G was defined to be the minimum number of times that vertices are dominated by vertices in a set S, given that each vertex must be dominated at least once. This can be stated as the following integer programming problem:

$$(6) \qquad R(G) = \min \sum_{i=1}^{n} (1 + deg(v_i))x_i$$

$$\text{subject to } N \cdot X \geq \vec{1}_n$$
$$\text{with } x_i \in \{0,1\}$$

Similarly, the efficient domination number $F(G)$ was defined to be the maximum number of vertices that can be dominated given that no vertex is dominated more than once. This can be stated as the following integer programming problem:

$$(7) \qquad F(G) = \max \sum_{i=1}^{n} (1 + deg(v_i))x_i$$

$$\text{subject to } N \cdot X \leq \vec{1}_n$$
$$\text{with } x_i \in \{0,1\}$$

Note that if a graph G has a solution to the matrix equation $N \cdot X = \vec{1}_n$, that is, if $F(G) = n$, then coding theorists would call G a perfect graph and the set of vertices for which $x_i = 1$ a perfect code. (Alternatively, we say that G is efficiently dominatable.)

Notice that, within the framework of linear and integer programming, at least eight problems can be defined by considering all combinations of $\leq$ or $\geq$, N or A, and $x_i \in \{0, 1\}$ or $[0, 1]$. More combinations are possible by considering $\sum_{i=1}^{n}(deg(v_i))x_i$ or $\sum_{i=1}^{n}(1 + deg(v_i))x_i$, and even more combinations are possible by considering other matrices than N or A, which can be associated with a graph G (see Chapter 4).

11.3   Domination Sequences

Recall from Chapter 3 the well-studied domination chain:

$$ir \leq \gamma \leq i \leq \beta \leq \Gamma \leq IR.$$

In this framework, we start with the property of being an independent set of vertices; this can be called the *seed* property (see [271]). We next observe that the maximality condition for an independent set is precisely the definition of a dominating set. Thus, every maximal independent set is a (minimal) dominating set and we get this much of the inequality chain:

$$\gamma \leq i \leq \beta \leq \Gamma.$$

Continuing in this manner, the minimality condition for a dominating set is precisely the definition of an irredundant set. Thus, every minimal dominating set is a (maximal) irredundant set, and we get the next extension to the inequality chain:

$$ir \leq \gamma \leq i \leq \beta \leq \Gamma \leq IR.$$

Continuing with this process, we can consider the maximality condition for an irredundant set. This yields the definition of what is called an external redundant set, and it can be shown that every maximal irredundant set is a minimal external redundant set. This leads to still another extension to the familiar inequality chain:

$$er \leq ir \leq \gamma \leq i \leq \beta \leq \Gamma \leq IR \leq ER.$$

Although, at this time, the next extension to this chain has not been developed, it would seem in principle that the derivation of longer inequality chains is possible. (Recall that the neighborhood knockout interior extension of Section 3.6.2 yields $i \leq nk \leq NK \leq \beta$.)

The minimality - maximality technique for generating this inequality chain, which starts with the initial seed property of being an independent set, provides a general framework for developing similar chains from other seed properties. Thus, almost any property of a subset of vertices in a graph can be considered (or for that matter, any property of subsets of a set). For example, we might be interested in sets of vertices S having the property that:

(i) S is a vertex cover;

(ii) S is a packing;

(iii) $\langle S \rangle$ is acyclic;

(iv) $\langle S \rangle$ is a regular graph of degree one, that is, a matching;

(v) $\langle S \rangle$ has a maximum degree one;

(vi) $\langle S \rangle$ in planar;

(vii) S is nearly perfect, that is, $|N(S) \cap S| \leq 1$, for every vertex $u \in V - S$;

(viii) $\langle S \rangle$ is a complete graph;

(ix) $\langle S \rangle$ has a perfect matching; or

(x) $\langle S \rangle$ has minimum degree one.

For each of these, as seed properties, an inequality sequence can in principle be generated. It remains to be seen what sorts of sequences result.

Of particular interest are questions such as:

(i) do chains of arbitrary length exist? or, for a given seed property, is the length of the generated inequality chain necessarily finite? is there a way to determine whether a given seed property will generate only a finite inequality chain or an arbitrarily long inequality chain?

(ii) how difficult is it to define the next pair of parameters in a given sequence as a function of the length of the sequence?

(iii) what is the complexity of computing values of parameters in such sequences? does this complexity depend inherently on the complexity of the seed property? could the complexity sooner or later become polynomial if the seed property is in **NP**? or conversely, could it become NP-complete if the seed property is in **P**?

11.4 Functions $f : V \rightarrow \{0, 1\}$

Given a graph $G = (V, E)$, consider defining various types of functions of the form $f : V \rightarrow \{0, 1\}$, where we impose various conditions that must be met. We say that a function f is *minimal* with respect to some condition if there does not exist a function of the same type, say g, $f \neq g$, for which $g(v) \leq f(v)$, for every $v \in V$. Similarly, a function f is *maximal* if there does not exist a function g of the same type, $f \neq g$, for which $g(v) \geq f(v)$, for every $v \in V$. For a vertex

$v \in V$, let $f(N[v])$ equal the sum of values of $f(v)$ for every vertex $w \in N[v]$. Given a function f, the weight of f, denoted $w(f)$, is the sum of the values $f(v)$ for all $v \in V$. We will be interested either in the value $\min w(f)$ or the value $\max w(f)$ over all functions satisfying a given condition. For example, consider the following conditions:

(i) for every vertex $v \in V$, $f(N[v]) \geq 1$ and f is minimal with respect to this property. Such a function is called a dominating function and the value $\min w(f) = \gamma(G)$, the domination number of G. Similarly, the value $\max w(f) = \Gamma(G)$, the upper domination number of G. Notice that if a set S is a dominating set of a graph $G = (V, E)$, then the characteristic function $\phi_S : V(G) \to \{0, 1\}$ is a dominating function.

(ii) for every vertex $v \in V$, $f(N(v)) \geq 1$ and f is minimal. Such a function is called a total dominating function and the value $\min w(f) = \gamma_t(G)$, the total domination number of G. Similarly, the value $\max w(f) = \Gamma_t(G)$, the upper total domination number of G.

(iii) for every vertex $v \in V$, $f(N[v]) \leq 1$ and f is maximal. Such a function is called a packing function and the value $\max w(f) = P(G)$, the packing number of G. Similarly, the value $\min w(f) = \rho(G)$, the lower packing number of G.

(iv) for every vertex $v \in V$, $f(N(v))) \leq 1$ and f is maximal. Such a function is called an open packing function and the value $\max w(f) = P^o(G)$, the open packing number of G. Similarly, the value $\min w(f) = \rho^o(G)$, the lower open packing number of G.

(v) for every edge $uv \in E$, $f(u) + f(v) \geq 1$ and f is minimal. Such a function is called a vertex covering function and the value $\min w(f) = \alpha_0(G)$, the vertex covering number of G. Similarly, the value $\max w(f) = \Lambda(G)$, the upper edge covering number of G.

(vi) for every edge $uv \in E$, $f(u) + f(v) \leq 1$ and f is maximal. Such a function is called a vertex independent function and the value $\max w(f) = \beta_0(G)$, the vertex independence number of G. Similarly, the value $\min w(f) = i(G)$, the independent domination number of G.

(vii) for every vertex $v \in V$, $f(v) = 1 - \min\{f(u) : u \in N[v]\}$. Such a function has been called a *stable labelling*. Although it is not immediately obvious, it can be proved that $\min w(f) = \alpha_0(G)$ and $\max w(f) = \Lambda(G)$.

(viii) for every vertex $v \in V$, $f(v) = 1 - \max\{f(u) : u \in N[v]\}$. Such a function has been called a *maximum stable labelling*. Although it is not immediately obvious, it can be proved that $\min w(f) = i(G)$ and $\max w(f) = \beta_0(G)$.

(ix) for every vertex $v \in V$, there exists a vertex $u \in N(v)$, such that $f(u) + f(v) \geq 1$ and f is minimal. It can be shown that $\min w(f) = \gamma(G)$ and $\max w(f) = \Gamma(G)$.

(x) for every vertex $v \in V$, there exists a vertex $u \in N(v)$, such that $f(u) + f(v) = 1$. Again it can be shown that $\min w(f) = \gamma(G)$. But in this case the value $\max w(f) = n - \gamma(G) = \Psi(G)$.

(xi) for every vertex $v \in V$, there exists a vertex $u \in N[v]$, such that $f(N[v]) = 1$. Such a function is called a perfect neighborhood function (see Section 3.5.4). This can be formulated somewhat differently as follows. Let $S = \{v : f(v) = 1\}$. A vertex w is called perfect if $f(N[w]) = 1$. A set S is called a perfect neighborhood set if every vertex $v \in V$ is either perfect or is adjacent to a perfect vertex. The perfect neighborhood number $\theta(G)$ of a graph G is the minimum cardinality of a perfect neighborhood set in G. Similarly, the upper perfect neighborhood number $\Theta(G)$ is the maximum cardinality of a perfect neighborhood set in G. Open perfect neighborhood functions have also been studied [831], that is, there exists a vertex $u \in N(v)$ such that $f(N(u)) = 1$.

11.5 Functions $f : V \to Y$

Generalizations of functions in Section 11.4, of the form $f : V \to \{0,1\}$ to the form $f : V \to Y$, for various sets Y of real numbers, yield other types of domination. Since these are discussed in some detail in Chapter 10, they will only be reviewed here.

(i) Let $Y = [0,1]$ be the closed unit interval between 0 and 1. Then $f : V \to [0,1]$ is a real-valued function. If for every vertex $v \in V$, $f(N[v]) \geq 1$, then we say that f is a fractional dominating function and the minimum weight of a fractional dominating function $\gamma_f(G)$ is the fractional domination number.

(ii) Let $Y = \{-1,1\}$. If for every vertex $v \in V$, $f(N[v]) \geq 1$, then we say that f is a signed dominating function and the minimum weight of a signed dominating function is called the signed domination number $\gamma_s(G)$ ([404, 627]).

(iii) Let $Y = \{-1,1\}$. If for at least half of the vertices $v \in V$, $f(N[v]) \geq 1$, then we say that f is a *majority dominating function* and the minimum weight of a majority dominating function is called the *majority domination number* $\gamma_{maj}(G)$ ([168, 617, 1158]).

(iv) Let $Y = \{-1,0,1\}$. If for every vertex $v \in V$, $f(N[v]) \geq 1$, then we say that f is a minus dominating function, and the minimum weight of a minus dominating function is called the minus domination number $\gamma^-(G)$ ([402, 403]).

(v) Let $Y = \{0, 1, 2, \cdots, k\}$. If for every vertex $v \in V$, $f(N[v]) \geq k$, then we say that f is a $\{k\}$-dominating function, and the minimum weight of a $\{k\}$-dominating function is called the $\{k\}$-domination number $\gamma_{\{k\}}(G)$ ([360, 371]).

11.6 Fundamental Dominating Sets or Conditions on the Dominating Set

The original definition of a dominating set states that a set $S \subseteq V$ is a dominating set of a graph G if for every vertex $v \in V - S$ there exists a vertex $u \in S$ such that u is adjacent to v. Historically, the next few types of dominating sets were all defined in terms of conditions on the subgraph induced by a dominating set S. Since most of these types of dominating sets have been discussed in earlier chapters, we will only review them here, with an eye toward the framework in which all of these types of dominating sets appear. Although more than 75 different types of dominating and related sets have been defined in the discrete mathematics literature (see Appendix), it has been suggested that only few of them are 'fundamental', that is, fundamental in the sense that (i) every nontrivial connected graph has at least one such dominating set, (ii) this type of dominating set is in a sense 'natural', and (iii) it is defined in terms of a condition on the dominating set S. With this in mind, the following types of dominating sets could be said to be fundamental:

(i) The second type of dominating set ever defined was an independent dominating set, which is a dominating set S such that $\langle S \rangle$ is a totally disconnected graph, that is, has no edges, or equivalently, such that $\Delta(\langle S \rangle) = 0$.

(ii) The third type of dominating set defined was a connected dominating set, which is a dominating set S such that $\langle S \rangle$ is a connected graph ([669, 1004]). Generalizations of connected domination can be obtained by imposing the condition that $\langle S \rangle$ is either 2-connected or 2-edge connected ([753, 757]).

(iii) The next type of dominating set defined was a total dominating set, which is a dominating set S such that $\langle S \rangle$ does not contain an isolated vertex, or equivalently, $\delta(\langle S \rangle) \geq 1$ ([28, 256]).

(iv) Another type of dominating set which could be said to be fundamental is a paired dominating set, which is a dominating set S such that $\langle S \rangle$ contains a perfect matching, that is, the vertices in S can be partitioned into pairs of adjacent vertices (see Section 6.6).

(v) Still another type of dominating set which could be said to be fundamental is an acyclic dominating set, which is a dominating set S such that $\langle S \rangle$ is an acyclic graph. This type of dominating set has not appeared in the literature, but is offered here as an example of a type of dominating set which can be defined within this particular framework of conditions on $\langle S \rangle$.

Still other types of dominating sets have been defined in terms of conditions on the dominating subgraph $\langle S \rangle$. These include the following:

(vi) clique domination: a dominating set S is a clique dominating set if $\langle S \rangle$ is a complete graph (see Section 6.5);

(vii) cycle domination: a dominating set S is a cycle dominating set if $\langle S \rangle$ has a hamiltonian cycle (see Section 6.7);

(viii) *path domination*: a dominating set S is a path dominating set if $\langle S \rangle$ has a hamiltonian path;

(ix) weakly connected domination: a dominating set S is a weakly connected dominating set if the subgraph weakly induced by S, $\langle S \rangle_W = (N[S], E_w)$, is connected, where E_w consists of the set of all edges having at least one vertex in S (see Section 3.5.3).

Note that while not every graph has a clique dominating set, a cycle dominating set, or a path dominating set, every connected graph does have a weakly connected dominating set.

11.7 Conditions on the Dominated Set $V - S$

As discussed in Chapter 7, a wide variety of dominating sets can be defined by imposing conditions on the complement $V - S$ of a dominating set S. We review here a number of these definitions, which are given by a statement which starts with "for every vertex $v \in V - S$" and then specifies some condition.

(i) strong domination: for every vertex $v \in V - S$, there exists a vertex $u \in S$ such that the $deg(u) \geq deg(v)$. Weak domination is defined by requiring that $deg(u) \leq deg(v)$ ([1003]).

(ii) *restrained domination*: for every vertex $v \in V - S$, there exists at least one other vertex $w \in V - S$ such that v is adjacent to w ([366, 367]).

(iii) k-domination: for every vertex $v \in V - S$, $|N(v) \cap S| \geq k$ (see Section 7.1).

(iv) perfect domination: for every vertex $v \in V - S$, $|N(v) \cap S| = 1$ ([270, 852, 1161]).

(v) odd domination: for every vertex $v \in V - S$, $|N(v) \cap S|$ is an odd number. Although this type of domination has not been studied in the literature, odd covers have. These are sets S which have the property that for every vertex $v \in V$, $|N[v] \cap S|$ is an odd number. Although it is not immediately obvious, it can be shown that every graph has an odd cover (see Section 7.2).

(vi) distance-k domination: for every vertex $v \in V - S$, there exists at least one vertex $u \in S$ such that $d(u, v) \leq k$ ([686]).

(vii) k-step domination: for every vertex $v \in V - S$, there exists at least one vertex $u \in S$ such that $d(u, v) = k$ ([217, 526]).

11.8   Degree Conditions on Sets S and $V - S$

Consider an arbitrary set S and its complement $V - S$ in a graph $G = (V, E)$. Consider also the subgraphs $\langle S \rangle$ and $\langle V - S \rangle$ and the bipartite subgraph $\langle S, V - S \rangle$ generated by all edges between S and $V - S$. Note that $G = \langle S \rangle \cup \langle V - S \rangle \cup \langle S, V - S \rangle$. Let us consider four values: (i) the degrees of vertices u in $\langle S \rangle$, (ii) the degrees of vertices $u \in S$ in $\langle S, V - S \rangle$ denoted $deg_u \langle S, V - S \rangle$, (iii) the degrees of vertices $v \in V - S$ in $\langle S, V - S \rangle$ denoted $deg_v \langle V - S, S \rangle$, and (iv) the degrees of vertices v in $\langle V - S \rangle$. As we will see, many types of domination can be defined in terms of various combinations of these four values. In Table 11.1, let X denote the situation that the value does not matter.

Clearly other combinations of these four values are worth considering. This particular framework is rich in possible models of domination for future study, and was first proposed, in a slightly different form, by Telle [1079].

11.9   Maximizing Functions Over a Set and Its Complement

Still another framework for studying domination has been proposed by Goddard and Henning [540]. Let $S \subseteq V$ and consider the subgraph $\langle S \rangle$ induced by S. Let $I(S)$ be the isolated vertices and $A(S)$ the nonisolated vertices in $\langle S \rangle$, let $B(S) = N[S] - S$ and let $Z(S) = V - S - B(S)$, see Figure 11.1.

For a set $S \subseteq V$ and for values i, a, d, and z, define:

$$f(S : i, a, d, z) = i|I(S)| + a|A(S)| + d|B(S)| + z|Z(S)|$$

and

$$f(G : i, a, d, z) = \max\{f(S : i, a, d, z) : S \subseteq V(G)\}.$$

Table 11.1: Degree conditions.

	$deg\langle S\rangle$	$deg\langle S, V-S\rangle$	$deg\langle V-S, S\rangle$	$deg\langle V-S\rangle$
S is a dominating set	X	X	≥ 1	X
S is an independent dominating set	$=0$	X	≥ 1	X
S is a total dominating set	≥ 1	X	≥ 1	X
S is a perfect dominating set	X	X	$=1$	X
there is a perfect matching between S and $V-S$	X	$=1$	$=1$	X
S is a restrained dominating set	X	X	≥ 1	≥ 1
S is a nearly perfect set	X	X	≤ 1	X
Both S and $V-S$ are dominating sets (dominating bipartition)	X	≥ 1	≥ 1	X
S is an k-domination set	X	X	$\geq k$	X
S is a k-restricted dominating set	X	$\leq k$	≥ 1	X
S is an independent set	0	X	X	X
S is a packing	0	X	≤ 1	X
S is a perfect code	0	X	$=1$	X
S is a total dominating set	≥ 1	X	≥ 1	X

Any set S for which this maximum is achieved is called a *realizing set* for G, i,a,d, and z. We can think of the function $f(G : i, a, d, z)$ as a maximum profit achievement in selecting a vertex set S, whereby vertices in $I(S)$ are worth \$ i each, vertices in $A(S)$ are worth \$ a, vertices in $B(S)$ are worth \$ d and vertices in $Z(S)$ are worth \$ z. Given this framework, the following instances of functions $f(G : i, a, d, z)$ produce several well-known domination and domination-related parameters, for every graph G:

(a) $f(G : -1, -1, 0, -\infty) = -\gamma(G)$;

(b) $f(G : -1, -\infty, 0, -\infty) = -i(G)$;

(c) $f(G : -\infty, -1, 0, -\infty) = -\gamma_t(G)$;

(d) $f(G : 1, -\infty, 0, 0) = \beta_0(G)$;

(e) $f(G : 0, -\infty, -1, -\infty) = -\alpha_0(G)$;

(f) $f(G : 0, 0, 1, \leq 0) = \Psi(G)$.

Clearly many combinations of parameters a, i, d, and z are possible, each of which can lead to interesting parameters of a graph.

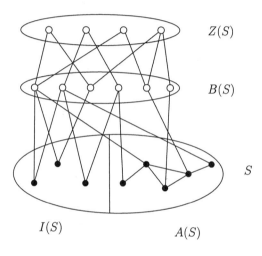

Figure 11.1: A set S and its boundary $B(S)$. ($B(S)$ and $Z(S)$ are not necessarily independent.)

11.10   Chessboard Concepts

The vertex independence number, the independent domination number , the domination number and the total domination number, although not defined as formal graphical parameters, all arise in the early mathematical studies of chessboard problems. This is documented by W. W. Rouse Ball, in his book *Mathematical Recreation and Problems of Past and Present Times* [72], published in 1892, and by P. J. Campbell, in his article entitled "Gauss and the eight queens problem: a study in miniature of the propagation of historical error", [179]. In this section, we review some of their observations about the origins of such chessboard problems.

From the book by W. W. Rouse Ball and H. S. M. Coxeter, *Mathematical Recreations and Essays*, Thirteenth Edition, Dover, 1987, page 166, we find the following passage:

"One of the classical problems connected with a chess-board is the determination of the number of ways in which eight queens can be placed on a chessboard (or, more generally, in which n queens can be placed on a board of n^2 cells) so that no queen can take any other. The was proposed originally by Franz Nauck in 1850."

Actually, Ball was in error here, as the problem was originally stated by a chessplayer, Max Bezzel in 1848 (this is discussed in Campbell's paper, mentioned above). But Dr. Franz Nauck is given credit, by Campbell, for being the first person to show that one can always place n nonattacking queens on an order n board. Thus, Nauck can be given credit for showing that the vertex independence number of the queens graph Q_n is n, that is, $\beta(Q_n) = n$.

Later in the same chapter, Ball comments:

"The Eight Queens Problem suggests the somewhat analogous question of finding the maximum number of kings - or more generally of pieces of one type - which can be put on a board so that no one can take any other, and the number of solutions possible in each case."

Thus, it is clear that Ball had the more general idea of independent sets in graphs and particularly of maximum independent sets.

Ball continues:

"Another problem of a somewhat similar character is the determination of the minimum number of kings - or more generally of pieces of one type - which can be put on a board so as to command or occupy all the cells."

Thus, Ball had the idea of domination in graphs, and particularly the domination number of a (chessboard) graph. As examples of this, Ball discusses what we might call the kings domination number, the queens domination number, the knights domination number, the bishops domination number and the rooks domination number of the order-8 board.

Ball continues:

"de Jaenisch proposed also the problem of the determination of the minimum number of queens which can be placed on a board of n^2 cells so as to command all the unoccupied cells, subject to the restriction that no queen shall attack the cell occupied by any other queen."

Thus, de Jaenisch should be given credit for the idea of independent domination $i(G)$ in graphs (see C. F. de Jaenisch, *Applications de l'Analyse Mathematique au Jeu des Echecs*, Leningrad, 1862 [347]). Ball discusses in his book the values of $i(Q_n)$ for $n = 4, 5, 6, 7$, and 8.

Finally, Ball adds this discussion:

"A problem of the same nature would be the determination of the minimum number of queens (or other pieces) which can be placed on a board so as to protect one another and command all the unoccupied cells."

Thus, Ball effectively defines the concept of the total domination number of a graph. He goes on to give the total domination number for the order-8 board for queens (5), bishops (10), knights (14) and rooks (8). It is interesting to note that he did not discuss the total domination number for kings.

One final problem mentioned by Ball has not received very much attention in the modern literature until very recently by an Italian computer scientist, Mario Velucchi (personal communication). Ball attributes to a Captain Turton the problem of placing n queens on an order-n board so as to attack or dominate the least number of cells.

11.11   Summary

We have seen in this chapter, 10 different mathematical frameworks in which the domination number of a graph naturally arises. Each of these frameworks places the domination number in a different perspective and suggests not only other parameters of a graph which are related in some way to the domination number, but generalizations of the domination number itself.

Each framework provides a unifying theory and a generalized viewpoint that enables one to

(1) identify and define new parameters,

(2) see relationships among these parameters, and

(3) develop insight into the computational problems involving these parameters.

We suggest that this is but the beginning, that is, other frameworks for the domination number undoubtably exist, each of which provides a richer understanding of the concept of domination in graphs. We also suggest that many other concepts, like vertex independence, vertex covering, and domatic number, can be studied in different mathematical frameworks.

EXERCISES

11.1 Many open research problems involve deciding which graph theorems extend to hypergraphs. Note that frequently many different extensions need to be considered. For example,

    (a) does one define the degree of a hypervertex to be the number of hyperedges containing it or the number of hypervertices adjacent to it?

    (b) how does one define the complementary hypergraph $\overline{H}$?

11.2 For each of the parameters defined in Section 11.2, identify the dual and complementary parameters.

11.3 The six-term sequence $ir \leq \gamma \leq i \leq \beta_0 \leq \Gamma \leq IR$ was extended outward to include er and ER and inward for nk and NK.

    (a) Does maintaining the maximality/minimality process for outward extensions result in infinitely many distinct parameters?

    (b) Can nk and NK be extended further inward?

11.4 Starting with vertex covering as a seed property, do minimality/maximality extensions, and note the complementarity to $ir \leq \gamma \leq i \leq \beta \leq \Gamma \leq IR$?

11.5 Prove the claims of Section 11.4 in (vii), (viii), (ix), and (x). For example, show that the minimum and maximum weights of stable labelling functions for G are $\alpha_0(G)$ and $\Lambda(G)$, respectively.

11.6 For each pair of parameters in Table 11.1 of Section 11.8, show if they are comparable or incomparable.

11.7 Verify (a) - (f) in Section 11.9.

Chapter 12

Domination Complexity and Algorithms

12.1 Introduction

With more than 200 research papers published on the algorithmic complexity of domination and related parameters of graphs, it is difficult to know what to cover in a relatively short chapter on the subject, especially since it could take quite a few pages just to cover the preliminaries of computational complexity and the many algorithm design paradigms that have been applied to domination problems. But consider this. It is well known and generally accepted that the problem of determining the domination number of an arbitrary graph is a difficult one. Since this problem has been shown to be NP-complete (see Chapter 1), it is generally thought to require exponential time in the order of the graph. Because of this, researchers have turned their attention to the study of classes of graphs for which the domination problem can be solved in polynomial time.

However, approximately 100 different classes of graphs have been considered. Just to define and discuss the basic properties of each of these classes of graphs would take 100 pages. Researchers have also defined and studied some 75 different types of domination and domination-related parameters of graphs (see the Appendix). Thus, in principle, there are about 7,500 different algorithmic complexity questions to be considered, specifically, can a given one of the 75 types of domination problems be solved in polynomial time, when restricted to one of the 100 different classes of graphs? And when you consider that each problem has both an unweighted and a weighted version, the number of possible questions doubles to 15,000. In addition, researchers have studied the algorithmic complexity of domination problems in line graphs, total graphs, and clique graphs of classes of graphs, thereby further increasing the number of algorithmic questions one could ask.

This chapter can therefore only scratch the surface of a very large field of study. It would seem reasonable, however, to present some of the basic results from which many of the others were inspired. It would also seem reasonable to split this chapter into two parts. The first part, Sections 12.2 and 12.3, considers complexity and NP-completeness results, which strongly suggest that the problems of determining the domination number and the values of related parameters for most classes of graphs are inherently exponential in nature. This part also reviews what is known, both in terms of NP-completeness results and polynomial algorithms, about eight of the standard domination parameters: the irredundance number $ir(G)$, the domination number $\gamma(G)$, the independent domination number $i(G)$, the vertex independence number $\beta_0(G)$, the upper domination number $\Gamma(G)$, the upper irredundance number $IR(G)$, the connected domination number $\gamma_c(G)$, and the total domination number $\gamma_t(G)$.

The second part, Sections 12.4 and 12.5, presents several domination algorithms, on trees, interval graphs, and permutation graphs, and presents a brief discussion of several other algorithmic aspects of domination in graphs.

Let us proceed first to the basic question: how difficult is it to compute the domination number of an arbitrary graph? In Chapter 1, we provided a brief discussion of NP-completeness and a proof that the decision problem DOMINATING SET is NP-complete. In the next section we will show that this domination problem is NP-complete when instances are restricted to either bipartite graphs or chordal graphs.

12.2   NP-Completeness of the Domination Problem

Recall from Chapter 1 that the basic complexity question concerning the decision problem for the domination number takes the following form:

DOMINATING SET
INSTANCE: A graph $G = (V, E)$ and a positive integer k
QUESTION: Does G have a dominating set of size $\leq k$?

David Johnson [524] was the first person to show that DOMINATING SET is NP-complete. The proof of the following theorem is given in Chapter 1.

Theorem 12.1 [524] *DOMINATING SET is NP-complete.*

Theorem 12.1 is the most basic complexity result concerning domination in graphs. It suggests that we should not expect to find a polynomial time algorithm for determining the domination number of an arbitrary graph. Thus, if we expect to be able to compute the domination number of a graph in polynomial time, we will have to restrict the instances to classes of graphs other than the class of arbitrary graphs. We have for our consideration a very large number of classes from which to choose, as indicated in the introduction to this chapter. DOMINATING SET remains NP-complete when instances are restricted to graphs in most of these classes, while for a relatively few classes we are able to compute the value of $\gamma(G)$ in polynomial time.

Bipartite graphs. Dewdney [355], Chang and Nemhauser [200], Bertossi [108] and several others have independently shown that DOMINATING SET remains NP-complete when instances are restricted to bipartite graphs. The following construction is due to Chang and Nemhauser, and is based on a reduction from DOMINATING SET for arbitrary graphs, which we have proved to be an element of **NPc**. Let $G = (V, E)$ be an arbitrary graph. By the VV^+ graph of G, we mean the bipartite graph $VV^+ = (V \cup \{x\}, V' \cup \{y\}, E^+)$, whose vertex set consists of two copies of V, denoted V and V', together with two special vertices, denoted x and y, and whose edges E^+ consist of: (i) edges uv' and $u'v$ for each edge $uv \in E(G)$, (ii) edges of the form uu' for each vertex $u \in V$, (iii) edges of the form $u'x$ for every vertex $u \in V$, and (iv) the one additional edge xy. This construction is illustrated in Figure 12.1.

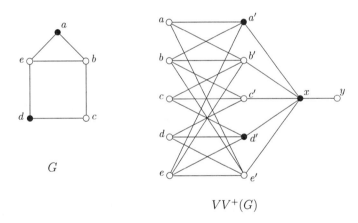

Figure 12.1: Reduction from DOMINATING SET to bipartite graphs.

For the VV^+ graph, one can prove that an arbitrary graph G has a dominating set of size $\leq k$ (for example, the shaded vertices in G in Figure 12.1) if and only if the corresponding bipartite graph VV^+ has a dominating set of size $\leq k + 1$ (for example, the shaded vertices in $VV^+(G)$ in Figure 12.1).

Chordal graphs. Booth and Johnson [136] were the first to show that DOMINATING SET remains NP-complete when restricted to chordal graphs. McRae [888] has provided the following simple proof of this, using a reduction from the following, well-known, NP-complete problem.

EXACT COVER BY 3-SETS

INSTANCE: A finite set $X = \{x_1, x_2, \cdots, x_{3q}\}$ of cardinality $3q$, for some positive integer q, and a set $C = \{C_1, C_2, \cdots, C_m\}$ of 3-element subsets of X. QUESTION: Does C contain an exact cover for X, that is, a subset $C' \subseteq C$ such that every element of X occurs in exactly one 3-element subset of C'?

Given an instance of EXACT COVER BY 3-SETS, we construct the following graph $G = (V, E)$. For each element $x_i \in X$, we create a vertex x_i in V. For each 3-element subset C_j in C, we create a path on three vertices, labelled u_j, v_j and w_j. We then add edges so that the set of vertices $U = \{u_1, u_2, \cdots, u_m\}$ forms a complete graph. We also add communication edges between the vertices labelled u_j and the three vertices corresponding to the 3-element subset C_j (see Figure 12.2, where the bold edges among the vertices labelled u_i indicate a complete graph K_5. Note that since the vertices in U form a complete subgraph, the graph G is a chordal graph. One can show that the set C contains an exact cover for X (for example, the set $\{C_4, C_5\}$) if and only if the chordal graph G has a dominating set of size $k = q + m$ (for example, the set of $k = 7$ shaded vertices). We omit the proof.

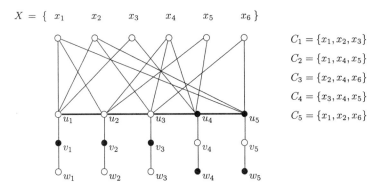

$X = \{\ x_1 \quad x_2 \quad x_3 \quad x_4 \quad x_5 \quad x_6\ \}$

$C_1 = \{x_1, x_2, x_3\}$

$C_2 = \{x_1, x_4, x_5\}$

$C_3 = \{x_2, x_4, x_6\}$

$C_4 = \{x_3, x_4, x_5\}$

$C_5 = \{x_1, x_2, x_6\}$

Figure 12.2: NP-completeness for chordal graphs.

Note that if we make the set V' in Figure 12.1 a clique, then the resulting graph, say G', is a split graph, that is, a graph $G' = (V', E')$ whose vertex set can be partitioned $V' = V_1' \cup V_2'$ so that $\langle V_1' \rangle$ is a complete graph and $\langle V_2' \rangle$ has no edges. The proof of NP-completeness for bipartite graphs then also works to establish the NP-completeness of DOMINATING SET for split graphs.

The following series of containment-relation chains among various classes of graphs indicates with either [**P**] or [**NPc**] whether DOMINATING SET is in the class **P** or is in **NPc** when restricted to the given family of graphs. Much of this material is taken from Corneil and Stewart [332].

trees[**P**] $\subset$ k-trees, fixed k [**P**] $\subset$ k-trees, arbitrary k [**NPc**] $\subset$ chordal [**NPc**]

trees [**P**] ⊂ block graphs [**P**] ⊂ directed path graphs [**P**] ⊂ undirected path graphs [**NPc**] ⊂ chordal graphs [**NPc**]

interval graphs [**P**] ⊂ trapezoid graphs [**P**] ⊂ cocomparability graphs [**P**] ⊂ perfect graphs [**NPc**]

split graphs [**NPc**] ⊂ chordal graphs [**NPc**] ⊂ weakly chordal graphs [**NPc**]

cographs [**P**] ⊂ 1-CUBs [**P**] ⊂ 2-CUBs [**NPc**] ⊂ CUBs [**NPc**]

interval graphs [**P**] ⊂ directed path graphs [**P**] ⊂ strongly chordal graphs [**P**] ⊂ chordal graphs [**NPc**]

cographs [**P**] ⊂ permutation graphs [**P**] ⊂ comparability graphs [**NPc**] ⊂ perfectly orderable graphs [**NPc**]

cographs [**P**] ⊂ P_4-reducible graphs [**P**] ⊂ permutation graphs [**P**] ⊂ trapezoid graphs [**P**]

trees [**P**] ⊂ chordal bipartite graphs [**NPc**] ⊂ semichordal graphs [**NPc**] ⊂ bipartite graphs [**NPc**] ⊂ comparability graphs [**NPc**]

trees [**P**] ⊂ generalized series-parallel graphs [**P**] ⊂ partial 2-trees [**P**] ⊂ partial k-trees, fixed k [**P**]

For several other classes of graphs, the complexity of DOMINATING SET has been considered; these include:

dually chordal graphs [**P**]
$k \times n$ grids, fixed k [**P**] ⊂ arbitrary grids [**NPc**],
cacti [**P**] ⊂ almost tree(k) [**P**],
BCH-graphs [**P**],
circular arc graphs [**P**], and
asteroidal triple-free graphs [**P**].

12.3 Complexity of Other Domination Related Parameters

Recall the domination chain discussed in Chapter 3:

$$ir(G) \leq \gamma(G) \leq i(G) \leq \beta_0(G) \leq \Gamma(G) \leq IR(G).$$

In this section we review the algorithmic complexity of each of these parameters, as well as that for the connected domination number $\gamma_c(G)$ and the total domination number $\gamma_t(G)$.

12.3.1   Irredundance number $ir(G)$

IRREDUNDANT SET
INSTANCE: A graph $G = (V, E)$ and a positive integer k
QUESTION: Does G have a maximal irredundant set of size $\leq k$?

Pfaff (see [940]) was the first to demonstrate the NP-completeness of IRREDUNDANT SET for arbitrary graphs. In fact, his construction is identical to that given in Figure 12.1. Laskar and Pfaff [830] have also shown that IRREDUNDANT SET is NP-complete when restricted to split graphs and thus, chordal graphs. McRae [888] showed that IRREDUNDANT SET remains NP-complete when restricted to line graphs, and Fricke, Hedetniemi, and Jacobs showed it to be NP-complete when restricted to k-regular graphs, for any fixed k [499].

In an impressive paper, Bern, Lawler, and Wong [103] constructed the first irredundance algorithm, an $O(n)$ algorithm, for determining the value of $ir(T)$ for an arbitrary tree. Their algorithm remains one of the most complex, polynomial-time, tree algorithms ever constructed, and consists largely of a 20-by-20 table of possible cases to consider between a vertex in a rooted tree and its parent. The Bern-Lawler-Wong algorithm was extended by Wimer to partial k-trees in his Ph.D. dissertation [1147].

Bertossi and Gori [112] published an $O(n^4)$ irredundance algorithm for weighted interval graphs, but it appeared to have a flaw, which was corrected by Chang, Hsing, and Peng [205]. Their algorithm runs in $O(mn^2 + n^3)$ time. The time complexity for irredundance in weighted interval graphs was later improved to $O(n^3)$ by Chang, Nagavamsi, and Pandu Rangan [209].

12.3.2   Domination number $\gamma(G)$

In the previous section we reviewed the NP-completeness results for the domination problem for quite a few classes of graphs. In this section we will provide some of the historical references.

The first domination algorithm was a linear algorithm for computing $\gamma(T)$ for an arbitrary unweighted tree, by Cockayne, Goodman, and Hedetniemi in 1975 [267]. An algorithm for the domination number of a tree had been published in 1966 by Daykin and Ng [346], however, it was found to contain a flaw. Shortly after the domination algorithm for trees had been constructed, the first NP-completeness result for DOMINATING SET was constructed by David Johnson, in 1975. This was followed by linear algorithm by Slater in 1976 [1031] for the R-domination number $\gamma_R(G)$ for any forest G, and a linear algorithm for the domination number of a line graph of a tree by Mitchell and Hedetniemi in 1977 [900]. This algorithm, in effect, computes the edge domination number of a tree, and was the first paper to show that for trees, $\gamma'(T) = i'(T)$. In 1978 Natarajan and White [912] published a linear algorithm for the weighted domination number of a tree, as did Farber in 1981 [430].

12.3.3   Independent domination number $i(G)$

Unlike the irredundance number, a fair amount of progress has been made in understanding the complexity of INDEPENDENT DOMINATING SET.

INDEPENDENT DOMINATING SET
INSTANCE: A graph $G = (V, E)$ and a positive integer k
QUESTION: Does G have an independent dominating set of size $\leq k$?

Garey and Johnson were the first to show that INDEPENDENT DOMINATING SET is NP-complete [524]. Corneil and Perl [329] showed that it remains NP-complete when restricted to bipartite graphs and comparability graphs; Yannakakis and Gavril [1156] showed that it is NP-complete when restricted to line graphs of graphs (equivalently, they established the NP-completeness of (independent) edge domination).

The first algorithm for computing $i(G)$ for any class of graphs was constructed for trees by Beyer, Proskurowski, Hedetniemi, and Mitchell in 1979 [114]. Somewhat surprisingly, Farber showed that the Minimum Weight INDEPENDENT DOMINATING SET can be solved in polynomial time when restricted to chordal graphs, and thus also when restricted to strongly chordal or interval graphs [431, 433]. Farber and Keil [434] have shown that the weighted domination and weighted independent domination problems can be solved in $O(n^3)$ time for permutation graphs. Their independent domination algorithm was improved from $O(n^3)$ to $O(n^2)$ by Brandstädt and Kratsch [153]. Kratsch and Stewart [802] extended the polynomial independent dominating algorithms for permutation and interval graphs to cocomparability graphs, which are complements of comparability graphs.

Elmallah and Stewart [415] studied k-polygon graphs, which are the intersection graphs of straight line chords inside a convex k-gon. They showed that for any fixed k, INDEPENDENT DOMINATING SET is polynomially solvable on k-polygon graphs.

12.3.4   Independence number $\beta_0(G)$

INDEPENDENT SET
INSTANCE: A graph $G = (V, E)$ and a positive integer k
QUESTION: Does G have an independent set of size $\leq k$?

This problem is shown to be NP-complete for arbitrary graphs in Garey and Johnson [524], even for cubic planar graphs. Kratochvíl and Nešetřil [796] showed that INDEPENDENT SET is NP-complete when restricted to the classes of 2-DIR and PURE-3-DIR graphs. Here, k-DIR refers to the class of intersection graphs of segments lying in at most k directions in the plane, and PURE-k-DIR is the class of intersection graphs of segments lying in at most k directions in the plane, with the added condition that any two parallel segments are disjoint.

For example, 1-DIR graphs are interval graphs and PURE-2-DIR graphs are bipartite graphs. INDEPENDENT SET was also shown to be NP-complete when restricted to k-regular graphs for any fixed $k \geq 3$, by Fricke, Hedetniemi, and Jacobs [499].

The first algorithm for computing the value of $\beta_0(G)$ for a class of graphs was due to Daykin and Ng in 1966 [346] for trees. This algorithm was generalized to weighted trees by Cockayne and Hedetniemi in 1976 [279] and to maximal outerplanar graphs by Mitchell in 1977 [898]. It is well known that a polynomial algorithm exists for solving this problem on bipartite graphs, since, by König's Theorem, the vertex independence number $\beta_0(G)$ equals the edge covering number $\alpha_1(G)$ for bipartite graphs (see Harary [576]), and by Gallai's theorem, $\alpha_1(G) + \beta_1(G) = n$. Since $\beta_1(G)$ can be computed for any graph in polynomial time, it follows that $\beta_0(G)$ can be computed in polynomial time for any bipartite graph.

Gavril [528] showed that the INDEPENDENT SET problem can be solved in polynomial time when restricted to chordal graphs.

12.3.5 Upper domination number $\Gamma(G)$

UPPER DOMINATING SET

INSTANCE: A graph $G = (V, E)$ and a positive integer k
QUESTION: Does G have a minimal dominating set of size $\leq k$?

This problem was shown to be NP-complete by Cheston, Fricke, Hedetniemi, and Jacobs [233]. Other than this, relatively little is known about the complexity of this problem.

However, Cockayne, Favaron, Payan, and Thomason [260] showed that for bipartite graphs, $\beta_0(G) = \Gamma(G) = IR(G)$. Thus, $\Gamma(G)$ can be computed for bipartite graphs in polynomial time. It was also shown, by Jacobson and Peters [738], that $\beta_0(G) = \Gamma(G) = IR(G)$ holds for all chordal graphs, and thus, $\Gamma(G)$ can be computed in polynomial time for chordal graphs.

Other than the fact that $\Gamma(G)$ can be computed in polynomial time for bipartite and chordal graphs, perhaps the only known algorithm for computing the value of $\Gamma(G)$ for a class of graphs is due to Hare, Hedetniemi, Laskar, Peters, and Wimer [597], who constructed a polynomial algorithm for generalized series-parallel graphs.

12.3.6 Upper irredundance number $IR(G)$

UPPER IRREDUNDANT SET

INSTANCE: A graph $G = (V, E)$ and a positive integer k
QUESTION: Does G have an irredundant set of size $\leq k$?

This problem was shown to be NP-complete by Fellows, Fricke, Hedetniemi, and Jacobs [465]. McRae [888] showed that UPPER IRREDUNDANT SET remains NP-complete when restricted to either line graphs or line graphs of bipartite graphs. Other than this, relatively little is known about the complexity of UPPER IRREDUNDANT SET.

However, as pointed out above, for both bipartite graphs and chordal graphs, $\beta_0(G) = \Gamma(G) = IR(G)$. Therefore, $IR(G)$ can be computed in polynomial time for these graphs.

12.3.7 Connected domination number $\gamma_c(G)$

CONNECTED DOMINATING SET
INSTANCE: A graph $G = (V, E)$ and a positive integer k
QUESTION: Does G have a connected dominating set of size $\leq k$?

Pfaff, Laskar, and Hedetniemi [940] showed that CONNECTED DOMINAT-ING SET is NP-complete for bipartite graphs, and Laskar and Pfaff [830] showed this problem to be NP-complete for split graphs. Since split graphs form a proper subclass of chordal graphs, it follows that CONNECTED DOMINAT-ING SET is NP-complete for chordal graphs. Müller and Brandstädt [905] showed that CONNECTED DOMINATING SET is NP-complete for chordal bipartite graphs, that is, bipartite graphs in which every cycle of length greater than four has a chord.

Keil, Laskar, and Manuel [769] considered *undirected path graphs* and *directed path graphs*, which are defined to be the intersection graphs of families of undirected paths in an undirected tree, and directed paths in a directed tree, respectively. They showed that CONNECTED DOMINATING SET is NP-complete for undirected path graphs of diameter three, but is polynomially solvable for directed path graphs of diameter three.

White, Farber, and Pulleyblank [1143] showed that CONNECTED DOMI-NATING SET is NP-hard for planar bipartite graphs, they constructed a poly-nomial algorithm for computing the connected domination number of any series-parallel graph and they showed that polynomial algorithms also exist for solving this problem on strongly chordal graphs and 2-trees. They made an interesting connection between connected dominating sets and Steiner trees. Let T be a subset of the vertices of a graph $G = (V, E)$. A *Steiner tree* of G on the ter-minal vertices T is a minimum cardinality subset S of V which contains T and for which $\langle S \rangle$ is a connected graph. White et al. showed that the Steiner tree problem is NP-hard for any class of chordal graphs for which CONNECTED DOMINATING SET is NP-hard.

Corneil and Perl [329] constructed a polynomial algorithm for computing the connected domination number of a cograph. This was generalized to an $O(n^2)$ algorithm for permutation graphs by Colbourn and Stewart [319], who also had an $O(n^3)$ algorithm for the weighted case. The complexity of connected domi-nation for permutation graphs was subsequently improved to $O(m + n \log n)$ by Arvind and Pandu Rangan [61], and later by Liang [845] to $O(n)$ in the unweighted case and to $O(m + n)$ in the weighted case. Liang also generalized these permutation algorithms to an $O(m + n)$ algorithm for trapezoid graphs, which form a superclass of both interval graphs and permutation graphs, but a subclass of cocomparability graphs. He also constructed an $O(m + n \log n)$ algorithm in the weighted case.

An even greater generalization is due to Breu and Kirkpatrick [154] and Kratsch and Stewart [802], who constructed an $O(n^3)$ algorithm for the connected domination number of a cocomparability graph and Chang [204], who showed that CONNECTED DOMINATING SET is NP-complete in the weighted case. All of this was generalized one step further by Balakrishnan, Rajaraman, and Pandu Rangan [70] and Corneil, Olariu, and Stewart [327] to asteroidal triple-free (AT-free) graphs.

Moscarini [904] defined a new class of chordal graphs called *doubly chordal*, which generalizes strongly chordal graphs, but is a proper subclass of chordal graphs. He constructed an $O(n^3)$ algorithm for computing the connected domination number of a doubly chordal graph.

Elmallah and Stewart [415] showed that for any fixed k, CONNECTED DOMINATING SET is polynomially solvable on k-polygon graphs. It is interesting to note that for a k-polygon graph with n vertices, their algorithm requires $O(n^{4k^2+3})$ time. Circle graphs are the intersection graphs of chords of a circle, and, as such, generalize k-polygon graphs. In fact, the union of k-polygon graphs over all k is the class of circle graphs. Keil [767] showed that CONNECTED DOMINATING SET is NP-complete for circle graphs.

Finally, D'atri and Moscarini [344] have shown that CONNECTED DOMINATING SET is solvable in polynomial time for distance-hereditary graphs. These are graphs having the property that for every pair of vertices u and v, the distance from u to v is the same in every connected, induced subgraph containing u and v. Yeh and Chang [1160] extended this algorithm to weighted distance-hereditary graphs.

12.3.8   Total domination number $\gamma_t(G)$

TOTAL DOMINATING SET
INSTANCE: A graph $G = (V, E)$ and a positive integer k
QUESTION: Does G have a total dominating set of size $\leq k$?

Pfaff (see [941]) was the first to demonstrate the NP-completeness of TOTAL DOMINATING SET, even when restricted to bipartite graphs. Laskar, Pfaff, Hedetniemi, and Hedetniemi [834] showed that TOTAL DOMINATING SET is NP-complete when restricted to undirected path graphs and constructed a linear algorithm for computing the total domination number of any tree. Also, Laskar and Pfaff [830] showed that TOTAL DOMINATING SET is NP-complete for split graphs, which implies its NP-completeness for chordal graphs. McRae [888] showed that TOTAL DOMINATING SET remains NP-complete when restricted to line graphs or line graphs of bipartite graphs.

Keil [766] was the first to construct a polynomial algorithm for solving the TOTAL DOMINATING SET problem on interval graphs. His algorithm had an $O(n + m)$ time complexity. This was subsequently improved by Bertossi and Gori [112], who constructed an $O(n \log n)$ algorithm for computing the total domination number of an interval graph.

Corneil and Stewart [332] and Brandstädt and Kratsch [153] showed that TOTAL DOMINATING SET is polynomially solvable on permutation graphs, and Chang [194] did the same for strongly chordal graphs. Kratsch and Stewart [802] constructed an $O(n^6)$ algorithm for the total domination number of a cocomparability graph. Kratsch [798] showed that TOTAL DOMINATING SET is polynomially solvable on asteroidal triple-free graphs. Keil, Laskar, and Manuel [769] showed that TOTAL DOMINATING SET is NP-hard for undirected path graphs of diameter three, but is polynomially solvable for directed path graphs of diameter three. Keil [767] showed that TOTAL DOMINATING SET is NP-complete for circle graphs, and finally, TOTAL DOMINATING SET was shown to be polynomially solvable for bipartite distance hereditary graphs by Yeh and Chang [1159] and subsequently, for arbitrary distance hereditary graphs, in linear time, by Chang, Wu, Chang, and Yeh [210]. In general, Stewart and Kratsch [801] showed that for many graph classes, the algorithm for DOMINATING SET can be used for TOTAL DOMINATING SET.

12.4 A Sample of Domination Algorithms

As we indicated in the previous section, quite a wide variety of polynomial domination and domination-related algorithms have been constructed. Within the limitations of the space allowed, we will present in this section a sampler of these algorithms, limited to three of the most well-studied families of graphs: trees, interval graphs and permutation graphs.

12.4.1 Trees

As previously mentioned, an impressive linear algorithm for computing $ir(T)$ was constructed by Bern, Lawler, and Wong in 1987 [103]. A linear algorithm for computing $\gamma(T)$ was given by Cockayne, Goodman, and Hedetniemi in 1975 [267]. A linear algorithm for computing $i(T)$ was constructed by Beyer, Hedetniemi, Proskurowski, and Mitchell in 1979 [114]. The first linear algorithm for computing $\beta_0(T)$ was constructed in 1966 by Daykin and Ng [346]. Since $\beta_0(G) = \Gamma(G) = IR(G)$ for trees, the Daykin and Ng algorithm suffices to compute $\Gamma(G)$ and $IR(G)$ for trees as well. A very simple $O(n)$ algorithm for computing $\beta_0(T)$ for an arbitrary unweighted tree can be found in [899]. It is trivial to compute $\gamma_c(T)$ for any tree T, it is simply the number of vertices which remain when all of the endvertices of a tree are deleted. And finally, as previously mentioned, Laskar, Pfaff, Hedetniemi, and Hedetniemi constructed the first linear algorithm for computing $\gamma_t(T)$ in 1984 [834].

We present one of the simplest of these algorithms, which computes the value of $\gamma(T)$ for an arbitrary unweighted tree T. It is taken from Mitchell, Cockayne, and Hedetniemi [899] and is a linear-time implementation of the algorithm in [267]. This algorithm actually solves a more general problem, as follows. Let the vertices of a tree $T = (V, E)$ be partitioned into three subsets, $V = V_1 \cup V_2 \cup V_3$, where V_1 consists of *free* vertices, V_2 consists of *bound* vertices, and V_3 consists of *required* vertices. An *optional dominating set* in T is any set of vertices D which contains all required vertices, that is, $V_3 \subseteq D$, and dominates all bound vertices. Notice that a bound vertex is either an element of D or is adjacent to a vertex in D. Free vertices need not be dominated and need not be in D, but can be used in D to dominate bound vertices. The *optional domination number* $\gamma_{opt}(T)$ is the minimum cardinality of a optional dominating set in T. Notice that if we specify that $V_1 = \emptyset$, $V_2 = V$ and $V_3 = \emptyset$, then $\gamma_{opt}(T) = \gamma(T)$. Thus, an algorithm for computing the value of $\gamma_{opt}(T)$ is sufficient to compute the value of $\gamma(T)$.

Figure 12.3 shows an example of a tree T and the data structure, called a Parent array, that is used to represent T once it has been rooted at an arbitrary vertex, labelled 1. Notice, in the Parent array, that the Parent of a vertex labelled i is given by Parent[i], and that the vertex labelled 1 has no Parent (indicated by Parent [1] = 0).

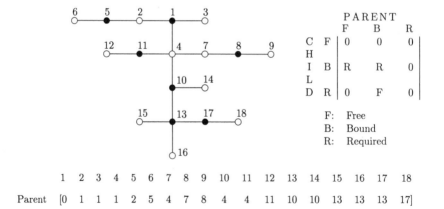

Figure 12.3: Domination in trees.

The algorithm for computing $\gamma(T)$ makes one pass over the array PARENT $[1..N]$, from N down to 2. Initially all vertices are labelled 'Bound'. As a vertex is encountered, say vertex i, its current LABEL $[\ i\ ]$ together with that of its PARENT $[\ i\ ]$ are used to possibly relabel PARENT $[\ i\ ]$, according to the table in Figure 12.3. If no relabelling is necessary, a 0 appears in the table, meaning 'do nothing'.

Algorithm TREE DOMINATION

INPUT: a tree T represented by an array PARENT $[\ 1..N\ ]$
OUTPUT: a minimum cardinality dominating set DOM of T, represented by the set of vertices i for which LABEL[i] = 'Required'.

Begin
    DOM $\leftarrow$ $\emptyset$;
    for $i = 1$ **to** N **do**
        LABEL$[\ i\ ]$ = 'Bound';
    od;
    for $j = N$ **to** 2 **by** -1 **do**
        if LABEL $[\ i\ ]$ = 'Bound' **then**
            LABEL $[$ PARENT $[\ i\ ]\ ]$ = 'Required';
        else if LABEL $[\ i\ ]$ = 'Required' **then**
            DOM $\leftarrow$ DOM $\cup\{i\}$;
            if LABEL $[$ PARENT $[\ i\ ]]$ = 'Bound' **then**
                LABEL $[$ PARENT $[i]]$ = 'Free';
            fi;
        fi;
        fi;
    od;

    /* Last vertex */
    if LABEL $[\ 1\ ]$ = 'Bound' **OR** LABEL $[1]$ = 'Required' **then**
        DOM $\leftarrow$ DOM $\cup$ $\{1\}$;
    fi; **End**.

The shaded vertices in Figure 12.3 are the vertices which receive the label 'Required' when Algorithm TREE DOMINATION is applied to the PARENT array in Figure 12.3. It is trivial to see that Algorithm TREE DOMINATION runs in $O(n)$ time, as it merely executes a simple **for**-loop, all of the statements within which can be executed in at most constant time. The correctness of this algorithm is based on the following theorem; we omit the straightforward proof.

Theorem 12.2 [267] *Let T be a tree whose vertices have been labelled 'Free' (V_1), 'Bound' (V_2) and 'Required'(V_3). Let v be an endvertex of T which is adjacent to a vertex u and let $T - v$ denote the tree which results from deleting v from T. Then*

(a) *if $v \in V_1$, then $\gamma_{opt}(T) = \gamma_{opt}(T - v)$.*

(b) *if $v \in V_2$ and T' is the tree which results from deleting v and relabelling u as 'Required', then $\gamma_{opt}(T) = \gamma_{opt}(T')$.*

(c) *if $v \in V_3$ and $u \in V_3$ then $\gamma_{opt}(T) = 1 + \gamma_{opt}(T - v)$.*

(d) *if $v \in V_3$ and $u \notin V_3$ and if T' if the tree which results from deleting v and relabelling u as 'Free', then $\gamma_{opt}(T) = 1 + \gamma_{opt}(T')$.*

12.4.2   Interval graphs

A graph $G = (V, E)$ is an *interval graph* if its vertices can be associated one-to-one with intervals on the real line in such a way that two vertices are adjacent if and only if their corresponding intervals have a nonempty intersection, that is, interval graphs are the intersection graphs of sets of intervals on the real line.

As we have seen earlier, the following containment relations hold for interval graphs:

interval graphs [**P**] $\subset$ directed path graphs [**P**] $\subset$ strongly chordal graphs [**P**] $\subset$ chordal graphs [**NPc**]

DOMINATING SET is NP-complete when restricted to chordal graphs, as are most domination related problems for chordal graphs (independent domination being an exception). But for interval graphs, directed path graphs, and strongly chordal graphs, DOMINATING SET can be solved in polynomial time. In fact, most domination and domination related problems have linear time algorithms when restricted to interval graphs. In this section we will take a look at several of these algorithms, using the elegant and simple presentation of Ramalingam and Pandu Rangan [962].

Figure 12.4 illustrates an interval graph with $n = 15$ vertices and its representation as the intersection graph of a set of 15 intervals. Without loss of generality, we can assume that the $2n$ endpoints of these n intervals are all distinct, and they occur at integer points on the line. Thus, for example, the input to an algorithm on interval graphs usually consists of a set of pairs of integers, for example, (1,3), (2,5), (4,7), (6,10), (9,11), (12,13), (14,15), (16,17), (18,19), (8,21), (22,23), (24,25), (26,27), (28,29), (20,30), in our example. In fact, most algorithms on interval graphs assume that the intervals are sorted and labelled in ascending order on their right endpoints, as illustrated in Figure 12.4. Notice that proper containment of intervals is permitted, for example, interval 7 is properly contained within interval 10.

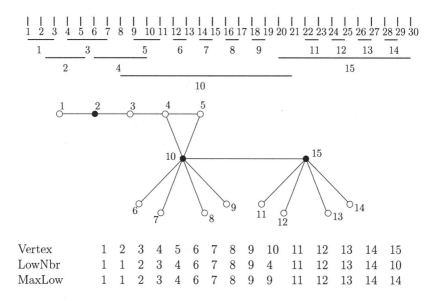

Vertex	1	2	3	4	5	6	7	8	9	10	11	12	13	14	15
LowNbr	1	1	2	3	4	6	7	8	9	4	11	12	13	14	10
MaxLow	1	1	2	3	4	6	7	8	9	9	11	12	13	14	14

Figure 12.4: Interval graph representation.

Most algorithms on interval graphs assume that the intervals, or vertices, have been numbered 1, 2, ..., n, and that one pass is made over these intervals from 1 to n. Let us define $V_i = \{1, 2, ..., i\}$ and $G_i = \langle V_i \rangle$ to be the subgraph of G induced by the vertices labelled 1, 2, ..., i. Notice that the graph G_i is obtained from G_{i-1} by adding vertex i and joining it to zero or more consecutive vertices at the right end of the sequence 1, 2, ..., $i-1$. Let us call this the left degree of vertex i. Thus, an interval graph can be completely specified by recording, for each vertex i, the number of consecutive vertices, to the immediate left in the sequence, to which it is adjacent. For example, the graph in Figure 12.4 can be completely specified as follows:

Vertex	1	2	3	4	5	6	7	8	9	10	11	12	13	14	15
Left degree	0	1	1	1	1	0	0	0	0	6	0	0	0	0	5

This ordering of the vertices of an interval graph, in fact, serves to characterize this class of graphs.

Theorem 12.3 [962] *A graph $G = (V, E)$ with n vertices is an interval graph if and only if its vertices can be labelled 1, 2, ..., n in such a way that for every $i < j < k$, $ik \in E$ implies that $jk \in E$.*

In the algorithms which follow we will need to know, for each vertex i, the smallest index $LowNbr(i)$ of a vertex adjacent to i; if vertex i is not adjacent to any vertices to its left, then $LowNbr(i) = i$. Notice that vertex i is not adjacent to vertices 1, 2, ..., $LowNbr(i) - 1$, but it is adjacent to every vertex between $LowNbr(i)$ and i. Suppose for some vertex t, that $LowNbr(t) = r$. Then consider the sequence of values, $LowNbr(s)$ for every integer s, $r \leq s \leq t$. Define $MaxLow(i) = \max\{LowNbr(s) : LowNbr(i) \leq s \leq i\}$. The values of $LowNbr(i)$ and $MaxLow(i)$ for each vertex in the graph G are given in Figure 12.4.

For each vertex i we define two more sets of vertices:

$$L(i) = \{MaxLow(i), \cdots, i\}$$

$$M(i) = \{j : j > i \text{ and } j \text{ is adjacent to } i\}.$$

Observe that in G_i, the closed neighborhood of vertex $MaxLow(i)$ satisfies $N[MaxLow(i)] \subseteq N[i]$ and $N[MaxLow(i)] \subseteq L(i) \cup M(i)$. Note also that the vertices in $L(i)$ form a clique in G_i.

Independent domination number

We are now ready to describe a simple algorithm for computing the independent domination number of an arbitrary interval graph, given the above-mentioned interval representation.

Let $ID(i)$ denote an independent dominating set in the graph G_i and let $MinID(i)$ denote the minimum weight of an independent dominating set in G_i. We observe that since $L(i)$ is a clique and since it contains a vertex which is not adjacent to any vertex in $V_i - L(i)$, any independent dominating set $ID(i)$ of G_i must contain exactly one vertex in $L(i)$, call it j. Furthermore, $ID(i) - \{j\}$ must independently dominate the vertices in $V(LowNbr(j) - 1)$. Thus, $ID(i)$ is an independent dominating set of V_i if and only if it is of the form $\{j\} \cup ID(LowNbr(j) - 1)$, for some vertex $j \in L(i)$. Therefore, we can compute the minimum weight of an independent dominating set in an interval graph G according to the following simple iteration.

Theorem 12.4 *The following algorithm computes the minimum weight of an independent dominating set, $MinID(n)$, in an interval graph:*

(i) $MinID(0) = \emptyset$;
(ii) **for** $i = 1$ **to** n **do**
$\qquad MinID(i) = \min\{\{j\} \cup MinID(LowNbr(j) - 1) : j \in L(i)\}$;
$\qquad$ **od**;

Notice that the iteration in Theorem 12.4 suffices to determine both a minimum independent dominating set and a minimum weight independent dominating set in an interval graph.

Let us illustrate this independent domination algorithm with the graph in Figure 12.4.

Vertex	1	2	3	4	5	6	7	8	9	10	11	12	13	14	15
LowNbr	1	1	2	3	4	5	7	8	9	4	11	12	13	14	10
MaxLow	1	1	2	3	4	6	7	8	9	9	11	12	13	14	14
$L(i)$	1	1	2	3	4	6	7	8	9	9	11	12	13	14	14
		2	3	4	5					10					15
$M(i)$	2	3	4	5	10	10	10	10	10	15	15	15	15	15	
				10											

$MinID(0) = \emptyset$
$MinID(1) = \{1\}$
$MinID(2) = \{2\}$
$MinID(3) = \{2\}$
$MinID(4) = \{3, 1\}$
$MinID(5) = \{5, 2\}$
$MinID(6) = \{6, 5, 2\}$
$MinID(7) = \{7, 6, 5, 2\}$
$MinID(8) = \{8, 7, 6, 5, 2\}$
$MinID(9) = \{9, 8, 7, 6, 5, 2\}$
$MinID(10) = \{10, 2\}$
$MinID(11) = \{11, 10, 2\}$
$MinID(12) = \{12, 11, 10, 2\}$
$MinID(13) = \{13, 12, 11, 10, 2\}$
$MinID(14) = \{14, 13, 12, 11, 10, 2\}$
$MinID(15) = \{14, 13, 12, 11, 10, 2\}$

Vertex independence number

The computation of the maximum cardinality of an independent set of vertices in an interval graph is as simple as changing 'min' to 'max' in Theorem 12.4.

Theorem 12.5 *The following algorithm computes the maximum weight of an independent set, $MaxI(n)$, in an interval graph:*

(i) $MaxI(0) = \emptyset$;
(ii) **for** $i = 1$ **to** n **do**
$$MaxI(i) = \max\{\{j\} \cup MaxI(LowNbr(j) - 1) : j \in L(i)\};$$
od;

Domination number

In order to compute the minimum weight of a dominating set in an interval graph, we let $D(i)$ denote a subset of V which dominates V_i. In this case we will permit a vertex not in V_i to be an element of $D(i)$. We next observe that for every i, there is a vertex in $L(i)$, say j, for which $N[j] \subseteq L(i) \cup M(i)$. Thus any dominating set $D(i)$ of V_i (in V) must include at least one vertex, say k, in $L(i) \cup M(i)$. Necessarily, the set $D(i) - \{k\}$ must dominate $V_{(LowNbr(k)-1)}$, since k suffices to dominate $V_i - V_{(LowNbr(k)-1)}$. Therefore, if $k \in L(i) \cup M(i)$, then $LowNbr(j) \leq i$, and we can conclude that a set $D(i) \subseteq V$ is a dominating set of V_i if and only if it is of the form $\{k\} \cup V_{(LowNbr(k)-1)}$, for some vertex $k \in L(i) \cup M(i)$.

Theorem 12.6 *The following algorithm computes the minimum weight of a dominating set, $MinD(n)$, in an interval graph:*

(i) $MinD(0) = \emptyset$;
(ii) **for** $i = 1$ **to** n **do**
$$MinD(i) = \min\{\{j\} \cup MinD(LowNbr(j) - 1) : j \in L(i) \cup M(i)\};$$
od;

Connected domination number

In order to compute the minimum weight of a connected dominating set in an interval graph, we let $CD(i)$ denote a connected dominating set of G_i which includes vertex i, and we let $MinCD(i)$ denote a minimum weight of a $CD(i)$. Certainly, if $LowNbr(i) = 1$, then vertex i dominates all of the vertices $1, 2, \ldots, i$; hence, we set $MinCD(i) = \{i\}$. If $LowNbr(i) > 1$ then every $CD(i)$ must include a vertex other than i, which is adjacent to i in G_i. So let j be a maximum vertex in $CD(i) - \{i\}$. We can assume that $LowNbr(j) < LowNbr(i)$, else vertex j is not needed to form a minimum weight connected dominating set (we are assuming that there are no negative weights). If $LowNbr(j) < LowNbr(i)$ then any vertex of G_j which is adjacent to vertex i must also be adjacent to vertex j, and therefore $CD(i) - \{i\}$ is a $CD(j)$. Therefore we have the basis of the following theorem.

Theorem 12.7 *The following algorithm computes the minimum weight of a connected dominating set, $MinCD(G)$ in a connected interval graph G:*

(i) **for** $i = 1$ **to** n **do**

   if $LowNbr(i) = 1$ **then**

     $MinCD(i) = \{i\}$;

   else if $LowNbr(i) > 1$ **then**

     $MinCD(i) = \min\{\{i\} \cup MinCD(j) : j < i, j \text{ is adjacent to } i,$
     $\text{and } LowNbr(j) < LowNbr(i)\}$;

     fi;

   fi;

  od;

(ii) $MinCD(G) = \min\{MinCD(i) : i \in L(n)\}$.

Total domination number

As before, let $TD(i) \subseteq V$ be a total dominating set of V_i and let $MinTD(i)$ be a minimum weight total dominating set of V_i. Let $PartialTD(i)$ be a subset of V that totally dominates the set $\{i\} \cup V_{(LowNbr(i)-1)}$, and let $MinPartialTD(i)$ denote a minimum weight $PartialTD(i)$.

As with the domination number algorithm, each $TD(i)$ must include at least one vertex in $L(i) \cup M(i)$, say vertex j. We have two cases to consider. If $j \in L(i)$, then it is necessary and sufficient that $TD(i) - \{j\}$ totally dominates $\{j\} \cup V_{(LowNbr(j)-1)}$. If $j \in M(i)$, then it is necessary and sufficient that $TD(i) - \{j\}$ totally dominates $V_{(LowNbr(j)-1)}$.

Similarly, every $PartialTD(i)$ includes at least one vertex adjacent to i, say k, so obviously $k \geq LowNbr(i)$. Therefore, it is necessary and sufficient that $PartialTD(i) - \{k\}$ totally dominate $V_{(\min\{LowNbr(i)-1, LowNbr(k)-1\})}$. The following theorem follows from these observations.

Theorem 12.8 *The following algorithm computes the minimum weight of a total dominating set, $MinTD(n)$, in a connected interval graph G:*

(i) $MinTD(0) = \emptyset$;

(ii) **for** $i = 1$ **to** n **do**

   $MinPartialTD(i) =$

     $\min\{\{j\} \cup MinTD(\min\{LowNbr(j) - 1, LowNbr(i) - 1\}) :$
     $j \text{ adjacent to } i\}$;

   $MinTD(i) =$

     $\min(\{\{j\} \cup MinPartialTD(j) : j \in L(i)\}$
     $\cup \{\{j\} \cup MinTD(LowNbr(j) - 1) : j \in M(i)\})$;

  od;

12.4.3   Permutation graphs

Let Π_n denote the set of all permutations of $\{1, 2, ..., n\}$. Let $\pi = (\pi(1),..., \pi(n)) \in \Pi_n$ be a permutation. If $\pi(i) = k$, we say that the position of k in the permutation π is i, that is, $\pi^{-1}(k) = i$. Each permutation π defines what is called an *inversion graph* $G(\pi) = (\{1, 2,... , n\}, E(\pi))$, where $ij \in E(\pi)$ if and only if $i < j$ and $\pi^{-1}(i) > \pi^{-1}(j)$. Figure 12.5 illustrates a permutation graph. Notice that if we draw a line between each integer i and its position in π , we create n lines, each with an associated integer. In this way, two vertices i and j are adjacent in $G(\pi)$ if and only if the corresponding lines cross. Notice that an independent set of vertices in $G(\pi)$ corresponds to an increasing subsequence of π, and a clique in $G(\pi)$ corresponds to a decreasing subsequence of π.

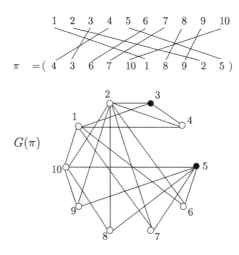

Figure 12.5: A permutation graph.

Permutation graphs were first introduced by Pnueli, Lempel, and Even in 1971 [947] and [427]. Since that time quite a few polynomial time algorithms have been constructed for solving problems on permutation graphs, that are NP-complete on arbitrary graphs. For example, Brandstädt and Kratsch [152] and Farber and Keil [434] constructed polynomial domination and independent domination algorithms, Brandstädt and Kratsch [152] and Colbourn and Stewart [319] constructed polynomial connected domination algorithms, and Brandstädt and Kratsch [152] and Corneil and Stewart [332] constructed polynomial total domination algorithms.

In this section we present a simple $O(n^2)$ algorithm, due to Brandstädt and Kratsch [153] for finding a minimum weight independent dominating set in a permutation graph G. We will assume here that the defining permutation π of the permutation graph is given as part of the input. Spinrad [1053] has shown that π can be constructed in $O(n^2)$ time, given the graph G. In more recent work, Spinrad and McConnell [882] designed a linear time algorithm to construct π.

This algorithm makes use of the observation that a set is an independent dominating set if and only if it is a maximal independent set. Since, as we have observed, maximal independent sets correspond in permutation graphs to maximal increasing subsequences in π, all that is necessary is to search for such a sequence in π of minimum weight. In particular, we will determine, for every i, $1 \leq i < n$, the minimum weight $MinID(i)$ of an independent dominating set in the subsequence $\pi(1)$, $\pi(2)$, ..., $\pi(i)$, which contains $\pi(i)$ as the rightmost element. We let $w(i)$ denote the weight of vertex i.

/* Stage 1 */

```
for i = 1 to n do
    LeftOfID(i) = 0;
    MinID(i) = w(π(i));
    for j = i − 1 to 1 step -1 do
        if π(j) < π(i) and LeftOfID(i) = 0 then
            MinID(i) = w(π(i)) + MinID(j);
            LeftOfID(i) = π(j);
        else if π(i) > π(j) > LeftOfID(i) > 0 then
            MinID(i) = min{MinID(i), w(π(i)) + MinID(j)};
            LeftOfID(i) = π(j);
            fi;
        fi;
    od;
od;
```

/* Stage 2 */

```
MaxElement = π(n);
MinID = MinID(n);
for i = n − 1 to 1 step -1 do
    if π(i) > MaxElement then
        MinID = min{MinID, MinID(i)};
        MaxElement = π(i);
    fi;
od;
```

We illustrate this algorithm with the permutation graph $G(\pi)$ in Figure 12.5, where $\pi = (4, 3, 6, 7, 10, 1, 8, 9, 2, 5)$ and all weights are equal to 1: $MinID[1..10] = (1, 1, 2, 3, 4, 1, 2, 3, 2, 2)$, and thus the minimum cardinality of an independent dominating set is 2, for example the set {3,5}, the shaded vertices in Figure 12.5.

12.5 Other Algorithmic Aspects

In this chapter we have discussed a variety of domination problems, most of which are NP-complete when instances are restricted to a wide variety of classes of graphs. On the other hand, we have seen that some domination problems have polynomial time solutions, particularly when restricted to such classes of graphs as trees, interval graphs and permutation graphs. In this section we briefly discuss several other aspects of domination complexity and algorithms.

12.5.1 Series-parallel graphs

The class of (two-terminal) series-parallel graphs has been well-studied in the field of electrical engineering, since they describe graphs of electrical circuits. This class of graphs can be defined in several different ways. The following is a recursive definition:

(i) The graph K_2 is a (two-terminal) series-parallel graph, in which one vertex is designated as a left terminal, and the other a right terminal.

(ii) If G_1, with terminals u_1 and v_1, and G_2, with terminals u_2 and v_2, are (two-terminal) series-parallel graphs, then so are the following:

(series connection) the graph $G_1 s G_2$, obtained from G_1 and G_2 by identifying v_1 with u_2, whose left terminal is u_1 and whose right terminal is v_2 (see Figure 12.6);

(parallel connection) the graph $G_1 p G_2$, obtained from G_1 and G_2 by identifying vertices u_1 and u_2 and calling the resulting vertex the left terminal, and identifying u_2 and v_2 and calling the resulting vertex the right terminal (see Figure 12.6).

(iii) no graph is a (two-terminal) series-parallel graph unless it can be constructed from copies of K_2 by a finite number of applications of the operations in (ii) above.

Series-parallel graphs were one of the first classes of graphs to be studied algorithmically, primarily because many of the standard NP-complete problems have polynomial solutions when restricted to these graphs. One of the first of

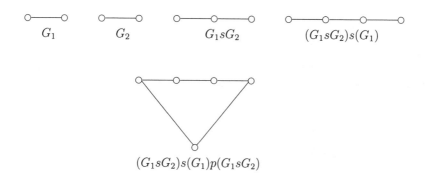

$$(G_1sG_2)s(G_1)p(G_1sG_2)$$

Figure 12.6: Series-parallel graphs.

these algorithms was a linear algorithm by Kikuno, Yoshida, and Kakuda [775] for solving the DOMINATING SET problem.

Takamizawa, Nishizeki, and Saito [1076] solved several other problems in linear time, including INDEPENDENT SET. Subsequently, linear algorithms were designed by Pfaff, Laskar, and Hedetniemi [941] for computing the values of $i(G)$ and $\gamma_t(G)$ for series-parallel graphs. As previously mentioned, Hare, Hedetniemi, Laskar, Peters, and Wimer [597] designed a linear algorithm for computing the value of $\Gamma(G)$ for generalized series-parallel graphs. We are not aware that anyone has designed algorithms for computing the values of $\gamma_c(G)$, $ir(G)$, or $IR(G)$ for series-parallel graphs.

12.5.2 Partial k-tree theory

The class of graphs called k-trees can be defined recursively as follows:

(i) A complete graph K_k on k vertices is a k-tree.

(ii) If G is a k-tree and H is a complete subgraph of G on k vertices, then the graph formed from G by adding a new vertex v, and joining v to each vertex in H, is a k-tree.

(iii) No graph is a k-tree unless it can be formed from a complete graph K_k by a finite number of applications of the operation in (ii).

A *partial k-tree* G is any graph which can be constructed from a k-tree H by deleting a subset of vertices and/or edges of H. It is not too difficult to see that the class of trees forms a subclass of the class of 1-trees (or that forests are partial 1-trees), and that the class of (two-terminal) series-parallel graphs is a subclass of the class of partial 2-trees.

Many researchers have independently observed that most problems which are NP-complete for the class of arbitrary graphs, have polynomial-time solutions when restricted to partial k-trees. This has become a rather large area of algorithmic research, which is much too large to be discussed in this text. The reader is referred to Volume of *Discrete Mathematics*, edited by Arnborg, Proskurowski, and Hedetniemi [43], which is devoted entirely to this topic.

12.5.3   Parallel algorithms

All of the algorithms presented in this chapter are sequential algorithms. They are designed to be executed on a computer having one processor and one random access memory. The advent of parallel computing, in which algorithms can be executed on computers with many, even thousands of, processors, using either a shared memory or a distributed memory architecture, has caused algorithm designers to completely rethink the design of their sequential algorithms. For example, algorithms which run in $O(n)$ sequential time can sometimes be redesigned to run in $O(\log n)$ parallel time. Other algorithms seem resistant to this kind of speedup, and no amount of additional processors or parallel memory seems to enable them to be executed any faster than linear time.

The parallel computation model which is most commonly used is the parallel random access machine (PRAM). Such a machine has a number of identical processors and a common, shared memory. In each unit of time, a processor can read from a memory cell, perform an ordinary arithmetic or logical computation, and write into a memory cell. In some PRAM models different processors are allowed to read from or write to a common memory cell concurrently; this is referred to as the CRCW PRAM. In other models, this is not allowed, for example, in the exclusive-read, exclusive-write, or EREW PRAM.

A good example of early work on the design of parallel algorithms for domination-type problems is that of He and Yesha [654]. They defined a very general, binary tree algebraic computation (BTAC) problem, which includes as special cases the problems of finding a minimum vertex cover, a maximum independent set, a maximum matching and a minimum distance-r dominating set. They designed efficient, parallel, EREW PRAM algorithms for solving the BTAC problem on trees in $O(\log n)$ time with $O(n)$ processors, and on (two-terminal) series-parallel graphs in $O(\log m)$ time with $O(m)$ processors.

The design and analysis of parallel algorithms has become a large field of study, which is, unfortunately, much too large and too complex to be discussed in this text. The interested reader is referred to any of a number of texts on the design and analysis of parallel algorithms.

12.5.4 Other publications on domination algorithms and complexity

We would be remiss if we failed to mention several other publications which discuss in some detail both algorithms and the complexity of domination problems on graphs. The first is the excellent survey paper by Corneil and Stewart [332] mentioned earlier. The second is an extensive chapter on this topic written by Kratsch in [645]. Kratsch's chapter also contains the most comprehensive bibliography yet compiled on domination algorithms and complexity. In addition, quite a few Ph.D. and M.S. dissertations have been written which contain results on domination algorithms and complexity. We include here just a partial listing of these, in historical order:

S. L. Mitchell. Linear algorithms on trees and maximal outerplanar graphs: design, complexity analysis and data structures study. PhD thesis, Univ. Virginia, 1977.

K. S. Booth. Dominating sets in chordal graphs. PhD thesis, Univ. Waterloo, Ontario, 1980.

M. Farber. Applications of linear programming duality to problems involving independence and domination. PhD thesis, Rutgers Univ., 1981.

G. J. Chang. k-Domination and graph covering problems. PhD thesis, Cornell Univ., 1982.

J. Pfaff. Algorithmic complexities of domination-related graph parameters. PhD thesis, Clemson Univ., 1984.

L. L. Kelleher. Domination in graphs and its application to social network theory. PhD thesis, Northeastern Univ., 1985.

M. B. Richey. Combinatorial optimization on series-parallel graphs: algorithms and complexity. PhD thesis, Georgia Institute of Technology, 1985.

L. Stewart. Permutation in graph structure and algorithms. PhD thesis, Univ. Toronto, 1985.

K. W. Peters. Theoretical and algorithmic results on domination and connectivity. PhD thesis, Clemson Univ., 1986.

E. El-Mallah. Decomposition and embedding problems for restricted networks. PhD thesis, Univ. Waterloo, 1987.

T. V. Wimer. Linear algorithms on K-terminal graphs. PhD thesis, Clemson Univ., 1987.

R. B. Borie. Recursively constructed graph families: membership and linear algorithms. PhD thesis, Georgia Institute of Technology, 1988.

K. Kilakos. On the complexity of edge domination. Master's thesis, Univ. New Brunswick, 1988.

T. W. (Haynes) Rice. On $k - \gamma$-insensitive Domination. PhD thesis, Univ. Central Florida, 1988.

D. L. Grinstead. Algorithmic templates and multiset problems in graphs. PhD thesis, Univ. Alabama in Huntsville, 1989.

E. O. Hare. Algorithms for grids and grid-like graphs. PhD thesis, Clemson Univ., 1989.

J. Haviland. Cliques and independent sets. PhD thesis, Univ. Cambridge, 1989.

P. Scheffler. Die Baumweite von Graphen als ein Mass für die Kompliziertheit algorithmischer Probleme. PhD thesis, Akademie der Wissenschaften der DDR, Berlin, 1989.

S. F. Hwang. Domination and related topics. PhD thesis, National Chiao Tung Univ., Hsinchu, Taiwan, 1991.

J. R. Carrington. Global domination of factors of a graph. PhD thesis, Univ. Central Florida, 1992.

T. Y. Chang. Domination numbers of grid graphs. PhD thesis, Univ. South Florida, 1992.

A. Majumdar. Neighborhood hypergraphs: a framework for covering and packing parameters in graphs. PhD thesis, Clemson Univ., 1992.

W. Piotrowski. Combinatorial optimization: scheduling, facility location, and domination. PhD thesis, North Dakota State Univ., 1992.

J. C. Schoeman. 'n Ry boonste en onderste dominaise -, onafhanklikheids- en onoorbodigheidsgetalle van 'n grafiek. Master's thesis, Rand Afrikaans Univ., 1992.

A. P. Burger. Domination in the Queen's graph. Master's thesis, Univ. South Africa, 1993.

P. J. P. Grobler. Functional generalisations of dominating sets in graphs. Master's thesis, Univ. South Africa, 1994.

C. Lee. On the domination number of a digraph. PhD thesis, Michigan State Univ., 1994.

A. A. McRae. Generalizing NP-completeness proofs for bipartite and chordal graphs. PhD thesis, Clemson Univ., 1994.

J. A. Telle. Vertex partitioning problems: characterization, complexity and algorithms on partial k-trees. PhD thesis, Univ. Oregon, 1994.

D. I. Carson. Computational aspects of some generalized domination parameters. PhD thesis, Univ. Natal, 1995.

C. B. Smart. Studies of graph based IP/LP parameters. PhD thesis, Univ. Alabama in Huntsville, 1996.

H. G. Yeh. Distance-hereditary graphs, combinatorial structures and algorithms. PhD thesis, National Chiao Tung Univ., Hsinchu, Taiwan, 1997.

EXERCISES

12.1 Complete the proofs that the decision problem DOMINATING SET is NP-complete for chordal graphs and for split graphs.

12.2 Construct linear time algorithms to compute the following parameters for trees T:

    (a) $\gamma_t(T)$, total domination.

    (b) $\gamma_p(T)$, paired-domination.

    (c) $\gamma_L(T)$, locating domination.

    (d) $PD(T)$, parity dimension.

    (e) $F(T)$ and $R(T)$, efficient domination and redundance, respectively.

    (f) $i(T)$ and $\beta_0(T)$.

12.3 Construct polynomial/linear time algorithms to compute the following parameters for interval graphs:

    (a) $\gamma_t(G)$.

    (b) $\gamma_p(G)$.

    (c) $\gamma_L(G)$.

    (d) $F(G)$ and $R(G)$.

(For exercises 12.4 - 12.6) Often enough one can get a fair amount of mileage out of a given theorem. In particular, one NP-completeness proof often suggests several others. Such is the case for the NP-completeness proofs of DOMINATING SET for arbitrary graphs in Chapter 1 and for bipartite graphs in Chapter 12.

12.4 Modify the proof in Chapter 1 that DOMINATING SET is NP-complete for arbitrary graphs to show that INDEPENDENT DOMINATING SET is NP-complete for arbitrary graphs.

12.5 Modify the graph G, which is constructed in the proof of the NP-completeness of DOMINATING SET in Chapter 1, so that the vertices which correspond to literals form a complete graph. Note that this modified graph is a chordal graph, say G'. We say that G' is obtained from G by clique-ing the literals. Use G' to show that each of the following decision problems are NP-complete, even when restricted to chordal graphs:

(a) DOMINATING SET.

(b) TOTAL DOMINATING SET.

(c) CONNECTED DOMINATING SET.

12.6 Modify the proof of the NP-completeness of DOMINATING SET for bipartite graphs, in Chapter 12, to show that each of the following problems are NP-complete, even when restricted to bipartite graphs:

(a) MINIMUM MAXIMAL IRREDUNDANT SET.

(b) CONNECTED DOMINATING SET.

(c) TOTAL DOMINATING SET.

12.7 In general it is reasonably straightforward to design linear algorithms for computing the exact values of various dominating parameters of trees. In Chapter 12 we presented a reasonably simple $O(n)$ algorithm for computing the value of $\gamma(T)$ for any tree T. Design similar algorithms for computing any of the following:

(a) the minimum weight of a dominating set in a tree T, in which each vertex is assigned an arbitrary integer weight.

(b) the independent domination number $i(T)$.

(c) the total domination number $\gamma_t(T)$.

(d) the strong domination number $\gamma_S(T)$.

(e) the vertex independence number $\beta_0(T)$.

12.8 A unicyclic graph G is a graph of order n and size n. It is reasonably easy to design linear algorithms for computing the values of many domination parameters of these graphs. Design a linear algorithm for computing the domination number of an arbitrary unicyclic graph.

12.9 A cactus is a connected graph every 2-connected component of which consists of a simple cycle of length ≥ 3. Design a linear algorithm for computing the domination number of an arbitrary cactus.

12.10 In Chapter 12 we presented linear algorithms for computing the domination number, independent domination number, vertex independence numer, connected domination number and total domination number of an arbitrary interval graphs. Design a polynomial algorithm for deciding if an arbitrary interval graph has a dominating clique.

12.11 "Let $i(T)$ be."

Appendix

The following pages contain a fairly comprehensive census of more than 75 models of dominating and related types of sets in graphs, which have appeared in the research literature over the past 20 years. For each type of domination we give the following information:

(i) a precise definition or equivalent definitions,

(ii) the generally accepted notation,

(iii) occasionally a comment or two about a noteworthy result or aspect of this type of domination, and

(iv) a citation to one or more of the first places where this type of domination appeared in the literature, or to a representative paper which discusses this type of domination.

In most cases we are interested in the optimization problems of finding either a maximum cardinality set and a minimum cardinality maximal set of a given type, or a minimum cardinality set and a maximum cardinality minimal set of a given type. In most of the following definitions, the X-domination number equals the minimum cardinality of an X-dominating set in G, and the upper X-domination number equals the maximum cardinality of a minimal X-dominating set in G. If the notation for the upper X-domination number is not specified, it means that this number has not been studied very much, if at all.

Following the presentation of the basic parameter γ, the remaining parameters are presented in loosely connected groupings. Many of these parameters could have been placed into more than one of these categories, and our order of presentation is somewhat random.

Types of Domination

Domination

A set S is a dominating set if

... for every vertex $u \in V - S$ there exists a vertex $v \in S$ such that u is adjacent to v.

... for every vertex $u \in V - S$, $d(u, S) = 1$.

... $N[S] = V$.

... the function $f : V \rightarrow \{0, 1\}$ with $S = \{v : f(v) = 1\}$ satisfies the condition that, for every $v \in V$, $f(N[v]) \geq 1$.

... the function $f : V \rightarrow \{0, 1\}$ with $S = \{v : f(v) = 1\}$ satisfies the condition that, for every $v \in V$, there exists a vertex $u \in N(v)$ such that $f(u) + f(v) \geq 1$.

The domination number $\gamma(G)$ of a graph G equals the minimum cardinality of a dominating set in G.

[95, 924]

The upper domination number $\Gamma(G)$ of a graph G equals the maximum cardinality of a minimal dominating set in G.

For an entire volume of topics on domination see [675].

Category A: For many applications it is not possible to use an arbitrary dominating set S. One possible form of restriction is based on considering $\langle S \rangle$ and/or $\langle V - S \rangle$.

Independent domination

A dominating set S is an independent dominating set if no two vertices in S are adjacent, that is, S is an independent set.

The independent domination number $i(G)$ of a graph G is the minimum cardinality of an independent dominating set. Thus, $i(G) = \min\{|S| : S$ dominates and $\Delta(\langle S \rangle) = 0\}$

[285, 433, 638]

It can also be shown that $i(G)$ equals the minimum cardinality of a maximal independent set of vertices in G. The upper independent domination number $\beta_0(G)$ equals the maximum cardinality of an independent set in G and is usually called the vertex independence number of G. The number $i(G)$ was also called the strong independence number by McFall and Nowakowski.

[114, 26, 126, 884, 433]

Total/open domination

A set S is a total dominating set, also called an open dominating set, if

... for every vertex $u \in V$ there exists a vertex $v \in S$, such that u is adjacent to v.

... $N(S) = V$.

... the function $f : V \rightarrow \{0,1\}$ with $S = \{v : f(v) = 1\}$ satisfies the condition that, for every $v \in V$, $f(N(v)) \geq 1$.

... S is a dominating set and $\delta(\langle S \rangle) \geq 1$.

The total (open) domination number of a graph G is $\gamma_t(G)$, and the upper total domination number is $\Gamma_t(G)$.

[256, 28, 766, 109]

Connected domination

A dominating set S is a connected dominating set if $\langle S \rangle$ is a connected subgraph of G. The connected domination number of a graph G is $\gamma_c(G)$.

[1131, 1004, 669, 533, 319]

Paired-domination

A dominating set is a paired-dominating set S if $\langle S \rangle$ has a perfect matching. Note that every graph without isolated vertices has a paired-dominating set. The paired-domination number of a graph G is denoted $\gamma_{pr}(G)$.

[652, 653, 603]

Weakly connected domination

The subgraph weakly induced by a set S of vertices is the graph $\langle S \rangle_w$ whose vertex set is $N[S]$ and whose edge set consists of those edges in $E(G)$ with at least one vertex, and possibly both, in S. A dominating set S is called a weakly connected dominating set if $\langle S \rangle_w$ is connected.

The weakly connected domination number is denoted $\gamma_w(G)$; while the upper weakly connected domination number is denoted $\Gamma_w(G)$.

Note that one can also define weakly connected independent sets, the weakly connected independent domination number $i_w(G)$ and the weakly connected independence number $\beta_w(G)$.

[393]

Set domination

Let $G = (V, E)$ be a connected graph. A set $S \subseteq V$ is a set dominating set if for every set $T \subseteq V - S$, there exists a nonempty set $R \subseteq S$ such that the subgraph $\langle R \cup T \rangle$ is connected. The set domination number is $\gamma_s(G)$.

[1001]

Point-set domination

Let $G = (V, E)$ be a connected graph. A set $S \subseteq V$ is a point-set dominating set if for every set $T \subseteq V - S$, there exists a vertex $v \in S$ such that
the subgraph $\langle T \cup \{v\} \rangle$ is connected. The point-set domination number is
denoted $\gamma_{ps}(G)$.

[999]

Order-k domination

Given a positive integer k, find the maximum number of vertices that can
be dominated by a set S of order k.

This problem is solved by an $O(n^2 k)$ algorithm for trees by Hsu.

[717]

$(b(v), c(v))$-domination

Given a graph G and nonnegative integers $(b(v), c(v))$ for every vertex
$v \in V$, we seek a subgraph H with a minimum number of components,
subject to the constraints that (i) each vertex v not in $V(H)$ is adjacent
to at least $b(v)$ vertices in H, and (ii) each vertex v in $V(H)$ has degree
at most $c(v)$.

Special cases of this problem are:

$(b(v) = 1, c(v) = 0)$: ordinary domination

$(b(v) = 1, c(v) = 1)$: K_1, K_2-domination

$(b(v) = \text{variable}, c(v) = 0)$: b-domination

$(b(v) = 1, c(v) = 2)$: path domination (in trees)

By choosing $b(v) = 0$ or $b(v) > deg(v)$, we can make v either a "free"
or "required" vertex, respectively, as necessary for optimal domination.
Lawler and Slater present a linear algorithm for $(b(v), c(v))$-domination in
trees.

[837]

Star partition number

The star partition number $\gamma^*(G)$ is the minimum order of a partition of
$V = \{V_1, V_2, \cdots, V_k\}$ such that for every i, $1 \leq i \leq k$, $\langle V_i \rangle$ is a star, $K_{1,t}$.
Clearly, if G has a star partition number of k then $\gamma(G) \leq k$, that is,
$\gamma(G) \leq \gamma^*(G)$.

[1130]

Restrained domination

A dominating set S is restrained if every vertex u in $V - S$ is adjacent to at least one other vertex in $V - S$.

The restrained domination number is denoted $\gamma_r(G)$.

[366]

Note that $n - \gamma_r(G)$ is the maximum cardinality of an enclaveless, isolate-free vertex set. This parameter is described in [1045].

2-connected

The minimum cardinality of a dominating set S such that $\langle S \rangle$ is 2-connected (respectively, 2-edge connected) is the 2-connected (respectively, 2-edge connected) domination number.

[753, 757]

Category B: One can consider restrictions based on what we allow $N(v) \cap S$ to be for each v outside the dominating set S, $v \in V - S$.

Weak and strong domination

A dominating set S is a strong dominating set if for every vertex $u \in V - S$ there is a vertex $v \in S$ with $deg(v) \geq deg(u)$ and u is adjacent to v.

A dominating set S is a weak dominating set if for every vertex $u \in V - S$ there is a vertex $v \in S$ with $deg(v) \leq deg(u)$ and u is adjacent to v.

The strong domination number is denoted $\gamma_S(G)$, and the weak domination number is denoted $\gamma_W(G)$.

[1003, 630]

---

k-domination

A set S is a k-dominating set if for every vertex $u \in V - S$, $|N(u) \cap S| \geq k$.

The k-domination number is denoted $\gamma_k(G)$.

[473, 474, 740, 264, 444]

Locating-domination

A dominating set S is called a locating-dominating set if for any two vertices $v, w \in V - S$, $N(v) \cap S \neq N(w) \cap S$. Thus, with a locating dominating set every vertex in $V - S$ is dominated by a distinct subset of the vertices of S.

The locating-domination number is denoted $\gamma_L(G)$.

[1035, 960, 469]

| Perfect domination |

A dominating set S is called a perfect dominating set if for every vertex $u \in V - S$, $|N(u) \cap S| = 1$.

Note that every graph G has at least the trivial perfect dominating set consisting of all vertices in V.

The perfect domination number is denoted $\gamma_p(G)$.

[P. M. Weichsel, Large subgraphs of hypercubes: Steiner systems and codes, manuscript]

[270, 466, 852, 1161]

| K_k-domination |

We say that two vertices u and v are K_k-adjacent if they are contained in a complete subgraph isomorphic to K_k in G. A set S of vertices is a K_k-dominating set if every vertex u in $V - S$ is K_k-adjacent to at least one vertex in S. A set S is a total K_k-dominating set if every vertex $u \in V$ is K_k-adjacent to at least one vertex in S.

The K_k-domination number is denoted $\gamma_{K_k}(G)$, and the total K_k-domination number is denoted $\gamma_{K_k}^t(G)$. Clearly, $\gamma_{K_2}(G) = \gamma(G)$ and $\gamma_{K_2}^t(G) = \gamma_t(G)$.

[681, 705]

Category C: Restrictions can be based on how much dominating set S dominates each $v \in V(G)$

| Global domination and factor domination |

A dominating set S is a global dominating set if S is also a dominating set in the complement of G.

The global domination number is denoted $\gamma_g(G)$, and the upper global domination number is denoted $\Gamma_g(G)$.

Sampathkumar has shown that for any tree T, $\gamma(T) \leq \gamma_g(T) \leq \gamma(T) + 1$.

[989]

Let $G = (V, E)$ be a graph having spanning subgraphs $G_i = (V, E_i)$, $1 \leq i \leq k$, where $E = \{E_1, E_2, \cdots, E_k\}$ is a partition of E. The subgraphs

G_i are called factors of G. A set of vertices S is called a factor dominating set if S is a dominating set of every subgraph G_i. The factor domination number is denoted $\gamma_{ft}(G)$. In the special case that there are two factors, a graph G and its complement, both of these definitions are the same.

[164]

The idea of global domination has been extended to global irredundance and global total domination.

A set S of vertices is called a global irredundant set if for every vertex $v \in S$, either $pn[v, S] \neq \emptyset$ in G or $pn[v, S] \neq \emptyset$ in the complement of G.

The global irredundance numbers are denoted $ir_g(G)$ and $IR_g(G)$.

Rall has shown that for trees (with two exceptions), and for graphs having diameter at least five, $\Gamma(G) = \Gamma_g(G)$.

Similarly, Dunbar and Laskar show that for graphs with diameter at least five, $\gamma_{gt}(G) = \gamma_t(G)$ and $\Gamma_{gt}(G) = \Gamma_t(G)$ for the global total domination and upper global total domination numbers.

They also show that for bipartite graphs, $IR(G) = IR_g(G)$ for upper global irredundance.

[407, 406, 959]

Redundance

Each vertex v dominates itself and its *deg* v neighbors. The influence of a vertex set S, which measures the total amount of domination done by S, is $I(S) = \sum_{v \in S}(1 + deg\ v)$. The redundance of a graph G is the minimum total amount of domination given that every vertex gets dominated at least once, $R(G) = \min\{I(S) : S$ is a dominating set $\}$.

[555, 558, 750]

The cardinality redundence $CR(G)$ is the minimum number of vertices dominated more than once by a dominating set S.

[749, 750]

---

k-tuple domination

A set S is an k-tuple dominating set if each vertex $v \in V$ is dominated by at least k vertices in S.

The difference between k-domination (defined above) and k-tuple domination is that with k-domination one must only dominate the vertices in $V - S$ k times, while with k-tuple domination every vertex $v \in V$ must be multiply dominated.

The minimum cardinality of a k-tuple dominating set is called the k-tuple domination number $\gamma_{\times k}(G)$.

When $k = 2$, the 2-tuple domination number is called the double domination number $dd(G)$.

[581]

$[1, k]$-domination

The $[1, k]$-domination number $\gamma_{[1,k]}(G)$ of a graph G is the minimum cardinality of a dominating set S such that for every vertex $u \in S$, $|N(u) \cap (V - S)| \le k$.

[988]

Capacity domination

Let $\pi = \{V_1, V_2, \cdots, V_t\}$ be a spanning star partition of the vertices of a graph G, that is, for every i, $1 \le i \le t$, $\langle V_i \rangle$ contains a spanning star with central vertex v_i. Suppose furthermore that $|V_i| \le k$. Then we say that π is a capacity-k spanning star partition. Let $S = \{v_1, v_2, \cdots, v_t\}$ be the central vertices of the stars in π. We say that S is a capacity-k dominating set of G, and $\gamma_{\text{cap-k}}(G)$ denotes the minimum order of such a set.

[Carson and Oellerman, manuscript 1995]

Private domination

Let S be a set of vertices in a graph $G = (V, E)$ and let $u \in S$. Define $pn[u, S] = N[u] - N[S - \{u\}]$. If $pn[u, S] \ne \emptyset$, then every vertex in $pn[u, S]$ is called a private neighbor of u (with respect to S), and we say that vertex u has a private neighbor.

A dominating set is called a private dominating set if every vertex $u \in S$ has a private neighbor. The private domination number $\Gamma_{pvt}(G)$ of a graph G equals the maximum cardinality of a private dominating set in G. It can be observed that the minimum cardinality of a private dominating set in G always equals $\gamma(G)$.

[657]

This concept has been generalized to private distance domination. A vertex u in a set S has a private n-neighbor if there exists a vertex $w \in V - S$, $u \ne w$, such that $d(u, w) \le n$ while $d(v, w) > n$ for all $v \in S - \{u\}$. A private distance-n dominating set is a distance-n dominating set S in which every vertex has a private n-neighbor.

[619]

K-minimal domination

For a positive integer k, a dominating set S is called k-minimal if the removal of a subset S' from S of cardinality $t \leq k$ followed by the addition of any subset S'' of cardinality $t - 1$ always results in a set which is not dominating.

The k-minimal domination number $\Gamma_k(G)$ is the minimum cardinality of a k-minimal dominating set.

Note that the following inequality chain holds:

$$\Gamma(G) = \Gamma_1(G) \geq \Gamma_2(G) \geq \cdots \geq \Gamma_k(G) \geq \cdots \geq \gamma(G).$$

[130, 296]

Parity-restricted domination

A very nice result first proved by Sutner is that for every graph G there is a vertex set S such that $|N[v] \cap S|$ is odd for every vertex $v \in V$. The minimum cardinality of such an "all-odd" set S is $\gamma_{odd}(G)$.

Let $f : V \to \{odd, even\}$ be a parity assignment for the vertices of G. Set S is an f-parity dominating set if, for each $v \in V$, $|N[v] \cap S|$ is odd or even corresponding to $f(v)$. The minimum cardinality of such a set S, when at least one such set exists, is $\gamma_{f-par}(G)$.

A variant that has not yet appeared in the literature is to set requirements on the parity of $|N(u) \cap S|$ just for each vertex $u \in V - S$.

If T is the linear transformation of the set of binary n-tuples $\mathcal{Z}_2^n$ with $T : \mathcal{Z}_2^n \to \mathcal{Z}_2^n$ defined by $T(X) = N \cdot X \pmod{2}$, then the null space $N(T)$ is the set of all-even parity sets. The parity dimension $PD(G) = k$ where $|N(T)| = 2^k$. We note that the number of all-odd sets S is also 2^k.

[1073, 33, 34, 32]

Set-restricted domination

With $V = \{v_1, v_2, \cdots, v_n\}$ to each v_i assign a set S_i of nonnegative integers, and we let $\mathcal{S} = \{S_1, S_2, \cdots, S_n\}$. If there exists a set $S \subseteq V$ such that $|N[v_i] \cap S| \in S_i$ for $1 \leq i \leq n$, then we let $\gamma_{\mathcal{S}}(G)$ denote the minimum cardinality of such a set.

If every $S_i = \{1, 2, \cdots, n\}$, then $\gamma_{\mathcal{S}}(G) = \gamma(G)$.

For k-tuple domination, every $S_i = \{k, k+1, \cdots, n\}$.

For Sutner's odd-parity problem, every $S_i = \{1, 3, 5, \cdots\}$.

For parity-restricted domination, each S_i is either $\{1, 3, 5, \cdots\}$ or $\{2, 4, 6, \cdots\}$.

If every $S_i = \{1\}$, then one is seeking an efficient-dominating set.

[33, 34]

Category D: Restrictions might be strong enough that appropriate dominating sets might not be obtainable for certain nontrivial, connected graphs.

Clique domination

A dominating set S is a clique dominating set if $\langle S \rangle$ is a complete graph; S is called a dominating clique.

[771, 333]

A chordal graph has a dominating clique if and only if it has diameter at most three. If a strongly chordal graph G has a dominating clique, then its size equals the domination number of G.

[800, 797]

Cycle domination

A dominating set S is a cycle dominating set if $\langle S \rangle$ has a Hamiltonian cycle; S is also called a dominating cycle.

[949, 64, 241, 318]

Path domination

A dominating set S is a path dominating set if $\langle S \rangle$ has a hamiltonian path; S is called a dominating path.

Note: another definition of path domination has also been given. Let $G = (V, E)$ be a graph, let P be a set of vertex-disjoint paths in G, and let $V(P)$ denote the set of vertices which lie on a path in P. We say that P is a path dominating set if for every vertex $u \in V - V(P)$ there is a vertex v on at least one path in P such that u is adjacent to v.

[489, 659, 1111]

Efficient domination

A dominating set S is called an efficient dominating set if for every vertex $u \in V$, $|N[u] \cap S| = 1$. Equivalently, a dominating set is efficient if the distance between any two vertices in S is at least three, that is, S is a packing.

It is worth noting that if a graph has an efficient dominating set, then all efficient dominating sets in G have the same cardinality, namely $\gamma(G)$.

[115, 77, 852, 466, 74]

Efficient open domination

> A set S of vertices is called a perfect total dominating set if for every vertex $v \in V$, $|N(v) \cap S| = 1$. Equivalently, a set S is called an efficient open dominating set if the open neighborhoods $N(v)$ for $v \in S$ forms a partition of V.

> [270, 526, 527, 660]

Bipartite domination

> Given a bipartite graph $G = (U, W, E)$, we seek a minimum cardinality subset $S \subseteq U$ which dominates all vertices in W. Such a set is called a W-dominating set.

> A subset $S \subseteq U$ is called an U-dominating set if for every vertex u in $U - S$ there exists a vertex $v \in S$ and a vertex $w \in W$ such that u and v are both adjacent to w.

> [670, 671]

Category E: The concept of a dominating set can be viewed as assigning a weight of one to the vertices in a set S and a weight of zero to the vertices in $V - S$. Then S is a dominating set if $|N[v] \cap S| \geq 1$ for each $v \in V$. This can be generalized by allowing weights other than just those in $\{0, 1\}$. If function f assigns the value $f(v)$ to each $v \in V$, the weight of f is $w(f) = f(V) = \sum_{v \in V} f(v)$.

$\{k\}$-dominating functions

> A function $g : V \to \{0, 1, 2, \cdots, k\}$ is called a $\{k\}$-dominating function if for every vertex $v \in V$, $g(N[v]) \geq k$. The $\{k\}$-domination number of G is given by $\gamma_{\{k\}}(G) = \min\{w(g) : g$ is a $\{k\}$-dominating function of $G\}$.

> [E. O. Hare, private communication, 1988]

> [360, 371]

Fractional domination

> A function $f : V \to [0, 1]$ is called a fractional dominating function if for every vertex $u \in V$, $f(N[u]) \geq 1$. The fractional domination number $\gamma_f(G)$ of a graph G equals the minimum weight of a fractional dominating function f on G. The upper fractional domination number $\Gamma_f(G)$ of a graph G equals the maximum weight of a minimal fractional dominating function on G.

> [232, 233, 371, 555, 664]

| Minus domination |

A function $f : V \to \{-1, 0, 1\}$ is called a minus dominating function if for every vertex $v \in V$, $f(N[v]) \geq 1$. The minus domination number $\gamma^-(G)$ equals the minimum weight of a minus dominating function f on G.

[392, 402, 403]

| Signed domination |

A function $f : V \to \{-1, 1\}$ is called a signed dominating function if for every vertex $v \in V$, $f(N[v]) \geq 1$. The signed domination number $\gamma_s(G)$ equals the minimum weight of a signed dominating function f on G.

[404, 627]

| Y-valued domination and efficient Y-domination |

Let Y be a subset of the reals. A function $f : V \to Y$ is a Y-dominating function if for every vertex $v \in V$, $f(N[v]) \geq 1$. If $f(N[v]) = 1$ for every vertex v then f is called an efficient Y-dominating function. In [74] it is shown that if the closed neighborhood matrix of a graph G is invertible, then G has an efficient Y-dominating function for some Y. It is also shown that G has an efficient Y-dominating function if and only if all equivalent Y-dominating functions have the same weight, where two Y-dominating functions f and f' are equivalent if and only if $f(N[v]) = f'(N[v])$ for every vertex $v \in V$.

The Y-domination number $\gamma_Y(G)$ is the minimum weight of a Y-dominating function. One can likewise define general Y-valued parameters, for example, involving independence, enclaveless, and covering concepts.

[74, 1042]

| Majority domination |

A function $f : V \to \{-1, 0, 1\}$ is called a majority dominating function if for at least half of the vertices $v \in V$, $f(N[v]) \geq 1$. The majority domination number $\gamma_{maj}(G)$ of a graph G equals the minimum weight $w(f)$ of a majority dominating function f on G.

[168, 617]

| k-sub domination |

For a positive integer k, a function $f : V \to \{-1, 0, 1\}$ is called a k-sub dominating function if for at least $1/k$ of the vertices, $f(N[v]) \geq 1$. The k-sub domination number $\gamma_{ksub}(G)$ equals the minimum weight $w(f)$ of a k-sub dominating function f on G.

[E. J. Cockayne and C. M. Mynhardt, private communication]

[167]

Resource allocation

Associate with every vertex $v \in V$ a positive integer requirement $r(v)$. We seek a function $f : V \to W = \{0, 1, 2, \cdots\}$ which has the property that for every vertex $v \in V$, $f(N[v]) \geq r(v)$ and the weight $w(f)$ is a minimum.

[664]

Closed neighborhood order domination

A function $f : V \to N = \{0, 1, 2, ...\}$ is a closed neighborhood order dominating function if for every vertex $v \in V$, $f(N[v]) \geq 1 + deg(v)$.

The closed neighborhood order domination number $W(G)$ is the minimum weight of a closed neighborhood order dominating function.

[701, 1037]

Category F: One can say that S is a dominating set if every vertex v is within a distance of one of S, $d(v, S) \leq 1$. We can consider domination from distances greater than one.

Distance-k domination

A set S is a distance-k dominating set if for every vertex $u \in V - S$, $d(u, S) \leq k$.

The distance-k domination number is denoted $\gamma_{\leq k}(G)$.

[200, 686, 719]

K-basis

For a positive integer k, a k-basis is a set of vertices S such that for every vertex $u \in V - S$, $d(u, S) \leq k$ and for every pair of vertices $v, w \in S$, $d(v, w) > k$.

[576]

$\Re$-domination

Given a graph G with vertex set $V = \{v_1, v_2, \cdots, v_n\}$ with a nonnegative integer r_i associated with v_i for $1 \leq i \leq n$, let $\Re = (r_1, r_2, \cdots, r_n)$. A vertex set S is an $\Re$-dominating set if for each v_i there is a vertex $u_i \in S$ such that the distance $d(v_i, u_i) \leq r_i$. The minimum cardinality of an $\Re$-dominating set is the $\Re$-domination number $\gamma_{\Re}(G)$.

[1031, 199]

If every $r_i = 1$, $\gamma_{\Re}(G) = \gamma(G)$.

If every $r_i = k$, $\gamma_{\Re}(G) = \gamma_{\leq k}(G)$ (distance-k domination).

If every $r_i \in \{\infty, 0, 1\}$, then we have optional domination, where ∞ indicates the vertex need not be dominated within any finite distance.

---

k-step and exact k-step domination

Two vertices u and v in a graph G for which the distance $d(u, v) = k$ are said to k-step dominate each other. The set $N_k(v)$ denotes the set of vertices that are k-step dominated by v. A set S of vertices is called a k-step dominating set if $\cup_{v \in V} N_k(v) = V$.

A k-step dominating set S such that the sets $N_k(v)$ are pairwise disjoint is called an exact k-step dominating set. If a graph G has an exact k-step dominating set, then G is called an exact k-step dominating graph.

[217]

Planetary domination

A sequence $s : l_1, l_2, \cdots, l_k$ $(k \leq n)$ of positive integers is called a planetary domination sequence for G if there are k distinct vertices $v_1, v_2, \cdots, v_k \in V$ such that every vertex of V is l_i-step dominated by v_i for some i $(1 \leq i \leq k)$. A planetary dominating sequence S for G is minimal if no proper subsequence of S is a planetary dominating sequence for G.

The planetary domination number $pl(G)$ is the length of a minimum planetary dominating sequence.

[31]

Optional domination

For a graph $G = (V, E)$ we partition the vertex set into three sets, $V = \{F, B, R\}$, where F is a set of free vertices, B is a set of bound vertices and R is a set of required vertices. We seek the minimum cardinality $\gamma_{opt}(G)$ of a set of vertices S which contains all vertices in R and dominates all vertices in B.

[267]

---

k-neighbor, r-domination

The (k, r)-domination problem is the problem of selecting a minimum cardinality vertex set S such that every vertex $v \in V - S$ is at distance $\leq r$ from at least k vertices in S.

If we impose the condition that the graph $\langle S \rangle$ is connected, then S is a connected (k, r)-dominating set.

If for each vertex in S there exists another vertex in S at a distance $\leq r$, then S is a total (k, r)-dominating set.

If for each vertex in S there exists another vertex in S adjacent to it, then S is called a reliable (k, r)-dominating set.

[758]

| k-neighborhood minus domination |

Function $f : V \rightarrow \{-1, 0, 1\}$ is a k-neighborhood minus dominating function if the sum of the function values over any closed k-neighborhood is at least one, $f(N_k[v]) \geq 1$ for every $v \in V$. The minimum weight of such a function is the k-neighborhood minus domination number $\gamma_k^-(G)$ of G.

[626]

Category G: A vertex v and an edge uv incident with it are said to cover each other, and two edges with a vertex in common dominate each other. Here we consider coverage and edge domination parameters.

| Mixed domination or total covers |

For simplicity, we say that a vertex dominates every edge to which it is incident as well as every vertex to which it is adjacent. Also, an edge uv dominates every edge to which it is adjacent, that is, has a vertex in common, and the two vertices u and v on the edge.

A set $S \subseteq V \cup E$ of vertices and/or edges is a mixed dominating set if it dominates every vertex and every edge of G.

The mixed domination number is denoted $\gamma_m(G)$.

[23, 570, 660, 24, 425]

| Vertex-edge, edge-vertex and mixed domination |

Define the vertex neighborhood $N[e]$ of an edge $e = uv$ to be $N[e] = N(u) \cup N(v)$. We say that a vertex v and an edge e weakly dominate each other if $v \in N[e]$. A vertex v and an edge e strongly dominate each other if $e \in \langle N[v] \rangle$. The weak vertex-edge domination number $\gamma_{01}(G)$ of a graph G is the minimum cardinality of a set of vertices weakly dominating all the edges of G. The weak edge-vertex domination number $\gamma_{10}(G)$ is the minimum cardinality of a set of edges weakly dominating all the vertices of G. The strong vertex-edge domination number $s\gamma_{01}(G)$ of a graph G

is the minimum cardinality of a set of vertices which strongly dominates every edge in G. Note that this number equals the neighborhood number $n_0(G)$, defined by Neeralagi and Sampathkumar.

The strong edge-vertex domination number $s\gamma_{10}(G)$ is the minimum cardinality of a set of edges which strongly dominates every vertex in G.

[829, 938, 994]

Edge domination

A set F of edges is an edge dominating set if for every edge $e \in E - F$ there exists an edge $f \in F$ such that e and f have a vertex in common.

The edge domination number is denoted $\gamma'(G)$ and the upper edge domination number is denoted $\Gamma'(G)$.

One can also define the independent edge domination number $i'(G)$ to equal the minimum cardinality of a maximal independent set of edges in G. It can be observed that for all graphs G, $\gamma'(G) = i'(G)$.

The maximum cardinality of an independent set of edges is the well known matching number, denoted $\beta_1(G)$.

One can also define the edge irredundance numbers, $ir'(G)$ and $IR'(G)$.

[716, 727, 900, 1156]

Efficient edge domination

An edge uv is said to dominate any edge ux or vx where $x \in V$, including the edge uv itself. An edge subset $E' \subseteq E$ is an efficient edge dominating set for G if each edge in E is dominated by exactly one edge in E'.

Efficient edge dominating sets correspond to efficient dominating sets in the line graph $L(G)$. Thus, in particular, if G has an efficient dominating set, then its cardinality is $\gamma'(G) = \gamma(L(G))$.

[549, 559]

Redundant coverage

A cover is a vertex set S such that, for each edge $uv \in E$, at least one of u and v is in S. That is, each edge must be covered by at least one vertex in S. Note that vertex v covers $deg\ v$ edges. Hence, the redundance-coverage number $R^{01}(G)$, which equals the minimum amount of coverage done by a vertex set that covers every edge at least once, satisfies $R^{01}(G) = \min\{\sum_{u \in S} deg\ u : S \text{ is a cover }\}$.

[1045, 396]

Domination-coverage

If we minimize the amount of coverage done by a dominating set rather than a cover, we have the domination-coverage number $\eta(G) = \min\{\sum_{u \in S} deg(u) : S$ is a dominating set$\}$.

[1045, 1048]

Category H: Domination parameters might involve more than one vertex set S. Several of these multiset parameters follow.

Domatic numbers

Mimicking the definition of the chromatic number of a graph G as the minimum cardinality of a partition of $V(G)$ into independent sets, Cockayne and Hedetniemi in 1977 [280] defined the domatic number $d(G)$ to be the maximum cardinality of a partition of $V(G)$ into dominating sets. In 1980, Cockayne, Dawes, and Hedetniemi [256] defined the total domatic number $d_t(G)$ to be the maximum cardinality of a partition of $V(G)$ into total dominating sets. The connected domatic number $d_c(G)$, introduced in 1984 by Hedetniemi and Laskar [669], is the maximum cardinality of a partition of $V(G)$ into connected dominating sets, where we require that G be a connected graph.

Notice, however, that the vertices of the cycle C_5 cannot be partitioned into independent dominating sets. But if there exists at least one partition of the vertices of a graph G into independent dominating sets, we say that G is idomatic and the idomatic number $id(G)$ equals the maximum cardinality of such a partition. This was defined by Cockayne and Hedetniemi [280] and Zelinka [1175].

Note also that since every paired-dominating set has even order, no graph G of odd order has a partition of $V(G)$ into paired-dominating sets.

As discussed in Section 8.3, a change of definition can incorporate both independent dominating sets and paired-dominating sets into the collection of domatic numbers as follows. We define: the domatic number $d(G)$, total domatic number $d_t(G)$, connected domatic number $d_c(G)$, independent domatic number $d_i(G)$, and paired-domatic number $d_{pr}(G)$, to equal the maximum number of pairwise disjoint dominating sets, total dominating sets, connected dominating sets, independent dominating sets, and paired-dominating sets in G, respectively.

The paired-domatic number was first defined and studied by Haynes and Slater in [652].

Among several other domatic numbers that have been studied we define only the following. The edge domatic number $d'(G)$ is the maximum number of disjoint edge dominating sets in a partition of $E(G)$. This was first defined and studied by Zelinka in 1983 [1179], who used the notation $ed(G)$.

The adomatic number defined by Cockayne and Hedetniemi [280] and studied by Zelinka in [1175] equals the minimum number of pairwise disjoint sets $S_1, S_2, ..., S_t$ that partition $V(G)$, where no S_i can be partitioned into two dominating sets.

Other types of domatic numbers, all due to Zelinka, which we will not define here include:

> (distance) k-domatic, [1180],
>
> k-ply domatic, [1181],
>
> semidomatic, [1182],
>
> antidomatic, [1165],
>
> complementarily domatic, [1189],
>
> total adomatic, [1201],
>
> edge-adomatic, [1202], and
>
> total edge-domatic, [1199].

Finally, we introduce here a new definition of the "adomatic number" which can be extended to other types of domination. The adomatic number $ad(G)$ is the minimum number of pairwise disjoint sets $S_1, S_2, \cdots, S_t$ such that

> (i) each S_i is a dominating set and,
>
> (ii) $(V - \cup_{j \neq i} S_j) \cup S_i$ does not contain two disjoint dominating sets.

| Iterated domination |

Consider a greedy algorithm which finds a minimal dominating set, say S_1. Remove S_1 from G, and once again find a minimal dominating set S_2 in the graph $G - S_1$. Remove S_2 and once again find a minimal dominating set in the graph $G - S_1 - S_2$. Repeat this process until no vertices remain. The minimum possible number of iterations/removals in this algorithm is called the iterated domination number $\gamma^*(G)$ of a graph G. The maximum number of iterations possible is called the upper iterated domination number $\Gamma^*(G)$.

One can also iteratively remove maximal independent sets of vertices, and maximal irredundant sets of vertices. This gives rise to the iterated independence numbers, $i^*(G)$ and $\beta^*(G)$, and the iterated irredundance numbers $ir^*(G)$ and $IR^*(G)$.

[662]

Least domination

A dominating set S is called a least dominating set if $\gamma(\langle S \rangle) \leq \gamma(\langle S' \rangle)$ for every other dominating set S' of G.

The least domination number is denoted $\gamma_l(G)$.

[990]

Domination-forcing number

For any $S \subseteq V$, an S-dominating set in G is a set $D \subseteq V$ such that each vertex in S is dominated by a vertex in D, that is, $S \subseteq N[D]$. The order of a smallest S-dominating set in G is denoted by $\gamma(S, G)$, and clearly $\gamma(S, G) \leq \gamma(G)$. A domination-forcing set S of G is a set with $\gamma(S, G) = \gamma(G)$, and the cardinality of a smallest domination-forcing set of G is the domination-forcing number $\gamma^\#(G)$.

[1046]

Minimum intersection of γ-sets

The minimum cardinality of the intersection of two minimum dominating sets in G is denoted by $M_\gamma(G)$. If G has a unique minimum dominating set D, then $M_\gamma(G) = \gamma(G) = |D|$.

[75, 76, 77, 556, 557]

Category I: One can think of combining with domination a second property such as independence, matching, or locating, resulting in parameters i, γ_{pr}, and γ_L. For prioritized multiproperty problems with domination as the priority property, one insists on having a γ-set and measures how close to the second property we can come. Several prioritized multiproperty parameters have been identified.

Prioritized(independence, domination)

The first prioritized multiproperty parameter that was defined is $PR(m, \gamma)$. In this case one seeks a γ-set S such that $\langle S \rangle$ is as independent as possible. Clearly, "as independent as possible" could mean many different things for $\langle S \rangle$, such as, minimizing the maximum degree or maximizing the number of components. $PR(m, \gamma)(G)$ is defined to be the minimum possible number of edges in $\langle S \rangle$ for a γ-set S. In particular, $PR(m, \gamma)(G) = 0$ if and only if $\gamma(G) = i(G)$.

[1020]

Prioritized(matching, domination)

The maximum cardinality of a matching in a subgraph induced by a minimum dominating set is denoted $PR(\beta_1, \gamma)(G)$. That is, $PR(\beta_1, \gamma)(G) = \max\{\beta_1(\langle S \rangle) : S$ is a γ-set$\}$.

[1038]

Prioritized(redundance, independence)

With independence as the priority property, one can optimize a secondary domination parameter such as redundance. The influence or "total redundance" of a (dominating) set S is $I(S) = \sum_{s \in S}(1 + deg\ s)$. The minimum total redundance of a maximum independent set S is $PR(R, \beta_0)(G) = \min\{I(S) : S$ is a β_0-set$\}$. The cardinality redundance of a (dominating) set S is $CR(S) = |\{v \in V : |N[v] \cap S| \geq 2\}|$, the number of vertices dominated more than once by S. The minimum cardinality redundance of a maximum independent set is $PR(CR, \beta_0)(G)$.

[749, 750]

Category J: We present here some other domination-related subset problems.

Kernels [in directed graphs]

A set of vertices S is called absorbant if for every vertex $u \in V - S$ there is a vertex $v \in S$ and a directed arc from u to v.

A kernel is an absorbant set which is also an independent set.

[99, 101, 389, 513, 863, 1012]

Independence

A set S of vertices is said to be independent if no two vertices in S are adjacent. The vertex independence number $\beta_0(G)$ is the maximum cardinality of an independent set in G.

A set F of edges is said to be independent if no two edges in F have a vertex in common; an independent set of edges is also called a matching.

The matching number $\beta_1(G)$ is the maximum cardinality of a matching/independent set of edges in G.

It is known that the minimum cardinality of a maximal independent set of vertices is the independent domination number $i(G)$.

The minimum cardinality of a maximal independent set of edges is called the minimaximal matching number.

The independence numbers of a graph are related to domination as follows:

$$\gamma(G) \leq i(G) \leq \beta_0(G) \leq \Gamma(G).$$

The independence number can also be seen to equal the maximum weight $w(f)$ of a function $f : V \rightarrow \{0,1\}$ which satisfies the condition that for every vertex $v \in V$, $f(v) = 1 - \max\{f(u) \mid u \in N(v)\}$.

Similarly, the independent domination number $i(G)$ can be seen to equal the minimum weight $w(f)$ of a function $f : V \rightarrow \{0,1\}$ which satisfies the condition that for every vertex $v \in V$, $f(v) = 1 - \max\{f(u) \mid u \in N(v)\}$.

Vertex covers/ edge covers

A vertex v is said to cover any edge containing v. A vertex cover is a set S of vertices which cover every edge in E. The vertex covering number $\alpha_0(G)$ is the minimum cardinality of a vertex cover.

The vertex covering number is also equal to the minimum weight $w(f)$ of a function $f : V \rightarrow \{0,1\}$ which satisfies the condition that for every edge $uv \in E$, $f(u) + f(v) \geq 1$. A function $f : V \rightarrow \{0,1\}$ satisfying the condition that for every vertex $u \in V$, $f(u) = 1 - \min\{f(v) : v \in N(u)\}$ is called a stable function. It can be shown that $\alpha_0(G)$ equals the minimum weight of a stable function.

It is well known by Gallai's Theorem that for any graph G with n vertices,

$$\alpha_0(G) + \beta_0(G) = n.$$

An edge uv is said to cover vertices u and v. An edge cover is a set F of edges which covers every vertex in V. The edge covering number $\alpha_1(G)$ is the minimum cardinality of an edge cover.

Again, by Gallai's Theorem, it is known that for any graph G with n vertices,

$$\alpha_1(G) + \beta_1(G) = n.$$

One can also define the upper vertex covering number $\Lambda(G)$ to equal the maximum cardinality of a minimal vertex cover. McFall and Nowakowski showed that:

$$i(G) + \Lambda(G) = n.$$

[884]

Finally, one can also define the upper edge covering number, $\Lambda_1(G)$, to equal the maximum cardinality of a minimal edge cover. The subsequent Gallai-type result is:

$$i_1(G) + \Lambda_1(G) = n,$$

where $i_1(G)$ is the minimum cardinality of a maximal matching in G. [284, 1042, 1045]

Irredundance numbers

A set S of vertices is said to be irredundant if for every vertex $v \in S$, $pn[v, S] = N[v] - N[S - \{v\}] \neq \emptyset$, that is, every vertex $v \in S$ has a private neighbor. The irredundance number $ir(G)$ equals the minimum cardinality of a maximal irredundant set in G. The upper irredundance number $IR(G)$ equals the maximum cardinality of an irredundant set in G.

A set S is said to be open irredundant if for every vertex $v \in S$, $pn(v, S] = N(v) - N[S - \{v\}] \neq \emptyset$. The open irredundance numbers are $oir(G)$ and $OIR(G)$. A set S is said to be open-open irredundant if for every vertex $v \in S$, $pn(v, S) = N(v) - N(S - \{v\}) \neq \emptyset$. The open-open irredundance numbers are $ooir(G)$ and $OOIR(G)$. A set S is said to be closed-open irredundant if for every vertex $v \in S$, $pn[v, S) = N[v] - N(S - \{v\}) \neq \emptyset$. The closed-open irredundance numbers are $coir(G)$ and $COIR(G)$.

The irredundance numbers are related to the domination numbers as follows:

$$ir(G) \leq \gamma(G) \leq i(G) \leq \beta_0(G) \leq \Gamma(G) \leq IR(G)$$

[285, 437, 438, 465, 658, 676]

Nearly perfect sets and strong stable sets

A set S of vertices is nearly perfect if for every vertex $v \in V - S$, $|N(v) \cap S| \leq 1$.

A set S of vertices is a strong stable set if for every vertex $v \in V$, $|N(v) \cap S| \leq 1$.

[394, 709]

2-packing number, upper 2-packing number

A set S is a 2-packing if for every pair of vertices $u, v \in S$, $d(u, v) > 2$.

Equivalently, a set S is a 2-packing if for every vertex $v \in V$, $|N[v] \cap S| \leq 1$.

The 2-packing number $\rho(G)$ equals the maximum cardinality of a 2-packing in G, and the lower 2-packing number $p_2(G)$ equals the minimum cardinality of a maximal 2-packing in G.

The 2-packing number $\rho(G)$ can also be seen to equal the maximum weight $w(f)$ of a function $f : V \to \{0, 1\}$ which satisfies the condition that for every vertex $v \in V$, $f(N[v]) \leq 1$. Such an f is called a packing function.

The minimum weight $w(f)$ of a maximal packing function equals $p_2(G)$.

We also note that if we change the definition above to read: for every vertex $v \in V$, $|N(v) \cap S| \leq 1$, then we define the open 2-packing numbers of G, denoted $\rho^o(G)$ and $p_2^o(G)$.

[893]

Maximum spanning forests/enclaveless sets

The maximum number of pendant or end-edges in a spanning forest is denoted $\varepsilon_F(G)$ for a graph G. For any graph G with n vertices, it is easy to see that:

$$\gamma(G) + \varepsilon_F(G) = n.$$

A vertex set S is enclaveless if it does not contain any closed neighborhood $N[v]$, that is, for each $v \in V$ one has $N[v] \cap (V - S) \neq \emptyset$. Hence, S is enclaveless if and only if $V - S$ is dominating. The maximum cardinality of an enclaveless set is $\Psi(G) = n - \gamma(G) = \varepsilon_F(G)$.

[920, 1032]

External redundance, upper external redundance

A subset S of vertices is an external redundant set if for all vertices $v \in V - S$, there exists a vertex $w \in S \cup \{v\}$ such that $pn[w, S \cup \{v\}] = \emptyset$ and if $w \in S$ then $pn[w, S] \neq \emptyset$.

The minimum and maximum cardinalities of a minimal external redundant set are denoted $er(G)$ and $ER(G)$, respectively.

It is to be noted that for any graph G:

$$er(G) \leq ir(G) \leq \gamma(G) \leq i(G) \leq \beta_0(G) \leq \Gamma(G) \leq IR(G) \leq ER(G).$$

[271, 268]

1-dependence number

A set S of vertices is called 1-dependent if the maximum degree of every vertex in $\langle S \rangle$ is at most one. Thus, $\langle S \rangle$ consists of disjoint copies of K_1 and/or K_2.

The 1-dependence number $\beta^1(G)$ equals the maximum cardinality of a 1-dependent set in G.

[444, 446, 474]

| Matchability number |

A set S of vertices is matchable if there exists a 1-1 function $f : S \rightarrow V - S$ such that for every vertex $v \in S$, v is adjacent to $f(v)$. The matchability number $\mu(G)$ of a graph G is the minimum cardinality of a maximal matchable set of vertices in G.

It is interesting to note that the matchability number lies in between the independent edge domination number and the matching number:

$$ir'(G) \leq \gamma'(G) = i'(G) \leq \mu(G) \leq \beta_1(G) \leq \Gamma'(G) \leq IR'(G).$$

[284]

[S. M. Hedetniemi, S. T. Hedetniemi, A. McRae and D. Parks, Matchings, matchability and 2-maximal matchings in graphs, submitted]

| Neighborhood number |

A set S of vertices is called a neighborhood set provided G is the union of the subgraphs induced by the closed neighborhoods of the vertices in S. The neighborhood number $n(G)$ of a graph G equals the minimum cardinality of a neighborhood set, and the upper neighborhood number, $N(G)$, equals the maximum cardinality of a minimal neighborhood set.

[995, 745]

| Efficiency |

The efficiency of a set S of vertices equals the number of vertices in $V - S$ which are dominated exactly once by a vertex in S, that is, $\varepsilon(S) = |\{v : |N(v) \cap S| = 1$. The efficiency of a graph equals the maximum efficiency of any subset $S \subseteq V$.

[105, 117]

| Perfect and open perfect neighborhood sets |

Let S be a set of vertices in a graph G. A vertex v is called perfect with respect to S if $|N[v] \cap S| = 1$. A set S is called a perfect neighborhood set of G if every vertex of G is either perfect or is adjacent to a perfect vertex.

The perfect neighborhood number $\theta(G)$ equals the minimum cardinality of a perfect neighborhood set in G, and the upper perfect neighborhood number $\Theta(G)$ equals the maximum cardinality of a perfect neighborhood set.

It is worth noting that for all graphs G, $\Theta(G) = \Gamma(G)$.

The open perfect neighborhood is defined similarly, but uses the condition that a vertex v is open perfect if $|N(v) \cap S| = 1$.

[495, 668]

| Strong or induced matching number |

A set F of edges is a strong matching if the subgraph induced by F consists of disjoint copies of K_2, that is, F defines a matching in that no two edges in F have a vertex in common, but furthermore, there are no edges in $E - F$ connecting two edges in F.

[178]

| Efficient domination |

The efficient domination number $F(G)$ is the maximum number of vertices that can be dominated, given that no vertex is dominated more than once. So $F(G) = \max\{I(S) : S \text{ is a packing}\}$.

[77]

| Y-valued parameters |

For a subset Y of the reals, a function $f : V \to Y$ is a Y-valued function. As previously defined, such a function is a Y-dominating function if $f(N[v]) \geq 1$ for every vertex $v \in V$. Likewise, other parameters can be generalized from $\{0, 1\}$ to general Y-valued parameters. For example, $f : V \to Y$ is a Y-independent function if $f(u) + f(v) \leq 1$ for every edge $uv \in E$. The Y-independence number $\beta_Y(G)$ is the maximum weight $w(f)$ of a Y-independent function, and the lower Y-independence number $i_Y(G)$ is the minimum weight of a maximal Y-independent function.

Other pairs of parameters can similarly be defined. For example, $f : V \to Y$ is a Y-covering if $f(u) + f(v) \geq 1$ for every $uv \in E$. Function $f : V \to Y$ is Y-enclaveless if $f(N[v]) \leq |N[v]| - 1 = |N(v)|$ for every $v \in V$. In the obvious way, we have parameters α_Y, Λ_Y, Ψ_Y, and ψ_Y.

[1042, 1043]

| Closed neighborhood order packing |

A function $f : V \to N = \{0, 1, 2, ...\}$ is a closed neighborhood order packing function if for every vertex $v \in V$, $f(N[v]) \leq 1 + deg(v)$. The closed neighborhood order packing number $P(G)$ is the maximum weight of a closed neighborhood order packing function.

[1037, 1040]

Fractional irredundance

A function $f : V \to [0,1]$ is an irredundant function if for every $v \in V$ with $f(v) > 0$, there exists a vertex $u \in N[v]$ such that $f(N[u]) = 1$. An irredundant function f is maximal if there does not exist an irredundant function $g \neq f$, with $g(v) \geq f(v)$ for every $v \in V$.

The fractional irredundance number of a graph G equals $ir_f(G) = inf\{w(f) : f$ is a maximal irredundant function$\}$.

It remains an open question whether $ir_f(T)$ can be computed in polynomial time for any tree T.

[500]

Strong and weak connecting sets

A path P with terminal vertices u and v is a connecting path through $V' \subseteq V$ if $V(P) - V' = \{u, v\}$ and $V(P) \cap V' \neq \emptyset$. The subset $V' \subseteq V$ is a strong connecting set (SCS) if every pair of vertices of $V - V'$ has a connecting path through V'. It is a weak connecting set (WCS) if every pair of non-adjacent vertices of $V - V'$ has a connecting path through V'. The minimum cardinality of an SCS is denoted by $\gamma_s(G)$ and the minimum cardinality of a WCS is denoted by $\gamma_w(G)$.

It can be shown that $\gamma_w(G) \leq \gamma_s(G) \leq \gamma_c(G) \leq \gamma_w(G) + 1$, where $\gamma_c(G)$ is the connected domination number.

[917]

Distance irredundance

(cf. private domination and private distance domination)

A set S is called distance-k irredundant if every vertex in S has a private distance-k neighbor.

[625]

Reinforcement

The reinforcement number $r(G)$ of a graph G equals the minimum number of edges which have to be added to G so that in the resulting graph G', $\gamma(G') < \gamma(G)$.

[782]

Bondage number

The bondage number $b(G)$ of a graph G is the minimum cardinality among all sets of edges $F \subseteq E$ for which $\gamma(G - F) > \gamma(G)$. The bondage number, therefore, is the minimum number of edges in a set whose removal from G renders every minimum dominating set a nondominating set.

[476]

Efficient edge coverage

The efficient covering number $F^{01}(G)$ of G equals the maximum number of edges covered by a vertex set S with no edge covered more than once. Equivalently, $F^{01}(G)$ is the maximum number of edges covered by an independent vertex set.

[396]

Bibliography

[1] H. L. Abbott and A. C. Liu. Bounds for the covering number of a graph. *Discrete Math.*, 25:281–284, 1979.

[2] B. Abramson and M. Yung. Divide and conquer under global constraints: a solution to the N-queens problem. *J. Parallel Distrib. Comput.*, 6:649–662, 1989.

[3] B. D. Acharya. The strong domination number of a graph and related concepts. *J. Math. Phys. Sci.*, 14:471–475, 1980.

[4] B. D. Acharya. On a relation between neighborhood number and Dilworth number. *Proc. Nat. Acad. Sci. India Sect. A*, 57:600–603, 1987.

[5] B. D. Acharya. Interrelations among the notions of independence, domination and full sets in a hypergraph. *Nat. Acad. Sci. Lett.*, 13:421–422, 1990.

[6] B. D. Acharya. Full sets in hypergraphs. *Sankhyā*, 54:1–6, 1992.

[7] B. D. Acharya and P. Gupta. On point-set domination in graphs. In N. M. Bujurke, editor, *Recent Developements in Mathematics*, pages 106–108. Karnatak Univ. Press, Dharwad, 1994.

[8] B. D. Acharya and P. Gupta. On point-set domination in graphs, II: Minimum psd-sets. Submitted, 1996.

[9] B. D. Acharya and P. Gupta. On point-set domination in graphs, III: Minimal psd-sets. Submitted, 1996.

[10] B. D. Acharya and P. Gupta. On point-set domination in graphs, IV: Separable graphs with unique minimum psd-sets. Submitted, 1996.

[11] B. D. Acharya and P. Gupta. A direct inductive proof of a conjecture due to E. Sampathkumar and L. Pushpa Latha on the weak domination number of a tree. Submitted, 1997.

[12] B. D. Acharya and P. Gupta. Domination in graphoidal covers of a graph. Submitted, 1997.

[13] B. D. Acharya and P. Gupta. On point-set domination in graphs, V: Independent psd-sets. Submitted, 1997.

[14] B. D. Acharya and H. B. Walikar. On graphs having unique minimum dominating sets. *Graph Theory Newsletter*, 8:2, 1979.

[15] B. D. Acharya and H. B. Walikar. Indominable graphs cannot be characterized by a finite family of forbidden subgraphs. *Nat. Acad. Sci. Lett.*, 4(1):23–25, 1981.

[16] G. S. Adhar and S. Peng. Parallel algorithms for cographs and parity graphs with applications. *J. Algorithms*, 11:252–284, 1990.

[17] G. S. Adhar and S. Peng. Parallel algorithms for finding connected, independent and total domination in interval graphs. Elsevier, Amsterdam, 1992.

[18] G. S. Adhar and S. Peng. Mixed domination in trees: a parallel algorithm. *Congr. Numer.*, 100:73–80, 1994.

[19] W. Ahrens. *Mathematische Unterhaltungen und Spiele.* Leipzig, 1910.

[20] M. Aigner. Some theorems on coverings. *Studia Sci. Math. Hungar.*, 5:303–315, 1970.

[21] S. Ainley. *Mathematical Puzzles.* G. Bell and Sons, Ltd., UK, 1977.

[22] J. Akiyama, G. Exoo, and F. Harary. Covering and packing in graphs III:Cyclic and acyclic invariants. *Math. Slovaca*, 30:405–417, 1980.

[23] Y. Alavi, M. Behzad, L. M. Lesniak-Foster, and E. A. Nordhaus. Total matchings and total coverings of graphs. *J. Graph Theory*, 1:135–140, 1977.

[24] Y. Alavi, J. Liu, J. Wang, and Z. Zhang. On total covers of graphs. *Discrete Math.*, 100:229–233, 1992.

[25] M. O. Albertson, L. Chan, and R. Haas. Independence and graph homomorphisms. *J. Graph Theory*, 17:581–588, 1993.

[26] R. B. Allan and R. C. Laskar. On domination and independent domination numbers of a graph. *Discrete Math.*, 23:73–76, 1978.

[27] R. B. Allan and R. C. Laskar. On domination and some related concepts in graph theory. *Congr. Numer.*, 21:43–58, 1978.

[28] R. B. Allan, R. C. Laskar, and S. T. Hedetniemi. A note on total domination. *Discrete Math.*, 49:7–13, 1984.

[29] N. Alon, M. R. Fellows, and D. R. Hare. Vertex transversals that dominate. *J. Graph Theory*, 21:21–31, 1996.

[30] N. Alon and J. H. Spencer. *The Probablistic Method.* John Wiley and Sons, Inc., 1992.

[31] G. Chartrand amd M. A. Henning and K. Schultz. On planetary domination numbers of graphs. Manuscript, 1997.

[32] A. T. Amin, L. H. Clark, and P. J. Slater. Parity dimension for graphs. Submitted, 1996.

[33] A. T. Amin and P. J. Slater. Neighborhood domination with parity restrictions in graphs. *Congr. Numer.*, 91:19–31, 1992.

[34] A. T. Amin and P. J. Slater. All parity realizable trees. *J. Combin. Math. Combin. Comput.*, 20:53–63, 1996.

[35] M. Anciaux-Mundeleer and P. Hansen. On kernels in strongly connected graphs. *Networks*, 7:263–266, 1977.

[36] B. Andreas. Graphs with unique maximal clumpings. *J. Graph Theory*, 2:19–24, 1978.

[37] S. Ao. *Independent Domination Critical Graphs.* Master's thesis, Univ. Victoria, 1994.

[38] S. Ao, E. J. Cockayne, G. MacGillivray, and C. M. Mynhardt. Domination critical graphs with higher independent domination numbers. *J. Graph Theory*, 22:9–14, 1996.

[39] S. Ao and G. MacGillivray. Hamiltonian properties of independent domination critical graphs. Submitted, 1996.

[40] V. I. Arnautov. The exterior stability number of a graph. *Diskret. Analiz*, 20:3–8, 1972.

[41] V. I. Arnautov. Estimation of the exterior stability number of a graph by means of the minimal degree of the vertices. *(Russian) Prikl. Mat. i Programmirovanie Vyp.*, 11:3–8, 126, 1974.

[42] S. Arnborg. Efficient algorithms for combinatorial problems on graphs with bounded decomposability - a survey. *BIT*, 25:2–23, 1985.

[43] S. Arnborg, S. T. Hedetniemi, and A. Proskurowski, editors. *Efficient Algorithms and Partial k-trees, Special Issue Discrete Math.*, volume 54. 1994.

[44] S. Arnborg, J. Lagergren, and D. Seese. Easy problems for tree-decomposable graphs. *J. Algorithms*, 12:308–340, 1991.

[45] S. Arnborg and A. Proskurowski. Problems on graphs with bounded decomposability. *Congr. Numer.*, 53:167–170, 1986.

[46] S. Arnborg and A. Proskurowski. Linear time algorithms for NP-hard problems restricted to partial k-trees. *Discrete Appl. Math.*, 23:11–24, 1989.

[47] L. Arseneau, A. Finbow, B. Hartnell, A. Hynick, D. MacLean, and L. O'Sullivan. On minimal connected dominating sets. *J. Combin. Math. Combin. Comput.*, 24:185–191, 1997.

[48] S. Arumugam and J. Paulraj Joseph. Domination in subdivision graphs. *J. Indian Math. Soc.* To appear.

[49] S. Arumugam and J. Paulraj Joseph. Uniform domination in graphs. *Internat. J. Management Systems*, 11:111–116, 1995.

[50] S. Arumugam and J. Paulraj Joseph. Graphs with equal domination and connected domination numbers. Submitted, 1997.

[51] S. Arumugam and R. Kala. Domination parameters of hypercubes. *J. Indian Math.* To appear.

[52] S. Arumugam and R. Kala. Domination parameters of star graphs. *Ars Combin.*, 44:93–96, 1996.

[53] S. Arumugam and A. Thuraiswamy. Total domatic number of a graph I. Submitted, 1995.

[54] S. Arumugam and A. Thuraiswamy. Mixed domination in graphs. Submitted, 1996.

[55] S. Arumugam and A. Thuraiswamy. Total domination in graphs. *Ars Combin.*, 43:89–92, 1996.

[56] S. Arumugam and S. Velammal. Edge domination in graphs. *Chinese J. Math.* To appear.

[57] S. Arumugam and S. Velammal. Maximum size of a connected graph with given domination parameters. *Ars Combin.* To appear.

[58] S. Arumugam and S. Velammal. Connected edge domination in graphs. Submitted, 1995.

[59] K. Arvind, H. Breu, M. S. Chang, D. G. Kirkpatrick, F. Y. Lee, Y. D. Liang, K. Madhukar, C. Pandu Rangan, and A. Srinivasan. Efficient algorithms in cocomparability and trapezoid graphs. Submitted, 1996.

[60] K. Arvind and C. Pandu Rangan. Efficient algorithms for domination problems on cocomparability graphs. Technical Report TR-TCS-909-18, Indian Instit. Tech., 1990.

[61] K. Arvind and C. Pandu Rangan. Connected domination and Steiner set on weighted permutation graphs. *Inform. Process. Lett.*, 41:215–220, 1992.

[62] K. Arvind and C. Pandu Rangan. Transitive reduction and efficient polylog algorithms on permutation graphs. Submitted, 1996.

[63] T. Asano. Dynamic programming on intervals. *Internat. J. Comput. Geom. Appl.*, 3:323–330, 1993.

[64] P. Ash and B. Jackson. Dominating cycles in bipartite graphs. In J. A. Bondy, editor, *Progress in Graph Theory*, pages 81–87, Waterloo, Ont., 1982. Academic Press (Toronto, Ont.).

[65] M. J. Atallah and S. R. Kosaraju. An efficient algorithm for maxdominance, with applications. *Algorithmica*, 4:221–236, 1989.

[66] M. J. Atallah, G. K. Manacher, and J. Urrutia. Finding a minimum independent dominating set in a permutation graph. *Discrete Appl. Math.*, 21:177–183, 1988.

[67] G. Bacsó and Z. Tuza. Dominating cliques in P_5-free graphs. *Period. Math. Hungar.*, 21:303–308, 1990.

[68] G. Bacsó and Z. Tuza. Domination properties and induced subgraphs. *Discrete Math.*, 111:37–40, 1993.

[69] B. S. Baker. Approximation algorithms for NP-complete problems on planar graphs. In *24th Ann. Symp. on the Foundations of Computer Science*, pages 265–273, 1983.

[70] H. Balakrishnan, A. Rajaraman, and C. Pandu Rangan. Connected domination and Steiner set on asteroidal triple-free graphs. In F. Dehne, J. R. Sack, N. Santoro, and S. Whitesides, editors, *Proc. Workshop on Algorithms and Data Structures (WADS'93)*, volume 709, pages 131–141, Montreal, Canada, 1993. Springler-Verlag, Berlin.

[71] E. Balas and M. W. Padberg. On the set covering problem. *Oper. Res.*, 20:1152–1161, 1972.

[72] W. W. Rouse Ball. *Mathematical Recreation and Problems of Past and Present Times.* MacMillan, London, 1892.

[73] D. W. Bange, A. E. Barkauskas, L. H. Host, and P. J. Slater. Efficient near-domination of grid graphs. *Congr. Numer.*, 58:43–52, 1987.

[74] D. W. Bange, A. E. Barkauskas, L. H. Host, and P. J. Slater. Generalized domination and efficient domination in graphs. *Discrete Math.*, 159:1–11, 1996.

[75] D. W. Bange, A. E. Barkauskas, and P. J. Slater. A constructive characterization of trees with two disjoint minimum dominating sets. *Congr. Numer.*, 21:101–112, 1978.

[76] D. W. Bange, A. E. Barkauskas, and P. J. Slater. Disjoint dominating sets in trees. Technical Report 78-1087J, Sandia Laboratories, 1978.

[77] D. W. Bange, A. E. Barkauskas, and P. J. Slater. Efficient dominating sets in graphs. In R. D. Ringeisen and F. S. Roberts, editors, *Applications of Discrete Mathematics*, pages 189–199. SIAM, Philadelphia, PA, 1988.

[78] R. Bar-Yehuda and S. Moran. On approximation problems related to the independent set and vertex cover problems. *Discrete Appl. Math.*, 9:1–10, 1984.

[79] R. Bar-Yehuda and U. Vishkin. Complexity of finding k-path-free dominating sets in graphs. *Inform. Process. Lett.*, 14:228–232, 1982.

[80] A. M. Barcalkin and L. F. German. The external stability number of the Cartesian product of graphs. *Bul. Akad. Stiince RSS Moldoven*, No. 1:5–8, 94, 1979.

[81] C. Barefoot, F. Harary, and K. F. Jones. What is the difference between the domination and independent domination numbers of a cubic graph? *Graphs Combin.*, 7:205–208, 1991.

[82] A. E. Barkauskas and L. H. Host. Finding efficient dominating sets in oriented graphs. *Congr. Numer.*, 98:27–32, 1993.

[83] D. Barraez, E. Flandrin, H. Li, and O. Ordaz. Dominating cycles in bipartite biclaw-free graphs. *Discrete Math.*, 146:11–18, 1995.

[84] S. B. Basapur and H. B. Walikar. Domatically cocritical graphs and some open problems. *Discrete Math.* To appear.

[85] S. B. Basapur and H. B. Walikar. Domatically cocritical graphs. In *Proc. Symp. Graph Theory Combin., Kochi, India*, pages 67–74, 1991.

[86] S. B. Basapur and H. B. Walikar. Domatically critical graphs. Submitted, 1997.

[87] D. Bauer, F. Harary, J. Nieminen, and C. L. Suffel. Domination alteration sets in graphs. *Discrete Math.*, 47:153–161, 1983.

[88] D. Bauer, E. Schmeichel, and H. J. Veldman. Some recent results on long cycles in tough graphs. In Y. Alavi, G. Chartrand, O. R. Oellermann, and A. J. Schwenk, editors, *Graph Theory, Combinatorics, and Applications, Proc. Sixth Quad. Conf. on the Theory and Applications of Graphs*, volume 1, pages 113–123, (Kalamazoo, MI 1988), 1991. Wiley.

[89] D. Bauer, E. Schmeichel, and H. J. Veldman. A note on dominating cycles in 2-connected graphs. *Discrete Math.*, 155:13–18, 1996.

[90] T. J. Bean, M. A. Henning, and H. C. Swart. On the integrity of distance domination in graphs. *Australas. J. Combin.*, 10:29–43, 1994.

[91] L. Beineke and M. A. Henning. Some extremal results on independent distance domination in graphs. *Ars Combin.*, 37:223–233, 1994.

[92] L. Beineke and M. A. Henning. Opinion functions in graphs. *Discrete Math.*, 167:167–178, 1997.

[93] M. A. Benedetti and F. M. Mason. On the characterization of domistable graphs (Italian; English summary). *Ann. Univ. Ferrara Sez. VII (N.S.)*, 27:1–11, 1981.

[94] C. Benzaken and P. L. Hammer. Linear separation of dominating sets in graphs. *Ann. Discrete Math.*, 3:1–10, 1978.

[95] C. Berge. *Theory of Graphs and its Applications.* Methuen, London, 1962.

[96] C. Berge. *Graphs and Hypergraphs.* North-Holland, Amsterdam, 1973.

[97] C. Berge. Nouvelles extensions du noyeau d'un graphe et des applications en theorie des jeux. *Publ. Econometriques*, 6:6–11, 1973.

[98] C. Berge. Some common properties for regularizable graphs, edge-critical graphs, and B-graphs. In *Lecture Notes in Computer Science, Graph Theory and Algorithms, Proc. Symp. Res. Inst. Electr. Comm. Tohoku Univ., Sendi, 1980*, volume 108, pages 108–123, Berlin, 1987. Springer.

[99] C. Berge and P. Duchet. Perfect graphs and kernels. *Bull. Inst. Math. Acad. Sinica*, 16:263–274, 1988.

[100] C. Berge and P. Duchet. Recent problems and results about kernels in directed graphs. In *Applications of Discrete Mathematics*, pages 200–204, Philadelphia, PA, 1988. SIAM.

[101] C. Berge and P. Duchet. Recent problems and results about kernels in directed graphs. *Discrete Math.*, 86:27–31, 1990.

[102] F. Berman, F. T. Leighton, P. Shor, and L. Synder. Generalized planar matching. Technical Report MIT/LCS/TM-273, MIT, 1985.

[103] M. W. Bern, E. L. Lawler, and A. L. Wong. Why certain subgraph computations require only linear time. In *Proc. 26th Ann. IEEE Symp. on the Foundations of Computer Science*, pages 117–125, Portland, OR, 1985.

[104] M. W. Bern, E. L. Lawler, and A. L. Wong. Linear-time computation of optimal subgraphs of decomposable graphs. *J. Algorithms*, 8:216–235, 1987.

[105] P. J. Bernhard, S. T. Hedetniemi, and D. P. Jacobs. Efficient sets in graphs. *Discrete Appl. Math.*, 44:99–108, 1993.

[106] B. Bernhardsson. Explicit solutions to the *N*-queens problem for all *N*. *SIGART Bull.*, 2:7, 1991.

[107] P. Bertolazzi and A. Sassono. A class of polynomially solvable set covering problems. *SIAM J. Discrete Math.*, 1:306–316, 1988.

[108] A. A. Bertossi. Dominating sets for split and bipartite graphs. *Inform. Process. Lett.*, 19:37–40, 1984.

[109] A. A. Bertossi. Total domination in interval graphs. *Inform. Process. Lett.*, 23:131–134, 1986.

[110] A. A. Bertossi. On the domatic number of interval graphs. *Inform. Process. Lett.*, 28:275–280, 1988.

[111] A. A. Bertossi and M. A. Bonuccelli. Some parallel algorithms on interval graphs. *Discrete Appl. Math.*, 16:101–111, 1987.

[112] A. A. Bertossi and A. Gori. Total domination and irredundance in weighted interval graphs. *SIAM J. Discrete Math.*, 1:317–327, 1988.

[113] A. A. Bertossi and S. Moretti. Parallel algorithms on circular-arc graphs. *Inform. Process. Lett.*, 33:275–281, 1990.

[114] T. A. Beyer, A. Proskurowski, S. T. Hedetniemi, and S. Mitchell. Independent domination in trees. *Congr. Numer.*, 19:321–328, 1977.

[115] N. Biggs. Perfect codes in graphs. *J. Combin. Theory Ser. B*, 15:289–296, 1973.

[116] A. Billionnet. Upper and lower bounds for the weight and cardinality of every maximal stable set of a graph. *Rev. Roumaine Math. Pures Appl.*, 29:641–656, 1984.

[117] J. R. S. Blair. The efficiency of AC graphs. *Discrete Appl. Math.*, 44:119–138, 1993.

[118] M. M. Blank. An estimate of the external stability number of a graph without suspended vertices. *Prikl. Mat. i Programmirovanie Vyp.*, 10:3–11, 149, 1973. Russian.

[119] M. Blidia. A parity digraph has a kernel. *Combinatorica*, 6:23–27, 1986.

[120] M. Blidia. Parity graphs with an orientation condition have a kernel. Submitted, 1996.

[121] M. Blidia, P. Duchet, and F. Maffray. Meyniel graphs are kernel-M-solvable. *Res. Rep. Univ. Paris*, 6, 1988.

[122] P. Blitch. *Domination in Graphs*. PhD thesis, Univ. South Carolina, 1983.

[123] A. Blokhuis and C. W. H. La. More coverings by rook domains. *J. Combin. Theory Ser. A*, 36:240–244, 1984.

[124] C. Bo and B. Liu. Some inequalities about connected domination number. *Discrete Math.*, 159:241–245, 1996.

[125] H. L. Bodlaender. Dynamic programming on graphs with bounded treewidth. In T. Lepisto and A. Salomaa, editors, *Lecture Notes in Comput. Sci., Proc. 15th Internat. Colloq. on Automata, Languages and Programming*, volume 317, pages 105–118, Heidelberg, 1988. Springer Verlag.

[126] B. Bollobás and E. J. Cockayne. Graph theoretic parameters concerning domination, independence and irredundance. *J. Graph Theory*, 3:241–250, 1979.

[127] B. Bollobás and E. J. Cockayne. On the domination number and minimum degree of a graph. Technical Report DM-276-IR, Dept. Mathematics, Univ. Victoria, 1983.

[128] B. Bollobás and E. J. Cockayne. On the irredundance number and maximum degree of a graph. *Discrete Math.*, 49:197–199, 1984.

[129] B. Bollobás and E. J. Cockayne. The irredundance number and maximum degree of a graph. *Discrete Math.*, 33:249–258, 1989.

[130] B. Bollobás, E. J. Cockayne, and C. M. Mynhardt. On generalised minimal domination parameters for paths. *Discrete Math.*, 86:89–97, 1990.

[131] B. Bollobás and A. G. Thomason. Uniquely partitionable graphs. *J. London Math. Soc.* 2, 3:403–410, 1977.

[132] J. A. Bondy and G. Fan. A sufficient condition for dominating cycles. *Discrete Math.*, 67:205–208, 1987.

[133] J. A. Bondy and U. S. R. Murty. *Graph Theory with Applications*. Macmillan, London and Elsevier, New York, 1976.

[134] M. A. Bonuccelli. Dominating sets and domatic number of circular arc graphs. *Discrete Appl. Math.*, 12:203–213, 1985.

[135] K. S. Booth. *Dominating Sets in Chordal Graphs*. PhD thesis, Univ. Waterloo, Ontario, 1980.

[136] K. S. Booth and J. H. Johnson. Dominating sets in chordal graphs. *SIAM J. Comput.*, 11:191–199, 1982.

[137] J. Borges and I. J. Dejter. On perfect dominating sets in hypercubes and their complements. *J. Combin. Math. Combin. Comput.*, 20:161–173, 1996.

[138] E. Boros and V. Gurevich. Perfect graphs are kernel solvable. *Discrete Math.*, 159:35–55, 1996.

[139] M. Borowiecki. On a minimaximal kernel of trees. *Discuss. Math*, 1:3–6, 1975.

[140] M. Borowiecki. On graphs with minimaximal kernels. *Prace Nauk. Inst. Mat. Politech. Wroclaw *. 17 Ser. Stud. i Materialy*, 13:3–7, 1977.

[141] M. Borowiecki. On the stable and absorbent sets of hypergraphs. In *Contributions to Graph Theory and its Applications, (Internat. Colloq. Oberhof)*, pages 33–39, 1977. German, Tech. Hochschule Ilmenau.

[142] M. Borowiecki and M. Kuzak. On the k-stable and k-dominating sets of graphs. In M. Borowiecki, Z. Skupien, L. Szamkolowicz, and G. Zielona, editors, *Graphs Hypergraphs and Block Systems, Proc. Symp. Zielona Gora*, 1976.

[143] A. Brandstädt. The computational complexity of feedback vertex set, Hamiltonian circuit, dominating set, Steiner tree, and bandwidth on special perfect graphs. *J. Inform. Process. Cybernet.*, 23:471–477, 1987.

[144] A. Brandstädt. Classes of bipartite graphs related to chordal graphs and domination of one colour class by the other. Technical Report N/88/10, Friedrich-Schiller-Univ., Jena, 1988.

[145] A. Brandstädt. On the domination problem for bipartite graphs. In R. Bodendieck and R. Henn, editors, *Topics in Combinatorics and Graph Theory*, pages 145–152. Physica, Heidelberg, 1990.

[146] A. Brandstädt. On improved bounds for permutation graph problems. In E. W. Mayr, editor, *Lecture Notes in Comput. Sci., Proc. WG'92*, volume 657, pages 1–10, Berlin, 1993. Springer Verlag.

[147] A. Brandstädt and H. Behrendt. Domination and the use of maximum neighbourhoods. Technical Report SM-DU-204, Univ. Duisburg, 1992.

[148] A. Brandstädt, V. D. Chepoi, and F. F. Dragan. Clique r-domination and clique r-packing problems on dually chordal graphs. Technical Report SM-DU-251, Univ. Duisburg, 1994.

[149] A. Brandstädt, V. D. Chepoi, and F. F. Dragan. The algorithmic use of hypertree structure and maximum neighbourhood orderings. In E. W. Mayr, G. Schmidt, and G. Tinhofer, editors, *Lecture Notes in Comput. Sci., 20th Internat. Workshop Graph-Theoretic Concepts in Computer Science (WG'94)*, volume 903, pages 65–80, Berlin, 1995. Springer-Verlag.

[150] A. Brandstädt and F. F. Dragan. A linear-time algorithm for connected r-domination and Steiner tree on distance-hereditary graphs. Technical Report SM-DU-261, Univ. Duisburg, 1994.

[151] A. Brandstädt, F. F. Dragan, V. D. Chepoi, and V. I. Voloshin. Dually chordal graphs. In *Lecture Notes in Comput. Sci., 19th Internat. Workshop Graph-Theoretic Concepts in Computer Science (WG'93)*, volume 790, pages 237–251, Berlin, 1993. Springer-Verlag.

[152] A. Brandstädt and D. Kratsch. On the restriction of some NP-complete graph problems to permutation graphs. In L. Budach, editor, *Lecture Notes in Comput. Sci., Proc. FCT'85*, volume 199, pages 53–62, Berlin, 1985. Springer-Verlag.

[153] A. Brandstädt and D. Kratsch. On domination problems on permutation and other graphs. *Theoret. Comput. Sci.*, 54:181–198, 1987.

[154] H. Breu and D. G. Kirkpatrick. Algorithms for dominating and Steiner set problems in cocomparability graphs. Manuscript, 1993.

[155] R. C. Brewster, E. J. Cockayne, and C. M. Mynhardt. Irredundant Ramsey numbers for graphs. *J. Graph Theory*, 13:283–290, 1989.

[156] R. C. Brewster, E. J. Cockayne, and C. M. Mynhardt. The irredundant Ramsey number $s(3,6)$. *Quaestiones Math.*, 13:141–157, 1990.

[157] R. C. Brigham and J. R. Carrington. Global domination. In T. W. Haynes, S. T. Hedetniemi, and P. J. Slater, editors, *Domination in Graphs: Advanced Topics*, chapter 11. Marcel Dekker, Inc., 1997.

[158] R. C. Brigham, J. R. Carrington, and R. P. Vitray. Connected graphs with maximum total domination number. Submitted, 1997.

[159] R. C. Brigham, P. Z. Chinn, and R. D. Dutton. Vertex domination-critical graphs. *Networks*, 18:173–179, 1988.

[160] R. C. Brigham and R. D. Dutton. A study of vertex domination-critical graphs. Technical Report M-2, Univ. Central Florida, 1984.

[161] R. C. Brigham and R. D. Dutton. A compilation of relations between graph invariants. *Networks*, 15:73–107, 1985.

[162] R. C. Brigham and R. D. Dutton. Neighborhood numbers, new invariants of undirected graphs. *Congr. Numer.*, 53:121–132, 1986.

[163] R. C. Brigham and R. D. Dutton. Bounds on the domination number of a graph. *Quart. J. Math. Oxford Ser. 2*, 41:269–275, 1990.

[164] R. C. Brigham and R. D. Dutton. Factor domination in graphs. *Discrete Math.*, 86:127–136, 1990.

[165] R. C. Brigham and R. D. Dutton. A compilation of relations between graph invariants, supplement I. *Networks*, 21:421–455, 1991.

[166] R. C. Brigham, R. D. Dutton, F. Harary, and T. W. Haynes. On graphs having equal domination and codomination numbers. *Utilitas Math.*, 50:53–64, 1996.

[167] I. Broere, J. E. Dunbar, and J. H. Hattingh. Minus k-subdomination in graphs. *Ars. Combin.* To appear.

[168] I. Broere, J. H. Hattingh, M. A. Henning, and A. A. McRae. Majority domination in graphs. *Discrete Math.*, 138:125–135, 1995.

[169] H. J. Broersma, T. Kloks, D. Kratsch, and H. Müller. Independent sets in asteroidal tripple-free graphs. Manuscript, 1996.

[170] H. J. Broersma and H. J. Veldman. Long dominating cycles and paths in graphs with large neighborhood unions. *J. Graph Theory*, 15:29–38, 1991.

[171] R. L. Brooks. On colouring the nodes of a network. *Proc. Cambridge Philos. Soc.*, 37:194–197, 1941.

[172] J. Brown, R. Nowakowski, and D. F. Rall. The ultimate categorical independence ratio of a graph. *SIAM J. Discrete Math.*, 9:290–300, 1996.

[173] R. D. Bryant, E. R. Fredricksen, and H. M. Fredricksen. Covering graphs. *Congr. Numer.*, 76:157–165, 1990.

[174] F. Buckley and F. Harary. *Distance in Graphs.* Addison-Wesley, 1989.

[175] A. P. Burger. *Domination in the Queen's Graph.* Master's thesis, Univ. South Africa, 1993.

[176] A. P. Burger, E. J. Cockayne, and C. M. Mynhardt. Domination and irredundance in the queen's graph. *Discrete Math.*, 163:47–66, 1997.

[177] A. P. Burger, C. M. Mynhardt, and E. J. Cockayne. Domination numbers for the queen's graph. *Bull. Inst. Combin. Appl.*, 10:73–82, 1994.

[178] K. Cameron. Induced matchings. *Discrete Appl. Math.*, 24:97–102, 1989.

[179] P. J. Campbell. Gauss and the eight queens problem: a study in miniature of the propagation of historical error. *Historia Math.*, 4:397–404, 1977.

[180] W. A. Carnielli. On covering and coloring problems for rook domains. *Discrete Math.*, 57:9–16, 1985.

[181] Y. Caro. On the k-domination and k-transversal numbers of graphs and hypergraphs. *Ars Combin.*, 29:49–55, 1990.

[182] Y. Caro and Y. Roditty. On the vertex-independence number and star decomposition of graphs. *Ars Combin.*, 20:167–180, 1985.

[183] Y. Caro and Y. Roditty. A note on the k-domination number of a graph. *Internat. J. Math. Sci.*, 13:205–206, 1990.

[184] Y. Caro and Y. Roditty. On the k-domination number of a graph. Submitted, 1996.

[185] Y. Caro and Z. Tuza. Improved lower bounds on k-independence. *J. Graph Theory*, 15:99–107, 1991.

[186] J. R. Carrington. *Global Domination of Factors of a Graph.* PhD thesis, Univ. Central Florida, Orlando, 1992.

[187] J. R. Carrington and R. C. Brigham. Factor domination. *Congr. Numer.*, 83:201–211, 1991.

[188] J. R. Carrington and R. C. Brigham. Global domination of simple factors. *Congr. Numer.*, 88:161–167, 1992.

[189] J. R. Carrington, F. Harary, and T. W. Haynes. Changing and unchanging the domination number of a graph. *J. Combin. Math. Combin. Comput.*, 9:57–63, 1991.

[190] D. I. Carson. *Computational Aspects of Some Generalized Domination Parameters.* PhD thesis, Univ. Natal, 1995.

[191] V. E. Cazanescu and S. Rudeanu. Independent sets and kernels in graphs and hypergraphs. *Ann. Fac. Sci. Univ. Nat. Zaire (Kinshasa) Sect. Math.-Phys.*, 4:37–66, 1978.

[192] C. Champetier. Kernels in some orientations of comparability graphs. *J. Combin. Theory Ser. B*, 47:111–113, 1989.

[193] G. J. Chang. *k-domination and Graph Covering Problems.* PhD thesis, Cornell Univ., Ithaca, NY, 1982.

[194] G. J. Chang. Labeling algorithms for domination problems in sun-free chordal graphs. *Discrete Appl. Math.*, 22:21–34, 1988/89.

[195] G. J. Chang. Total domination in block graphs. *Oper. Res. Lett.*, 8:53–57, 1989.

[196] G. J. Chang. The domatic number problem. *Discrete Math.*, 125:115–122, 1994.

[197] G. J. Chang, M. Farber, and Z. Tuza. Algorithmic aspects of neighborhood numbers. *SIAM J. Discrete Math.*, 6(1):24–29, 1993.

[198] G. J. Chang and G. L. Nemhauser. The *k*-domination and *k*-stability problems on graphs. Technical Report TR-540, School Oper. Res. Industrial Eng., Cornell Univ., 1982.

[199] G. J. Chang and G. L. Nemhauser. *R*-domination of block graphs. *Oper. Res. Lett.*, 1(6):214–218, 1982.

[200] G. J. Chang and G. L. Nemhauser. The *k*-domination and *k*-stability problems in sun-free chordal graphs. *SIAM J. Algebraic Discrete Methods*, 5:332–345, 1984.

[201] G. J. Chang and G. L. Nemhauser. Covering, packing and generalized perfection. *SIAM J. Algebraic Discrete Methods*, 6:109–132, 1985.

[202] G. J. Chang, C. Pandu Rangan, and S. R. Coorg. Weighted independent perfect domination on cocomparability graphs. *Discrete Appl. Math.*, 63(3):215–222, 1995.

[203] M. S. Chang. Efficient algorithms for the domination problems on interval graphs and circular-arc graphs. In *IFIP Transactions A-12, Proc. IFIP 12th World Congress*, volume 1, pages 402–408, 1992.

[204] M. S. Chang. Weighted domination on cocomparability graphs. In *Lecture Notes in Comput. Sci., Proc. ISAAC'95*, volume 1004, pages 121–131, Berlin, 1995. Springer-Verlag.

[205] M. S. Chang, F. H. Hsing, and S. L. Peng. Irredundance in weighted interval graphs. In *Proc. National Computer Symp.*, pages 128–137, Taipei, Taiwan, 1993.

[206] M. S. Chang and C. Hsu. On minimum intersection of two minimum dominating sets of interval graphs. Submitted, 1996.

[207] M. S. Chang and Y. C. Liu. Polynomial algorithms for the weighted perfect domination problems on chordal and split graphs. *Inform. Process. Lett.*, 48:205–210, 1993.

[208] M. S. Chang and Y. C. Liu. Polynomial algorithms for weighted perfect domination problems on interval and circular-arc graphs. *J. Inform. Sci. Engineering*, 10:549–568, 1994.

[209] M. S. Chang, P. Nagavamsi, and C. Pandu Rangan. Weighted irredundance of interval graphs. Manuscript, 1996.

[210] M. S. Chang, S. Wu, G. J. Chang, and H. G. Yeh. Domination in distance-hereditary graphs. 1996. Submitted.

[211] T. Y. Chang. *Domination Numbers of Grid Graphs*. PhD thesis, Univ. South Florida, 1992.

[212] T. Y. Chang and W. E. Clark. The domination numbers of the $5 \times n$ and $6 \times n$ grid graphs. *J. Graph Theory*, 17:81–107, 1993.

[213] T. Y. Chang, W. E. Clark, and E. O. Hare. Domination numbers of complete grid graphs I. *Ars Combin.*, 38:97–111, 1994.

[214] Y. Chang, M. S. Jacobson, C. L. Monma, and D. B. West. Subtree and substar intersection numbers. *Discrete Appl. Math.*, 44:205–220, 1993.

[215] C. Chao. On the kernels of graphs. Technical Report RC-685, IBM, 1962.

[216] G. Chartrand, H. Gavlas, F. Harary, and R. C. Vandell. The forcing domination number of a graph. *J. Combin. Math. Combin. Comput.* To appear.

[217] G. Chartrand, F. Harary, M. Hossain, and K. Schultz. Exact 2-step domination in graphs. *Math. Bohem.* To appear.

[218] G. Chartrand, M. A. Henning, and K. Schultz. On planetary domination numbers of graphs. Submitted, 1996.

[219] G. Chartrand and L. Lesniak. *Graphs & Digraphs*. Wadsworth and Brooks/Cole, Monterey, CA, third edition, 1996.

[220] G. Chartrand and J. Mitchem. Graphical theorems of the Nordhaus-Gaddum class. In *Lecture Notes in Math., Recent Trends in Graph Theory*, volume 186, pages 55–61. Springer-Verlag, Berlin, 1971.

[221] G. Chartrand and S. Schuster. On the independence numbers of complementary graphs. *Trans. New York Acad. Sci., Series II*, 36:247–251, 1974.

[222] G. Chartrand, D. W. VanderJagt, and B. Q. Yue. Orientable domination in graphs. *Congr. Numer.*, 119:51–63, 1996.

[223] G. Chaty and J. L. Szwarcfiter. Enumerating the kernels of a directed graph with no odd circuits. *Inform. Process. Lett.*, 51:149–153, 1994.

[224] B. Chen and H. Fu. Edge domination in complete partite graphs. *Discrete Math.*, 132:29–35, 1994.

[225] C. Chen, A. Harkat-Benhamdine, and H. Li. Distance-dominating cycles in quasi-claw-free graphs. Submitted, 1996.

[226] G. Chen, J. H. Hattingh, and C. C. Rousseau. Asymptotic bounds for irredundant and mixed Ramsey numbers. *J. Graph Theory*, 17:193–206, 1993.

[227] G. Chen and M. Jacobson. On a problem in domination. Submitted, 1992.

[228] G. Chen, W. Piotrowski, and W. Shreve. A partition approach to Vizing's conjecture. *J. Graph Theory*, 21:103–111, 1996.

[229] G. Chen and C. C. Rousseau. The independent Ramsey number $s(3, 7)$. *J. Graph Theory*, 19:263–270, 1995.

[230] W. Chengde. On the sum of two parameters concerning independence and irredundance in graphs. *J. Math. (PRC)*, 5:201–204, 1985.

[231] A. A. Chernyak and Zh. A. Chernyak. Pseudodomishold graphs. *Discrete Math.*, 84:193–196, 1990.

[232] G. Cheston and G. H. Fricke. Classes of graphs for which upper fractional domination equals independence, upper domination, and upper irredundance. *Discrete Appl. Math.*, 55:241–258, 1994.

[233] G. A. Cheston, G. H. Fricke, S. T. Hedetniemi, and D. P. Jacobs. On the computational complexity of upper fractional domination. *Discrete Appl. Math.*, 27:195–207, 1990.

[234] G. A. Cheston, E. O. Hare, S. T. Hedetniemi, and R. C. Laskar. Simplicial graphs. *Congr. Numer.*, 67:105–113, 1988.

[235] K. B. Chilakamarri and P. Hamburger. On a class of kernel-perfect and kernel-perfect-critical graphs. *Discrete Math.*, 118:253–257, 1993.

[236] F. R. K. Chung, R. L. Graham, E. J. Cockayne, and D. J. Miller. On the minimum dominating pair number of a class of graphs. *Caribbean J. Math.*, 1:73–76, 1982.

[237] V. Chvátal. On the computational complexity of finding a kernel. Technical Report CRM-300, Centre de Recherches Mathematiques, Univ. Montreal, 1973.

[238] V. Chvátal and W. Cook. The discipline number of a graph. *Discrete Math.*, 86:191–198, 1990.

[239] V. Chvátal and L. Lovász. Every directed graph has a semikernel. In C. Berge and D. K. Ray-Chaudhuri, editors, *Lecture Notes in Math., Hypergraph Seminar*, volume 411, page 75. Springer, Berlin, 1974.

[240] V. Chvátal and P. J. Slater. A note on well-covered graphs. *Ann. Discrete Math.*, 55:179–182, 1993.

[241] B. N. Clark, C. J. Colbourn, and P. Erdös. A conjecture on dominating cycles. *Congr. Numer.*, 47:189–197, 1985.

[242] B. N. Clark, C. J. Colbourn, and D. S. Johnson. Unit disk graphs. *Discrete Math.*, 86:165–177, 1990.

[243] L. Clark. Perfect domination in random graphs. *J. Combin. Math. Combin. Comput.*, 14:173–182, 1993.

[244] W. E. Clark and L. A. Dunning. Sharp upper bounds for the domination number of graphs of given order and minimum degree. Submitted, 1997.

[245] W. E. Clark, D. Fisher, B. Shekhtman, and S. Suen. Some upper bounds for the domination number. Submitted, 1996.

[246] W. E. Clark and B. Shekhtman. Covering by complements of subspaces. *Linear and Multilinear Algebra*, 40:1–13, 1995.

[247] W. E. Clark and B. Shekhtman. On the domination matrices of the $\mathcal{C}$-analogues of Kneser graphs. *Congr. Numer.*, 107:193–197, 1995.

[248] W. E. Clark and B. Shekhtman. Covering by complements of subspaces, II. *Proc. Amer. Math. Soc.*, 125(1):251–254, 1997.

[249] W. E. Clark and B. Shekhtman. Domination numbers of q-analogues of Kneser graphs. *Bull. Inst. Combin. Appl.*, 19:83–92, 1997.

[250] W. E. Clark and B. Shekhtman. On the domination numbers of certain analogues of Kneser graphs. Submitted, 1997.

[251] E. J. Cockayne. Domination in undirected graphs - a survey. In Y. Alavi and D. R. Lick, editors, *Theory and Applications of Graphs in America's Bicentennial Year*, pages 141–147. Springer-Verlag, Berlin, 1978.

[252] E. J. Cockayne. Disjoint cliques in coset graphs. *Congr. Numer.*, 23:275–279, 1980.

[253] E. J. Cockayne. Chessboard domination problems. *Discrete Math.*, 86:13–20, 1990.

[254] E. J. Cockayne. Irredundance in the queen's graph. Submitted, 1994.

[255] E. J. Cockayne. Bibliography on irredundance in graphs. Personal communication, 1997.

[256] E. J. Cockayne, R. M. Dawes, and S. T. Hedetniemi. Total domination in graphs. *Networks*, 10:211–219, 1980.

[257] E. J. Cockayne, G. Exoo, J. H. Hattingh, and C. M. Mynhardt. The irredundant Ramsey number $s(4, 4)$. *Utilitas Math.*, 41:119–128, 1992.

[258] E. J. Cockayne, O. Favaron, H. Li, and G. MacGillivray. On the product of independent domination numbers of a graph and its complement. *Discrete Math.*, 90(3):313–317, 1991.

[259] E. J. Cockayne, O. Favaron, and C. M. Mynhardt. Universal maximal packing functions of graphs. *Discrete Math.*, 159:57–68, 1996.

[260] E. J. Cockayne, O. Favaron, C. Payan, and A. G. Thomason. Contributions to the theory of domination, independence and irredundance in graphs. *Discrete Math.*, 33:249–258, 1981.

[261] E. J. Cockayne, G. H. Fricke, S. T. Hedetniemi, and C. M. Mynhardt. Extremum aggregates of hypergraph functions: existence and interpolation theorems. Submitted, 1994.

[262] E. J. Cockayne, G. H. Fricke, S. T. Hedetniemi, and C. M. Mynhardt. Properties of minimal dominating functions of graphs. *Ars Combin.*, 41:107–115, 1995.

[263] E. J. Cockayne, G. H. Fricke, and C. M. Mynhardt. On a Nordhaus-Gaddum type problem for independent domination. *Discrete Math.*, 138:199–205, 1995.

[264] E. J. Cockayne, B. Gamble, and B. Shepherd. An upper bound for the k-domination number of a graph. *J. Graph Theory*, 9:533–534, 1985.

[265] E. J. Cockayne, B. Gamble, and B. Shepherd. Domination parameters for the bishops graph. *Discrete Math.*, 58:221–227, 1986.

[266] E. J. Cockayne and G. Geldenhuys. Partial dominating functions for trees and cycles. *J. Combin. Math. Combin. Comput.*, 17:175–190, 1995.

[267] E. J. Cockayne, S. E. Goodman, and S. T. Hedetniemi. A linear algorithm for the domination number of a tree. *Inform. Process. Lett.*, 4:41–44, 1975.

[268] E. J. Cockayne, P. J. P. Gröbler, S. T. Hedetniemi, and A. A. McRae. What makes an irredundant set maximal? *J. Combin. Math. Combin. Comput.* To appear.

[269] E. J. Cockayne, E. O. Hare, S. T. Hedetniemi, and T. V. Wimer. Bounds for the domination number of grid graphs. *Congr. Numer.*, 47:217–228, 1985.

[270] E. J. Cockayne, B. L. Hartnell, S. T. Hedetniemi, and R. Laskar. Perfect domination in graphs. *J. Combin. Inform. System Sci.*, 18:136–148, 1993.

[271] E. J. Cockayne, J. H. Hattingh, S. M. Hedetniemi, S. T. Hedetniemi, and A. A. McRae. Using maximality and minimality conditions to construct inequality chains. *Discrete Math.* To appear.

[272] E. J. Cockayne, J. H. Hattingh, J. Kok, and C. M. Mynhardt. Mixed Ramsey numbers and irredundant Turán numbers for graphs. *Ars Combin.*, 29C:57–68, 1990.

[273] E. J. Cockayne, J. H. Hattingh, and C. M. Mynhardt. The irredundant Ramsey number $s(3,7)$. *Utilitas Math.*, 39:145–160, 1991.

[274] E. J. Cockayne, T. W. Haynes, and S. T. Hedetniemi. Extremal graphs for inequalities involving domination parameters. Submitted, 1996.

[275] E. J. Cockayne, S. M. Hedetniemi, S. T. Hedetniemi, and C. M. Mynhardt. Irredundant and perfect neighborhood sets in trees. Submitted, 1997.

[276] E. J. Cockayne and S. T. Hedetniemi. Independence graphs. In *Proc. Fifth S.E. Conf. on Combinatorics, Graph Theory, and Computing*, pages 471–491. Utilitas Math., Winnipeg, 1974.

[277] E. J. Cockayne and S. T. Hedetniemi. Optimal domination in graphs. *IEEE Trans. Circuits and Systems*, 22:855–857, 1975.

[278] E. J. Cockayne and S. T. Hedetniemi. Disjoint independent dominating sets in graphs. *Discrete Math.*, 15:213–222, 1976.

[279] E. J. Cockayne and S. T. Hedetniemi. A linear algorithm for the maximum weight of an independent set in a tree. In *Proc. Seventh S.E. Conf. on Combinatorics, Graph Theory and Computing*, pages 217–228. Utilitas Math., Winnipeg, 1976.

[280] E. J. Cockayne and S. T. Hedetniemi. Towards a theory of domination in graphs. *Networks*, 7:247–261, 1977.

[281] E. J. Cockayne and S. T. Hedetniemi. Disjoint cliques in regular graphs of degree seven and eight. *J. Combin. Theory Ser. B*, 24(2):233–237, 1978.

[282] E. J. Cockayne and S. T. Hedetniemi. On group graphs with disjoint cliques. In *Problémes combinatoires et theorie des graphes*, Orsay, France, 1978.

[283] E. J. Cockayne and S. T. Hedetniemi. On the diagonal queens domination problem. *J. Combin. Theory Ser. A*, 42:137–139, 1986.

[284] E. J. Cockayne, S. T. Hedetniemi, and R. C. Laskar. Gallai theorems for graphs, hypergraphs and set systems. *Discrete Math.*, 72:35–47, 1988.

[285] E. J. Cockayne, S. T. Hedetniemi, and D. J. Miller. Properties of hereditary hypergraphs and middle graphs. *Canad. Math. Bull.*, 21:461–468, 1978.

[286] E. J. Cockayne, S. T. Hedetniemi, and P. J. Slater. Matchings and transversals in hypergraphs, domination and independence in trees. *J. Combin. Theory Ser. B*, 26:78–80, 1979.

[287] E. J. Cockayne, C. W. Ko, and F. B. Shepherd. Inequalities concerning dominating sets in graphs. Technical Report DM-370-IR, Dept. Math., Univ. Victoria, 1985.

[288] E. J. Cockayne, G. MacGillivray, and C. M. Mynhardt. Generalized maximal independence parameters for paths and cycles. *Quaestiones Math.*, 13:123–139, 1990.

[289] E. J. Cockayne, G. MacGillivray, and C. M. Mynhardt. A linear algorithm for 0-1 universal minimal dominating functions of trees. *J. Combin. Math. Combin. Comput.*, 10:23–31, 1991.

[290] E. J. Cockayne, G. MacGillivray, and C. M. Mynhardt. Convexity of minimal dominating functions and universals of graphs. *Bull. Inst. Combin. Appl.*, 5:37–48, 1992.

[291] E. J. Cockayne, G. MacGillivray, and C. M. Mynhardt. Convexity of minimal dominating functions of trees–II. *Discrete Math.*, 125:137–146, 1994.

[292] E. J. Cockayne, G. MacGillivray, and C. M. Mynhardt. Convexity of minimal dominating functions of trees. *Utilitas Math.*, 48:129–144, 1995.

[293] E. J. Cockayne and C. M. Mynhardt. Minimality and convexity of dominating and related functions in graphs: a unifying theory. *Utilitas Math.* To appear.

[294] E. J. Cockayne and C. M. Mynhardt. On a conjecture concerning irredundant and perfect neighborhood sets in graphs. *J. Combin. Math. Combin. Comput.* To appear.

[295] E. J. Cockayne and C. M. Mynhardt. On triangle-free graphs with unequal upper domination and irredundance numbers. *Bull. Inst. Combin. Appl.* To appear.

[296] E. J. Cockayne and C. M. Mynhardt. k-minimal domination numbers of cycles. *Ars Combin.*, 23A:195–206, 1987.

[297] E. J. Cockayne and C. M. Mynhardt. On the product of upper irredundance numbers of a graph and its complement. *Discrete Math.*, 76:117–121, 1989.

[298] E. J. Cockayne and C. M. Mynhardt. On the irredundant Ramsey number $s(3,3,3)$. *Ars Combin.*, 29C:189–202, 1990.

[299] E. J. Cockayne and C. M. Mynhardt. On the product of k-minimal domination numbers of a graph and its complement. *J. Combin. Math. Combin. Comput.*, 8:118–122, 1990.

[300] E. J. Cockayne and C. M. Mynhardt. Independence and domination in 3-connected cubic graphs. *J. Combin. Math. Combin. Comput.*, 10:173–182, 1991.

[301] E. J. Cockayne and C. M. Mynhardt. Domination sequences of graphs. *Ars. Combin.*, 33:257–275, 1992.

[302] E. J. Cockayne and C. M. Mynhardt. Convexity of minimal dominating functions of trees- A survey. *Quaestiones Math.*, 16(3):301–317, 1993.

[303] E. J. Cockayne and C. M. Mynhardt. The sequence of upper and lower domination, independence and irredundance numbers of a graph. *Discrete Math.*, 122:89–102, 1993.

[304] E. J. Cockayne and C. M. Mynhardt. The irredundant Ramsey number $s(3,3,3) = 13$. *J. Graph Theory*, 18:595–604, 1994.

[305] E. J. Cockayne and C. M. Mynhardt. A characterization of universal minimal total dominating functions in trees. *Discrete Math.*, 141:75–84, 1995.

[306] E. J. Cockayne and C. M. Mynhardt. On a generalisation of signed dominating functions of graphs. *Ars. Combin.*, 43:235–245, 1996.

[307] E. J. Cockayne and C. M. Mynhardt. Convexity of extremal domination-related functions of graphs. In T. W. Haynes, S. T. Hedetniemi, and P. J. Slater, editors, *Domination in Graphs: Advanced Topics*, chapter 5. Marcel Dekker, Inc., 1997.

[308] E. J. Cockayne and C. M. Mynhardt. Domination and irredundance in cubic graphs. *Discrete Math.*, 167/168:205–214, 1997.

[309] E. J. Cockayne and C. M. Mynhardt. Domination-related parameters. In T. W. Haynes, S. T. Hedetniemi, and P. J. Slater, editors, *Domination in Graphs: Advanced Topics*, chapter 14. Marcel Dekker, Inc., 1997.

[310] E. J. Cockayne and C. M. Mynhardt. Irredundance and maximum degree in graphs. *Combin. Prob. Comput.*, 6:153–157, 1997.

[311] E. J. Cockayne, C. M. Mynhardt, and B. Yu. Universal minimal total dominating functions in graphs. *Networks*, 24:83–90, 1994.

[312] E. J. Cockayne, C. M. Mynhardt, and B. Yu. Total dominating functions in trees: minimality and convexity. *J. Graph Theory*, 19(1):83–92, 1995.

[313] E. J. Cockayne and F. D. K. Roberts. Computation of minimum independent dominating sets in graphs. Technical Report DM-76-IR, Dept. Math., Univ. Victoria, 1974.

[314] E. J. Cockayne and F. D. K. Roberts. Computation of dominating partitions. *INFOR.*, 15:94–106, 1977.

[315] E. J. Cockayne and P. H. Spencer. On the independent queens covering problem. *Graphs Combin.*, 4:101–110, 1988.

[316] C. J. Colbourn, J. M. Keil, and L. K. Stewart. Finding minimum dominating cycles in permutation graphs. *Oper. Res. Lett.*, 4:13–17, 1985.

[317] C. J. Colbourn, P. J. Slater, and L. K. Stewart. Locating-dominating sets in series-parallel networks. *Congr. Numer.*, 56:135–162, 1987.

[318] C. J. Colbourn and L. K. Stewart. Dominating cycles in series-parallel graphs. *Ars Combin.*, 19A:107–112, 1985.

[319] C. J. Colbourn and L. K. Stewart. Permutation graphs: connected domination and Steiner trees. *Discrete Math.*, 86:179–189, 1990.

[320] D. G. Corneil. The complexity of generalized clique packing. *Discrete Appl. Math.*, 12:233–239, 1985.

[321] D. G. Corneil and J. Fonlupt. Stable set bonding in perfect graphs and parity graphs. *J. Combin. Theory Ser. B*, 59:1–14, 1993.

[322] D. G. Corneil and J. M. Keil. A dynamic programming approach to the dominating set problem on k-trees. *SIAM J. Algebraic Discrete Methods*, 8:535–543, 1987.

[323] D. G. Corneil, S. Olariu, and L. Stewart. Asteroidal triple-free graphs. *SIAM J. Discrete Math.* To appear.

[324] D. G. Corneil, S. Olariu, and L. Stewart. Linear time algorithms for dominating pairs in asteroidal triple-free graphs. *SIAM J. Comput.* To appear.

[325] D. G. Corneil, S. Olariu, and L. Stewart. Asteroidal triple-free graphs. In *Lecture Notes in Comput. Sci., Proc. 19th Internat. Workshop on Graph-Theoretic Concepts in Computer Science (WG'93)*, volume 790, pages 211–224, Berlin, 1994. Springer-Verlag.

[326] D. G. Corneil, S. Olariu, and L. Stewart. Computing a dominating pair in an asteroidal triple-free graph in linear time. In *Proc. 4th Algorithms and Data Structures Workshop, LNCS 955*, volume 955, pages 358–368. Springer, 1995.

[327] D. G. Corneil, S. Olariu, and L. Stewart. A linear time algorithm to compute dominating pairs in asteroidal triple-free graphs. In *Lecture Notes in Comput. Sci., Proc. 22nd Internat. Colloq. on Automata, Languages and Programming (ICALP'95)*, volume 944, pages 292–302, Berlin, 1995. Springer-Verlag.

[328] D. G. Corneil, S. Olariu, and L. Stewart. A linear time algorithm to compute a dominating path in an AT-free graph. *Inform. Process. Lett.*, 54:253–257, 1995.

[329] D. G. Corneil and Y. Perl. Clustering and domination in perfect graphs. *Discrete Appl. Math.*, 9:27–39, 1984.

[330] D. G. Corneil, Y. Perl, and L. Stewart Burlingham. A linear recognition algorithm for cographs. *SIAM J. Comput.*, 14:926–934, 1985.

[331] D. G. Corneil, Y. Perl, and L. Stewart. Cographs: recognition, application and algorithms. *Congr. Numer.*, 43:249–258, 1984.

[332] D. G. Corneil and L. Stewart. Dominating sets in perfect graphs. *Discrete Math.*, 86:145–164, 1990.

[333] M. B. Cozzens and L. L. Kelleher. Dominating cliques in graphs. *Discrete Math.*, 86:101–116, 1990.

[334] C. Croitoru. On stables in graphs. In *Proc. Third Colloq. On Operations Research (cluj-Napoca), (Univ. Babes-Bolyai, cluj-Napoca, 1978)*, pages 55–60, 1979.

[335] C. Croitoru and E. Suditu. Perfect stables in graphs. *Inform. Process. Lett.*, 17:53–56, 1983.

[336] R. W. Cutler. Edge domination of $G \times Q_n$. *Bull. Inst. Combin. Appl.*, 15:69–79, 1995.

[337] E. Dahlhaus and P. Damaschke. The parallel solution of domination problems on chordal and strongly chordal graphs. *Discrete Appl. Math.*, 52:261–273, 1994.

[338] E. Dahlhaus, J. Kratochvíl, P. Manual, and M. Miller. Transversal partitioning in balanced hypergraphs. *Discrete Appl. Math.* To appear.

[339] P. Damaschke. Irredundance number versus domination number. *Discrete Math.*, 89:101–104, 1991.

[340] P. Damaschke, H. Müller, and D. Kratsch. Domination in convex and chordal bipartite graphs. *Inform. Process. Lett.*, 36:231–236, 1990.

[341] P. Dankelmann, P. J. Slater, V. Smithdorf, and H. C. Swart. A note on domination, total domination, and connected domination triples of graphs. Submitted, 1996.

[342] S. K. Das, N. Deo, and S. Prasad. Reverse binary digraphs. *Congr. Numer.*, 71:53–66, 1990.

[343] A. D'Atri. Cross graphs, Steiner trees and connected domination. Technical Report R147, Consiglio Nazionale delle Ricerche, Instituto do Analisi ed Informatica, 1986.

[344] A. D'Atri and M. Moscarini. Distance-hereditary graphs, Steiner trees, and connected domination. *SIAM J. Comput.*, 17:521–538, 1988.

[345] R. W. Dawes. Neighborhood covers for trees. Technical Report ISSN-0836-0227, Dept. Computing and Information Science, Queen's Univ., 1990. External Report.

[346] D. E. Daykin and C. P. Ng. Algorithms for generalized stability number of tree graphs. *J. Austral. Math. Soc.*, 6:89–100, 1966.

[347] C. F. De Jaenisch. *Applications de l'Analuse mathematique an Jen des Echecs*. Petrograd, 1862.

[348] W. F. De La Vega. Kernels in random graphs. *Discrete Math.*, 82:213–217, 1990.

[349] C. De Simone and A. Sassano. Stability number of bull- and chair-free graphs. *Discrete Applied Math.*, 41:121–129, 1993.

[350] N. Dean and J. Zito. Well-covered graphs and extendability. *Discrete Math.*, 126:67–80, 1994.

[351] I. J. Dejter and J. Pujol. Perfect domination and symmetry in hypercubes. *Congr. Numer.*, 111:18–32, 1995.

[352] I. J. Dejter and P. M. Weichsel. Twisted perfect dominating subgraphs of hypercubes. *Congr. Numer.*, 94:67–78, 1993.

[353] O. Demirors, N. Rafraf, and M. Tanik. Obtaining N-queens solutions from magic squares and constructing magic squares from N-queens solutions. *J. Recreational Math.*, 24:272–280, 1992.

[354] A. P. Deshpande and H. B. Walikar. Domatically critical graphs. *J. Graph Theory*. To appear.

[355] A. K. Dewdney. Fast turing reductions between problems in NP; chapter 4; reductions between NP-complete problems. Technical Report 71, Dept. Computer Science, Univ. Western Ontario, 1981.

[356] R. Diestel. Dominating functions and topological graph minors. In *Graph Structure Theory (Seattle, WA, 1991)*, pages 461–476. Amer. Math. Soc., Providence, RI, 1993. Series: Contemp. Math., 147.

[357] R. Diestel, S. Shelah, and J. Steprans. Dominating functions and graphs. *J. London Math. Soc. 2*, 49:16–24, 1994.

[358] G. Ding. Stable sets versus independent sets. *Discrete Math.*, 117:73–87, 1993.

[359] P. Dolan and M. Halsey. Random edge domination. *Discrete Math.*, 112:259–260, 1993.

[360] G. S. Domke. *Variations of Colorings, Coverings and Packings of Graphs.* PhD thesis, Clemson Univ., 1988.

[361] G. S. Domke, J. E. Dunbar, T. W. Haynes, D. J. Knisley, and L. R. Markus. Independence, domination and generalized maximum degree. *Congr. Numer.* To appear.

[362] G. S. Domke, J. E. Dunbar, T. W. Haynes, D. J. Knisley, and L. R. Markus. Independence, domination and uniform maximum degree. In *Graph Theory, Combinatorics, Algorithms and Applications, Proc. Eighth Internat. Conf. on Graph Theory, Combinatorics, Algorithms and Applications*, (Kalamazoo, MI). Wiley. To appear.

[363] G. S. Domke, J. E. Dunbar, and L. Markus. Gallai-type theorems and domination parameters. *Discrete Math.* To appear.

[364] G. S. Domke, G. H. Fricke, R. C. Laskar, and A. Majumdar. Fractional domination and related parameters. In T. W. Haynes, S. T. Hedetniemi, and P. J. Slater, editors, *Domination in Graphs : Advanced Topics*, chapter 3. Marcel Dekker, Inc., 1997.

[365] G. S. Domke and J. H. Hattingh. The bondage and reinforcement numbers of fractional domination for complete multipartite graphs. Submitted, 1997.

[366] G. S. Domke, J. H. Hattingh, S. T. Hedetniemi, R. C. Laskar, and L. R. Markus. Restrained domination in graphs. Submitted, 1997.

[367] G. S. Domke, J. H. Hattingh, M. A. Henning, and L. R. Markus. Restrained domination in graphs with minimum degree two. Submitted, 1996.

[368] G. S. Domke, J. H. Hattingh, R. C. Laskar, and L. R. Markus. A constructive characterization of trees with equal domination and restrained domination numbers. Submitted, 1997.

[369] G. S. Domke, S. T. Hedetniemi, and R. C. Laskar. Fractional packings, coverings and irredundance in graphs. *Congr. Numer.*, 66:227–238, 1988.

[370] G. S. Domke, S. T. Hedetniemi, R. C. Laskar, and R. B. Allan. Generalized packings and coverings of graphs. *Congr. Numer.*, 62:259–270, 1988.

[371] G. S. Domke, S. T. Hedetniemi, R. C. Laskar, and G. H. Fricke. Relationships between integer and fractional parameters of graphs. In Y. Alavi, G. Chartrand, O. R. Oellermann, and A. J. Schwenk, editors, *Graph Theory, Combinatorics, and Applications, Proc. Sixth Quad. Conf. on the Theory and Applications of Graphs*, volume 2, pages 371–387, (Kalamazoo, MI 1988), 1991. Wiley.

[372] G. S. Domke and R. C. Laskar. The bondage and reinforcement numbers of γ_f for some graphs. *Discrete Math.* To appear.

[373] R. G. Downey, M. R. Fellows, and V. Raman. The complexity of irredundant sets parameterized by size. Submitted, 1997.

[374] F. F. Dragan. Dominating and packing in triangulated graphs. *Metody Diskret. Analiz.*, 1(51):17–36, 1991.

[375] F. F. Dragan. Domination in Helly graphs without quadrangles. *Cybernet. Systems Anal.*, 6:47–57, 1993.

[376] F. F. Dragan. HT-graphs: centers, connected r-domination and Steiner trees. *Comput. Sci. J. Moldova (Kishinev)*, 1:64–83, 1993.

[377] F. F. Dragan. Dominating cliques in distance-hereditary graphs. In *Lecture Notes in Comput. Sci., Algorithm Theory - SWAT/94: 4th Scandinavian Workshop on Algorithm Theory*, volume 824, pages 370–381, Berlin, 1994. Springer-Verlag.

[378] F. F. Dragan and A. Brandstädt. Dominating cliques in graphs with hypertree structure. In E. M. Schmidt and S. Skyum, editors, *Lecture Notes in Comput. Sci., Internat. Symp. on Theoretical Aspects of Computer Science (STACS'94)*, volume 775, pages 735–746, Berlin, 1994. Springer-Verlag.

[379] F. F. Dragan and A. Brandstädt. r-dominating cliques in graphs with hypertree structure. *Discrete Math.*, 162:93–108, 1996.

[380] F. F. Dragan and F. Nicolai. r-domination problems on homogeneously orderable graphs. In H. Reichel, editor, *Lecture Notes in Comput. Sci., Proc. Fundamentals of Computation Theory, FCT'95*, volume 965, pages 201–210, Berlin, 1995. Springer-Verlag.

[381] F. F. Dragan, C. F. Prisacaru, and V. D. Chepoi. Location problems in graphs and the Helly property. *Discrete Math. (Moscow)*, 4(4):67–73, 1992. in Russian.

[382] F. F. Dragan and V. I. Voloshin. Incidence graphs of biacyclic hypergraphs. *Discrete Appl. Math.* To appear.

[383] P. Duchet. Graphes noyau-parfaits. *Ann. Discrete Math.*, 9:93–101, 1980.

[384] P. Duchet. Two problems in kernel theory. *Ann. Discrete Math.*, 9:302, 1980.

[385] P. Duchet. Parity graphs are kernel-M-solvable. *J. Combin. Theory Ser. B*, 43:121–126, 1987.

[386] P. Duchet. A sufficient condition for a graph to be kernel-perfect. *J. Graph Theory*, 11:81–86, 1987.

[387] P. Duchet and H. Meyniel. A note on kernel-critical graphs. *Discrete Math.*, 33:93–101, 1980.

[388] P. Duchet and H. Meyniel. On Hadwiger's number and the stability number. *Ann. Discrete Math.*, 13:71–74, 1982.

[389] P. Duchet and H. Meyniel. Kernels in directed graphs: a poison game. *Discrete Math.*, 115:273–276, 1993.

[390] H. E. Dudeney. *Amusements in Mathematics.* Thomas Nelson, London, 1907.

[391] J. E. Dunbar, J. Ghoshal, and R. P. Grimaldi. Unit parallelogram graphs. *Congr. Numer.*, 112:162–172, 1995.

[392] J. E. Dunbar, W. Goddard, S. T. Hedetniemi, M. A. Henning, and A. A. McRae. The algorithmic complexity of minus domination in graphs. *Discrete Appl. Math.*, 68:73–84, 1996.

[393] J. E. Dunbar, J. W. Grossman, J. H. Hattingh, S. T. Hedetniemi, and A. A. McRae. On weakly-connected domination in graphs. *Discrete Math.*, 167/168:261–269, 1997.

[394] J. E. Dunbar, F. C. Harris, Jr., S. M. Hedetniemi, S. T. Hedetniemi, A. A. McRae, and R. C. Laskar. Nearly perfect sets in graphs. *Discrete Math.*, 138:229–246, 1995.

[395] J. E. Dunbar, J. H. Hattingh, R. C. Laskar, and L. Markus. Induced clique domination in graphs. *Ars Combin.* To appear.

[396] J. E. Dunbar, J. H. Hattingh, A. A. McRae, and P. J. Slater. Efficient coverage by edge sets in graphs. *Utilitas Math.* To appear.

[397] J. E. Dunbar and T. W. Haynes. Domination in inflated graphs. *Congr. Numer.*, 118:143–154, 1996.

[398] J. E. Dunbar, T. W. Haynes, and M. A. Henning. The domatic number of a graph and its complement. *Congr. Numer.* To appear.

[399] J. E. Dunbar, T. W. Haynes, and M. A. Henning. Nordhaus-Gaddum type results for the domatic number of a graph. In *Graph Theory, Combinatorics, and Applications, Proc. 8th Internat. Conf. on Graph Theory, Combinatorics, Algorithms and Applications*, (Kalamazoo, MI). Wiley. To appear.

[400] J. E. Dunbar, T. W. Haynes, and M. A. Henning. The codomatic number of a cubic graph. Submitted, 1997.

[401] J. E. Dunbar, T. W. Haynes, U. Teschner, and L. Volkmann. Bondage, insensitivity, and reinforcement. In T. W. Haynes, S. T. Hedetniemi, and P. J. Slater, editors, *Domination in Graphs : Advanced Topics*, chapter 17. Marcel Dekker, Inc., 1997.

[402] J. E. Dunbar, S. T. Hedetniemi, M. A. Henning, and A. A. McRae. Minus domination in graphs. *Comput. Math. Appl.* To appear.

[403] J. E. Dunbar, S. T. Hedetniemi, M. A. Henning, and A. A. McRae. Minus domination in regular graphs. *Discrete Math.*, 149:311–312, 1996.

[404] J. E. Dunbar, S. T. Hedetniemi, M. A. Henning, and P. J. Slater. Signed domination in graphs. In Y. Alavi and A. J. Schwenk, editors, *Graph Theory, Combinatorics, and Applications, Proc. 7th Internat. Conf. Combinatorics, Graph Theory, Applications*, volume 1, pages 311–322. John Wiley & Sons, Inc., 1995.

[405] J. E. Dunbar and R. C. Laskar. Universal and global irredundancy in graphs. *J. Combin. Math. Combin. Comput.*, 12:179–185, 1992.

[406] J. E. Dunbar, R. C. Laskar, and T. Monroe. Global irredundant sets in graphs. *Congr. Numer.*, 85:65–72, 1991.

[407] J. E. Dunbar, R. C. Laskar, and C. Wallis. Some global parameters of graphs. *Congr. Numer.*, 89:187–191, 1992.

[408] J. E. Dunbar, L. R. Markus, and D. F. Rall. Graphs with two sizes of minimal dominating sets. *Congr. Numer.*, 111:115–128, 1995.

[409] R. D. Dutton and R. C. Brigham. An extremal problem for the edge domination insensitive graphs. *Discrete Appl. Math.*, 20:113–125, 1988.

[410] I. Dvořáková-Rulićová. Perfect codes in regular graphs. *Comment. Math. Univ. Carol.*, 29:79–83, 1988.

[411] J. Edmonds and D. R. Fulkerson. Bottleneck extrema. *J. Combin. Theory Ser. B*, 8:299–306, 1970.

[412] E. V. Egiazaryan. The domination number of graphs (Russian, Armenian summary). *Appl. Math.*, 2:67–74, 1983. Erevan Univ., Erevan.

[413] M. Eisenstein, C. M. Grinstead, B. Hahne, and D. Van Stone. The queen domination problem. *Congr. Numer.*, 91:189–193, 1992.

[414] M. El-Zahar and C. M. Pareek. Domination number of products of graphs. *Ars Combin.*, 31:223–227, 1991.

[415] E. S. Elmallah and L. K. Stewart. Domination in polygon graphs. *Congr. Numer.*, 77:63–76, 1990.

[416] E. S. Elmallah and L. K. Stewart. Independence and domination in polygon graphs. *Discrete Appl. Math.*, 44:65–77, 1993.

[417] D. Ene. Algorithms for the detection of the minimal externally stables sets in a graph, and applications. *Stud. Cerc. Mat.*, 23:1009–1016, 1971.

[418] C. Erbas and M. M. Tanik. Storage schemes for parallel memory systems and the N-queens problem. In *Proc. 15th Ann. ASME ETCE Conf., Computer Applications Symp., Houston, TX*, 1992.

[419] C. Erbas, M. M. Tanik, and Z. Aliyazicioglu. Linear congruence equations for the solutions of the N-queens problem. *Inform. Process. Lett.*, 41:301–306, 1992.

[420] P. Erdös. Applications of probability to combinatorial problems. In *Proc. Colloq. on Combin. Methods in Probability Theory*, pages 90–92, 1962.

[421] P. Erdös. On a problem in graph theory. *Math. Gaz.*, 47:220–223, 1963.

[422] P. Erdös, R. Faudree, A. Gyarfas, and R. H. Schelp. Domination in colored complete graphs. *J. Graph Theory*, 13:713–718, 1989.

[423] P. Erdös and J. H. Hattingh. Asymptotic bounds for irredundant Ramsey numbers. *Quaestiones Math.*, 16:319–331, 1993.

[424] P. Erdös, M. A. Henning, and H. C. Swart. The smallest order of a graph with domination number equal to two and with every vertex contained in a K_n. *Ars Combin.*, 35A:217–224, 1993.

[425] P. Erdös and A. Meir. On total matching numbers and total covering numbers of complementary graphs. *Discrete Math.*, 19:229–233, 1977.

[426] P. Erdös and S. Schuster. Existence of complementary graphs having specified edge domination numbers. *Combin. Inform. System Sci.*, 16:7–10, 1991.

[427] S. Even, A. Pnueli, and A. Lempel. Permutation graphs and transitive graphs. *J. Assoc. Comput. Mach.*, 19(3):400–410, 1972.

[428] B. J. Falkowski and L. Schmitz. A note on the queens problem. *Inform. Process. Lett.*, 23:39–46, 1986.

[429] M. Farber. *Applications of Linear Programming Duality to Problems Involving Independence and Domination.* PhD thesis, Simon Fraser Univ., B.C., 1981.

[430] M. Farber. Domination and duality in weighted trees. *Congr. Numer.*, 33:3–13, 1981.

[431] M. Farber. Independent domination in chordal graphs. *Oper. Res. Lett.*, 1:134–138, 1982.

[432] M. Farber. Characterizations of strongly chordal graphs. *Discrete Math.*, 43:173–189, 1983.

[433] M. Farber. Domination, independent domination, and duality in strongly chordal graphs. *Discrete Appl. Math.*, 7:115–130, 1984.

[434] M. Farber and J. M. Keil. Domination in permutation graphs. *J. Algorithms*, 6:309–321, 1985.

[435] A. M. Farley, S. T. Hedetniemi, and A. Proskurowski. Partitioning trees: matching, domination and maximum diameter. *Internat. J. Comput. Inform. Sci.*, 10:55–61, 1981.

[436] A. M. Farley and A. Proskurowski. Broadcasting in trees with multiple originators. *SIAM J. Algebraic Discrete Methods 2*, 4:381–386, 1981.

[437] A. M. Farley and A. Proskurowski. Computing the maximum order of an open irredundant set in a tree. *Congr. Numer.*, 41:219–228, 1984.

[438] A. M. Farley and N. Schacham. Senders in broadcast networks: open irredundancy in graphs. *Congr. Numer.*, 38:47–57, 1983.

[439] R. Faudree, O. Favaron, and H. Li. Independence, domination, irredundance, and forbidden pairs. Technical Report 988, Univ. de Paris Sud, Orsay, 1995.

[440] R. J. Faudree, E. Flandrin, and Z. Ryjáček. Claw-free graphs - a survey. *Discrete Math.*, 164:87–147, 1997.

[441] R. J. Faudree, R. J. Gould, M. S. Jacobson, and L. Lesniak. Lower bounds for lower Ramsey numbers. *J. Graph Theory*, 14:723–730, 1990.

[442] R. J. Faudree, R. H. Schelp, and W. E. Shreve. The domination number for the product of graphs. *Congr. Numer.*, 79:29–33, 1990.

[443] O. Favaron. Very well covered graphs. *Discrete Math.*, 42:177–187, 1982.

[444] O. Favaron. On a conjecture of Fink and Jacobson concerning k-domination and k-dependence. *J. Combin. Theory Ser. B*, 39:101–102, 1985.

[445] O. Favaron. Stability, domination and irredundance in a graph. *J. Graph Theory*, 10:429–438, 1986.

[446] O. Favaron. k-domination and k-dependence in graphs. *Ars. Combin.*, 25C:159–167, 1988.

[447] O. Favaron. A note on the open irredundance in a graph. *Congr. Numer.*, 66:316–318, 1988.

[448] O. Favaron. Two relations between the parameters of independence and irredundance. *Discrete Math.*, 70:17–20, 1988.

[449] O. Favaron. A bound on the independent domination number of a tree. *Internat. J. Graph Theory*, 1:19–27, 1992.

[450] O. Favaron. A note on the irredundance number after vertex-deletion. *Discrete Math.*, 121:51–54, 1993.

[451] O. Favaron. Irredundance in inflated graphs. Submitted, 1996.

[452] O. Favaron. Least domination in a graph. *Discrete Math.*, 150:115–122, 1996.

[453] O. Favaron. Signed domination in regular graphs. *Discrete Math.*, 158:287–293, 1996.

[454] O. Favaron and J. L. Fouquet. On m-centers in p_t-free graphs. *Discrete Math.*, 125:147–152, 1994.

[455] O. Favaron, G. H. Fricke, D. Pritikin, and J. Puech. Irredundance and domination in kings graphs. Submitted, 1997.

[456] O. Favaron and D. Kratsch. Ratios of domination parameters. In *Advances in Graph Theory*, pages 173–182. Vishwa, Gulbarga, 1991.

[457] O. Favaron, H. Li, and M. D. Plummer. Some results on k-covered graphs. Submitted, 1996.

[458] O. Favaron and C. M. Mynhardt. On equality in an upper bound for domination parameters of graphs. *J. Graph Theory*, 24:221–231, 1997.

[459] O. Favaron and J. Puech. Irredundance in grids. *Discrete Math.* To appear.

[460] O. Favaron and J. Puech. Irredundant and perfect neighborhood sets in graphs and claw-free graphs. Submitted, 1997.

[461] O. Favaron, D. Sumner, and E. Wojcicka. The diameter of domination k-critical graphs. *J. Graph Theory*, 18(7):723–734, 1994.

[462] O. Favaron, F. Tian, and L. Zhang. Independence and hamiltonicity in 3-domination-critical graphs. *J. Graph Theory*, 25(3):173–184, 1997.

[463] M. R. Fellows. Applications of some domination-related categories of graphs. Manuscript, 1986.

[464] M. R. Fellows. Transversals of vertex partitions in graphs. *SIAM J. Discrete Math.*, 3:206–215, 1990.

[465] M. R. Fellows, G. H. Fricke, S. T. Hedetniemi, and D. Jacobs. The private neighbor cube. *SIAM J. Discrete Math.*, 7(1):41–47, 1994.

[466] M. R. Fellows and M. N. Hoover. Perfect domination. *Australas. J. Combin.*, 3:141–150, 1991.

[467] D. Fernandez-Baca and G. J. Slutzki. Solving parametric problems on trees. *J. Algorithms*, 10:381–402, 1989.

[468] A. Finbow and B. L. Hartnell. A game related to covering by stars. *Ars Combin.*, 16-A:189–198, 1983.

[469] A. Finbow and B. L. Hartnell. On locating dominating sets and well-covered graphs. *Congr. Numer.*, 65:191–200, 1988.

[470] A. Finbow, B. L. Hartnell, and R. Nowakowski. Well-dominated graphs: a collection of well-covered ones. *Ars. Combin.*, 25A:5–10, 1988.

[471] A. Finbow, B. L. Hartnell, and R. Nowakowski. A characterization of well covered graphs of girth 5 or greater. *J. Combin. Theory Ser. B*, 57:44–68, 1993.

[472] A. Finbow, B. L. Hartnell, and C. Whitehead. A characterization of graphs of girth eight or more with exactly two sizes of maximal independent sets. *Discrete Math.*, 125:153–167, 1994.

[473] J. F. Fink and M. S. Jacobson. n-domination in graphs. In Y. Alavi and A. J. Schwenk, editors, *Graph Theory with Applications to Algorithms and Computer Science*, pages 283–300, (Kalamazoo, MI 1984), 1985. Wiley.

[474] J. F. Fink and M. S. Jacobson. On n-domination, n-dependence and forbidden subgraphs. In Y. Alavi and A. J. Schwenk, editors, *Graph Theory with Applications to Algorithms and Computer Science*, pages 301–311, (Kalamazoo, MI 1984), 1985. Wiley.

[475] J. F. Fink, M. S. Jacobson, L. F. Kinch, and J. Roberts. On graphs having domination number half their order. *Period. Math. Hungar.*, 16:287–293, 1985.

[476] J. F. Fink, M. S. Jacobson, L. F. Kinch, and J. Roberts. The bondage number of a graph. *Discrete Math.*, 86:47–57, 1990.

[477] D. C. Fisher. The 2-packing number of complete grid graphs. *Ars Combin.*, 36:261–270, 1993.

[478] D. C. Fisher, J. R. Lundgren, S. K. Merz, and K. B. Reid. The domination and competition graphs of a tournament. *J. Graph Theory*. To appear.

[479] D. C. Fisher, J. R. Lundgren, S. K. Merz, and K. B. Reid. Domination graphs of tournaments and digraphs. *Congr. Numer.*, 108:97–107, 1995.

[480] D. C. Fisher, J. R. Lundgren, S. K. Merz, and K. B. Reid. Connected domination graphs of tournaments. Submitted, 1996.

[481] D. C. Fisher, P. A. McKenna, and E. D. Boyer. Hamiltonicity, diameter, domination, 2-packing, spectrum, and biclique partitions of Mycielskians. Manuscript, 1995.

[482] D. C. Fisher, J. Ryan, G. S. Domke, and A. Majumdar. Fractional domination of strong direct products. *Discrete Appl. Math.*, 50:89–91, 1994.

[483] P. Flach and L. Volkmann. Estimations for the domination number of a graph. *Discrete Math.*, 80:145–151, 1990.

[484] H. Fleischner. Uniqueness of maximal dominating cycles in 3-regular graphs and of hamiltonian cycles in 4-regular graphs. *J. Graph Theory*, 18(5):449–459, 1994.

[485] R. Forcade. Smallest maximal matchings in the graph of the d-dimensional cube. *J. Combin. Theory Ser. B*, 14:153–156, 1973.

[486] A. Fraenkel. Planar kernel and grundy with $d \leq 3$, $d_{out} \leq 2$, $d_{in} \leq 2$ are NP-complete. *Discrete Appl. Math.*, 3:257–262, 1981.

[487] A. Fraenkel. Even kernels. *Electronic J. Combin.*, 1:5–13, 1994.

[488] A. Fraenkel. Combinatorial game theory foundations applied to digraph kernels. *Electronic J. Combin.*, 4(2):10–17, 1997.

[489] P. Fraisse. Dominating cycles and paths. *Notices Amer. Math. Soc.*, 5:231, 1984.

[490] P. Fraisse. A note on distance dominating cycles. *Discrete Math.*, 71:89–92, 1988.

[491] C. N. Frangakis. A backtracking algorithm to generate all kernels of a directed graph. *Internat. J. Comput. Math.*, 10:35–41, 1981-82.

[492] A. Frank. Kernel systems of directed graphs. *Acta Sci. Math. (Szeged)*, 4:63–76, 1979.

[493] G. H. Fricke. Upper domination on double cone graphs. *Congr. Numer.*, 72:199–207, 1990.

[494] G. H. Fricke, E. O. Hare, D. P. Jacobs, and A. Majumdar. On integral and fractional total domination. *Congr. Numer.*, 77:87–95, 1990.

[495] G. H. Fricke, T. W. Haynes, S. M. Hedetniemi, S. T. Hedetniemi, and M. A. Henning. Perfect neighborhoods in graphs. Submitted, 1996.

[496] G. H. Fricke, S. M. Hedetniemi, S. T. Hedetniemi, A. A. McRae, C. K. Wallis, M. S. Jacobson, W. W. Martin, and W. D. Weakley. Combinatorial problems on chessboards: a brief survey. In Y. Alavi and A. J. Schwenk, editors, *Graph Theory, Combinatorics, Algorithms and Applications*, volume 1, pages 507–528, (Kalamazoo, MI 1992), 1995. Wiley.

[497] G. H. Fricke, S. T. Hedetniemi, and M. A. Henning. Asymptotic results on distance independent domination in graphs. *J. Combin. Math. Combin. Comput.*, 17:160–174, 1995.

[498] G. H. Fricke, S. T. Hedetniemi, and M. A. Henning. Distance independent domination in graphs. *Ars Combin.*, 41, 1995. 33–44.

[499] G. H. Fricke, S. T. Hedetniemi, and D. P. Jacobs. Independence and irredundance in k-regular graphs. *Ars Combin.* To appear.

[500] G. H. Fricke, S. T. Hedetniemi, and D. P. Jacobs. Maximal irredundant functions. *Discrete Appl. Math.*, 68(3):267–277, 1996.

[501] G. H. Fricke, M. A. Henning, O. R. Oellermann, and H. C. Swart. An efficient algorithm to compute the sum of two distance domination parameters. *Discrete Appl. Math.*, 68:85–91, 1996.

[502] Y. Fu. Dominating set and converse dominating set of a directed graph. *Amer. Math. Monthly*, 75:861–863, 1968.

[503] S. Fujita, T. Kameda, and M. Yamashita. A resource assignment problem on graphs. In *Proc. 6th Internat. Symp. on Algorithms and Computation*, pages 418–427, Cairns, Australia, December 1995.

[504] S. Fujita, T. Kameda, and M. Yamashita. A study on r-configurations – a resource arrangement problem. Submitted, 1996.

[505] J. Fulman. A note on the characterization of domination perfect graphs. *J. Graph Theory*, 17:47–51, 1993.

[506] J. Fulman. A generalization of Vizing's theorem on domination. *Discrete Math.*, 126:403–406, 1994.

[507] J. Fulman, D. Hanson, and G. MacGillivray. Vertex domination-critical graphs. *Networks*, 25:41–43, 1995.

[508] H. Galeana-Sánchez. A counterexample to a conjecture of Meyniel on kernel-perfect graphs. *Discrete Math.*, 41:105–107, 1982.

[509] H. Galeana-Sánchez. A theorem about a conjecture of H. Meyniel on kernel-perfect graphs. *Discrete Math.*, 59:35–41, 1986.

[510] H. Galeana-Sánchez. On the existence of $(k, 1)$-kernels in digraphs. *Discrete Math.*, 85:99–102, 1990.

[511] H. Galeana-Sánchez. b_1- and b_2-orientable graphs in kernel theory. *Discrete Math.*, 143:269–274, 1995.

[512] H. Galeana-Sánchez and L. V. Neumann-Lara. Extending kernel perfect digraphs to kernel-perfect critical digraphs. *Discrete Math.* To appear.

[513] H. Galeana-Sánchez and L. V. Neumann-Lara. On kernels and semikernels of digraphs. *Discrete Math.*, 48:67–76, 1984.

[514] H. Galeana-Sánchez and L. V. Neumann-Lara. On kernel-perfect critical digraphs. *Discrete Math.*, 59:257–265, 1986.

[515] H. Galeana-Sánchez and L. V. Neumann-Lara. Orientations of graphs in kernel theory. *Discrete Math.*, 87:271–280, 1991.

[516] H. Galeana-Sánchez and L. V. Neumann-Lara. New extensions of kernel perfect digraphs to kernel imperfect critical digraphs. *Graphs Combin.*, 10:329–336, 1994.

[517] H. Galeana-Sánchez, L. Pastrana-Ramirez, and H. A. Rincon-Mejia. Semikernels, quasikernels, and Grundy functions in line digraphs. *SIAM J. Discrete Math.*, 4:80–83, 1991.

[518] T. Gallai. Über extreme Punkt-und Kantenmengen. *Ann. Univ. Sci. Budapest, Eotvos Sect. Math.*, 2:133–138, 1959.

[519] B. Gamble, B. Shepherd, and E. J. Cockayne. Domination of chessboards by queens on a column. *Ars Combin.*, 19:105–118, 1985.

[520] M. Gardner. Mathematical games. *Sci. Amer.*, pages 114–115, May 1972. (cf. also *Sci. Amer.*, June, 1972, p.117).

[521] M. Gardner. Mathematical games. *Sci. Amer.*, 20, Feb. 1978. (cf. also *Sci. Amer.*, March 1978, p. 27).

[522] M. Gardner. Mathematical games. *Sci. Amer.*, pages 22–32, June 1981.

[523] M. Gardner. Wheels, life and other mathematical amusements. *Freeman*, 1983.

[524] M. R. Garey and D. S. Johnson. *Computers and Intractability: A Guide to the Theory of NP-Completeness.* Freeman, New York, 1979.

[525] D. K. Garnick and N. A. Nieuwejaar. Total domination of the $m \times n$ chessboard by kings, crosses, and knights. *Ars. Combin.*, 41:65–75, 1995.

[526] H. Gavlas and K. Schultz. Open domination in graphs. Submitted, 1996.

[527] H. Gavlas, K. Schultz, and P. J. Slater. Efficient open domination in graphs. *Scientia.* To appear.

[528] F. Gavril. Algorithms for minimum colorings, maximum clique, minimum coverings by cliques, and maximum independent set of a chordal graph. *SIAM J. Comput.*, 1:180–187, 1972.

[529] J. P. Georges and M. D. Halsey. Edge domination and graph structure. *Congr. Numer.*, 76:127–144, 1990.

[530] D. Gernert. Forbidden and unavoidable subgraphs. *Ars Combin.*, 27:165–175, 1989.

[531] D. Gernert. Inequalities between the domination number and the chromatic number of a graph. *Discrete Math.*, 76:151–153, 1989.

[532] D. Gernert. Some new relations between graph invariants. In *Series: Methods Oper. Res., XII Symp. on Operations Research*, volume 58, pages 183–189, Passau (1987), 1989. Athenaum/Hain/Hanstein, Königstein.

[533] J. Ghoshal, R. Laskar, and D. Pillone. Connected domination and C-irredundance. *Congr. Numer.*, 107:161–171, 1995.

[534] J. Ghoshal, R. Laskar, and D. Pillone. Topics on domination in directed graphs. In T. W. Haynes, S. T. Hedetniemi, and P. J. Slater, editors, *Domination in Graphs: Advanced Topics*, chapter 15. Marcel Dekker, Inc., 1997.

[535] P. B. Gibbons and J. A. Webb. Some new results for the queens domination problem. *Australian J. Combin.*, 1997.

[536] J. Gimbel and M. A. Henning. Bounds on an independent distance domination parameter. *J. Combin. Math. Combin. Comput.*, 20:193–205, 1996.

[537] J. Gimbel, M. Mahêo, and C. Virlouvet. Double total domination of graphs. *Discrete Math.*, 165/166:343–351, 1997.

[538] J. Ginsburg. Gauss's arithmetization of the problem of 8 queens. *Scripta Math.*, 5(1):63–66, 1938.

[539] W. Goddard and M. A. Henning. Real and integer domination in graphs. Submitted, 1995.

[540] W. Goddard and M. A. Henning. Generalized domination and independence in graphs. Submitted, 1997.

[541] W. Goddard, M. A. Henning, and H. C. Swart. Some Nordhaus-Gaddum-type results. *J. Graph Theory*, 16(3):221–231, 1992.

[542] W. Goddard, O. Oellermann, P. J. Slater, and H. Swart. Bounds on the total redundance and efficiency of a graph. *Ars Combin.* To appear.

[543] M. Goldberg and T. Spencer. An efficient parallel algorithm that finds independent sets of guaranteed size. *SIAM J. Discrete Math.*, 6(3):443–459, 1993.

[544] M. Goljan, J. Kratochvíl, and P. Kučera. String graphs. Technical Report MR 88c:05088, Academia, Prague, 1986.

[545] S. W. Golomb and L. D. Baumert. Backtrack programming. *J. Assoc. Comput. Mach.*, 12(4):516–524, 1965.

[546] S. W. Golomb and E. C. Posner. Rook domains, Latin squares, affine planes, and error-distributing codes. *IEEE Trans. Inform. Theory*, IT-10:196–208, 1964.

[547] M. C. Golumbic and P. L. Hammer. Stability in circular arc graphs. *J. Algorithms*, 9:314–320, 1988.

[548] M. C. Golumbic and R. C. Laskar. Irredundancy in circular arc graphs. *Discrete Appl. Math.*, 44:79–89, 1993.

[549] R. L. Graham and J. H. Spencer. A constructive solution to a tournament problem. *Canad. Math. Bull.*, 14:45–48, 1971.

[550] S. Gravier and A. Khelladi. On the domination number of cross products of graphs. *Discrete Math.*, 145:273–277, 1995.

[551] J. R. Griggs, C. M. Grinstead, and D. R. Guichard. The number of maximal independent sets in a connected graph. *Discrete Math.*, 68:221–220, 1988.

[552] J. R. Griggs and J. P. Hutchinson. On the r-domination number of a graph. *Discrete Math.*, 101:65–72, 1992.

[553] C. M. Grinstead, B. Hahne, and D. Van Stone. On the queen domination problem. *Discrete Math.*, 86:21–26, 1990.

[554] D. L. Grinstead. *Algorithmic Templates and Multiset Problems in Graphs.* PhD thesis, Univ. Alabama, Huntsville, 1989.

[555] D. L. Grinstead and P. J. Slater. Fractional domination and fractional packing in graphs. *Congr. Numer.*, 71:153–172, 1990.

[556] D. L. Grinstead and P. J. Slater. On minimum dominating sets with minimum intersection. *Discrete Math.*, 86:239–254, 1990.

[557] D. L. Grinstead and P. J. Slater. On the minimum intersection of minimum dominating sets in series-parallel graphs. In Y. Alavi, G. Chartrand, O. R. Oellermann, and A. J. Schwenk, editors, *Graph Theory, Combinatorics, and Applications VI, Proc. Sixth Quad. Internat. Conf. on the Theory and Applications of Graphs*, volume 1, pages 563–584, (Kalamazoo, MI 1988), 1991. Wiley.

[558] D. L. Grinstead and P. J. Slater. A recurrence template for several parameters in series-parallel graphs. *Discrete Appl. Math.*, 54:151–168, 1994.

[559] D. L. Grinstead, P. J. Slater, N. A. Sherwani, and N. D. Holmes. Efficient edge domination problems in graphs. *Inform. Process. Lett.*, 48:221–228, 1993.

[560] P. J. P. Gröbler. *Functional Generalisations of Dominating Sets of Graphs.* Master's thesis, Univ. South Africa, 1994.

[561] P. J. P. Gröbler and C. M. Mynhardt. Extremum aggregates of minimal 0-dominating functions of graphs. *Quaestiones Math.*, 19:291–313, 1996.

[562] J. W. Grossman. Spanning star trees in regular graphs. *Graphs Combin.* To appear.

[563] J. W. Grossman. Dominating sets whose closed stars form spanning trees. *Discrete Math.*, 169:83–94, 1997.

[564] R. K. Guha and T. W. Haynes. Some remarks on k–insensitive graphs in network system design. *Sankhyā*, 54:177–187, 1992.

[565] G. Gunther, B. L. Hartnell, L. Markus, and D. F. Rall. Graphs with unique minimum dominating sets. *Congr. Numer.*, 101:55–63, 1994.

[566] G. Gunther, B. L. Hartnell, and D. F. Rall. Critical β^- stable graphs. *Congr. Numer.*, 86:55–64, 1992.

[567] G. Gunther, B. L. Hartnell, and D. F. Rall. Graphs whose vertex independence number is unaffected by single edge addition or deletion. *Discrete Appl. Math.*, 46:167–172, 1993.

[568] G. Gunther, B. L. Hartnell, and D. F. Rall. A restricted domination problem. *Congr. Numer.*, 94:215–222, 1993.

[569] G. Gunther, B. L. Hartnell, and D. F. Rall. Star-factors and k-bounded total domination. *Networks*, 27:197–201, 1996.

[570] R. P. Gupta. Independence and covering numbers of line graphs and total graphs. In F. Harary, editor, *Proof Techniques in Graph Theory*, pages 61–62, New York, 1969. Academic Press.

[571] Y. Gurevich, L. Stockmeyer, and U. Vishkin. Solving NP-hard problems on graphs that are almost trees and an application to facility location problems. Technical Report RC9348, IBM, Yorktown Heights, NY, 1982.

[572] G. Gutin and V. È. Zverovich. Upper domination and upper irredundance perfect graphs. *Discrete Math.* To appear.

[573] P. Hansen. Upper bounds for the stability number of a graph. *Rev. Roumaine Math. Pures Appl.*, 24:1195–1199, 1979.

[574] P. Hansen. Bornes et algorithmes pour les stables d'un graph (Bounds and algorithms for the stable vertices of a graph). In *A Look at Graph Theory (Proc. Colloq., Cerisy) (French)*, pages 39–53. Polytech. Romandes, Lausanne, 1980.

[575] D. Hanson. Hamilton closures in domination critical graphs. *J. Combin. Math. Combin. Comput.*, 13:121–128, 1993.

[576] F. Harary. *Graph Theory*. Addison-Wesley, Reading, MA, 1969.

[577] F. Harary. Conditional graph theory III: Independence numbers. *Bull. Greek Math. Soc.*, 28B:1–8, 1987.

[578] F. Harary and M. Bezhad. On the problem of characterizing digraphs with solutions and kernels. *Sociometry*, 20:205–215, 1957.

[579] F. Harary, J. P. Hayes, and H. Wu. A survey of the theory of hypercube graphs. *Comput. Math. Appl.*, 15:277–289, 1988.

[580] F. Harary and T. W. Haynes. Conditional graph theory IV: Dominating sets. *Utilitas Math.* To appear.

[581] F. Harary and T. W. Haynes. Double domination in graphs. *Ars Combin.* To appear.

[582] F. Harary and T. W. Haynes. The k-tuple domatic number of a graph. *Math. Slovaca.* To appear.

[583] F. Harary and T. W. Haynes. Nordhaus-Gaddum inequalities for domination in graphs. *Discrete Math.*, 155:99–105, 1996.

[584] F. Harary, T. W. Haynes, and M. Lewinter. On the codomination number of a graph. *Proyecciones*, 12:149–153, 1993.

[585] F. Harary, T. W. Haynes, and P. J. Slater. Efficient and excess domination in graphs. *J. Combin. Math. Combin. Comput.* To appear.

[586] F. Harary and M. Livingston. Characterization of trees with equal domination and independent domination numbers. *Congr. Numer.*, 55:121–150, 1986.

[587] F. Harary and M. Livingston. Caterpillars with equal domination and independence domination numbers. In V. R. Kulli, editor, *Recent Studies in Graph Theory*, pages 149–154. Vishwa Internat. Publ., Gulbarga, India, 1989.

[588] F. Harary and M. Livingston. Independent domination in hypercubes. Manuscript, 1995.

[589] F. Harary and M. Richardson. A matrix algorithm for solutions and r-bases of a finite irreflexive relation. *Naval Res. Logist. Quart.*, 6:307–314, 1959.

[590] F. Harary and S. Schuster. Interpolation theorems for the independence and domination numbers of spanning trees. *Ann. Discrete Math.*, 41:221–228, 1989. North-Holland, Amsterdam-New York.

[591] E. O. Hare. *Algorithms for Grids and Grid-like Graphs.* PhD thesis, Clemson Univ., 1989.

[592] E. O. Hare. k-weight domination and fractional domination of $P_m \times P_n$. *Congr. Numer.*, 78:71–80, 1990.

[593] E. O. Hare. Fibonacci numbers and fractional domination of $P_m \times P_n$. *Fibonacci Quart.*, 32:69–73, 1994.

[594] E. O. Hare and W. R. Hare. k-packing of $P_m \times P_n$. *Congr. Numer.*, 84:33–39, 1991.

[595] E. O. Hare and W. R. Hare. Domination in graphs similar to grid graphs. *Congr. Numer.*, 97:143–154, 1993.

[596] E. O. Hare, W. R. Hare, and S. T. Hedetniemi. Algorithms for computing the domination number of $k \times n$ complete grid graphs. *Congr. Numer.*, 55:81–92, 1986.

[597] E. O. Hare, S. Hedetniemi, R. C. Laskar, K. Peters, and T. Wimer. Linear-time computability of combinatorial problems on generalized-series-parallel graphs. In D. S. Johnson, T. Nishizeki, A. Nozaki, and H. S. Wilf, editors, *Discrete Algorithms and Complexity, Proc. Japan-US Joint Seminar*, pages 437–457, Kyoto, Japan, 1987. Academic Press, New York.

[598] E. O. Hare and S. T. Hedetniemi. A linear algorithm for computing the knight's domination number of a $K \times N$ chessboard. *Congr. Numer.*, 59:115–130, 1987.

[599] M. Harminic. Solutions and kernels of a directed graph. *Math. Slovaca*, 32:263–267, 1982.

[600] Y. H. Harris Kwong and H. J. Straight. An extremal problem involving neighborhood number. Manuscript, 1995.

[601] B. L. Hartnell. On determining the 2-packing and domination numbers of the Cartesian product of certain graphs. *Ars Combin.* To appear.

[602] B. L. Hartnell. Some problems on minimum dominating sets. *Congr. Numer.*, 19:317–320, 1977.

[603] B. L. Hartnell and S. Fitzpatrick. Paired-domination. *Discuss. Math.- Graph Theory.* To appear.

[604] B. L. Hartnell, L. K. Jorgensen, P. D. Vestergaard, and C. Whitehead. Edge stability of the k-domination number of trees. *Bull. Inst. Combin. Appl.* To appear.

[605] B. L. Hartnell and D. F. Rall. On Vizing's conjecture. *Congr. Numer.*, 82:87–96, 1991.

[606] B. L. Hartnell and D. F. Rall. A characterization of trees in which no edge is essential to the domination number. *Ars Combin.*, 33:65–76, 1992.

[607] B. L. Hartnell and D. F. Rall. Bounds on the bondage number of a graph. *Discrete Math.*, 128:173–177, 1994.

[608] B. L. Hartnell and D. F. Rall. A characterization of graphs in which some minimum dominating set covers all the edges. *Czech. Math. J.*, 45:221–230, 1995.

[609] B. L. Hartnell and D. F. Rall. A characterization of graphs in which some minimum dominating set covers all the edges. *Czech. Math. J.*, 45:221–230, 1995.

[610] B. L. Hartnell and D. F. Rall. Vizing's conjecture and the one-half argument. *Discuss. Math. - Graph Theory*, 15:205–216, 1995.

[611] B. L. Hartnell and D. F. Rall. A bound on the size of a graph with given order and bondage number. Submitted, 1997.

[612] B. L. Hartnell and D. F. Rall. Domination in cartesian products: Vizing's conjecture. In T. W. Haynes, S. T. Hedetniemi, and P. J. Slater, editors, *Domination in Graphs: Advanced Topics*, chapter 7. Marcel Dekker, Inc., 1997.

[613] B. L. Hartnell and D. F. Rall. On graphs in which every minimal total dominating set is minimum. Submitted, 1997.

[614] A. S. Hasratian and N. K. Khachatrian. Stable properties of graphs. *Discrete Math.*, 90:143–152, 1991.

[615] M. Hassan. The p-center problem in a unicyclic graph. *Acta Math. Univ. Comenian.*, 50/51:15–26, 1987. (1988).

[616] J. H. Hattingh. On irredundant Ramsey numbers for graphs. *J. Graph Theory*, 14(4):437–441, 1990.

[617] J. H. Hattingh. Majority domination and its generalizations. In T. W. Haynes, S. T. Hedetniemi, and P. J. Slater, editors, *Domination in Graphs: Advanced Topics*, chapter 4. Marcel Dekker, Inc., 1997.

[618] J. H. Hattingh, G. Chen, and C. C. Rousseau. Asymptotic bounds for irredundant and mixed Ramsey numbers. *J. Graph Theory*, 17:193–206, 1993.

[619] J. H. Hattingh, J. Heidema, and E. Jonck. Private distance domination. Submitted, 1995.

[620] J. H. Hattingh and M. A. Henning. On strong domination in graphs. *J. Combin. Math. Combin. Comput.* To appear.

[621] J. H. Hattingh and M. A. Henning. The complexity of upper distance irredundance. *Congr. Numer.*, 91:107–115, 1992.

[622] J. H. Hattingh and M. A. Henning. A characterization of block graphs that are well-k-dominated. *J. Combin. Math. Combin. Comput.*, 13:33–38, 1993.

[623] J. H. Hattingh and M. A. Henning. The ratio of the distance irredundance and domination numbers of a graph. *J. Graph Theory*, 18(1):1–9, 1994.

[624] J. H. Hattingh and M. A. Henning. Distance irredundance in graphs. In *Graph Theory, Combinatorics, Algorithms and Applications, Proc. Seventh Quad. Internat. Conf. on the Theory and Applications of Graphs*, volume 1, pages 529–542, (Kalamazoo, MI 1992), 1995. Wiley.

[625] J. H. Hattingh, M. A. Henning, and J. C. Schoeman. Distance irredundance in graphs: complexity issues. *Ars Combin.* To appear.

[626] J. H. Hattingh, M. A. Henning, and P. J. Slater. Three-valued k-neighbourhood domination in graphs. *Australas. J. Combin.*, 9:233–242, 1994.

[627] J. H. Hattingh, M. A. Henning, and P. J. Slater. On the algorithmic complexity of signed domination in graphs. *Australas. J. Combin.*, 12, 1995. 101–112.

[628] J. H. Hattingh, M. A. Henning, and E. Ungerer. Partial signed domination in graphs. *Ars Combin.* To appear.

[629] J. H. Hattingh, M. A. Henning, and J. L. Walters. On the computational complexity of upper distance fractional domination. *Australas. J. Combin.*, 7:133–144, 1993.

[630] J. H. Hattingh and R. C. Laskar. On weak domination in graphs. *Ars Combin.* To appear.

[631] J. H. Hattingh, A. A. McRae, and E. Ungerer. The minus bondage number of a graph. In *Graph Theory, Combinatorics, Algorithms, and Applications, Proc. Eighth Quad. Internat. Conf. on the Theory and Applications of Graphs*. To appear.

[632] J. H. Hattingh, A. A. McRae, and E. Ungerer. Minus k-subdomination in graphs III. Submitted, 1997.

[633] J. H. Hattingh and E. Ungerer. Minus k-subdomination in graphs II. *Discrete Math.* To appear.

[634] J. H. Hattingh and E. Ungerer. On the signed k-subdomination number of trees. In *Graph Theory, Combinatorics, Algorithms and Applications, Proc. Eighth Quad. Internat. Conf. on the Theory and Applications of Graphs.* To appear.

[635] J. H. Hattingh and E. Ungerer. The signed and minus k-subdomination numbers of comets. *Discrete Math.* To appear.

[636] J. Haviland. *Cliques and Independent Sets.* PhD thesis, Univ. Cambridge, 1989.

[637] J. Haviland. On minimum maximal independent sets of a graph. *Discrete Math.*, 94:95–101, 1991.

[638] J. Haviland. Independent domination in regular graphs. *Discrete Math.*, 143:275–280, 1995.

[639] L. Hayes, K. Schultz, and J. Yates. Universal dominating sequences of graphs. Submitted, 1996.

[640] P. Hayes. A problem of chess queens. *J. Recreational Math.*, 24(4):264–271, 1992.

[641] T. W. Haynes. Domination in graphs: a brief overview. *J. Combin. Math. Combin. Comput.*, 24:225–237, 1997.

[642] T. W. Haynes, R. C. Brigham, and R. D. Dutton. Extremal 2-2-insensitive graphs. *Congr. Numer.*, 67:158–166, 1988.

[643] T. W. Haynes, R. C. Brigham, and R. D. Dutton. Extremal graphs domination insensitive to the removal of k edges. *Discrete Appl. Math.*, 44:295–304, 1993.

[644] T. W. Haynes, R. K. Guha, R. D. Dutton, and R. C. Brigham. The G-network and its inherent fault tolerant properties. *Internat. J. Computer Math.*, 31:167–175, 1990.

[645] T. W. Haynes, S. T. Hedetniemi, and P. J. Slater, editors. *Domination in Graphs: Advanced Topics.* Marcel Dekker, Inc., New York, NY, 1997.

[646] T. W. Haynes and M. A. Henning. Domination critical graphs with respect to relative complements. Submitted, 1995.

[647] T. W. Haynes and M. A. Henning. The domatic number of factors of graphs. Submitted, 1997.

[648] T. W. Haynes and M. A. Henning. Forbidden subgraph domination. Manuscript, 1997.

[649] T. W. Haynes and D. J. Knisley. Generalized maximum degree and totally regular graphs. *Utilitas Math.* To appear.

[650] T. W. Haynes and L. M. Lawson. Applications of E-graphs in network design. *Networks*, 23:473–479, 1993.

[651] T. W. Haynes, L. M. Lawson, R. C. Brigham, and R. D. Dutton. Changing and unchanging of the graphical invariants: minimum degree, maximum degree, node independence number, maximum clique size and edge independence number. *Congr. Numer.*, 72:239–252, 1990.

[652] T. W. Haynes and P. J. Slater. Paired-domination and the paired-domatic number. *Congr. Numer.*, 109:67–72, 1995.

[653] T. W. Haynes and P. J. Slater. Paired-domination in graphs. Submitted, 1995.

[654] X. He and Y. Yesha. Efficient parallel algorithms for r-dominating set and p-center problems on trees. *Algorithmica*, 5:129–145, 1990.

[655] S. M. Hedetniemi. Optimum domination in unicyclic graphs. Technical Report CIS-TR-80-14, Dept. Computer and Information Science, Univ. Oregon, 1980.

[656] S. M. Hedetniemi, S. T. Hedetniemi, and M. A. Henning. The algorithmic complexity of perfect neighborhoods in graphs. *J. Combin. Math. Combin. Comput.* To appear.

[657] S. M. Hedetniemi, S. T. Hedetniemi, and D. P. Jacobs. Private domination: theory and algorithms. *Congr. Numer.*, 79:147–157, 1990.

[658] S. M. Hedetniemi, S. T. Hedetniemi, and D. P. Jacobs. Total irredundance in graphs: theory and algorithms. *Ars Combin.*, 35A:271–284, 1993.

[659] S. M. Hedetniemi, S. T. Hedetniemi, and R. C. Laskar. Domination in trees: models and algorithms. In Y. Alavi, G. Chartrand, L. Lesniak, D. R. Lick, and C. E. Wall, editors, *Graph Theory with Applications to Algorithms and Computer Science*, pages 423–442. Wiley, New York, 1985.

[660] S. M. Hedetniemi, S. T. Hedetniemi, R. C. Laskar, A. A. McRae, and A. Majumdar. Domination, independence and irredundance in total graphs: a brief survey. In Y. Alavi and A. J. Schwenk, editors, *Graph Theory, Combinatorics, Algorithms and Applications, Proc. Seventh Quad. Internat. Conf. on the Theory and Applications of Graphs*, volume 2, pages 671–683, (Kalamazoo, MI 1992), 1995. Wiley.

[661] S. M. Hedetniemi, S. T. Hedetniemi, A. McRae, and D. Parks. Matchings, matchability, and 2-maximal matchings in graphs. Submitted, 1997.

[662] S. M. Hedetniemi, S. T. Hedetniemi, A. A. McRae, D. Parks, and J. A. Telle. Iterated coloring. Technical Report 134, Univ. Bergen, 1997.

[663] S. M. Hedetniemi, S. T. Hedetniemi, and R. Reynolds. Combinatorial problems on chessboards: II. In T. W. Haynes, S. T. Hedetniemi, and P. J. Slater, editors, *Domination in Graphs: Advanced Topics*, chapter 6. Marcel Dekker, Inc., 1997.

[664] S. M. Hedetniemi, S. T. Hedetniemi, and T. V. Wimer. Linear time resource allocation algorithms for trees. Technical Report URI-014, Dept. Mathematical Sciences, Clemson Univ., 1987. presented at Southeastern Conf. on Combinatorics, Graph Theory and Computing (Boca Raton, FL, 1987).

[665] S. T. Hedetniemi. Hereditary properties of graphs. *J. Combin. Theory*, 14:16–27, 1973.

[666] S. T. Hedetniemi. A max-min relationship between matchings and domination in graphs. *Congr. Numer.*, 40:23–34, 1983.

[667] S. T. Hedetniemi, D. P. Jacobs, and R. C. Laskar. Inequalities involving the rank of a graph. *J. Combin. Math. Combin. Comput.*, 6:173–176, 1989.

[668] S. T. Hedetniemi, D. P. Jacobs, R. C. Laskar, and D. Pillone. Open perfect neighborhood sets in graphs. Submitted, 1997.

[669] S. T. Hedetniemi and R. C. Laskar. Connected domination in graphs. In B. Bollobás, editor, *Graph Theory and Combinatorics*, pages 209–218. Academic Press, London, 1984.

[670] S. T. Hedetniemi and R. C. Laskar. A bipartite theory of graphs: I. *Congr. Numer.*, 55:5–14, 1986.

[671] S. T. Hedetniemi and R. C. Laskar. A bipartite theory of graphs: II. *Congr. Numer.*, 64:137–146, 1988.

[672] S. T. Hedetniemi and R. C. Laskar. Recent results and open problems in domination theory. In R. Ringeisen and F. Roberts, editors, *Applications of Discrete Mathematics*, pages 205–218. SIAM, Philadelphia, PA, 1988.

[673] S. T. Hedetniemi and R. C. Laskar. On irredundance, dimension and rank in posets. In M. F. Capobianco, M. Guan, D. F. Hsu, and F. Tian, editors, *Graph Theory and its Applications, East and West, Proc. First China-U.S.A. Internat. Graph Theory Conf.*, volume 576, pages 235–240. Ann. New York Acad. Sci., 1989.

[674] S. T. Hedetniemi and R. C. Laskar. Bibliography on domination in graphs and some basic definitions of domination parameters. *Discrete Math.*, 86:257–277, 1990.

[675] S. T. Hedetniemi and R. C. Laskar, editors. *Topics on Domination*, volume 48. North Holland, New York, 1990.

[676] S. T. Hedetniemi, R. C. Laskar, and J. Pfaff. Irredundance in graphs: a survey. *Congr. Numer.*, 48:183–193, 1985.

[677] S. T. Hedetniemi, R. C. Laskar, and J. Pfaff. A linear algorithm for finding a minimum dominating set in a cactus. *Discrete Appl. Math.*, 13:287–292, 1986.

[678] S. T. Hedetniemi, A. A. McRae, and D. A. Parks. Complexity results. In T. W. Haynes, S. T. Hedetniemi, and P. J. Slater, editors, *Domination in Graphs: Advanced Topics*, chapter 9. Marcel Dekker, Inc., 1997.

[679] P. Heggernes and J. A. Telle. Partitioning graphs into generalized dominating sets. In *Proc. of XV Internat. Conf. of the Chilean Computer Society*, pages 241–252, 1995.

[680] M. A. Henning. Distance perfect neighborhoods in graphs. *Utilitas Math.* To appear.

[681] M. A. Henning. K_n-domination sequences of graphs. *J. Combin. Math. Combin. Comput.*, 10:161–172, 1991.

[682] M. A. Henning. Irredundance perfect graphs. *Discrete Math.*, 142:107–120, 1995.

[683] M. A. Henning. Domination in graphs: a survey. In G. Chartrand and M. Jacobson, editors, *Surveys in Graph Theory*, volume 116, pages 139–172. *Congr. Numer.*, 1996.

[684] M. A. Henning. Domination in regular graphs. *Ars Combin.*, 43:263–271, 1996.

[685] M. A. Henning. A characterization of edge-minimal connected graphs on n vertices with minimum degree two and restrained domination number at least $(n-1)/2$. Submitted, 1997.

[686] M. A. Henning. Distance domination in graphs. In T. W. Haynes, S. T. Hedetniemi, and P. J. Slater, editors, *Domination in Graphs: Advanced Topics*, chapter 12. Marcel Dekker, Inc., 1997.

[687] M. A. Henning. Dominating functions in graphs. In T. W. Haynes, S. T. Hedetniemi, and P. J. Slater, editors, *Domination in Graphs: Advanced Topics*, chapter 2. Marcel Dekker, Inc., 1997.

[688] M. A. Henning. Packing in graphs. Submitted, 1997.

[689] M. A. Henning. Research in graph theory. *South African J. Sci.*, 93:15–18, 1997.

[690] M. A. Henning. Upper bounds on the lower open packing number of a tree. Submitted, 1997.

[691] M. A. Henning and H. R. Hind. Strict majority functions in graphs. Submitted, 1997.

[692] M. A. Henning and G. Kubicka. Real domination in graphs. *J. Combin. Math. Combin. Comput.* To appear.

[693] M. A. Henning, O. R. Oellermann, and H. C. Swart. Distance domination critical graphs. *J. Combin. Inform. System Sci.* To appear.

[694] M. A. Henning, O. R. Oellermann, and H. C. Swart. Diversity of generalized domination in graphs. *Discrete Math.* To appear.

[695] M. A. Henning, O. R. Oellermann, and H. C. Swart. Bounds on distance domination parameters. *J. Combin. Inform. System Sci.*, 16:11–18, 1991.

[696] M. A. Henning, O. R. Oellermann, and H. C. Swart. Relationships between distance domination parameters. *Math. Pannon.*, 5(1):69–79, 1994.

[697] M. A. Henning, O. R. Oellermann, and H. C. Swart. Relating pairs of distance domination parameters. *J. Combin. Math. Combin. Comput.*, 18:233–244, 1995.

[698] M. A. Henning, O. R. Oellermann, and H. C. Swart. The diversity of domination. *Discrete Math.*, 161:161–173, 1996.

[699] M. A. Henning, O. R. Oellermann, and H. C. Swart. Local edge-domination critical graphs. *Discrete Math.*, 161:175–184, 1996.

[700] M. A. Henning and P. J. Slater. Open packing in graphs. *J. Combin. Math. Combin. Comput.* To appear.

[701] M. A. Henning and P. J. Slater. Closed neighborhood order dominating functions. Submitted, 1995.

[702] M. A. Henning and P. J. Slater. Inequalities relating domination parameters in cubic graphs. *Discrete Math.*, 158:87–98, 1996.

[703] M. A. Henning and H. C. Swart. Bounds on a generalized total domination parameter. *J. Combin. Math. Combin. Comput.*, 6:143–153, 1989.

[704] M. A. Henning and H. C. Swart. Bounds on a generalized domination parameter. *Quaestiones Math.*, 13(2):237–257, 1990.

[705] M. A. Henning and H. C. Swart. Bounds relating generalized domination parameters. *Discrete Math.*, 120:93–105, 1993.

[706] A. Hertz and D. de Werra. On the stability number of AH-free graphs. *J. Graph Theory*, 17:53–63, 1993.

[707] H. Hitotumatsu and K. Noshita. A technique for implementing backtrack algorithms and its applications. *Inform. Process. Lett.*, 8(4):174–175, 1979.

[708] D. S. Hochbaum. On the fractional solution to the set covering problem. *SIAM J. Algebraic Discrete Methods*, 4:221–222, 1983.

[709] D. S. Hochbaum and D. B. Shmoys. A best possible heuristic for the k-center problem. *Math. Oper. Res.*, 10:180–184, 1985.

[710] E. J. Hoffman, J. C. Loessi, and R. C. Moore. Constructions for the solution of the m queens problem. *Math. Mag.*, 42:66–72, 1969.

[711] D. H. Hollander. An unexpected two-dimensional space-group containing seven of the twelve basic solutions to the eight queens problem. *J. Recreational Math.*, 6:287–291, 1973.

[712] L. Holley, Y. Lai, and B. QuanYue. Orientable step domination in graphs. *Congr. Numer.*, 117:129–143, 1996.

[713] I. J. Holyer and E. J. Cockayne. On the sum of cardinalities of extremum maximal independent sets. *Discrete Math.*, 26:243–246, 1979.

[714] J. N. Hooker, R. S. Garfinkel, and C. K. Chen. Finite dominating sets for network location problems. *Oper. Res.*, 39:100–118, 1991.

[715] G. Hopkins and W. Staton. Graphs with unique maximum independent sets. *Discrete Math.*, 57:245–251, 1985.

[716] J. D. Horton and K. Kilakos. Minimum edge dominating sets. *SIAM J. Discrete Math.*, 6(3):375–387, 1993.

[717] W. Hsu. On the domination numbers of trees. Technical report, Dept. Industrial Eng. and Management Sci., Northwestern Univ., 1980.

[718] W. Hsu. On the maximum coverage problem on trees. Technical report, Dept. Industrial Eng. and Management Sci., Northwestern Univ., 1981.

[719] W. Hsu. The distance-domination numbers of trees. *Oper. Res. Lett.*, 1:96–100, 1982.

[720] W. Hsu and K. Tsai. Linear time algorithms on circular-arc graphs. *Inform. Process. Lett.*, 40:123–129, 1991.

[721] M. Hujter. The irredundance and domination numbers are equal in domistable graphs. *MTA Számítástechnikai és Automatizálási Kutató Intézete*, pages 90–126, 1990. Budapest.

[722] A. B. Huseby. Domination theory and the Crapo B-invariant. *Networks*, 1:135–149, 1989.

[723] S. F. Hwang. *Domination and related topics*. PhD thesis, National Chiao Tung Univ., 1991.

[724] S. F. Hwang and G. J. Chang. k-Neighbor covering and independence problem. *SIAM J. Discrete Math.* To appear.

[725] S. F. Hwang and G. J. Chang. The k-neighbor domination problem in block graphs. *European J. Oper. Res.*, 52:373–377, 1991.

[726] S. F. Hwang and G. J. Chang. Edge domatic numbers of complete n-partite graphs. *Graphs Combin.*, 10:241–248, 1994.

[727] S. F. Hwang and G. J. Chang. The edge domination problem. *Discuss. Math.–Graph Theory*, 15:51–57, 1995.

[728] O. H. Ibarra and Q. Zheng. Some efficient algorithms for permutation graphs. *J. Algorithms*, 16:453–469, 1994.

[729] T. Ichiishi. Comparative cooperative game theory. *Internat. J. Game Theory*, 19:139–152, 1990.

[730] R. W. Irving. On approximating the minimum independent dominating set. *Inform. Process. Lett.*, 37(4):197–200, 1991.

[731] J. Ivanco and B. Zelinka. Domination in Kneser graphs. *Math. Bohem.*, 118:147–152, 1993.

[732] A. H. Jackson and R. P. Pargas. Solutions to the $N \times N$ knights cover problem. *Rec. Math.*, 23(4):255–267, 1991.

[733] B. Jackson, H. Li, and Y. Zhu. Dominating cycles in regular 3-connected graphs. *Discrete Math.*, 102:163–176, 1991.

[734] H. Jacob. Kernels in graphs with a clique-cutset. *Discrete Math.*, 156:265–267, 1996.

[735] M. S. Jacobson and L. F. Kinch. On the domination number of products of graphs: I. *Ars. Combin.*, 18:33–44, 1984.

[736] M. S. Jacobson and L. F. Kinch. On the domination of the products of graphs II: Trees. *J. Graph Theory*, 10:97–106, 1986.

[737] M. S. Jacobson and K. Peters. Complexity questions for n-domination and related parameters. *Congr. Numer.*, 68:7–22, 1989.

[738] M. S. Jacobson and K. Peters. Chordal graphs and upper irredundance, upper domination and independence. *Discrete Math.*, 86:59–69, 1990.

[739] M. S. Jacobson and K. Peters. A note on graphs which have upper irredundance equal to independence. *Discrete Appl. Math.*, 44:91–97, 1993.

[740] M. S. Jacobson, K. Peters, and D. Rall. On n-irredundance and n-domination. *Ars Combin.*, 29B:151–160, 1990.

[741] F. Jaeger and C. Payan. Relations du type Norhaus-Gaddum pour le nombre d'absorption d'un graphe simple. *C. R. Acad. Sci. Paris*, 274:728–730, 1972.

[742] S. R. Jayaram. The domatic number of a graph. *Nat. Acad. Sci. Lett.*, 10:23–25, 1987.

[743] S. R. Jayaram. Line domination in graphs. *Graphs Combin.*, 3:357–363, 1987.

[744] S. R. Jayaram. Maximum/minimum neighbourhood number of a graph. *J. Math. Phys. Sci.*, 24:37–44, 1990.

[745] S. R. Jayaram, Y. H. Harris Kwong, and H. J. Straight. Neighbourhood sets in graphs. *Indian J. Pure Appl. Math.*, 22:259–268, 1991.

[746] D. S. Johnson. The NP-completeness column: an ongoing guide. *J. Algorithms*, 5:147–160, 1984.

[747] D. S. Johnson. The NP-completeness column: an ongoing guide. *J. Algorithms*, 6:291–305,434–451, 1985.

[748] S. M. Johnson. A new lower bound for coverings by rook domains. *Utilitas Math.*, 1:121–140, 1972.

[749] T. W. Johnson and P. J. Slater. Maximum independent, minimally c-redundant sets in graphs. *Congr. Numer.*, 74:193–211, 1990.

[750] T. W. Johnson and P. J. Slater. Maximum independent, minimally redundant sets in series-parallel graphs. *Quaestiones Math.*, 16:351–370, 1993.

[751] C. Jordan. Sur les assemblages de lignes. *J. Reine Angew. Math.*, 70:185–190, 1869.

[752] J. Paulraj Joseph and S. Arumugam. Domination and connectivity in graphs. *Internat. J. Management Systems*, 8(3):233–236, 1992.

[753] J. Paulraj Joseph and S. Arumugam. On connected cutfree domination in graphs. *Indian J. Pure Appl. Math.*, 23:643–647, 1992.

[754] J. Paulraj Joseph and S. Arumugam. On the connected domatic number of a graph. *J. Ramanujan Math. Soc.*, 9(1):69–77, 1994.

[755] J. Paulraj Joseph and S. Arumugam. Domination in graphs. *Internat. J. Management Systems*, 11:177–182, 1995.

[756] J. Paulraj Joseph and S. Arumugam. Domination and colouring in graphs. Submitted, 1996.

[757] J. Paulraj Joseph and S. Arumugam. On 2-edge connected domination in graphs. *Internat. J. Management Systems*, 12:131–138, 1996.

[758] D. S. Joshi, S. Radhakrishnan, and N. Chandrasekharan. The k-neighbor, r-domination problem on interval graphs. Submitted, 1993.

[759] S. Judd. Domination and perpetual evasiveness. *Congr. Numer.*, 91:117–127, 1992.

[760] H. A. Jung and P. Fraisse. Longest cycle and independent sets in k-connected graphs. In V. R. Kulli, editor, *Recent Studies in Graph Theory*. Vishwa International, Gulbarga, 1989.

[761] J. G. Kalbfleisch, R. G. Stanton, and J. D. Horton. On covering sets and error-correcting codes. *J. Combin. Theory Ser. A*, 11:233–250, 1971.

[762] P. P. Kale and N. V. Deshpande. On line independence, domination, irredundance and neighbourhood numbers of a graph. *Indian J. Pure Appl. Math.*, 21:695–698, 1990.

[763] M. Kano and N. Tokushige. Binding numbers and f-factors of graphs. *J. Combin. Theory Ser. B*, 54:213–221, 1992.

[764] O. Kariv and S. L. Hakimi. An algorithmic approach to network location problems I: the p-centers. *SIAM J. Appl. Math.*, 37:513–538, 1979.

[765] O. Kariv and S. L. Hakimi. An algorithmic approach to network location problems II: the p-medians. *SIAM J. Appl. Math.*, 37:539–560, 1979.

[766] J. M. Keil. Total domination in interval graphs. *Inform. Process. Lett.*, 22:171–174, 1986.

[767] J. M. Keil. The complexity of domination problems in circle graphs. *Discrete Appl. Math.*, 42:51–63, 1993.

[768] J. M. Keil, R. C. Laskar, and P. D. Manuel. Domination problems in directed path graphs. Manuscript, 1997.

[769] J. M. Keil, R. C. Laskar, and P. D. Manuel. The vertex clique cover problem and some related problems in chordal graphs. Submitted, 1997.

[770] J. M. Keil and D. Schaefer. An optimal algorithm for finding dominating cycles in circular-arc graphs. *Discrete Appl. Math.*, 36:25–34, 1992.

[771] L. L. Kelleher. *Domination in Graphs and its Application to Social Network Theory*. PhD thesis, Northeastern Univ., 1985.

[772] L. L. Kelleher and M. B. Cozzens. Dominating sets in social network graphs. *Math. Social Sci.*, 16:267–279, 1988.

[773] T. Kikuno, N. Yoshida, and Y. Kakuda. Dominating set in planar graphs. Technical Report AL79-9, Inst. Electronics and Comm. Eng. of Japan, 1979. pages 21–30.

[774] T. Kikuno, N. Yoshida, and Y. Kakuda. The NP-completeness of the dominating set problem in cubic planar graphs. *IECE Japan*, E64:443–444, 1980.

[775] T. Kikuno, N. Yoshida, and Y. Kakuda. A linear algorithm for the domination number of a series-parallel graph. *Discrete Appl. Math.*, 5:299–311, 1983.

[776] K. Kilakos. *On the Complexity of Edge Domination*. Master's thesis, Univ. New Brunswick, 1988.

[777] S. Klavzar and N. Seifter. Dominating cartesian products of cycles. *Discrete Appl. Math.*, 59(2):129–136, 1995.

[778] D. J. Kleitman and D. B. West. Spanning trees with many leaves. *SIAM J. Discrete Math.*, 4:99–106, 1991.

[779] E. Koch and M. A. Perles. Covering efficiency of trees and *k*-trees. *Congr. Numer.*, 18:391–420, 1976.

[780] E. Köhler. Connected domination on tradezoid graphs in $O(n)$ time. Manuscript, 1996.

[781] E. G. Köhler. *Domination Problems in Three-dimensional Chessboard Graphs*. Master's thesis, Clemson Univ., 1994.

[782] J. Kok and C. M. Mynhardt. Reinforcement in graphs. *Congr. Numer.*, 79:225–231, 1990.

[783] M. C. Kong. On computing maximum *k*-independent sets. *Congr. Numer.*, 95:47–60, 1993.

[784] D. König. *Theorie der Endlichen und Unendlichen Graphen*. Chelsea, New York, 1950.

[785] A. V. Kostochka. Sequences of dominating sets. *Math Pannon.*, 1:51–54, 1990.

[786] A. V. Kostochka. The independent domination number of a cubic 3–connected graph can be much larger that its domination number. *Graphs Combin.*, 9:235–237, 1993.

[787] M. Kraitchik. *Mathematical Recreations*. Norton, New York, 1942.

[788] J. Kratochvíl. One-perfect codes over self-complementary graphs. *Comment. Math. Univ. Carolin.*, 26:589–595, 1985.

[789] J. Kratochvíl. Perfect codes in general graphs. In *Colloquia Math. Soc. J. Bolyai 52, Proc. VIIth Hungarian Colloq. on Combin.*, pages 357–364, Eger, 1987.

[790] J. Kratochvíl. Perfect codes and two-graphs. *Comment. Math. Univ. Carolin.*, 30:755–760, 1989.

[791] J. Kratochvíl. Regular codes in regular graphs are difficult. *Discrete Math.*, 133:191–205, 1994.

[792] J. Kratochvíl. Perfect codes over graphs. *J. Combin. Theory Ser. B*, 40:224–228, 1996.

[793] J. Kratochvíl and M. Křivánek. On the computational complexity of codes in graphs. In *Proc. MFCS Karlovy Vary 1988, Lecture Notes in Comp. Sci. 324*, pages 396–404. Springer-Verlag, Berlin, 1988.

[794] J. Kratochvíl, J. Maly, and J. Matous. On the existence of perfect codes in a random graph. In *Proc. Random Graphs '87*, pages 141–149. Wiley, 1990.

[795] J. Kratochvíl, P. Manuel, and M. Miller. Generalized domination in chordal graphs. *Nordic J. Comput.*, 2:41–50, 1995.

[796] J. Kratochvíl and J. J. Nešetřil. Independent set and clique problems in intersection-defined classes of graphs. *Comment. Math. Univ. Carolin.*, 31(1):1–9, 1990.

[797] D. Kratsch. Finding dominating cliques efficiently, in strongly chordal graphs and undirected path graphs. *Discrete Math.*, 86:225–238, 1990.

[798] D. Kratsch. Domination and total domination in asteroidal triple-free graphs. Technical Report Math/Inf/96/25, F.-Schiller-Univ., Jena, 1996.

[799] D. Kratsch. Algorithms. In T. W. Haynes, S. T. Hedetniemi, and P. J. Slater, editors, *Domination in Graphs: Advanced Topics*, chapter 8. Marcel Dekker, Inc., 1997.

[800] D. Kratsch, P. Damaschke, and A. Lubiw. Dominating cliques in chordal graphs. *Discrete Math.*, 128:269–275, 1994.

[801] D. Kratsch and L. Stewart. Total domination and transformation. *Inform. Process. Lett.* To appear.

[802] D. Kratsch and L. Stewart. Domination on cocomparability graphs. *SIAM J. Discrete Math.*, 6(3):400–417, 1993.

[803] M. S. Krishnamoorthy and K. Murthy. On the total dominating set problem. *Congr. Numer.*, 54:265–278, 1986.

[804] V. R. Kulli. On n-total domination number in graphs. In *Graph Theory, Combinatorics, Algorithms, and Applications*, pages 319–324. SIAM, Philadelphia, PA, 1991.

[805] V. R. Kulli and D. K. Patwari. On the total edge domination number of a graph. In A. M. Mathai, editor, *Proc. of the Symp. on Graph Theory and Combinatorics (Cochin, 1991), Centre Math. Sci., Trivandrum*, pages 75–81, 1991. Series: Publication, 21.

[806] V. R. Kulli and S. C. Sigarkanti. Inverse domination in graphs. *Nat. Acad. Sci. Lett.*, 14:473–475, 1991.

[807] V. R. Kulli and S. C. Sigarkanti. Further results on the neighborhood number of a graph. *Indian J. Pure Appl. Math.*, 23(8):575–577, 1992.

[808] V. R. Kulli, S. C. Sigarkanti, and N. D. Soner. Entire domination in graphs. In V. R. Kulli, editor, *Advances in Graph Theory*, pages 237–243. Vishwa, Gulbarga, 1991.

[809] M. Kwasnik. Charakteristische Funktion, k-Grundy Funktion, Ordinal-Funktion und k-Kern. *Zeszyty Naukowe WSInz.*, Nr. 55, 1980. Matematyka-Fizyka, Zielona Gora.

[810] M. Kwasnik. On the $(k, 1)$-kernels. In *Lecture Notes in Math., Graph Theory, Lagow*, pages 114–121. Springer-Verlag, 1981.

[811] M. Kwasnik. Die Kerne in der Summe und Komposition der Graphen. Manuscript, 1994.

[812] M. Kwasnik. The generalization of Richardson's theorem. Manuscript, 1996.

[813] M. Kwasnik. On $(k, 1)$-kernels of exclusive disjunction, Cartesian sum and normal point product of two directed graphs. Manuscript, 1996.

[814] M. Kwasnik, A. Wloch, and I. Wloch. Some remarks about $(k, 1)$-kernels in directed and undirected graphs. *Discuss. Math.*, 122:29–37, 1993.

[815] Y. H. H. Kwong and H. J. Straight. An extremal problem involving neighborhood number. Manuscript, 1996.

[816] P. Kys. Remarks on domatic number. *Acta Math. Univ. Comenian*, 46/47:237–241, 1985/1986.

[817] J. M. Laborde. On the domatic number of the n-cube and a conjecture of Zelinka. *European J. Combin.*, 8:175–177, 1987.

[818] J. Lalani, R. Laskar, and S. T. Hedetniemi. Graphs and posets: some common parameters. *Congr. Numer.*, 71:205–215, 1990.

[819] D. E. Lampert and P. J. Slater. The acquisition number of a graph. *Congr. Numer.*, 109:203–210, 1995.

[820] D. E. Lampert and P. J. Slater. Interior parameters in $ir \leq \gamma \leq i \leq \ldots \leq \beta \leq \gamma \leq ir$. *Congr. Numer.*, 122:129–143, 1996.

[821] D. E. Lampert and P. J. Slater. The expected number of surviving vertices under parallel knockouts. Submitted, 1997.

[822] D. E. Lampert and P. J. Slater. On the complexity of acquisition and consolidation parameters. Submitted, 1997.

[823] D. E. Lampert and P. J. Slater. Parallel knockouts in the complete graph. Submitted, 1997.

[824] M. Larsen. The problem of kings. *Electron. J. Combin.*, 2, 1995.

[825] L. C. Larson. A theorem about primes proved on a chessboard. *Math. Mag.*, 50(2):69–74, 1977.

[826] R. Laskar, A. Majumdar, G. Domke, and G. Fricke. A fractional view of graph theory. *Sankhyā*, 54:265–279, 1992.

[827] R. Laskar and K. Peters. Connected domination of complementary graphs. Technical Report 429, Dept. Mathematical Sciences, Clemson Univ., 1983.

[828] R. Laskar and K. Peters. Domination and irredundance in graphs. Technical Report 434, Dept. Mathematical Sciences, Clemson Univ., 1983.

[829] R. Laskar and K. Peters. Vertex and edge domination parameters in graphs. *Congr. Numer.*, 48:291–305, 1985.

[830] R. Laskar and J. Pfaff. Domination and irredundance in split graphs. Technical Report 430, Dept. Mathematical Sciences, Clemson Univ., 1983.

[831] R. Laskar, D. Pillone, S. T. Hedetniemi, and D. P. Jacobs. Open perfect neighborhood sets in graphs. Manuscript, 1996.

[832] R. Laskar and H. B. Walikar. On domination related concepts in graph theory. In *Lecture Notes in Mathematics, Combinatorics and Graph Theory (Calcutta, 1980)*, volume 885, pages 308–320. Springer, Berlin, 1981.

[833] R. Laskar and C. Wallis. Chessboard related designs and domination parameters. *R. C. Bose Memorial issue in J. Statist. Plann. Inference*. To appear.

[834] R. C. Laskar, J. Pfaff, S. M. Hedetniemi, and S. T. Hedetniemi. On the algorithmic complexity of total domination. *SIAM J. Algebraic Discrete Methods*, 5:420–425, 1984.

[835] R. C. Laskar and C. K. Wallis. Domination parameters of graphs of three-class association schemes and variations of chessboard graphs. *Congr. Numer.*, 100:199–213, 1994.

[836] C. Lautemann. Decomposition trees: structured graph representation and efficient algorithms. In *Lecture Notes in Comput. Sci.*, volume 229, pages 28–39. Springer, Berlin-New York, 1988.

[837] E. L. Lawler and P. J. Slater. A linear time algorithm for finding an optimal dominating subforest of a tree. In *Graph Theory with Applications to Algorithms and Computer Science*, pages 501–506, (Kalamazoo, MI 1984), 1985. Wiley.

[838] L. M. Lawson, T. W. Haynes, and J. W. Boland. Domination from a distance. *Congr. Numer.*, 103:89–96, 1994.

[839] M. Le Breton. The stability set of voting games: classification and genericity results. *Internat. J. Game Theory*, 19:111–127, 1990.

[840] C. Lee. *On the Domination Number of a Digraph*. PhD thesis, Michigan State Univ., 1994.

[841] J. Lee, M. Y. Sohn, and H. K. Kim. On minus domination numbers of graphs. Submitted, 1997.

[842] L. Lesniak-Foster and J. E. Williamson. On spanning and dominating circuits in graphs. *Canad. Math. Bull.*, 20:215–220, 1977.

[843] Y. D. Liang. Permutation graphs: connected domination and Steiner trees revisited. Manuscript, 1992.

[844] Y. D. Liang. Domination in trapezoid graphs. *Inform. Process. Lett.*, 52:309–315, 1994.

[845] Y. D. Liang. Steiner set and connected domination in trapezoid graphs. *Inform. Process. Lett.*, 56:101–108, 1995.

[846] Y. D. Liang, C. Rhee, S. K. Dall, and S. Lakshmivarahan. A new approach for the domination problem on permutation graphs. *Inform. Process. Lett.*, 37:219–224, 1991.

[847] N. Linial, D. Pleg, and U. Yu. Rabinovich. Neighborhood packings, covers, and polling in graphs and compact metric spaces. Submitted, 1995.

[848] C. L. Liu. *Introduction to Combinatorial Mathematics*. McGraw-Hill, New York, 1968.

[849] J. Liu. Maximal independent sets in bipartite graphs. *J. Graph Theory*, 17:495–507, 1993.

[850] J. Liu and Q. Yu. A dominating property of i-center in p_t-free graphs. *Ars Combin.*, 39:121–127, 1995.

[851] J. Liu and H. Zhou. Dominating subgraphs in graphs with some forbidden structures. *Discrete Math.*, 135:163–168, 1994.

[852] M. Livingston and Q. F. Stout. Perfect dominating sets. *Congr. Numer.*, 79:187–203, 1990.

[853] M. Livingston and Q. F. Stout. Constant time computation of minimum dominating sets. *Congr. Numer.*, 105:116–128, 1994.

[854] E. Loukakis. Two algorithms for determining a minimum independent dominating set. *Internat. J. Comput. Math.*, 15:213–229, 1984.

[855] L. Lovász. On the ratio of optimal integral and fractional covers. *Discrete Math.*, 13:383–390, 1975.

[856] T. L. Lu, P. H. Ho, and G. J. Chang. The domatic number problem in interval graphs. *SIAM J. Discrete Math.*, 3:531–536, 1990.

[857] E. Lucas. La solution complete du probleme des huit reines. *Revue Scientifique*, 9:948–953, 1880.

[858] E. Lucas. Le probleme des huites reines au jeu des echecs. *Recreations Mathematiques, Paris*, pages 57–86, 1882.

[859] C. Lund and M. Yannakakis. On the hardness of approximating minimization problems. In *Proc. 25th ACM Symp. on the Theory of Computing*, pages 286–293, 1993.

[860] K. L. Ma. *Partition Algorithm for the Dominating Set Problem.* Master's thesis, Concordia Univ., 1990.

[861] K. L. Ma and C. W. H. Lam. Partition algorithm for the dominating set problem. *Congr. Numer.*, 81:69–80, 1991.

[862] G. MacGillivray and K. Seyffarth. Domination numbers of planar graphs. *J. Graph Theory*, 22:213–229, 1996.

[863] F. Maffray. On kernels in i-triangulated graphs. *Discrete Math.*, 61:247–251, 1986.

[864] F. Maffray. Kernels in perfect line-graphs. *J. Combin. Theory Ser. B*, 55:1–8, 1992.

[865] K. Maghout. Sur la determination des nombres de stabilite et du nombre chromatique d'un graphe. *C. R. Acad. Sci. Paris*, 248:3522–3523, 1959.

[866] A. Majumdar. *Neighborhood Hypergraphs: A Framework for Covering and Packing Parameters in Graphs.* PhD thesis, Clemson Univ., 1992.

[867] A. Majumdar and R. Laskar. Covering and packing parameters in hypergraphs defined by neighborhoods in graphs. *J. Combin. Inform. System Sci.*, 17:9–18, 1992.

[868] G. K. Manacher and T. A. Mankus. Incorporating negative-weight vertices in certain vertex-search graph algorithms. *Inform. Process. Lett.*, 42:293–294, 1992.

[869] G. K. Manacher and T. A. Mankus. Finding a domatic partition of an interval graph in time $O(n)$. *SIAM J. Discrete Math.*, 9:167–172, 1996.

[870] M. V. Marathe, H. B. Hunt III, and S. S. Ravi. Efficient approximation algorithms for domatic partition and on-line coloring of circular arc graphs. *Discrete Appl. Math.*, 64(2):135–149, 1996.

[871] M. Marchioro and A. Morgana. Structure and recognition of domishold graphs. *Discrete Math.*, 50:239–251, 1984.

[872] D. Marcu. A method for finding the kernel of a digraph. *An. Univ. Bucuresti Mat.*, 27:41–43, 1978.

[873] D. Marcu. On the existence of a kernel in a strong connected digraph (Romanian summary). *Bul. Inst. Politehn. Iasi. Sect. I*, 25:35–37, 1979.

[874] D. Marcu. Some remarks concerning the kernels of a strong connected digraph. *Al. I. Cuza Iasi Sect. I a Math. (N. S.)*, 26:417–418, 1980.

[875] D. Marcu. Generalized kernels with considerations to the tournament digraphs. *Mathematica*, 24:57–63, 1982.

[876] D. Marcu. A new upper bound for the domination number of a graph. *Quart. J. Math. Oxford Ser. 2*, 36:221–223, 1985.

[877] D. Marcu. An upper bound on the domination number of a graph. *Math. Scand.*, 59:41–44, 1986.

[878] D. Marcu. On finding the domination number of a graph. *Anal. Numer. Theor. Approx.*, 18:77–79, 1989.

[879] D. Marcu. Note on the domination number of a graph. *Demonstratio Math.*, 26:103–105, 1993.

[880] J. D. Masters, Q. F. Stout, and D. M. Van Wieren. Unique domination in cross-product graphs. *Congr. Numer.*, 118:49–71, 1996.

[881] K. L. McAvaney. The number and stability indices of C_n-trees. In *Lecture Notes in Mathematics, Combinatorial Mathematics IV (Proc. Fourth Australian Conf., Univ. Adelaide, 1975)*, volume 560, pages 142–148. Springer, Berlin, 1976.

[882] R. M. McConnell and J. P. Spinrad. Modular decomposition and transitive orientation. Manuscript, 1995.

[883] W. McCuaig and B. Shepherd. Domination in graphs with minimum degree two. *J. Graph Theory*, 13:749–762, 1989.

[884] J. D. McFall and R. Nowakowski. Strong independence in graphs. *Congr. Numer.*, 29:639–656, 1980.

[885] J. McHugh and Y. Perl. Best location of service centers in a treelike network under budget constraints. *Discrete Math.*, 86:199–214, 1990.

[886] M. McKay. *A Study of G_2 Graphs*. Master's thesis, Clemson Univ., 1989.

[887] T. A. McKee. Intersection graphs and cographs. *Congr. Numer.*, 78:223–230, 1990.

[888] A. A. McRae. *Generalizing NP-Completeness Proofs for Bipartite and Chordal Graphs*. PhD thesis, Clemson Univ., 1994.

[889] A. A. McRae and S. T. Hedetniemi. Finding n-independent dominating sets. *Congr. Numer.*, 85:235–244, 1991.

[890] N. Megiddo. Are the vertex cover and the dominating set problems equally hard? Technical Report RJ 5783, Stanford Univ., 1987.

[891] N. Megiddo and U. Vishkin. On finding a minimum dominating set in a tournament. *Theor. Comput. Sci.*, 61:307–316, 1988.

[892] N. Megiddo, E. Zemel, and S. Louis Hakimi. The maximum coverage location problem. *SIAM J. Algebraic Discrete Methods*, 4:253–261, 1983.

[893] A. Meir and J. W. Moon. Relations between packing and covering numbers of a tree. *Pacific J. Math.*, 61:225–233, 1975.

[894] A. Meir and J. W. Moon. Packing and covering constants for certain families of trees. *J. Graph Theory*, 1:157–174, 1977.

[895] A. Meir and J. W. Moon. Packing and covering constants for families of trees: II. *Trans. Amer. Soc. Math.*, 233:167–178, 1977.

[896] A. Meir and J. W. Moon. Packing and covering constants for recursive trees. In *Lecture Notes in Mathematics, Theory and Applications of Graphs*, volume 642, pages 403–411. Springer, Berlin, 1978.

[897] B. C. Messick. *Finding Irredundant Sets in Queen's Graphs*. Master's thesis, Clemson Univ., 1995.

[898] S. L. Mitchell. *Linear Algorithms on Trees and Maximal Outerplanar Graphs: Design, Complexity Analysis and Data Structures Study*. PhD thesis, Univ. Virginia, 1977.

[899] S. L. Mitchell, E. J. Cockayne, and S. T. Hedetniemi. Linear algorithms on recursive representations of trees. *J. Comput. System Sci.*, 18(1):76–85, 1979.

[900] S. L. Mitchell and S. T. Hedetniemi. Edge domination in trees. *Congr. Numer.*, 19:489–509, 1977.

[901] S. L. Mitchell and S. T. Hedetniemi. Independent domination in trees. *Congr. Numer.*, 29:639–656, 1979.

[902] S. L. Mitchell, S. T. Hedetniemi, and S. Goodman. Some linear algorithms on trees. *Congr. Numer.*, 14:467–483, 1975.

[903] Z. Mo and K. Williams. (r, s)-domination in graphs and directed graphs. *Ars Combin.*, 29:129–141, 1990.

[904] M. Moscarini. Doubly chordal graphs, Steiner trees, and connected domination. *Networks*, 23:59–69, 1993.

[905] H. Müller and A. Brandstädt. The NP-completeness of STEINER TREE and DOMINATING SET for chordal bipartite graphs. *Theoret. Comput. Sci.*, 53:257–265, 1987.

[906] C. M. Mynhardt. Domination, independence and irredundance in graphs: a survey. Technical Report 13/85, Univ. South Africa, 1985.

[907] C. M. Mynhardt. Generalised maximal independence and clique numbers of a graph. *Quaestiones Math.*, 11:383–398, 1988.

[908] C. M. Mynhardt. Lower Ramsey numbers for graphs. *Discrete Math.*, 91:69–75, 1991.

[909] C. M. Mynhardt. On the difference between the domination and independent domination number of cubic graphs. In Y. Alavi, G. Chartrand, O. R. Oellermann, and A. J. Schwenk, editors, *Graph Theory, Combinatorics, and Applications, Proc. Sixth Quad. Conf. on the Theory and Applications of Graphs*, volume 2, pages 939–947, (Kalamazoo, MI 1988), 1991. Wiley.

[910] C. M. Mynhardt. Irredundant Ramsey numbers for graphs: a survey. *Congr. Numer.*, 86:65–79, 1992.

[911] B. A. Nadel. Representation selection for constraint satisfaction: a case study using the N-queens. *IEEE Expert*, 5(3):16–23, 1990.

[912] K. S. Natarajan and L. J. White. Optimum domination in weighted trees. *Inform. Process. Lett.*, 7:261–265, 1978.

[913] F. Nauck. Briefwechsel mit allen für alle. *Illustrirte Zeitung*, 15(377):182, Sept. 21, 1850.

[914] F. Nauck. Schach: eine in das Gebiet der Mathematik fallende Aufgabe von Herrn Dr. Nauck in Schleusingen. *Illustrirte Zeitung*, 14(365):416, June 29, 1850.

[915] P. Naur. An experiment on program development. *BIT*, 12:347–365, 1971.

[916] P. S. Neeralagi. Complete domination and neighbourhood numbers of a graph. *J. Math. Phy. Sci.*, 27(4):295–303, 1993.

[917] R. E. Newman-Wolfe, R. D. Dutton, and R. C. Brigham. Connecting sets in graphs - a domination related concept. *Congr. Numer.*, 67:67–76, 1988.

[918] F. Nicolai and T. Szymzcak. Domination and homogeneous sets – a linear time algorithm for distance-hereditary graphs. Technical Report SM-DU-336, Univ. Duisburg, 1996.

[919] J. Nieminen. A method to determine a minimal dominating set of a graph G_0. *IEEE Trans. Circuits and Systems*, CAS-21:12–14, 1974.

[920] J. Nieminen. Two bounds for the domination number of a graph. *J. Inst. Math. Applics.*, 14:183–187, 1974.

[921] R. G. Nigmatullin. The largest number of kernels in graphs with n vertices. *Kazan. Gos. Univ. Ucen. Zap.*, 130:75–82, 1970.

[922] E. A. Nordhaus and J. W. Gaddum. On complementary graphs. *Amer. Math. Monthly*, 63:175–177, 1956.

[923] R. J. Nowakowski and D. Rall. Associative graph products and their independence, domination and coloring numbers. *Discuss. Math.- Graph Theory*, 16:53–79, 1996.

[924] O. Ore. *Theory of Graphs*. Amer. Math. Soc. Colloq. Publ., 38 (Amer. Math. Soc., Providence, RI), 1962.

[925] C. Papadimitriou and M. Yannakakis. Optimization, approximation and complexity classes. In *Proc. 20th ACM Symp. on the Theory of Computing*, pages 229–234, 1988.

[926] A. K. Parekh. Analysis of a greedy heuristic for finding small dominating sets in graphs. *Inform. Process. Lett.*, 39:237–240, 1991.

[927] M. Paris. Note: The diameter of edge domination critical graphs. *Networks*, 24:261–262, 1994.

[928] M. Paris. The vertices of edge domination critical graphs. Manuscript, 1995.

[929] M. Paris, D. Sumner, and E. Wojcicka. Edge-domination critical graphs with cut-vertices. Submitted, 1997.

[930] C. Payan. Sur le nombre d'absorption d'un graphe simple. *Cahiers Centre Études Rech. Opér. (2.3.4)*, 171, 1975.

[931] C. Payan. A class of threshold and domishold graphs: equistable and equidominating graphs. *Discrete Math.*, 29:47–52, 1980.

[932] C. Payan. Remarks on cliques and dominating sets in graphs. Manuscript, undated.

[933] C. Payan and N. H. Xuong. Domination-balanced graphs. *J. Graph Theory*, 6:23–32, 1982.

[934] S. L. Peng. Parallel algorithms for cographs and parity graphs with applications. *J. Algorithms*, 11:252–284, 1990.

[935] S. L. Peng and M. S. Chang. A simple linear time algorithm for the domatic partition problem on strongly chordal graphs. *Inform. Process. Lett.*, 43:297–300, 1992.

[936] S.G. Penrice. Dominating sets with small clique covering number. Manuscript, 1993.

[937] S.G. Penrice. Clique-like dominating sets in perfect graphs. *Congr. Numer.*, 110:77–82, 1995.

[938] K. W. Peters. *Theoretical and Algorithmic Results on Domination and Connectivity*. PhD thesis, Clemson Univ., 1986.

[939] J. Pfaff. *Algorithmic Complexities of Domination-related Graph Parameters*. PhD thesis, Clemson Univ., 1984.

[940] J. Pfaff, R. Laskar, and S. T. Hedetniemi. NP-completeness of total and connected domination, and irredundance for bipartite graphs. Technical Report 428, Dept. Mathematical Sciences, Clemson Univ., 1983.

[941] J. Pfaff, R. Laskar, and S. T. Hedetniemi. Linear algorithms for independent domination and total domination in series-parallel graphs. *Congr. Numer.*, 45:71–82, 1984.

[942] V. Phillippe. Quasi-kernels of minimum weakness in a graph. *Discrete Math.*, 20:187–192, 1977.

[943] M. R. Pinter. Planar regular one-well-covered graphs. *Congr. Numer.*, 91:159 – 187, 1992.

[944] W. Piotrowski. *Combinatorial Optimization: Scheduling, Facility Location, and Domination.* PhD thesis, North Dakota State Univ., 1992.

[945] M. Plummer. Some covering concepts in graphs. *J. Combin. Theory*, 8:91–98, 1970.

[946] M. Plummer. Well-covered graphs: a survey. *Quaestiones Math.*, 16:253–287, 1993.

[947] A. Pnueli, A. Lempel, and S. Even. Transitive orientation of graphs and identification of permutation graphs. *Canad. J. Math.*, 23:160–175, 1971.

[948] E. Popescu. On the structure of preordered sets. *Bull. Math. Soc. Sci. Math. R. S. Roumanie (N.S.)*, 33:125–135, 1989. 2,81.

[949] A. Proskurowski. Minimum dominating cycles in 2-trees. *Internat. J. Comput. Inform. Sci.*, 8:405–417, 1979.

[950] A. Proskurowski and M. M. Syslo. Minimum dominating cycles in outerplanar graphs. *Internat. J. Comput. Inform. Sci.*, 10:127–139, 1981.

[951] A. Proskurowski and J. A. Telle. Algorithms for vertex partitioning problems on partial k-trees. *SIAM J. Discrete Math.*, 10(4), 1997.

[952] J. Puech. Irredundant and independent perfect neighborhood sets in graphs. Submitted.

[953] J. Puech. Stability, domination, and irredundance in W_{AR} graphs. *Ars Combin.* To appear.

[954] J. Puech. Irredundance perfection and P_6-free graphs. Submitted, 1997.

[955] J. Puech. Lower domination parameters in inflated trees. Submitted, 1997.

[956] L. Pushpa Latha. The global point-set domination number of a graph. *Indian J. Pure Appl. Math.*, 28(1):47–51, 1997.

[957] D. F. Rall. Domatically critical and domatically full graphs. *Discrete Math.*, 86:81–87, 1990.

[958] D. F. Rall. A fractional version of domatic number. *Congr. Numer.*, 74:100–106, 1990.

[959] D. F. Rall. Dominating a graph and its complement. *Congr. Numer.*, 80:89–95, 1991.

[960] D. F. Rall and P. J. Slater. On location-domination numbers for certain classes of graphs. *Congr. Numer.*, 45:97–106, 1984.

[961] G. Ramalingam and C. Pandu Rangan. Total domination in interval graphs revisited. *Inform. Process. Lett.*, 27:17–21, 1988.

[962] G. Ramalingam and C. Pandu Rangan. A unified approach to domination problems in interval graphs. *Inform. Process. Lett.*, 27:271–274, 1988.

[963] B. Randerath and L. Volkmann. Characterization of graphs with equal domination and covering number. *Discrete Math.* To appear.

[964] B. Randerath and L. Volkmann. Simplicial graphs and relationships to different graph invariants. *Ars. Combin.* To appear.

[965] B. Randerath and L. Volkmann. A characterization of well covered block-cactus graphs. *Australas. J. Combin.*, 9:307–314, 1994.

[966] B. Randerath and L. Volkmann. Relationships between different graph invariants and simplicial graphs. Submitted, 1996.

[967] A. S. Rao and C. Pandu Rangan. Optimal parallel algorithms on circular-arc graphs. *Inform. Process. Lett.*, 33:147–156, 1989.

[968] A. S. Rao and C. Pandu Rangan. Linear algorithm for domatic number problem on interval graphs. *Inform. Process. Lett.*, 33:29–33, 1989-90.

[969] G. Ravindra. Well covered graphs. *J. Combin. Inform. System Sci.*, 2:20–21, 1977.

[970] B. Reed. Paths, stars, and the number three. Submitted, 1996.

[971] E. Regener and M. Vo. New bounds on dominating numbers of grid graphs. *Congr. Numer.*, 56, 1987.

[972] M. Reichling. A simplified solution for the N queens problem. *Inform. Process. Lett.*, 25:253–355, 1987.

[973] C. Rhee, Y. D. Liang, S. K. Dhall, and S. Lakshmivaranhan. An $O(n+m)$ algorithm for finding a minimum-weight dominating set in a permutation graph. *SIAM J. Comput.*, 25:401–419, 1996.

[974] T. Haynes Rice. An extremal problem for 2-2-insensitive domination. Technical Report TR-88-07, Univ. Central Florida, 1988.

[975] T. Haynes Rice. *On k-γ-insensitive Domination.* PhD thesis, Univ. Central Florida, 1988.

[976] T. Haynes Rice and R. K. Guha. A multi-layered G-network for massively parallel computation. In *Proc. Second Symp. on the Frontiers of Massively Paralled Computation*, pages 519–520, 1989.

[977] M. Richardson. Extension theorems for solutions of irreflexive relations. *Proc. Nat. Acad. Sci.*, 39:649–655, 1953.

[978] M. Richardson. Solutions of irreflexive relations. *Ann. Math.*, 58:573–590, 1953.

[979] M. Richardson. Relativization and extension of solutions of irreflexive relations. *Pacific J. Math.*, 5:551–584, 1955.

[980] I. Rivin and R. Zabih. A dynamic programming solution to the N-queens problem. *Inform. Process. Lett.*, 41:253–256, 1992.

[981] E. R. Rodemich. Coverings by rook domains. *J. Combin. Theory*, 9:117–128, 1970.

[982] J. S. Rohl. A faster lexicographic N queens algorithm. *Inform. Process. Lett.*, 17:231–233, 1983.

[983] P. Rowlinson. Dominating sets and eigenvalues of graphs. *Bull. London Math. Soc.*, 26:248–254, 1994.

[984] S. Rudeanu. Notes sur l'existence et l'uniciteé des noyaux d'un graphe. I. *Rev. Francaise Rech. Opérationnelle*, 8:345–352, 1964.

[985] S. Rudeanu. Notes sur l'existence et l'uniciteé des noyaux d'un graphe. II. *Rev. Francaise Rech. Opérationnelle*, 10:301–310, 1966.

[986] S. Rudeanu. On solving Boolean equations in the theory of graphs. *Rev. Roumaine Math. Pures Appl.*, 11:653–664, 1966.

[987] W. Rytter and T. Szymacha. Parallel algorithms for a class of graphs generated recursively. *Inform. Process. Lett.*, 30:225–231, 1989.

[988] E. Sampathkumar. $(1, k)$-domination in a graph. *J. Math. Phys. Sci.*, 22:613–619, 1988.

[989] E. Sampathkumar. The global domination number of a graph. *J. Math. Phys. Sci.*, 23:377–385, 1989.

[990] E. Sampathkumar. The least point covering and domination numbers of a graph. *Discrete Math.*, 86:137–142, 1990.

[991] E. Sampathkumar. On some new domination parameters of a graph—a survey. In *Centre Math. Sci., Trivandrum, Proc. Symp. on Graph Theory and Combinatorics (Cochin, 1991)*, pages 7–13. (Series: Publication, 21), 1991.

[992] E. Sampathkumar. Generalizations of independence and chromatic numbers of a graph. *Discrete Math.*, 115:245–251, 1993.

[993] E. Sampathkumar. Domination parameters of a graph. In T. W. Haynes, S. T. Hedetniemi, and P. J. Slater, editors, *Domination in Graphs: Advanced Topics*, chapter 10. Marcel Dekker, Inc., 1997.

[994] E. Sampathkumar and S. S. Kamath. Mixed domination in graphs. *Sankhyā (Special Volume)*, 54:399–402, 1992.

[995] E. Sampathkumar and P. S. Neeralagi. The neighborhood number of a graph. *Indian. J. Pure Appl. Math.*, 16:126–132, 1985.

[996] E. Sampathkumar and P. S. Neeralagi. The line-neighborhood number of a graph. *J. Pure Appl. Math.*, 17(2):142–149, 1986.

[997] E. Sampathkumar and P. S. Neeralagi. Domination and neighbourhood critical, fixed, free and totally free points. *Sankhyā (Special Volume)*, 54:403 – 407, 1992.

[998] E. Sampathkumar and P. S. Neeralagi. Independent, perfect and connected neighborhood numbers of a graph. *J. Combin. Inform. System Sci.*, 19:139–145, 1994.

[999] E. Sampathkumar and L. Pushpa Latha. Point-set domination number of a graph. *Indian J. Pure Appl. Math.*, 24:225–229, 1993.

[1000] E. Sampathkumar and L. Pushpa Latha. The global set domination number of a graph. *Indian J. Pure Appl. Math.*, 25(10):1053–1057, 1994.

[1001] E. Sampathkumar and L. Pushpa Latha. Set domination in graphs. *J. Graph Theory*, 18(5):489–495, 1994.

[1002] E. Sampathkumar and L. Pushpa Latha. Dominating strength and weakness of a graph. Manuscript, 1995.

[1003] E. Sampathkumar and L. Pushpa Latha. Strong weak domination and domination balance in a graph. *Discrete Math.*, 161:235–242, 1996.

[1004] E. Sampathkumar and H. B. Walikar. The connected domination number of a graph. *J. Math. Phys. Sci.*, 13:607–613, 1979.

[1005] E. Sampathkumar and H. B. Walikar. On a conjecture of Cockayne and Hedetniemi. *J. Comput. Inform. System Sci.*, 7(3):1–4, 1982.

[1006] L. A. Sanchis. Bounds related to domination in graphs with minimum degree two. *J. Graph Theory.* To appear.

[1007] L. A. Sanchis. Maximum number of edges in connected graphs with a given domination number. *Discrete Math.*, 87:65–72, 1991.

[1008] R. S. Sankaranarayana. *Well-covered Graphs: Some New Sub-classes and Complexity Results.* PhD thesis, Univ. Alberta, 1994.

[1009] P. Scheffler. Linear-time algorithms for NP-complete problems restricted to partial k-trees. Technical Report 03/87, IMATH, Berlin, 1987.

[1010] F. Scheid. Some packing problems. *Amer. Math. Monthly*, 67:231–235, 1960.

[1011] G. Schmidt and T. Strohlein. Kernel in bipartite graphs (German summary). In *Proc. Eighth Conf. on Graph Theoretic Concepts in Computer Science (WG 82) Neunkirchen*, pages 251–256, Hauser, Munich, 1982.

[1012] G. Schmidt and T. Strohlein. On kernels of graphs and solutions of games: a synopsis based on relations and fixedpoints. *SIAM J. Algebraic Discrete Methods*, 6:54–65, 1985.

[1013] P. Schmidt and T. Haynes. On a graph transformation where nodes are replaced by complete subgraphs. *Congr. Numer.*, 78:99–107, 1990.

[1014] J. C. Schoeman. *'n Ry Boonste en Onderste Dominaise -, Onafhanklikheids- en Onoorbodigheidsgetalle van 'n Grafiek.* Master's thesis, Rand Afrikaans Univ., 1992.

[1015] W. Schöne. Algorithmen zur Bestimmung intern und extern stabiler Mengen und der chromatischen Zahl eines endlichen Graphen. *Wiss. Z. Techn. Hochsch. Karl Marx-Stadt*, 17:269–280, 1975.

[1016] F. Schuh. *The Master Book of Mathematical Recreations.* Dover, New York, 1959.

[1017] S. Schuster. Edge dominating numbers of complementary graphs. In Y. Alavi, G. Chartrand, O. R. Oellermann, and A. J. Schwenk, editors, *Graph Theory, Combinatorics and Applications, Proc. Sixth Internat. Conf. on the Theory and Application of Graphs*, volume 2, pages 1053–1060, (Kalamazoo, MI 1988), 1991. Wiley.

[1018] D. Seese. Tree-partite graphs and the complexity of algorithms. In *Lecture Notes in Computer Science, FCT 85*, volume 199, pages 412–421. Springer, Berlin, 1985.

[1019] N. Seifter. Domination and independent domination numbers of graphs. *Ars Combin.*, 38:119–128, 1994.

[1020] W. J. Selig and P. J. Slater. Minimum dominating, optimally independent vertex sets in graphs. In Y. Alavi, G. Chartrand, O. R. Oellermann, and A. J. Schwenk, editors, *Graph Theory, Combinatorics, and Applications, Proc. Sixth Internat. Conf. on the Theory and Applications of Graphs*, volume 2, pages 1061–1073, (Kalamazoo, MI 1988), 1991. Wiley.

[1021] S. C. Shee. On stabilities and chromatic number of a loop graph. *Nanta Math.*, IV:8–14, 1970.

[1022] B. N. Shneyder. Algebra of sets applied to solution of graph theory problems. *Izv. Akad. Nauk SSSR Tehn. Kibernet. (1970) 109-114*, pages 521–526, 1970. (Russian); translated as Engrg. Cybernetics 8 (3) (1970)(1971).

[1023] W. Siemes, J. Topp, and L. Volkmann. On unique independent sets in graphs. *Discrete Math.*, 131:279–285, 1994.

[1024] H. G. Singh. A parallel algorithm for the domination number of complete grid graphs on the floating point systems T-20 hypercube. Technical report, Dept. Computer Science, Clemson Univ., 1987.

[1025] H. G. Singh and R. P. Pargas. A parallel implementation for the domination number of a grid graph. *Congr. Numer.*, 59:297–311, 1987.

[1026] F. V. Sirokov and V. A. Signaevskii. Minimal coverings of a finite set I (Russian); connected coverings. *Diskret Analiz.*, 21:72–94, 1972.

[1027] F. V. Sirokov and V. A. Signaevskii. Minimal coverings of a finite set II (Russian); connected coverings. *Diskret Analiz.*, 22:57–78, 1973. 81; cf. Dokl. Akad. Nauk SSSR 207 (1972) 1066-1069, MR 47 3216.

[1028] M. Skowrońska and M. M. Syslo. Dominating cycles in Halin graphs. *Discrete Math.*, 86:215–224, 1990.

[1029] P. J. Slater. Irreducible point independence numbers and independence graphs. *Cong. Numer.*, 10:647–660, 1974.

[1030] P. J. Slater. Leaves of trees. *Congr. Numer.*, 14:549–559, 1975.

[1031] P. J. Slater. R-domination in graphs. *J. Assoc. Comput. Mach.*, 23:446–450, 1976.

[1032] P. J. Slater. Enclaveless sets and MK-systems. *J. Res. Nat. Bur. Standards*, 82:197–202, 1977.

[1033] P. J. Slater. A characterization of SOFT hypergraphs. *Canad. Math. Bull.*, 21:335–337, 1978.

[1034] P. J. Slater. A constructive characterization of trees with at least k disjoint maximum matchings. *J. Combin. Theory Ser. B*, 25:326–338, 1978.

[1035] P. J. Slater. Domination and location in acyclic graphs. *Networks*, 17:55–64, 1987.

[1036] P. J. Slater. Dominating and reference sets in a graph. *J. Math. Phys. Sci.*, 22:445–455, 1988.

[1037] P. J. Slater. Closed neighborhood order domination and packing. *Congr. Numer.*, 97:33–43, 1993.

[1038] P. J. Slater. Maximum matching among minimum dominating sets. *Ars Combin.*, 35A:239–249, 1993.

[1039] P. J. Slater. Locating dominating sets and locating-dominating sets. In Y. Alavi and A. Schwenk, editors, *Graph Theory, Combinatorics, and Applications, Proc. Seventh Quad. Internat. Conf. on the Theory and Applications of Graphs*, pages 1073–1079. John Wiley & Sons, Inc., 1995.

[1040] P. J. Slater. Packing into closed neighborhoods. *Bull. Inst. Combin. Appl.*, 13:23–34, 1995.

[1041] P. J. Slater. The complementation theorem. Submitted, 1996.

[1042] P. J. Slater. Generalized graph parameters I: Gallai theorems. *Bull. Inst. Combin. Appl.*, 17:27–37, 1996.

[1043] P. J. Slater. An overview of graphical weighted subset problems. *Graph Theory Notes of New York*, 31:44–47, 1996.

[1044] P. J. Slater. The automorphism class theorem for LP-based graphical parameters. *Bull. Inst. Combin. Appl.*, 20:19–26, 1997.

[1045] P. J. Slater. LP-duality, complementarity, and generality of graphical subset parameters. In T. W. Haynes, S. T. Hedetniemi, and P. J. Slater, editors, *Domination in Graphs: Advanced Topics*, chapter 1. Marcel Dekker, Inc., 1997.

[1046] P. J. Slater, V. Smithdorf, and H. Swart. Domination- and covering-forcing sets of graphs. Submitted, 1995.

[1047] C. B. Smart. *Studies of Graph Based IP/LP Parameters*. PhD thesis, Univ. Alabama in Huntsville, 1996.

[1048] C. B. Smart and P. J. Slater. Minimum coverage by dominating sets in graphs. *J. Combin. Math. Combin. Comput.* To appear.

[1049] C. B. Smart and P. J. Slater. Complexity results for closed neighborhood order parameters. *Congr. Numer.*, 112:83–96, 1995.

[1050] C. B. Smart and P. J. Slater. Closed neighborhood order parameters: complexity, algorithms, and inequalities. Submitted, 1996.

[1051] R. Sosic and J. Gu. A polynomial time algorithm for the N-queens problem. *SIGART Bull.*, 2(2):7–11, 1990.

[1052] J. Spinrad. Transitive orientation in $O(n^2)$ time. In *Proc. 15th Ann. ACM Symp. on Theory of Computing*, pages 457–466, 1983.

[1053] J. Spinrad. On comparability and permutation graphs. *SIAM J. Comput.*, 14:658–670, 1985.

[1054] J. Spinrad, A. Brandstädt, and L. Stewart. Bipartite permutation graphs. *Discrete Appl. Math.*, 18:279–292, 1987.

[1055] R. Sridhar and S. S. Iyengar. Efficient parallel algorithms for domination problems on strongly chordal graphs. Manuscript, 1990.

[1056] R. A. Srinivasa and C. Pandu Rangan. Linear algorithm for domatic number problem on interval graphs. *Inform. Process. Lett.*, 33:29–33, 1989.

[1057] A. Srinivasan and C. Pandu Rangan. Efficient algorithms for the minimum weighted dominating clique problem on permutation graphs. *Theoret. Comput. Sci.*, 91:1–21, 1991.

[1058] A. Stacey. Universal minimal total dominating functions of trees. *Discrete Math.*, 140:287–290, 1995.

[1059] L. Stewart. *Permutation in Graph Structure and Algorithms*. PhD thesis, Univ. Toronto, 1985.

[1060] L. S. Stewart and E. O. Hare. Fractional domination of $P_m \times P_n$. *Congr. Numer.*, 91:35–42, 1992.

[1061] H. S. Stone and J. M. Stone. Efficient search techniques - an empirical study of the N-queens problem. *IBM J. Res. Develop.*, 31:464–474, 1987.

[1062] C. Stracke and L. Volkmann. A new domination conception. *J. Graph Theory*, 17:315–323, 1993.

[1063] D. Studer. *Induced Paired Domination*. Master's thesis, East Tennessee State Univ., 1996.

[1064] D. Studer, T. W. Haynes, and L. M. Lawson. Strongly paired-domination in graphs. Submitted, 1997.

[1065] D. P. Sumner. The nucleus of a point determining graph. *Discrete Math.*, 14:91–97, 1976.

[1066] D. P. Sumner. Critical concepts in domination. *Discrete Math.*, 86:33–46, 1990.

[1067] D. P. Sumner and P. Blitch. Domination critical graphs. *J. Combin. Theory Ser. B*, 34:65–76, 1983.

[1068] D. P. Sumner and J.I. Moore. Domination perfect graphs. *Notices Amer. Math. Soc.*, 26:A–569, 1979.

[1069] D. P. Sumner and E. Wojcicka. Domination k-critical graphs. Manuscript, 1994.

[1070] D. P. Sumner and E. Wojcicka. Cut-vertices in k-edge-domination-critical graphs. Manuscript, 1996.

[1071] D. P. Sumner and E. Wojcicka. Graphs critical with respect to the domination number. In T. W. Haynes, S. T. Hedetniemi, and P. J. Slater, editors, *Domination in Graphs: Advanced Topics*, chapter 16. Marcel Dekker, Inc., 1997.

[1072] L. Sun. On three conjectures of B. Zelinka in graph theory. *Kexue Tongbao (Science Bulletin). A Monthly Journal of Science. English Edition*, 33(12):969–970, 1988. Kexue Tongbao Chinese Academy of Science.

[1073] K. Sutner. Linear cellular automata and the Garden-of-Eden. *Math. Intell.*, 12(2):49–53, 1989.

[1074] M. M. Syslo. NP-complete problems on some tree-structured graphs: a review. In M. Nagl and J. Perl, editors, *Proc. WG'83 (Workshop on Graph Theoretic Concepts in Computer Science, Univ. Osnabruck, Trauner Verlag, Linz*, pages 342–353, 1984.

[1075] E. Szekeres and G. Szekeres. On a problem of Schütte and Erdös. *Math. Gaz.*, 49:290–293, 1965.

[1076] K. Takamizawa, T. Nishizeki, and N. Saito. Linear-time computability of combinatorial problems on series-parallel graphs. *J. Assoc. Comput. Mach.*, 29:623–624, 1882.

[1077] J. A. Telle. Characterization of domination-type parameters in graphs. *Congr. Numer.*, 94:9–16, 1993.

[1078] J. A. Telle. Complexity of domination-type problems in graphs. *Nordic J. Comput.*, 1:157–171, 1994.

[1079] J. A. Telle. *Vertex Partitioning Problems: Characterization, Complexity and Algorithms on Partial k-trees.* PhD thesis, Univ. Oregon, 1994.

[1080] J. A. Telle and A. Proskurowski. Efficient sets in partial k-trees. *Discrete Appl. Math.*, 44:109–117, 1993.

[1081] J. A. Telle and A. Proskurowski. Practical algorithms on partial k-trees with an application to domination-type problems. In F. Dehne, J. R. Sack, N. Santoro, and S. Whitesides, editors, *Lecture Notes in Comput. Sci., Proc. Third Workshop on Algorithms and Data Structures (WADS'93)*, volume 703, pages 610–621, Montréal, 1993. Springer-Verlag.

[1082] U. Teschner. New results about the bondage number of a graph. *Discrete Math.* To appear.

[1083] U. Teschner. A counterexample to a conjecture on the bondage number of a graph. *Discrete Math.*, 122:393–395, 1993.

[1084] U. Teschner. *Die Bondagezahl eines Graphen.* PhD thesis, RWTH, Aachen, 1995.

[1085] U. Teschner. The bondage number of a graph G can be much greater than $\Delta(G)$. *Ars Combin.*, 43:81–87, 1996.

[1086] U. Teschner and L. Volkmann. On the bondage number of cactus graphs. Submitted, 1995.

[1087] L. I. Tjagunov. Some properties of the structure of the set of kernels of symmetric graph (Russian). *Trudy Inst. Mat. i Meh, Ural. Naucn. Centr Akad. Nauk SSSR 27; Metody Mat. Programmiv. i Prilozon*, pages 56–60, 126–127, 1979.

[1088] N. Tokushige. Binding number and minimum degree for k-factors. *J. Graph Theory*, 13:607–617, 1989.

[1089] I. Tomescu. Almost all digraphs have a kernel. In M. Karoński, J. Jaworski, and A. Ruciński, editors, *Random Graphs*, volume 2, pages 325–340. Wiley, Chichester, 1990.

[1090] R. W. Topor. Fundamental solutions of the eight queens problem. *BIT*, 22:42–52, 1982.

[1091] J. Topp. Graphs with unique minimum edge dominating sets and graphs with unique maximum independent sets of vertices. *Discrete Math.* To appear.

[1092] J. Topp and P. D. Vestergaard. Some classes of well-covered graphs. Technical Report R-93-2009, Dept. Math. and Computer Science, Aalborg Univ., 1993.

[1093] J. Topp and P. D. Vestergaard. α_k- and γ_k-stable graphs. Submitted, 1997.

[1094] J. Topp and L. Volkmann. On domination and independence numbers of graphs. *Results Math.*, 17:333–341, 1990.

[1095] J. Topp and L. Volkmann. Well covered and well dominated block graphs and unicyclic graphs. *Math. Pannon.*, 1:55–66, 1990.

[1096] J. Topp and L. Volkmann. Characterization of unicyclic graphs with equal domination and independence numbers. *Discuss. Math.*, 11:27–34, 1991.

[1097] J. Topp and L. Volkmann. On graphs with equal domination and independent domination numbers. *Discrete Math.*, 96:75–80, 1991.

[1098] J. Topp and L. Volkmann. On packing and covering numbers of graphs. *Discrete Math.*, 96:229–238, 1991.

[1099] J. Topp and L. Volkmann. Some classes of well covered graphs. Technical report, Inst. Electronic Systems, Dept. of Math. and Computer Science, Aalborg Univ., 1993. Submitted.

[1100] J. Topp and L. Volkmann. Some upper bounds for the product of the domination number and the chromatic number of a graph. *Discrete Math.*, 118:289–292, 1993.

[1101] J. Topp and L. Volkmann. Well irredundant graphs. Technical report, Inst. Electronic Systems, Dept. of Math. and Computer Science, Aalborg Univ., 1993. Submitted.

[1102] J. Topp and L. Volkmann. Totally equimatchable graphs. *Discrete Math.*, 164:285–290, 1997.

[1103] K. Tsai and W. L. Hsu. Fast algorithms for the dominating set problem on permutation graphs. *Algorithmica*, 9:109–117, 1993.

[1104] C. Tsouros and M. Satratzemi. Tree search algorithms for the dominating vertex set problem. *Internat. J. Computer Math.*, 47:127–133, 1993.

[1105] Z. Tuza. Covering all cliques of a graph. *Discrete Math.*, 86:117–126, 1990.

[1106] Z. Tuza. Small n-dominating sets. *Math. Pannon.*, 5:271–273, 1994.

[1107] E. Ungerer. *Aspects of signed and minus domination in graphs.* PhD thesis, Rand Afrikaans University, 1996.

[1108] C. Van Nuffelen. Rank and domination number. In *Graphs and Other Combinatorial Topics (Prague, 1982)*, pages 209–211. Teubner-Texte zur Mathematik 59 (Teubner, Leipzig), 1983.

[1109] D. Van Wieren, M. Livingston, and Q. F. Stout. Perfect dominating sets on cube-connected cycles. *Congr. Numer.*, 97:51–70, 1993.

[1110] L. P. Varvak. Generalization of a kernel of a graph. *Ukrainian Math. J.*, 25:78–81, 1973.

[1111] H. J. Veldman. Existence of dominating cycles and paths. *Discrete Math.*, 43:281–296, 1983.

[1112] H. J. Veldman. On dominating and spanning circuits in graphs. *Discrete Math.*, 124:229–239, 1994.

[1113] P. D. Vestergaard and B. Zelinka. Cut-vertices and domination in graphs. Technical report, Inst. Electronic Systems, Dept. of Math. and Computer Science, Aalborg Univ., Denmark, 1993.

[1114] M. G. Vinnicenko. A lower bound for the number of edges of a graph with prescribed internal and external stability numbers. *Dopovidi Akad. Nauk Ukrain RSR Ser. A*, pages 487–489, 571, 1973. Russian.

[1115] V. G. Vizing. The Cartesian product of graphs. *Vychisl. Sistemy*, 9:30–43, 1963.

[1116] V. G. Vizing. A bound on the external stability number of a graph. *Dokl. Akad. Nauk SSSR*, 164:729–731, 1965.

[1117] V. G. Vizing. Some unsolved problems in graph theory. *Uspekhi Mat. Nauk*, 23(6 (144)):117–134, 1968.

[1118] L. Volkmann. Simple reduction theorems for finding minimum coverings and minimum dominating sets. *Contempory Methods in Graph Theory, Bibliographisches Inst., Mannheim*, pages 667–672, 1990.

[1119] L. Volkmann. On graphs with equal domination and covering numbers. *Discrete Math.*, 51:211–217, 1994.

[1120] L. Volkmann. Irredundance number versus domination number for special graphs. Submitted, 1995.

[1121] L. Volkmann. On graphs with equal domination and edge independence numbers. *Ars Combin.*, 41:45–56, 1995.

[1122] L. Volkmann. A reduction principle concerning minimum dominating sets in graphs. Submitted, 1995.

[1123] L. Volkmann. *Fundamente der Grphentheorie.* Springer, 1996.

[1124] L. Volkmann and V. É. Zverovich. A proof of Favaron's conjecture on irredundance perfect graphs. Submitted.

[1125] L. Volkmann and V. É. Zverovich. A proof of Favaron's conjecture and a disproof of Henning's conjecture on irredundance perfect graphs. In *Fifth Twente Workshop on Graphs and Combin. Optimization*, pages 215–217, Enschede, 1997.

[1126] J. von Neumann and O. Morgenstern. *Theory of Games and Economic Behaviour.* Princeton Univ. Press, 1953.

[1127] H. Vu Dinh. On the length of longest dominating cycles in graphs. *Discrete Math.*, 121:211–222, 1993.

[1128] R. Wagner and R. Geist. Crippled queens placement problem. *Sci. Comput. Programming*, 4:221–248, 1984.

[1129] H. B. Walikar. Deficiency of non-d-critical graphs. *Discrete Math.* To appear.

[1130] H. B. Walikar. On the star-partition number of a graph. Manuscript, 1979.

[1131] H. B. Walikar. *Some Topics in Graph Theory (Contributions to the Theory of Domination in Graphs and its Applications).* PhD thesis, Karnatak Univ., 1980.

[1132] H. B. Walikar. Domatically critical and cocritical graphs. In S. Arumugam, B. D. Acharya, and E. Samathkumar, editors, *Graph Theory and Its Applications*, pages 223–234. Tata-MacGraw-Hill Publishing Co. Ltd., New Delhi, 1997.

[1133] H. B. Walikar and B. D. Acharya. Domination critical graphs. *Nat. Acad. Sci. Lett.*, 2:70–72, 1979.

[1134] H. B. Walikar, B. D. Acharya, and E. Sampathkumar. Recent developments in the theory of domination in graphs. In *MRI Lecture Notes in Math., Mahta Research Instit., Allahabad*, volume 1, 1979.

[1135] H. B. Walikar, E. Sampathkumar, and B. D. Acharya. Two new bounds for the domination number of a graph. Technical Report 14, MRI, 1979.

[1136] C. K. Wallis. *Domination Parameters of Line Graphs of Designs and Variations of Chessboard Graphs*. PhD thesis, Clemson Univ., 1994.

[1137] C. Wang. On the sum of two parameters concerning independence and irredundance in a graph. *Discrete Math.*, 69:199–202, 1988.

[1138] Y. Wang. On the bondage number of a graph. *Discrete Math.*, 159:291–294, 1996.

[1139] W. D. Weakley. Domination in the queen's graph. In Y. Alavi and A. J. Schwenk, editors, *Graph Theory, Combinatorics, Algorithms, and Applications, Proc. Seventh Quad. Internat. Conf. on the Theory and Application of Graphs*, volume 2, pages 1223–1232, (Kalamazoo, MI 1992), 1995. Wiley.

[1140] W. D. Weakley. Minimum dominating sets in hypercubes. Submitted, 1996.

[1141] K. Weber. Domination number for almost every graph. *Rostock. Math. Kolloq.*, 16:31–43, 1981.

[1142] P. M. Weichsel. Dominating sets in n-cubes. *J. Graph Theory*, 18(5):479–488, 1994.

[1143] K. White, M. Farber, and W. Pulleyblank. Steiner trees, connected domination and strongly chordal graphs. *Networks*, 15:109–124, 1985.

[1144] H. S. Wilf. The problem of the kings. *Electron. J. Combin.*, 2:1–7, 1995.

[1145] T. V. Wimer. Linear algorithms for the dominating cycle problems in series-parallel graphs, partial k-trees and Halin graphs. *Congr. Numer.*, 56:289–298, 1986.

[1146] T. V. Wimer. An $O(n)$ algorithm for domination in k-chordal graphs. Manuscript, 1986.

[1147] T. V. Wimer. *Linear Algorithms on K-terminal Graphs*. PhD thesis, Clemson Univ., 1987.

[1148] T. V. Wimer, S. T. Hedetniemi, and R. Laskar. A methodology for constructing linear graph algorithms. *Congr. Numer.*, 50:43–60, 1985.

[1149] E. Wojcicka. Hamiltonian properties of domination-critical graphs. *J. Graph Theory*, 14:205–215, 1990.

[1150] E. S. Wolk. A note on 'The comparability graph of a tree'. In *Proc. American Math. Soc., 16*, pages 17–20, 1965.

[1151] S. Xu. Some parameters of a graph and its complement. *Discrete Math.*, 65:197–207, 1987.

[1152] S. Xu. Relations between parameters of a graph. *Discrete Math.*, 89:65–88, 1991.

[1153] Y. F. Xue and Z. Q. Chen. Hamiltonian cycles in domination-critical graphs. *Nanjing Univ. J. Natur. Sci.*, 27:58–62, 1991.

[1154] Y. F. Xue and Z. Q. Chen. Dominating trails and pancyclicity of linear graphs. *Nei Mongol Daxue Xuebao, Ziran Kexue, J. Inner Mongolia Teach. Univ. Natur. Sci.*, 23:166–173, 1992.

[1155] A. M. Yaglom and I. M. Yaglom. Challenging mathematical problems with elementary solutions. In *Volume 1: Combinatorial Analysis and Probability Theory*, San Francisco, 1964. Holden-Day.

[1156] M. Yannakakis and F. Gavril. Edge dominating sets in graphs. *SIAM J. Appl. Math.*, 38:264–272, 1980.

[1157] H. G. Yeh. *Distance-Hereditary Graphs: Combinatorial Structures and Algorithms.* PhD thesis, National Chiao Tung Univ., 1997.

[1158] H. G. Yeh and G. J. Chang. Algorithmic aspect of majority domination. *Taiwanese J. Math.* To appear.

[1159] H. G. Yeh and G. J. Chang. Linear-time algorithms for bipartite distance-hereditary graphs. Submitted, 1997.

[1160] H. G. Yeh and G. J. Chang. Weighted connected domination and Steiner trees in distance-hereditary graphs. Submitted, 1997.

[1161] C. Yen and R. C. T. Lee. The weighted perfect domination problem. *Inform. Process. Lett.*, 35(6):295–299, 1990.

[1162] B. Yu. *Convexity of Minimal Total Dominating Functions of Graphs.* Master's thesis, Univ. Victoria, 1992.

[1163] B. Yu. Convexity of minimal total dominating functions in graphs. In *Lecture Notes in Comput. Sci., Proc. First Ann. Internat. Conf. on Computing and Combinatorics*, volume 959, pages 357–365, 1995.

[1164] B. Yu. Convexity of minimal total dominating functions in graphs. *J. Graph Theory*, 24:313–321, 1997.

[1165] B. Zelinka. Antidomatic number of a graph. *Arch. Math. Brno.* To appear.

[1166] B. Zelinka. Domatic numbers of directed graphs. *Czech. Math. J.* To appear.

[1167] B. Zelinka. Graphs with the domatic number χ_0. *Czech. Math. J.* To appear.

[1168] B. Zelinka. Total domatic number of graph. *Proc. Math. Liberec.* To appear.

[1169] B. Zelinka. Uniquely domatic regular domatically full graphs. *Czech. Math. J.* To appear.

[1170] B. Zelinka. Domatically critical graphs. *Czech. Math. J.*, 30:486–489, 1980.

[1171] B. Zelinka. On domatic numbers of graphs. *Math. Slovaca*, 31:91–95, 1981.

[1172] B. Zelinka. Some remarks on domatic numbers of graphs. *Časopis Pěst. Mat.*, 106:373–375, 1981.

[1173] B. Zelinka. Domatic numbers of cube graphs. *Math. Slovaca*, 32:117–119, 1982.

[1174] B. Zelinka. Domatische zahlen der graphen (Domatic numbers of graphs). In *Rostock. Math. Kolloq. 21, Discrete Mathematics and its Applications to Mathematical Cybernetics, Part 3 (Rostock, 1981)*, pages 69–72, 1982.

[1175] B. Zelinka. Adomatic and idomatic numbers of graphs. *Math. Slovaca*, 33:99–103, 1983.

[1176] B. Zelinka. Domatic number and bichromaticity of a graph. In *Lecture Notes in Mathematics, Graph Theory, Lagow 1981*, volume 1018, pages 278–285. Springer, Berlin, 1983.

[1177] B. Zelinka. Domatic number and degrees of vertices of a graph. *Math. Slovaca*, 33:145–147, 1983.

[1178] B. Zelinka. Domatically cocritical graphs. *Časopis Pěst. Mat.*, 108:82–88, 1983.

[1179] B. Zelinka. Edge-domatic number of a graph. *Czech. Math. J.*, 33:107–110, 1983.

[1180] B. Zelinka. On k-domatic numbers of graphs. *Czech. Math. J.*, 33:309–313, 1983.

[1181] B. Zelinka. On k-ply domatic numbers of graphs. *Math. Slovaca*, 34:313–318, 1984.

[1182] B. Zelinka. Semidomatic numbers of directed graphs. *Math. Slovaca*, 34:371–374, 1984.

[1183] B. Zelinka. Bichromaticity and domatic number of a bipartite graph. *Časopis Pěst. Mat.*, 110:113–115, 207, 1985.

[1184] B. Zelinka. Domatic numbers of uniform hypergraphs. *Arch. Math. Brno.*, 21:129–134, 1985.

[1185] B. Zelinka. Odd graphs. *Arch. Math. Brno.*, 21:181–188, 1985.

[1186] B. Zelinka. The bigraph decomposition number of a graph. *Časopis Pěst. Mat.*, 113:56–59, 1986.

[1187] B. Zelinka. Connected domatic number of a graph. *Math. Slovaca*, 36:387–392, 1986.

[1188] B. Zelinka. Domatic number and linear arboricity of cacti. *Math. Slovaca*, 36:49–54, 1986.

[1189] B. Zelinka. Complementarily domatic number of a graph. *Math. Slovaca*, 38:27–32, 1988.

[1190] B. Zelinka. *Some Numerical Invariants of Graphs, Czech.* PhD thesis, Prague, 1988.

[1191] B. Zelinka. Total domatic number of cacti. *Math. Slovaca*, 38:207–214, 1988.

[1192] B. Zelinka. Total domatic number and degrees of vertices of a graph. *Math. Slovaca*, 39:7–11, 1989.

[1193] B. Zelinka. Domatic numbers of lattice graphs. *Czech. Math. J.*, 40:113–124, 1990.

[1194] B. Zelinka. Domination in cubic graphs. In *Topics in Combinatorics and Graph Theory*. Physica-Verlag, Heidelberg, 1990. 727-735.

[1195] B. Zelinka. Edge-domatically full graphs. *Math. Slovaca*, 40:359–365, 1990.

[1196] B. Zelinka. Regular totally domatically full graphs. *Discrete Math.*, 86:71–79, 1990.

[1197] B. Zelinka. Domination in cubes. *Math. Slovaca*, 41:17–19, 1991.

[1198] B. Zelinka. Edge-domatic numbers of cacti. *Math. Bohem.*, 116:91–95, 1991.

[1199] B. Zelinka. Total edge-domatic number of a graph. *Math. Bohem.*, 116:96–100, 1991.

[1200] B. Zelinka. Domatic number of a graph and its variants (extended abstract). In M. Fiedler and J. Nešetřil, editors, *Fourth Czech. Symp. on Combin., Graphs, and Complexity*. Elsevier, 1992.

[1201] B. Zelinka. Adomatic and total adomatic numbers of graphs. Manuscript, 1995.

[1202] B. Zelinka. Edge-adomatic numbers of graphs. Manuscript, 1995.

[1203] B. Zelinka. Graphs with the domatic number χ. Manuscript, 1995.

[1204] B. Zelinka. Semidomatic and total semidomatic numbers of directed graphs. Manuscript, 1995.

[1205] B. Zelinka. Some remarks on domination in cubic graphs. *Discrete Math.*, 158:249–255, 1996.

[1206] B. Zelinka. Domatic numbers of graphs and their variants: a survey. In T. W. Haynes, S. T. Hedetniemi, and P. J. Slater, editors, *Domination in Graphs: Advanced Topics*, chapter 13. Marcel Dekker, Inc., 1997.

[1207] J. Žerovnik and J. Oplerova. A counterexample to a conjecture of Barefoot, Harary, and Jones. *Graphs Combin.*, 9:205–207, 1993.

[1208] H. Zhang. Self-complementary symmetric graphs. *J. Graph Theory*, 16:1–5, 1992.

[1209] I. E. Zverovich and V. É. Zverovich. A characterization of domination perfect graphs. *J. Graph Theory*, 15(2):109–114, 1991.

[1210] I. E. Zverovich and V. É. Zverovich. A note on domatically critical and cocritical graphs. *Czech. Math. J.*, 41:278–281, 1991.

[1211] I. E. Zverovich and V. É. Zverovich. Disproof of a conjecture in domination theory. *Graphs Combin.*, 10:389–396, 1994.

[1212] I. E. Zverovich and V. É. Zverovich. An induced subgraph characterization of domination perfect graphs. *J. Graph Theory*, 20(3):375–395,
1995.

[1213] I. E. Zverovich and V. É. Zverovich. A semi-induced subgraph characterization of upper domination perfect graphs. Submitted, 1997.

[1214] V. É. Zverovich. A solution of one problem on domatically cocritical
graphs. Technical Report 2293-B88, Vestnik Bel. Gos. Univ., 1988. (in
Russian).

[1215] V. É. Zverovich. On domination and irredundance in graphs. Technical
Report 4378-B89, Vestnik Bel. Gos. Univ., 1989. (in Russian).

[1216] V. É. Zverovich. Criticality and inequalities in domination theory of
graphs. *Izvest. AN BSSR Ser. Fiz.-Mat. Nauk*, 4:37–42, 1990. (in Russian).

[1217] V. É. Zverovich. Domination-complete graphs. *Math. Notes*, 48:920–922,
1990.

[1218] V. É. Zverovich. *Graph-Theoretic Invariants Connected with Neighborhoods of Vertex Sets.* PhD thesis, Belarus State Univ., 1992.

[1219] V. É. Zverovich. The ratio of the irredundance number and the domination number for a class of graphs. Manuscript, 1996.

[1220] V. É. Zverovich and I. E. Zverovich. Some classes of minimaximal graphs.
Technical Report 5070-B88, Vestnik Bel. Gos. Univ., 1988. (in Russian).

[1221] V. É. Zverovich and I. E. Zverovich. Domistable graphs. Technical Report
956-B89, Vestnik Bel. Gos. Univ., 1989. (in Russian).

[1222] V. É. Zverovich and I. E. Zverovich. A characterization of perfect
dominating-clique graphs. *Izvest. AN BSSR Ser. Fiz.-Mat. Nauk*, 2:107–
109, 1991. (in Russian).

Notation List

Symbol	Meaning
$a(G)$	acquisition number
$b(G)$	bondage number
$B(S)$	boundary of set S
$c(G)$	circumference
$C(G)$	center
$coir(G)$	closed-open irredundance number
$COIR(G)$	upper closed-open irredundance number
$CR(G)$	cardinality redundance number
$d(G)$	domatic number
$d_c(G)$	connected domatic number
$d_i(G)$	independent domatic number
$d_{pr}(G)$	paired domatic number
$d_t(G)$	total domatic number
$d(v_i, v_j)$	distance from v_i to v_j
$d(v, S)$	distance from vertex v to set S
$dd(G)$	double domination number
$deg(v)$	degree of vertex v
$diam(G)$	diameter
$dist(v)$	distance of vertex v
$ecc(S)$	eccentricity of set S
$er(G)$	external redundance number
$ER(G)$	upper external redundance number
$F(G)$	efficient domination number
$F_f(G)$	fractional efficient domination number
$g(G)$	girth
$\overline{G}$	complement of G
G^k	kth power of G
$Gr(G)$	Grundy coloring number
$G - v$	deletion of a vertex
$G - uv$	deletion of an edge
$i(G)$	independent domination number
$i'(G)$	independent edge domination number
$i^*(G)$	iterated independent domination number
$i_m(G)$	mixed independent domination number
$i_S(G)$	independent strongdomination number
$i_w(G)$	weakly connected independent domination number
$I(S)$	influence of set S

Symbol	Meaning
$id(G)$	idomatic number
$ir(G)$	irredundance number
$IR(G)$	upper irredundance number
$ir'(G)$	edge irredundance number
$IR'(G)$	upper edge irredundance number
$ir^*(G)$	iterated irredundance number
$IR^*(G)$	upper iterated irredundance number
$ir_f(G)$	fractional irredundance number
$IR_f(G)$	upper fractional irredundance number
$ir_m(G)$	mixed irredundance number
$IR_m(G)$	upper mixed irredundance number
$ir_{\{k\}}(G)$	$\{k\}$-irredundance number
$IR_{\{k\}}(G)$	upper $\{k\}$-irredundance number
$L(G)$	line graph
$M(G)$	median, middle graph
$N(v)$	open neighborhood of v
$N[v]$	closed neighborhood of v
$nk(G)$	knockout number
$NK(G)$	upper knockout number
NP	class of non-deterministic polynomial decision problems
NPc	class of NP-complete decision problems
$OB(S)$	open boundary of set S
$oir(G)$	open irredundance number
$OIR(G)$	upper open irredundance number
$ooir(G)$	open-open irredundance number
$OOIR(G)$	upper open-open irredundance number
P	class of deterministic polynomial decision problems
$P(G)$	closed neighborhood order packing number
$P_f(G)$	fractional closed neighborhood order packing number
$PD(G)$	parity dimension
$pn[u, S]$	private neighbor set of vertex u with respect to set S
$PR(m, \gamma)$	prioritized independent domination number
$r(G)$	reinforcement number
$R(G)$	redundance number
$R_f(G)$	fractional redundance number
$rad(G)$	radius of G
$\langle S \rangle$	subgraph induced by set S
$T(G)$	total graph
$w(f)$	weight of function f
$W(G)$	closed neighborhood order domination number
$W_f(G)$	fractional closed neighborhood order domination number

Symbol	Meaning
$\alpha_0(G)$	vertex covering number
$\Lambda_0(G)$	upper vertex covering number
$\alpha_1(G)$	edge covering number
$\Lambda_1(G)$	upper edge covering number
$\beta_0(G)$	vertex independent number
$\beta_1(G)$	matching number, edge independence number
$\beta_k(G)$	k-dependence number
$\beta_m(G)$	mixed independence number
$\beta_w(G)$	weakly connected independence number
$\beta^*(G)$	iterated independence number
$\gamma(G)$	domination number
$\Gamma(G)$	upper domination number
$\gamma'(G)$	edge domination number
$\Gamma'(G)$	upper edge domination number
$\gamma^\#(G)$	domination forcing number
$\gamma^*(G)$	iterated domination number
$\gamma^*(G)$	star partition number
$\Gamma^*(G)$	upper iterated domination number
$\gamma^-(G)$	minus domination number
$\Gamma^-(G)$	upper minus domination number
$\gamma_{1,k}(G)$	$(1,k)$-domination number
$\gamma_c(G)$	connected domination number
$\gamma_{cap}(G)$	capacity domination number
$\gamma_{cl}(G)$	clique domination number
$\gamma_{cy}(G)$	cycle domination number
$\gamma_f(G)$	fractional domination number
$\gamma_f^0(G)$	open fractional domination number
$\Gamma_f(G)$	upper fractional domination number
$\gamma_{ft}(G)$	factor domination number
$\gamma_g(G)$	global domination number
$\gamma_k(G)$	k-domination number
$\Gamma_k(G)$	k-minimal domination number
$\gamma_{ksub}(G)$	$ksub$-domination number
$\gamma_{\{k\}}(G)$	$\{k\}$-domination number
$\Gamma_{\{k\}}(G)$	upper $\{k\}$-domination number
$\gamma_{\leq k}(G)$	distance-k domination number
$\gamma_{\times k}(G)$	k-tuple domination number
$\gamma_l(G)$	least domination number
$\gamma_L(G)$	locating domination number
$\gamma_m(G)$	mixed domination number
$\Gamma_m(G)$	upper mixed domination number

Symbol	Meaning
$\gamma_{maj}(G)$	majority domination number
$\gamma_{opt}(G)$	optional domination number
$\gamma_p(G)$	perfect domination number
$\gamma_{pr}(G)$	paired domination number
$\gamma_{ps}(G)$	point-set domination number
$\Gamma_{pvt}(G)$	private domination number
$\gamma_r(G)$	restrained domination number
$\gamma_s(G)$	signed domination number, set domination number
$\Gamma_s(G)$	upper signed domination number
$\gamma_S(G)$	strong domination number
$\gamma_t(G)$	total domination number
$\Gamma_t(G)$	upper total domination number
$\gamma_w(G)$	weakly connected domination number
$\Gamma_w(G)$	upper weakly connected domination number
$\gamma_W(G)$	weak domination number
$\gamma_Y(G)$	Y-domination number
$\chi(G)$	chromatic number
$\delta(G)$	minimum degree of a vertex in G
$\Delta(G)$	maximum degree of a vertex in G
$\varepsilon_F(G)$	max. pendant edges in spanning forest
$\varepsilon_T(G)$	max. pendant edges in spanning tree
ϕ_S	characteristic function of set S
$\kappa(G)$	vertex connectivity
$\lambda(G)$	edge connectivity
$\mu(G)$	matchability number
$\theta(G)$	perfect neighborhood number
$\Theta(G)$	upper perfect neighborhood number
$\rho(G)$	packing number
$\rho_f(G)$	fractional packing number
$\psi(G)$	lower enclaveless number
$\Psi(G)$	enclaveless number

Index